*An Illustrated Guide
to Theoretical Ecology*

An Illustrated Guide to Theoretical Ecology

Ted J. Case

University of California, San Diego

New York   Oxford
OXFORD UNIVERSITY PRESS
2000

Oxford University Press

Oxford New York
Athens Auckland Bangkok Bogotá Buenos Aires Calcutta
Cape Town Chennai Dar es Salaam Delhi Florence Hong Kong Istanbul
Karachi Kuala Lumpur Madrid Melbourne Mexico City Mumbai
Nairobi Paris São Paulo Singapore Taipei Tokyo Toronto Warsaw

and associated companies in
Berlin Ibadan

Published by Oxford University Press, Inc.,
198 Madison Avenue, New York, New York, 10016
http://www.oup-usa.org
1-800-334-4249

Library of Congress Cataloging-in-Publication Data

```
Case, Ted J.
    An illustrated guide to theoretical ecology / Ted J. Case.
       p.   cm.
    Includes bibliographical references.
    ISBN 978-0-19-508512-9 (pbk. : alk. paper)
    1. Ecology--Mathematical models.   I. Title.
   QH541.15.M3C36   1999
   577',01--dc21
```
99-32309
CIP

ISBN 978-0-19-508512-9 (paper)

9 8 7 6 5 4

Printed in the United States of America
on acid-free paper

To Benita
and
Nathan

Contents

Preface

Increasingly our world requires that we make informed decisions to manage economically valuable natural resources, save threatened species, control agricultural pests, and improve the health of air and water. Yet an understanding of these problems and how precisely to achieve our goals is inevitably a quantitative and inferential subject. Recent national trends indicate that the mathematical preparation of undergraduates is deteriorating, and so for many academic ecologists the job of teaching the principles of ecology is becoming more challenging. This book is a response to that problem. My goal is to explain ecological theory in a user-friendly way. Some students seem to learn best through pictorial presentations while others are more facile with the symbolic logic of algebra and calculus. The deepest level of understanding seems to come with the joint application of both frames of understanding. I have attempted to give students the basic tools they need to appreciate the complexities of ecological systems and to analyze simple quantitative ecological problems. Mathematical techniques and tricks are only introduced after a question has been raised and the student is motivated to solve it. In this way, I attempt to provide students with a visceral feel for how and why these tools work and how they can be applied to the workings of the natural world.

The book builds sophisticated theoretical results from first principles and tries to gently make these results accessible to students through a step-by-step development of equations paired with illustrations. This book, therefore, could be used in advanced classes as a supplemental self-guided tutorial. In this context, it can serve as a companion to any of the standard texts in ecology. My hope here is that most instructors, even those not mathematically inclined, can use this primer to supplement their exposition of the more analytical aspects of ecology—those parts that seem to cause students the most trouble. Such a use would then free instructor's lecture time from being dominated by models. Because I want to ensure that this primer can be woven into a class using any of the major texts, I have spent some time explaining why equations may look different from one text to another. It has been my experience that many students get hung up on the symbolism and notation of mathematics, trying to memorize the meaning of r, K, etc., while forgetting the logical role that these symbols play in the conceptual development. I have tried to add exercises and notes of caution so that this pitfall can be avoided.

Instructors who prefer to emphasize quantitative and modeling approaches should find the book rigorous enough to be used as a stand-alone text in introductory theoretical ecology classes for undergraduates. In this case the material could be profitably supplemented with additional readings presenting more advanced mathematical techniques and extensions.

The coverage is not thorough in the sense of covering all aspects of ecology—particularly lacking is ecosystem-level dynamics and physiological ecology, but the primer does go into a bit of depth in the areas that seem essential for understanding population dynamics and community ecology. With Theodosius Dobzhansky, I, too, believe that nothing in biology makes sense except in the light of evolution. Consequently, the book has more evolutionary ecology and natural selection than most. The coverage, however, will not substitute for a course on population genetics. I assume that students have had a high-school level of algebra and trigonometry, and a first course in college calculus. In spite of these modest assumption about background skills, the students are introduced to some of the more complicated areas in ecology. Areas that are more challenging than others are labeled "Advanced." These may be skipped without losing continuity. Appendices at the end of the book allow students to conveniently refer back to formulas and analytical tools. Different mathematical tools are introduced and applied to ecological problems as follows:

Chapter 1.   Discrete time difference equations; continuous time differential equations; the log transformation.
Chapter 2.   Discrete and continuous diffusion; the formula for a mean and variance; stochastic processes.
Chapter 3.   Matrices and vectors; the eigenvalue problem.
Chapter 4.   Infinite sums and improper integrals.
Chapter 5.   Taylor's expansion; time lags; local stability of an equilibrium point; limit cycles; chaos; Lyapunov exponents.
Chapter 6.   Correlograms.
Chapter 7.   Optimization.
Chapter 8.   Partial derivatives.
Chapter 9.   Boundary (invasion) analysis.
Chapter 12. Coupled differential equations; zero-isoclines; phase space.
Chapter 13. Multidimensional stability analysis; Jacobian matrix.
Chapter 16. Reaction–dispersal equations.

I have tried to be reasonably rigorous but gentle. This makes for some redundancy—a few steps are inserted in derivations that might be obvious to the more advanced students or a figure is added when, for some, it is not needed. All this clearly makes for a somewhat longer text than a more terse and compact presentation would allow. At the same time, I have made a serious attempt to relate models to real data and provide examples and solved problems. One might argue that such "research"-related topics should be reserved for graduate students, and I accept this as fair criticism. Yet without introducing students to the methods of science as well as its accumulated body of knowledge, I feel that we do not adequately explain the job of science.

The order of chapters is generally by level of complication. The book is a primer in the very real sense that chapters and exercises build on each other and follow one another in a pedagogical order. The book begins with single-species geometric growth, then density-dependent growth, and life history theory. Multispecies interactions follow. Chapters develop different tools and techniques that are used in later chapters; thus it would be difficult to skip around in the book. Appendix 1, "Preparation" should be read first. It has two goals: to review elementary mathematical forms with an emphasis on visualization, and to present the semantics of model building. The other appendices are supplemental. Because third- and even fourth-year undergraduates these days are not exposed to matrix algebra, I present enough of these subjects to prepare students for the material in the chapters; some of this is delegated to the appendices for easy referral. Each chapter has questions and problems that make good homework assignments. It is my experience that students begin to understand this material only by working through problems and seeing simulations and examples.

Also, I have developed simulation models to accompany many of the models in the text. These are written in the systems modeling software Stella® (1999, High Performance Systems, Hanover, New Hampshire) and in a few cases Matlab® (The Math Works). These models are available for student simulation through the website: **http://www.nceas.ucsb.edu/BookCase.** A run-time version of Stella® that will allow users to run the models and change parameters, can be downloaded free from the High Performance Systems website: **http://www.hps-inc.com.** During lectures I present these simulations using computer projection. My colleague Mike Gilpin has developped

Java aplets for ecology including several from this book. He is adding to these continually and they are available at: **http://nemesis.webjump.com.** Another neat approach is Ecobeaker for Macs or PCs developed by Eli Meir. These are individual-based models. The web site is **http://www.ecobeaker.com.**

When I began working on Chapter 16, I found it impossible to improve much upon study designs that Mike Gilpin had prepared to explain island biogeography and metapopulation dynamics to undergraduates. I therefore enlisted his help in writing this chapter.

I am very grateful to Barry Noon, Martin Cody, John Donnell, Benita Epstein, Mike Gilpin, Lloyd Goldwasser, Bob Holt, Kathy Klingenberg, Russ Lande, Peter Kareiva, and Mark Taper for their comments and suggestions on the manuscript. I particularly want to thank Trevor Price for critiquing every chapter. Each of these people helped identify errors and contributed several terrific suggestions for improving the book. The book was finished during my sabbatical leave in 1998–1999 at the National Center for Ecological Analysis and Synthesis at Santa Barbara. I thank the center and staff for facillitating my stay. I also want to thank the many students who took my classes over the years in general ecology and community ecology at the University of California of San Diego. Various graduate students also made substantial contributions: Emilio Bruno, Doug Bolger, Paul Griffin, Kathy Hanley, Adam Richman, Andy Suarez, Mark Taper, and Barb Taylor. Peter Abrams deserves credit for being the first (to my knowledge) to discover that Strobeck's (1973) three-competitor unstable equilibrium point, in fact, emitted a limit cycle (Chapter 14). While he never published this result, it filtered through the ecological community and I eventually rediscovered it. Finally, a special thanks goes to my wife and colleagues for their support and encouragement.

An Illustrated Guide
to Theoretical Ecology

1 Exponential and Geometric Population Growth

A **population** is a group of individuals that belong to a single species and live in some defined area. Ecologists strive to understand the causes of variation in the sizes of populations and to predict trends in these numbers over time and from place to place. A population increases in size by births and the immigration of individuals into the population from outside. Deaths and emigration decrease population size. These inputs and outputs can be set up as an equation where the subscript t indicates a discrete point in time. The population at time $t + 1$ is N_{t+1} and its size is given by

$$N_{t+1} = N_t + \textbf{births} - \textbf{deaths} + \textbf{immigration} - \textbf{emigration.}$$

The births, deaths, immigration and emigration in this equation are those that occurred during the time interval between t and $t + 1$. If time is measured in years, then this equation says that the number of individuals at a specific time next year $(t + 1)$ will equal the number that are present at the same time this year, plus all the positive and negative changes to population size that occurred during the year. Figure 1.1 summarizes these flows.

However, this diagram doesn't capture the long-term dynamics of population growth. Each new birth has the potential itself to give birth after that individual reaches sexual maturity. Thus the birth component of the input/output diagram forms a positive feedback loop with population size. The more births, the larger the population size becomes, and, as population size becomes larger, more individuals produce new young, and so on. Figure 1.2 modifies the diagram shown in Figure 1.1 to illustrate this positive feedback loop involving births.

This diagram still ignores many realistic complications: for example, different ages or different genotypes may have different rates of death, reproduction, and movement. These parameters may also vary over time and space and with population size. The rest of the book is devoted to adding and investigating these complications one at a time. For now, however, it is important for you to appreciate the inherent potential of exponential growth. The ability of populations to increase in a multiplicative way has tremendous ramifications for ecology, evolution, and economics. A pair of rabbits might produce 2 pairs, which can produce 4 pairs, which can produce 8 pairs, and so on. The larger the population size, the greater is its potential to grow still larger during the next time period (i.e. $N_{t+1} - N_t$, $N_{t+2} - N_{t+1}$, . . . , $N_{t+n} - N_{t+n-1}$). The end of each interval completes a **time step.**

MODELING POPULATION GROWTH

Population biologists trying to predict future population growth take different approaches to the mathematics of this growth. The first approach, **"discrete time,"** as the name suggests, divides time into discrete chunks or intervals and forms equations that describe the growth of the population from one time step to the next. This method produces equations for population growth called **difference** equations; an example is

1

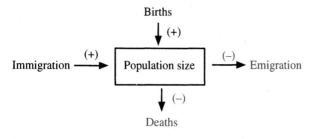

Figure 1.1
General factors increasing (+) and decreasing (–)
population size.

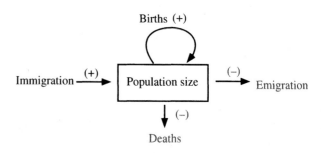

Figure 1.2
Births form a positive feedback loop with population size.

the opening equation of this chapter. The other approach considers time as a **continuous** variable and uses the calculus of differential equations to perform smooth, continuous projections over time. These different approaches are not based simply on mathematical convenience or an esoteric desire for completeness. We all know that time is continuous, nevertheless some organisms respond to time's arrow in a discontinuous, pulsed manner. Their breeding seasons may occur during brief intervals of the year. For other organisms, reproduction may go on continuously over the year, and for these species the mathematics of continuous time provides a more realistic description of their response. Most species are probably somewhere between these two extremes, but we could model them using either method as long as we recognize the minor errors involved. (The short Preparation section in Appendix 1 presents an overview of the strategy and syntax of model development in population biology.)

Box 1.1 shows side by side the basic equations for simple unlimited population growth in discrete time and continuous time. In both methods, we seek a **solution** that is an expression for population size at any arbitrary time in the future based on the population size that presently exists. In the discrete time approach, growth is formulated as

$$N_{t+1} = RN_t + N_t.$$

This equation says that the population size, N, one time step (e.g., one year) later is equal to the present size of the population (at time t) plus some increment that represents the net change. This net change is given by a per capita rate R that then multiplies the number of individuals present at time t; R may be positive or negative. Of course, R probably will vary over time, space, and individuals. However, the case in which it is constant is interesting and a good demonstration of the mathematical tools used in population projection. The goal is to find an expression for N, not for a single time step into the future, but rather for many time steps, say T. Box 1.1 shows that this expression is $N_T = N_0\lambda^T$.

In the continuous time approach, the *instantaneous rate* of population change (dN/dt) is given as a function of the population size at the time t, $N(t)$, and a parameter r. (Later we determine the precise relationship between r and R in Box 1.2.) Note the differences in notation used in deriving the discrete time formula (use of subscripts and superscripts) and in deriving the continuous time formula (use of function notation).

For both methods we obtained **solutions** (Eqs. 1.3a and 1.3b) that give the population size any arbitrary time into the future, based on a knowledge of the initial population size, N_0, a parameter describing population growth rate (either r or λ), and the growth model (either Eq. 1.1a or 1.1b).

Three examples of exponential population growth in continuous time are shown in Figure 1.3. All of these populations begin with a single inseminated female and, as time goes on, all approach infinity but at different rates.

When we plot the natural log of N (written ln N), rather than N, the curves in Figure 1.3 become straight lines in Figure 1.4. Logarithms, regardless of the base chosen, convert multiplicative to additive processes (recall that $\ln(ab) = \ln a + \ln b$). The slope of each line equals $r = \ln \lambda$.

Box 1.1 Unrestricted Growth: Two Different Approaches When the Rate of Growth Is Constant

Discrete Time Using a Difference Equation (Geometric Growth)

The starting expression is

$$N_{t+1} = RN_t + N_t,$$

where R is the net discrete (or geometric) per capita rate of growth. Collecting terms gives

$$N_{t+1} = (R + 1) N_t, \tag{1.1a}$$

which we may also write as

$$N_{t+1} = \lambda N_t,$$

where $\lambda = (R + 1)$ is the discrete (or geometric) per capita rate of growth; its units are per time period.

Similarly, for two time steps,

$$N_{t+2} = \lambda N_{t+1}$$
$$= \lambda \lambda N_t$$
$$= \lambda^2 N_t$$

and, for any arbitrary number of time steps into the future (say, T time steps),

$$N_{t+T} = \lambda^T N_t. \tag{1.2a}$$

If we start with N_0 individuals at time $t = 0$, then at time T the number of individuals is

$$N_T = N_0 \lambda^T. \tag{1.3a}$$

Equation (1.3a) is the "solution" for discrete time because it is a formula giving N for any arbitrary time period into the future.

Continuous Time Using a Differential Equation (Exponential Growth)

The starting expression is

$$\frac{dN}{dt} = rN(t). \tag{1.1b}$$

Initial conditions specify the beginning time ($t = 0$) and initial population size

$$N \text{ (at } t = 0) = N(0).$$

In Eq. (1.1b), r is the intrinsic (or exponential) per capita rate of growth; its units are per time period.

To solve Eq. (1.1b) with its initial conditions, we separate the differentials and integrate both sides. Then we evaluate the integral from $t = 0$ to $t = T$:

$$\int_{N(0)}^{N(T)} \frac{dN(t)}{N} = r \int_0^T dt. \tag{1.2b}$$

From the integral formulas of calculus, the left-hand side of Eq. (1.2b) becomes

$$\ln N(T) - \ln N(0)$$

and the right-hand side of Eq. (1.2b) becomes

$$rT - r0 = rT.$$

After exponentiation of both sides,

$$\frac{N(T)}{N(0)} = e^{rT}.$$

Finally, rearranging yields

$$N(T) = N(0) \, e^{rT}. \tag{1.3b}$$

Equation (1.3b) is the "solution" for continuous time because it is a formula giving $N(t)$ for any arbitrary time T into the future.

Comparing Eqs. (1.3a) and (1.3b), we see that $e^r = \lambda$.

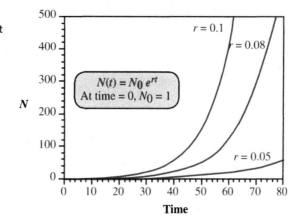

Figure 1.3
Exponential growth for three different values of the intrinsic growth rate, r.

$N(t) = N_0 e^{rt}$
At time = 0, $N_0 = 1$

$r = 0.1$
$r = 0.08$
$r = 0.05$

(Graph: N vs Time, Time axis from 0 to 80, N axis from 0 to 500)

Figure 1.4
The three curves in Figure 1.3 plotted on a logarithmic scale for *N*. All populations begin at the same starting number of one inseminated female ln 1 = 0.

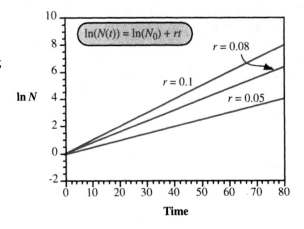

Figure 1.5
The effect of different initial population sizes on the course of exponential growth: (a) *r* > 0 and (b) *r* < 0.

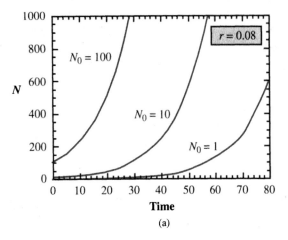

(a)

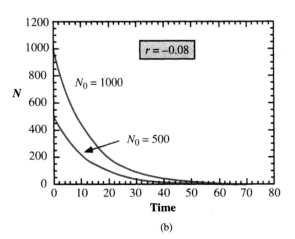

(b)

Figure 1.6
The results in Figure 1.5 plotted on a natural log scale. Ultimately the populations in (a) approach infinity and those in (b) approach zero.

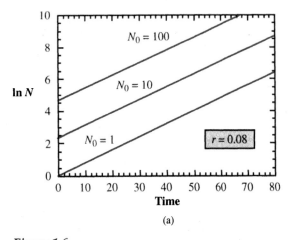

(a)

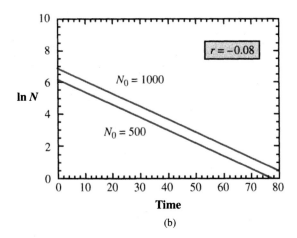

(b)

Exercise: Use Eq. (1.3b) and the definition of the slope of a straight line to show that the slope of each line in Figure 1.4 is $r = \ln \lambda$.

Figure 1.5 shows the effect of varying the initial population size.

After conversion of N to a natural log scale, this plot produces the straight lines shown in Figure 1.6.

The lines for the different starting population sizes are parallel on the log scale. In Figure 1.6(a) it takes the $N_0 = 1$ population 75 years to reach 403 ($\ln 403 \approx 6$). Increasing the initial population size tenfold ($N_0 = 10$) reduces this time but only by about one-third, to about 46 years. Thus doubling the initial population size does *not* necessarily halve the time it takes for a population to reach some threshold size.

Question: One hundred rabbits are introduced onto a small island at the beginning of 1995. If this population has a geometric growth rate λ of 1.3/year on the island, what will be the population size at the beginning of 2000? At the beginning of 2005?

Answer: In this example, N_0 is the population size in 1995; label it $N_{1995} = 100$. Applying Eq. (1.3a), the number of rabbits in 2000 is

$$N_{2000} = N_{1995}(1.3^5) = (100)(3.713) \approx 371 \text{ rabbits.}$$

Or, equivalently, Eq. (1.3b) could be applied. After we take the natural log of both sides, that equation becomes

$$\ln N_{2000} = \ln(N_{1995}) + rt.$$

As $r = \ln \lambda$, this expression becomes

$$\ln N_{2000} = 4.605 + (0.2623)(5) = 5.9168.$$

The antilog of 5.9168 is $N_{2000} \approx 371$ rabbits.
The population size in 2005 is

$$N_{2005} = N_{1995}(1.3^{10}) = (100)(13.786) \approx 1379 \text{ rabbits.}$$

This quantitative equivalency between the dynamics of populations in discrete time and continuous time, with the substitution $e^r = \lambda$, does not extend to more complicated population dynamic models, as we show in Chapter 5 when we discuss density dependence.

A graphical technique helps illustrate the sequential change in the number of individuals over time by creating a "ladder" between N_{t+1} and N_t. The black diagonal line in Figure 1.7 is the line of equality $N_{t+1} = N_t$. Start the population with some number N_0 at time 0 and follow the dashed arrow up to the geometric growth line (slope = λ). This intersection gives the population size at $t + 1$. To translate this value into the starting number for the next iteration, move from that point horizontally to the diagonal line $N_{t+1} = N_t$. The population sizes for successive times $t = 1$, 2, and 3 are projected as dots on the N_t axis. Because we already have an analytical formula for the growth of the population (Eq. 1.3a), this graphical method isn't really necessary. However, these diagrams can be generalized to situations where the growth rate, λ, is changing over time (Chapter 5), so there is some advantage to introducing the method for a simpler case.

Real populations do not often grow exponentially or geometrically for very long. Yet, species introduced to new places where resources are initially plentiful and their natural enemies are lacking can exhibit approximately exponential growth, at least temporarily. Figure 1.8 shows the growth of the U.S. population from 1790 to 1995. At first glance, this increase looks like it is close to being exponential. However, after converting population numbers to a natural log scale and then finding a best-fit linear regression line through the points, we see more clearly that the growth rate has been tapering off, as shown in Figure 1.9.

Figure 1.7
A graphical iteration method for determining discrete population growth. The initial population size is N_0. (a) $\lambda = 1.3$. The population increases geometrically. (b) $\lambda = 0.8$. The population decreases geometrically.

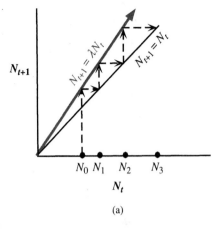

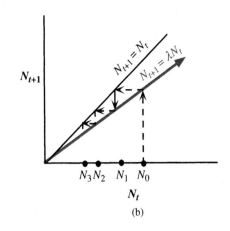

(a)

(b)

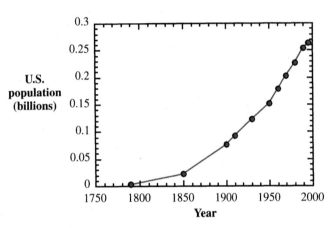

Figure 1.8
Population growth in the United States from 1790 to the present.

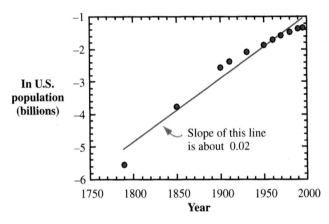

Figure 1.9
The data in Figure 1.8 plotted on a natural log scale. A regression line is shown through the points. It has a slope of about 0.02, meaning that the average rate of growth is about 2% per year. Note, however, that the actual points show a decline in growth rate over the past several decades.

The U.S. population has been growing at an average rate of about 2% per year, although current rates are less than 1%. The logarithmic plot reveals the decline in growth rate with time; note that the points are better described by a curve with a decreasing slope than by the straight line of constant r shown.

To give you a better understanding of geometric growth, in Figure 1.10 we show a simulation of a population of water lilies covering a pond. The open red squares represent the pond on selected days. The population of lily pads on those days is represented by the solid red areas. The black horizontal bars portray elapsed time. The simulation begins at day 0 with a single lily pad. Relative to the scale of the entire pond, this single lily pad is so small as to be invisible. The number of lily pads doubles every day ($\lambda = 2$/day) until the pads become visible on day 15 (lower left-hand corner of window). Ultimately the pond can hold 100 million lily pads. It is completely full at day 30 but was just half full 1 day before, on day 29. The point is that **geometric growth can sneak up on you.**

With a population doubling time of just 1 day, population size will get out of hand quickly, despite a very low initial number. However, the same kind of explosive growth occurs even with longer doubling times. The average growth rate in the United States since colonial times is only about 2%/year, translating to a doubling time of about 35 years. (See the section Solved Problems for this doubling time calculation.) The U.S. Geological Survey has produced a video showing the growth of urban and suburban areas in the Baltimore/Washington, D.C., area during this period. Figure 1.11 shows

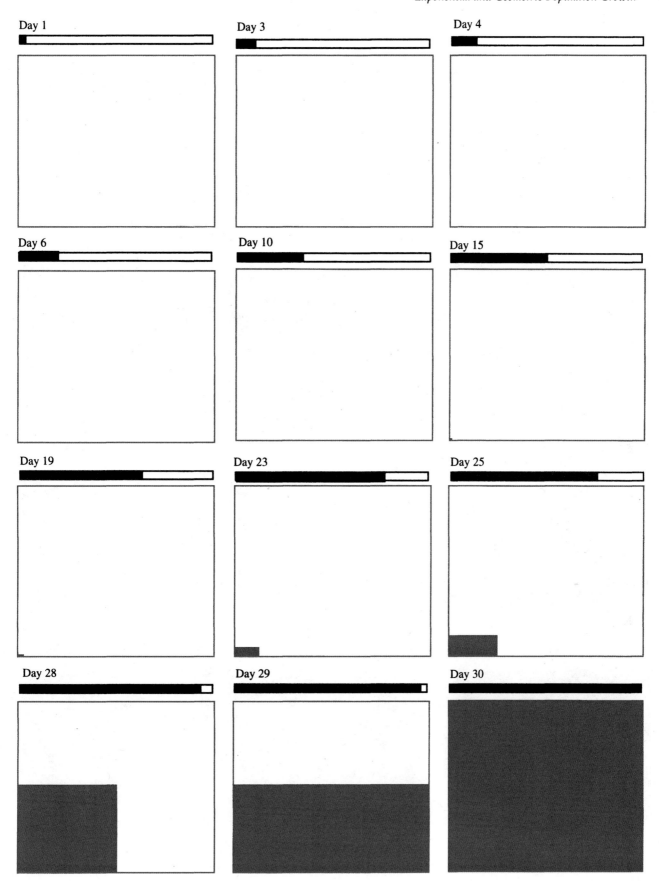

Figure 1.10
Simulation of the growth of lily pads in a pond.

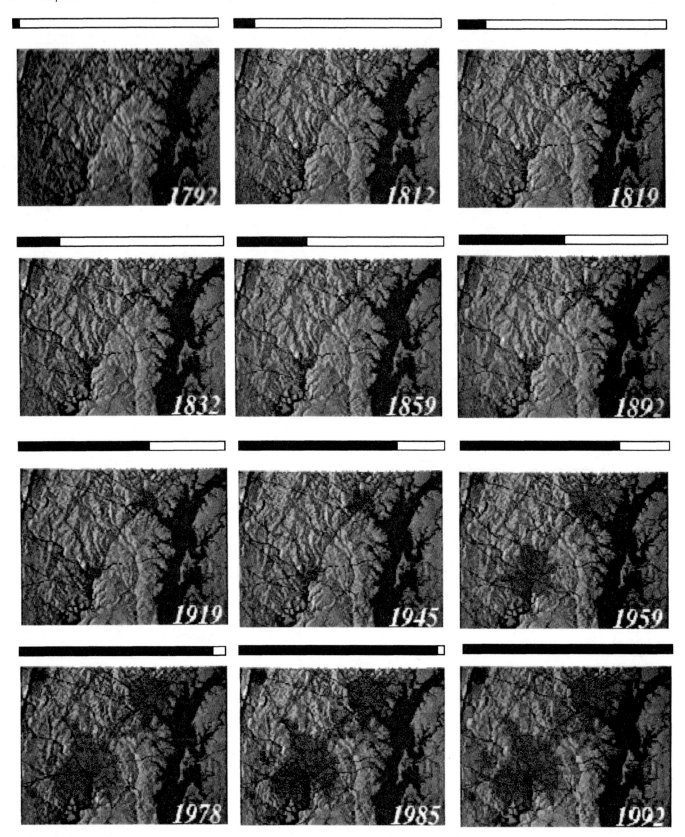

Figure 1.11
Historical growth of the Baltimore/Washington, D.C., urban area (shown in red). Adapted from a U.S. Geological Survey video.

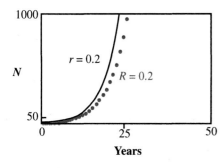

Figure 1.12
Comparison of exponential growth and geometric growth when $r = R = 0.2$/year.

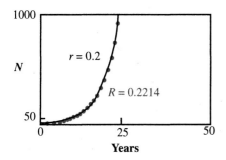

Figure 1.13
Exponential growth with $r = 0.2$ and discrete growth with $R = 0.2214$ give virtually identical time courses.

maps based on the same relative sequence of time slices as that for the lily pond simulation. Urbanized areas are shown in red, indicating clearly that development accelerated after World War II until the mid 1980s when it appears to have slowed.

EXPONENTIAL GROWTH: *R* VERSUS *r*

In financial circles R is called the *yield* or *annual yield* if expressed on a yearly basis, whereas r corresponds to what is called the *annual rate*. Financial institutions usually post both the annual yield and the annual rate in their advertisements. For example, a newspaper advertisement for a bank proclaims the following interest rates for its money market investments.

Annual yield	**Annual rate**
7.51%	7.25%

Note that the two rates are slightly different. To see how they are related, we plotted growth, N, over 25 years for $R = 0.2$/year and for $r = 0.2$/year, as shown in Figure 1.12.

The two rates of growth shown in Figure 1.12 are not identical. The growth of the population (or money if we're talking about a financial investment) is larger for $r = 0.2$ than for $R = 0.2$. This difference makes sense because discrete growth *compounds* only once a year while continuous growth compounds much more frequently—in fact, instantaneously.

We may then reason that, if $R = 0.2$/year in discrete time yields a slower rate of growth than $r = 0.2$/year in continuous time, we need to make R larger if we want to superimpose its growth curve on that for $r = 0.2$. But how much larger?

You can immediately get an insight into this question by examining the two expressions at the bottom of Box 1.1. For geometric growth $N_T = N_0 \lambda^T$, and for exponential growth $N_T = N_0 e^{rT}$. These two expressions are identical after the substitution $\lambda = e^r$. And, since $R = \lambda - 1$,

$$R = e^r - 1. \qquad (1.4)$$

Now let's work through this example. If $r = 0.2$, then we will get an identical curve, starting with the same initial numbers, if

$$R = e^{0.2} - 1 = 0.2214.$$

Thus in financial terms the annual yield would be 0.2214 (or 22.14%) if the annual rate was 0.20 (or 20.0%); \$1000.00 invested today would produce \$1221.40 one year later. The two curves are plotted in Figure 1.13.

The discrepancy between r and R decreases as r approaches 0. When the absolute value of r is much less than 1, we can apply a convenient approximation formula: $e^r \approx 1 + r$. So by Eq. (1.4), $R \approx 1 + r - 1 = r$. For example, for $r = 0.2$, the substitution yields $R = 1.02$, which is very close to the true value of 1.0202.

Exercise: Verify from an advertisement for money market interest rates that the two rates converge according to Eq. (1.4).

A word of caution: Here we have used r for the instantaneous growth rate and R for the discrete rate of population growth, but there is nothing sacred about this particular symbolism and it is not uniform in ecology texts. You will always be able to discern the meaning behind the symbols by looking at the equation, discrete or continuous time, in which they appear. A common alternative is to call r the **Malthusian parameter** and abbreviate it as *m*. Thomas Malthus was one of the first to realize the tremendous social implications of exponential growth for human beings. Charles Darwin attributed Malthus's 1798 book, *An Essay on the Principles of Population as It Affects*

Box 1.2 (Advanced)   *Exponential Growth: Recovering the Continuous from the Discrete*

In discrete geometric growth, there is an implicit time lag of one time step (or one generation) each iteration. The population size is projected forward (or backward) in time, one step at a time. For example, in terms of yearly interest rates this time lag is like the compounding of interest on an investment only at the end of each year. If interest is compounded twice a year, then the time lag is reduced by half. With this modification, Eq. (1.1a) for discrete growth becomes

$$N_{t+1} = N_t \left(1 + \frac{R}{2}\right)^2,$$

where, as before, $t + 1$ means 1 year later than year t. Similarly if interest is compounded q times a year, then

$$N_{t+1} = N_t \left(1 + \frac{R}{q}\right)^q.$$

To reduce notational complexity, we can substitute the symbol x for R/q, leading to an expression for which calculus can provide a ready solution:

$$N_{t+1} = N_t(1 + x)^{R/x}. \qquad \text{(a)}$$

From calculus we can use the identity that defines the base of the natural logarithms e:

$$\text{In the limit as } x \to 0, \quad (1 + x)^{1/x} \to e. \qquad \text{(b)}$$

This equation is read as: In the limit, as x gets smaller and smaller, eventually approaching 0, then $(1 + x)^{1/x}$ comes closer and closer to the number e. For example, when $= 0.1$, $(1 + x)^{1/x} = 1.1^{10} = 2.593$; and when $x = 0.01$, $(1 + x)^{1/x} = 1.01^{100} = 2.705$, thus approaching $e = 2.7183$. . . .

This is where our trick came in handy, allowing us to convert to something simpler. We also use the fact that $(1 + x)^{R/x} = [(1 + x)^{1/x}]^R$ and then substitute this limit of Eq. (b) into Eq. (a), to get

$$N_{t+1} = N_t e^R.$$

In other words, if the interest rate R is compounded continually rather than annually, the total amount of money after 1 year equals $(e^R)(N_t)$ rather than $RN_t + N_t = \lambda N_t$; the latter is smaller. For example, if $R = 8\%$/year, then $\lambda = 1.08$/year—but if this same rate was compounded instantaneously we would have $e^R = 1.083$, or 8.3% interest at the end of a year.

---

the *Future Improvement of Society,* to helping him crystallize his thinking on the theory of evolution by natural selection. Box 1.2 presents a more formal connection between r and R.

PROBLEM (ADVANCED)

Exponential Growth in the Consumption Rate of Resources

Question: If the present worldwide rate of consumption of petroleum increases by 2% every year, how long will it take to use up the earth's petroleum?

Answer: It is important to recognize that in this problem, unlike the problems considered so far in this chapter, the rate of change (in this case of petroleum consumption) is itself growing exponentially. Let's call the rate of petroleum consumption at time t, $C(t)$. Then we can express $C(t)$ as a function of present consumption ($C(0)$), at initial time 0:

$$C(t) = C(0)e^{rt}.$$

The rate constant, r, in units of per year can be determined by the condition $C(1) = 1.02C(0)$. Thus

$$e^{r(1)} = 1.02$$

and because $\ln(1.02) = 0.0198$,

$$r = 0.0198.$$

If $M(t)$ is the amount of petroleum remaining at time t, then the rate of change of $M(t)$ is

$$\frac{dM}{dt} = -C(t) = -C(0)e^{0.0198t}$$

Consumption depletes petroleum, so the minus sign is necessary on the right-hand side of the equation. This equation can be integrated to yield an expression for $M(t)$ versus t. If we call the integration variable t', then we may write this integration as

$$\int_{M(0)}^{M(t)} dM = -\int_0^t C(0)e^{rt'}dt'. \tag{1.5}$$

Evaluating the integrals in Eq. (1.5) gives

$$M(t) - M(0) = \frac{-C(0)}{r}e^{rt} + \frac{C(0)}{r}. \tag{1.6}$$

[Do you see where the r's came from in the denominators of Eq. (1.6)? The indefinite integral of e^t, is just e^t but the indefinite integral of e^{rt} is e^{rt}/r.]

After collecting and rearranging terms, we have

$$M(t) = M(0) + \frac{C(0)}{r}(1 - e^{rt}). \tag{1.7}$$

The petroleum stock will be entirely consumed when $M(t)$ is 0. Let's call that particular time, when petroleum is depleted, T. (The capital T is not to be confused with the lowercase t, which we are using as the symbol for "generic" time.) Hence we defined T by setting $M(t)$ in Eq. (1.7) equal to 0, or

$$0 = M(0) + \frac{C(0)}{r}(1 - e^{rT}). \tag{1.8}$$

After collecting terms to get T on the left-hand side of the equals sign, we get

$$e^{rT} = 1 + \frac{rM(0)}{C(0)}.$$

Taking the log of both sides and dividing by r, we get

$$T = \frac{\ln\left[1 + \dfrac{rM(0)}{C(0)}\right]}{r}. \tag{1.9}$$

[Note that generally the logarithm of a sum of terms cannot be reduced to an expression involving the sum of the logarithms of the separate terms.]

As of 1995, the rate of petroleum consumption, $C(0)$, was about 24.163 billion barrels (bbl) per year and reserves were estimated at $M(0) = 1111$ billion barrels. Substituting these values and $r = 0.0198$/year into eq (1.9), gives the final answer:

$$T = \frac{\ln\left[1 + \dfrac{(0.0198 / \text{year})(1111 \text{ bbl})}{24.163 \text{ bbl} / \text{year}}\right]}{0.0198 / \text{year}} = 32.2 \text{ years}.$$

If these estimates are accurate—and if the rates of consumption continue to increase exponentially—the world's petroleum reserves will be depleted by about 2027.

SOLVED PROBLEMS

1. How long would it take for a pair of individuals to produce the world population today (about 6 billion people) at the present rate of population growth ($r \approx 2\%$ per year)?

Solution:

Call the population size today N_{now} and assume a continuous time model:

$$N_{now} = N_0 e^{rt}.$$

Take the natural log of both sides:

$$\ln N_{now} = \ln N_0 + rt \text{ and } t = \frac{\ln N_{now} - \ln N_0}{r}.$$

The present population, N_{now}, is 6 billion, r is about 2% per year, and $N_0 = 2.0$. Hence

$$t = \frac{\ln 6{,}000{,}000{,}000 - \ln 2}{0.02}$$

$$= \frac{22.51 - 0.693}{0.02}$$

$$= 1091 \text{ years.}$$

2. What is the *doubling time* of this population?

Solution:

We denote the time until the population doubles as t_d, and it occurs when

$$2N_0 = N_0 e^{rt_d}$$

Taking the natural log of both sides, we get

$$\ln[(2)N_0] = \ln(N_0) + rt_d$$

or

$$\ln 2 + \ln N_0 - \ln N_0 = rt_d.$$

Rearranging to solve for the doubling time gives

$$t_d = \frac{\ln 2}{r}. \tag{1.10}$$

This formula gives the doubling time of an exponentially growing population if r is known. In this case, Eq. (1.10) yields

$$t_d = \frac{0.693}{0.02} = 34.7 \text{ years.}$$

Note that for exponential growth the doubling time is independent of initial population size. A population growing at 2% per year will double about every 35 years. A population growing at 1% per year will double about every 69 years.

QUESTIONS

1. The population growth curves for two populations (A and B) are plotted in Figure 1.14. Both populations begin with 4 individuals. One population is growing continuously through time with $r = 0.25$/week, and the other population is growing discretely with $R = 0.25$/week (i.e., $\lambda = 1.25$/week). Which is

    a. Discrete? _____

       Continuous? _____

    b. What is the doubling time for population A? _____
    c. What would the value of R need to be to exactly superimpose curve B on curve A? _____

    d. What would the value of r need to be to exactly superimpose curve A on curve B? _____

    e. Use the initial conditions $t_0 = 0$ and $N_0 = 10$ to write the growth equations for the two populations shown:

$$N_A(t) =$$

$$N_B(t_0 + n) =$$

2. A certain moth has a constant intrinsic per capita growth rate of $r = 0.4$ per generation ($\lambda = e^r = 1.492$/generation). If there are 10,000 moths now and the generation time is 10 days, how many moths will there be in 30 days?

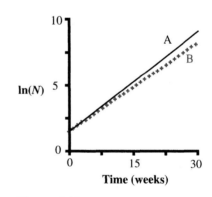

Figure 1.14

3. Another moth breeds only in the summer and survives the winter in the egg stage. A local population increased from 10,000 individuals to 12,000 individuals after one year. Predict the population size after two years, assuming that the per capita growth rate remains constant.

4. Mountain goats were introduced into the Olympic Peninsula in Washington state. The population increased from 12 individuals in 1929 to 1175 individuals in 1983 (Houston et al. 1994). Calculate r for this time frame, assuming a continuous and constant growth rate.

What is the doubling time of this population? Assuming that r remains constant in the future, in what year will the mountain goat population reach 10,000?

5. Mary is going to have an outdoor party in 15 days. She wants to have her backyard pond covered in water lilies before this party, so she goes to the nursery to buy some water lilies. Mary gives the clerk the dimensions of her pool and the clerk, knowing the geometric growth rate of the water lilies that he stocks, calculates that if she purchases a single water lily, it will produce a population of $N = 10,000$ that will completely cover the surface of Mary's pool in 30 days. Mary reasons that if she buys two water lilies instead of one, she can meet her goal of having the pond surface covered in 15 days. Is anything wrong with her logic? If so, how many lilies will she need to buy to meet her goal?

2

Spatial, Temporal, and Individual Variation in Birth and Death Rates

As you have seen in Chapter 1, populations have the capacity to either grow explosively or decline to extinction in short order. We know from experience, however, that natural populations are not climbing to infinity or collapsing to zero within the time frames predicted by exponential growth. Nature is more complicated than that. A population grows slower in some years than in others and in some places better than in others. This temporal and spatial variation means that populations are unlikely to grow at a constant rate. In this chapter we meld such variation into our models for population growth. We explore these models with an eye toward seeing if such variability can prevent explosions and/or collapses, or simply delay them.

MOVEMENTS AND POPULATION GROWTH

Real populations occupy physical space and individuals move across this space. If we could visualize the position of each individual—say through a high resolution aerial photography—we would have a snapshot of the population's distribution at that instant. If this snapshot captured the entire species, it would also depict the **species' geographic range.** While individuals move many times over their lifetimes, the borders of the geographic range may stay relatively constant, except on a much longer time scale. How might this range expand over time? Consider the simplest case where all space is essentially identical in terms of its ability to support population growth; that is, the space is **uniform** in topography and habitat type. Imagine that we inoculate some part of that space with a few individuals of some focal species; we watch the population size grow and as it grows we watch the individuals move across space. We ultimately want to know how quickly the population spreads. To answer this question we need to back up and first examine the rate of spread of a population where individuals only move, but do not die or give birth. Then we will superimpose the birth and death processes that give rise to exponential growth (from Chapter 1).

Random Walks and Diffusion

A starting place to think about the small-scale movement of individuals is simply to assume a blind random walk. Figure 2.1 shows the trails of two individuals following such a walk on a plane (i.e., in two dimensions). Each time period, an individual moves to one of the eight neighboring cells at random.

For most animals a random walk will probably be a poor approximation to their actual movements. Real animals will change their rate of movement over time, in different regions of space, or because of different experiences that they have had. Importantly, real organisms certainly do not choose which direction to head just by chance alone. All these complications produce deviations from a random walk model. Nevertheless, a random walk is a good starting place to examine the effect of the simplest kinds of movement on population dynamics.

Figure 2.1
Two examples of a simulated random walk. Each time the individual lands on a new pixel, it is colored black to indicate the path.

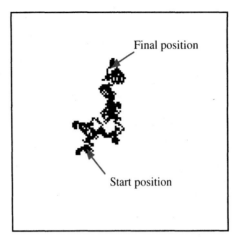 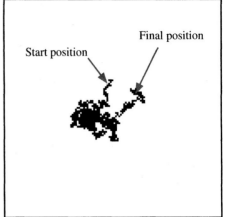

The population consequence of many individuals simultaneously undergoing random walks in continuous time and across continuous space is called **diffusion.** Figure 2.2 illustrates this movement.

The exact position of each particle cannot be predicted since their movements are truly random. Nevertheless, it is possible to derive statistical properties of the collection of particles as they move around.

- The center of this circle should, on average, not change over time because movements are equally likely in each direction. Hence the expected value of all the x and y positions of all the particles will not change over time (again assuming a very large number of particles).

- The particles spread out over time. The radius of the circle contains 90% of all the particles in Figure 2.2, but this particular percentage is arbitrary. The radius is proportional to the expected standard deviation of the particles' positions in the x and y directions. Later we examine this rate of spread in more detail.

- Finally, particle density (number of individuals per grid cell) quickly takes on the shape of a sample from a normal distribution in two-dimensional space (Figure 2.3). This is called a **bivariate normal** distribution because it is normal or Gaussian in both variables (directions) x and y.

Now we make some of these arguments a bit more quantitative. Imagine that space is simply linear (i.e., one-dimensional) and uniform (i.e., with no spatial heterogeneity). A long linear and stagnant stream provides a convenient analogy. To make the scene more intuitive, you can think of the individuals as N dye particles introduced into the center of the stream. For purposes of simulating this on a computer, we divide this stream into 20 connected grid cells numbered from 1 to 20. Each time period a fraction d of the particles in each cell moves to the left and another fraction d moves to the right (Figure 2.4)

To see the effects of diffusion, imagine that at time 0 all the individuals are confined to the central two cells and that each of these cells has 1000 individuals. All the other cells on either side of the central two cells have zero individuals; thus the total initial population is $N = 2000$. At each time period, one-tenth of the individuals in each cell move to the left and another tenth move to the right ($d = 0.1$). We have in tabular form the following numbers across space for three successive time periods:

Time 0:			1000	1000		
Time 1:		100	900	900	100	
Time 2:	10	170	820	820	170	10

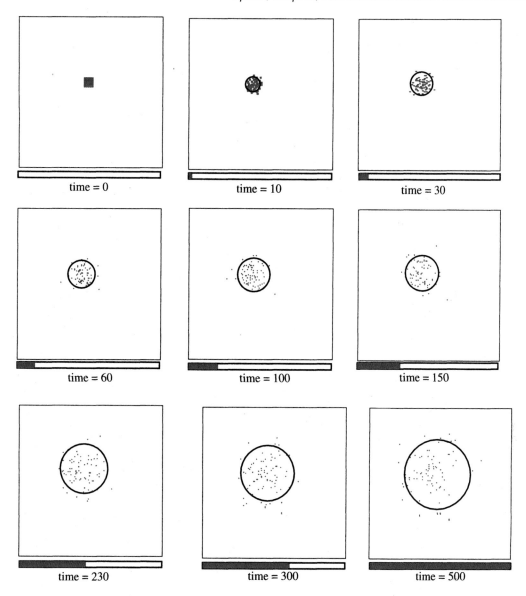

Figure 2.2
Diffusion: Independent random walks. At time 0, 70 particles clustered in the center of the box, begin individual random walks. As for Figure 2.1, space here is really a very fine grid where each grid cell is a single pixel. The particles occupy a single pixel at any one time and move from one pixel to the next each time step. The direction of movement is random: all eight directions have equal probability. The configuration of the particles in space is shown for several successive time steps up to time 500. Since the time intervals between subfigures are not equal, the bars below provide an index of the elapsed time from the beginning of the simulation. The circle in each subfigure encloses 90% of the particles. The center of this circle is at the mean position $(\bar{x}, \bar{y})$ of all the particles at time $t = 0$ (i.e., $\bar{x} = \Sigma\, x_i/70$ and $\bar{y} = \Sigma\, y_i/70$).

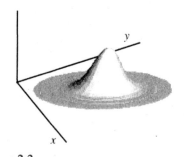

Figure 2.3
Density. A bivariate (x and y) normal distribution, where x and y are uncorrelated and have equal variance.

Exercise: In the preceding "pyramid" of numbers, where did the number $N_x = 170$ come from at time 2?

Solution: The number in that same cell in the previous time step ($t = 1$) was $N_x = 100$:

$$\xleftarrow{\;10\;}\ \boxed{100}\ \xrightarrow{\;10\;}$$
$$\xrightarrow[90]{}\qquad\qquad \xleftarrow[0]{}$$

$$100 - 20 + 90 = 170.$$

Figure 2.4
Discrete Diffusion. The height of the bars measures the density $N_{x,t}$ in each cell x at time t. Some individuals move each time period to neighboring cells. In this example, the initial numbers in the cells are not equal.

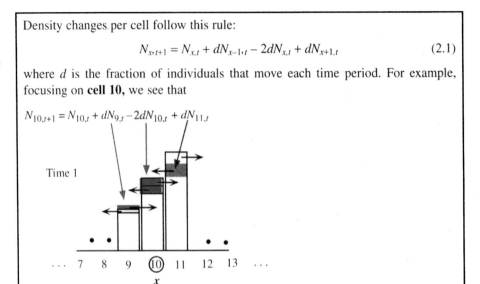

Density changes per cell follow this rule:

$$N_{x,t+1} = N_{x,t} + dN_{x-1,t} - 2dN_{x,t} + dN_{x+1,t} \qquad (2.1)$$

where d is the fraction of individuals that move each time period. For example, focusing on **cell 10,** we see that

$$N_{10,t+1} = N_{10,t} + dN_{9,t} - 2dN_{10,t} + dN_{11,t}$$

Time 1

$$\cdots \quad 7 \quad 8 \quad 9 \quad ⑩ \quad 11 \quad 12 \quad 13 \quad \cdots$$
$$x$$

Figure 2.5
Diffusion without population growth. Two thousand particles are introduced into the central two cells of a long thin stream with no directional flow. A fraction, $d = 0.1$, of the particles moves to each neighboring cell each time period. The distribution of numbers per cell are shown in five-step intervals up to $t = 20$. The red dotted line at $N = 40$ individuals serves as a marker to follow the tail of the distribution as it spreads out from the center. As time goes to infinity, the distribution becomes flat at 100 individuals per cell.

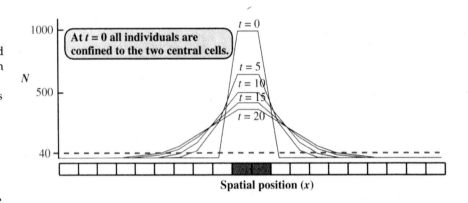

At $t = 0$ all individuals are confined to the two central cells.

Spatial position (x)

Figure 2.5 shows the continuation of this process. The abundance curve for each time period shows N_i in the center point of each cell. A special rule is invoked at the extreme end positions 1 and 20: simply let dN_1 individuals in cell 1 move to the right and dN_{20} individuals in cell 20 move to the left. This creates a **reflecting boundary** since no individuals fall off the edge.

The $N = 2000$ individuals become redistributed over space; the curve describing their abundance across space becomes normally distributed as time goes on. Every time step, the variance of this distribution curve increases. This "flattening" of the curve is a feature of **diffusion.** The greater the fraction of individuals in each cell that moves every time period, d, the sooner the particles will be distributed evenly across the stream. Also as time goes on, the *rate* that the curve expands decreases. Look at the successive distances from one curve to the next, along the x axis in the vicinity of the tails (the red dotted line). Define the radial distance moved by an individual at time t in a one- or two-dimensional random walk as the straight line (**Euclidean**) distance from that individual's position at time t to its initial position at time 0. This radial distance ignores the path of movement over this time interval, that is, the actual path that the individual took to arrive at its present position. It can be shown that, for an infinite number of particles, the mean radial distance averaged across all individuals soon (once the distribution curve takes on the bell shape) increases at a rate directly proportional to the **square root** of time. That is,

Radial distance moved by an individual in a random walk $\propto \sqrt{\text{time}}$, where $\propto$ means "is proportional to". In order for the average individual to wander twice as far, it will take 4 times as much time. In order for the individual to wander 10 times as far, it takes 100 times as long.

A species range will expand because of individual movements. For the moment, let's define a species range as the length (or area if we are working in two dimensions) that contains some fixed proportion of the all the individuals-say, 95%. We can be more precise about how the range changes with time by looking at the equation describing the curves in Figure 2.5, once they take on the Gaussian shape and when space and time are continuous, rather than discrete. (A derivation of this result is in Berg (1983) and Pielou (1969)). We have

$$\mathcal{N}(x,t) = \frac{N_0}{\sqrt{4\pi Dt}} \exp\left(\frac{-(x-\bar{x})^2}{4Dt}\right),$$

where D is a constant that is related to what we have been calling d (more about this later), t is time, and N_0 is the initial number of individuals. Compare this with the standard formula for a Gaussian curve with area 1:

$$N(x;\bar{x},\sigma) = \frac{1}{\sigma\sqrt{2\pi}} \exp\left(\frac{-(x-\bar{x})^2}{2\sigma^2}\right), \tag{2.2}$$

where σ^2 is the variance. By comparing the two formulas, we see that, if we set

$$\sigma^2 = 2Dt,$$

we can get N to look just like $\mathcal{N}$. Thus the variance across space for one-dimensional diffusion increases linearly with time, which implies that the standard deviation grows in proportion to the *square root* of time, or

$$\sigma = \sqrt{2Dt}.$$

Now, the integral (or area) under a one-dimensional normal curve from -1 standard deviation to $+1$ standard deviation is about 68% of the total area, and the area under ±2 standard deviations contains about 95%. Hence, if we define our geographic range as the radius of space that contains 95% of the individuals, then

$$\text{Range} = 2\sigma = 2\sqrt{2Dt}.$$

We conclude that the geographic range expands linearly with the square root of time. It also increases with the square root of D, not D itself.

A remarkable feature of diffusion is that this same linear relationship between average distance moved and the square root of time does not depend upon the size of space and is true for diffusion in two dimensions, three dimensions, or even higher dimensions. Thus for a group of individuals diffusing in two dimensions, **the mean displacement (the radius of the particles' boundary from the center of release) expands across space at a rate proportional to the square root of time.** This means that the geographic *area* occupied by the particles (i.e., the area of the circle that contains a fixed proportion of individuals, as in Figure 2.2) is proportional to the square of the radius and thus increases linearly with time, or

$$\text{Area occupied in two dimensions} \propto \text{radius}^2 \propto \sqrt{\text{time}}\sqrt{\text{time}} = \text{time}.$$

Coming back to the particles enclosed by the circle in Figure 2.2, the radius of the circle increases with the square root of time and thus the area of the circle increases linearly with time.

Fick's Law (Advanced)

We have illustrated diffusion in discrete time and discrete space. **Fick's law** describes the properties of diffusion for continuous time and space (Figure 2.6). The parameter D in Eq. (2.3b) is called the **diffusion coefficient** and has units of distance^2 per unit time. For a small molecule in water at about room temperature, $D \approx 10^{-5}$ cm^2/sec. This molecule would diffuse about 1 cm in 14 hr. One physical interpretation of D is related to

Figure 2.6
Fick's law of diffusive movements.

For discrete space and discrete time, the iteration equation for density change in one dimension is

$$\Delta N_x = N_{x,t+1} - N_{x,t} = -2dN_{x,t} + dN_{x-1,t} + dN_{x+1,t}. \tag{2.3a}$$

We express this change on the right in order of the cells from left to right as

$$= d[(N_{x-1,t} - N_{x,t}) - (N_{x,t} - N_{x+1,t})]$$

$$= \text{(the difference from } x - 1 \text{ to } x) - \text{(the difference from } x \text{ to } x + 1).$$

Note that ΔN_x is then a difference between two differences or, in other words, the *change of a change.*

As both time and space become continuous, this formula becomes:

$$\frac{dN(x,t)}{dt} = D\frac{\partial^2 N(x,t)}{\partial^2 x}. \tag{2.3b}$$

The rate that the density at spatial position x is changing at time t D is the diffusion parameter.   . . . is proportional to the second derivitive of the change in density across space at point x and at time t.

A solution to Eq. (2.3b) for N_0 individuals released at point $\bar{x}$ is the normal distribution of Eq. (2.2):

$$\mathcal{N}(x,t) = \frac{N_0}{\sqrt{4\pi Dt}}\exp\left(\frac{-(x-\bar{x})^2}{4Dt}\right)$$

For two-dimensional space (spatial dimensions x and y), Eq. (2.3b) becomes

$$\frac{dN(x,y,t)}{dt} = D\left[\frac{\partial^2 N(x,y,t)}{\partial^2 x} + \frac{\partial^2 N(x,y,t)}{\partial^2 y}\right],$$

with a solution for the radius r from the point of release $(\bar{x}, \bar{y})$ of

$$N(r,t) = \frac{N_0}{4\pi Dt}\exp\left(\frac{-r^2}{4Dt}\right).$$

the mean squared displacement (i.e., Euclidean distance) of the particles $\overline{\delta^2}$ per unit time, or

$$D = \frac{\overline{\delta^2}}{4t} \quad \text{(in two dimensions)} \tag{2.4a}$$

$$D = \frac{\overline{\delta^2}}{2t} \quad \text{(in one dimension)} \tag{2.4b}$$

That is, the recipe for D is: take the Euclidean distance of each individual from its position at an earlier time, δ_i, square each of these distances, δ_i^2, take the average of these squares, and then divide by either $4t$ or $2t$.

There is one last important connection we can make involving D. We have discussed the bivariate normal distribution of particles over space, which is a solution to the diffusion equation. The variance of any random variable δ is defined as variance $= E(\delta^2) - (E(\delta))^2$, where $E(\ )$ means the expected value, the mean. Since we are assuming that movements in each direction are equally likely, $E(\delta) = 0$ because movements to the left, on average, cancel out movements to the right. Thus the variance is just $E(\delta^2)$, which is the mean squared displacement, that is, $\overline{\delta^2}$. We can also connect $\overline{\delta^2}$ with the variance equation, Eq. (2.2); namely, $\overline{\delta^2} = \sigma^2$. The reason for the difference in the constant (2 versus 4) in Eqs. (2.4a) and (2.4b) is that in two dimensions the distance moved has both a Δx and a Δy component. Thus the square of the Euclidean distance $\delta_{x,y}$ is

$$\delta_{x,y}^2 = (\Delta x)^2 + (\Delta y)^2.$$

Considerations of symmetry show that $(\Delta x)^2 = (\Delta y)^2$. Consequently, if the variance (after one time step) in one dimension is defined to be $2D$, then the variance in two dimensions is twice that, or $4D$, since the variance of a sum of independent variables is the sum of the individual variances.

Since D is a kind of variance, we now understand why its units are distance^2 per time unit. We can also derive a rough approximation of the diffusion coefficient D for continuous time and space from the fraction dispersing each time step in discrete space and time, which we have been calling d. Consider the following table that calculates the mean squared displacement along one dimension for individuals moving out of cell 0 to cells -1 and $+1$ in a single time step.

Distance moved, δ	Distance squared, δ^2	Probability	Probability times distance squared
-1	1	d	d
0	0	$1 - 2d$	0
1	1	d	d

Thus the mean squared displacement is the sum of the terms in the last column which is $\delta^2 = 2d$ and substituting this into Eq. (2.4b), we find that $D = d$. This is not a mathematically rigorous derivation because it glosses over the complications that arise from going to continuous space and time, but it does illustrate the point that D and d are closely related. More formally, for one-dimensional space, D is,

$$D \cong d\frac{\Delta x^2}{\Delta t}$$

And D goes to d as long as Δx^2 and Δt decrease at about the same rate as they both go to zero.

For most students, the fact that Fick's law says that the growth rate dN/dt at each position is proportional to the second derivative in Eq. (2.3b) seems strange, but a little thought shows that this relationship has to be the case. It is clear from Eq. (2.3a) that the net change in density at position x at time t is 0 unless there is some differential in density between that point and its neighbors. If $N_{x,t} = N_{x-1,t} = N_{x+1,t}$, then $N_{x,t+1} = N_{x,t}$. However, a differential in density can still give zero net change if it is linear across space. Berg (1983) provides a delightful proof of this relationship. We illustrate it with an example based on applying the discrete space random walk of Eq. (2.3a), as illustrated in Figure 2.7.

Movements Plus Geometric Growth

With the understanding of simple diffusion given by Figures 2.6 and 2.7, we can return to the simple case of one-dimensional space and now add geometric population growth with discrete rate λ in each cell. The complete dynamics may be modeled as a difference equation for each cell x:

$$N_{x,t+1} = \lambda N_{x,t} - 2dN_{x,t} + dN_{x-1,t} + dN_{x+1,t}.$$

Figure 2.8 shows the result of a simulation in which the dispersal fraction d is again 0.1 and $\lambda = 1.1$/time step in each of the 20 cells. Thus, if each cell was a closed population, the cell's population would increase by 10% each time step.

The two shaded cells in the center receive 100 individuals each at $t = 0$; all the other cells have zero individuals. The population changes in each cell are due to both population growth and movements to and from immediate neighbors. The result is displayed each 5 time steps. The average individual's displacement from its initial position still increases with the square root of time, but now individuals give birth while they are wandering across space. Each of their offspring then begins a new random walk from its birthplace. The population of individuals in each cell is shown as the successive curves for different elapsed times; these curves do not flatten out, as they do for pure diffusion without population growth, but instead become amplified as time advances.

Focus on the **cell $x = 1$.** Its change in density over time is described by

$$N_{1,t+1} = N_{1,t} + dN_{0,t} - 2dN_{1,t} + dN_{2,t}$$

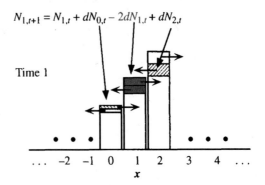

Time 1

If the slope in density is constant over space, then $N_{x,t}$ can be described by applying the formula for a straight line with slope a and y intercept b, $N_{x,t} = ax + b$. The cell densities at time 1 are:

x	$N_{x,1}$
0	b
1	$a + b$
2	$2a + b$

So cell 1, at time $t = 1$ contains $N_{1,1} = a + b$ individuals. But at time 2, cell 1 will also contain $a + b$:

$$N_{1,2} = a + b - 2d(a + b) + db + d(2a + b) = a + b$$

initial density in $x = 1$

emigration from $x = 1$ to $x = 0$ and 2

immigration from $x = 0$

immigration from $x = 2$

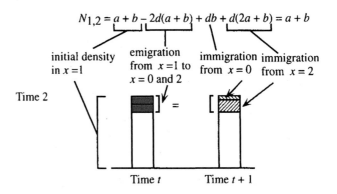

Time 2

Time t Time $t + 1$

Conclusion: No net change in local density occurs if the slope of density is constant across space. Net density changes per cell require that the slope vary across space. Since the slope itself is the first derivative, dN_x/dx, its rate of change over space is the second derivative, d^2N_x/d^2x.

Figure 2.7
Density changes over time for diffusion are given by the second derivative of density over space.

Consequently, as shown by the dashed line across Figure 2.8 at the initial number $N_0 = 100$, successively more peripheral cells reach this threshold abundance as time goes on. Moreover, if you look carefully at these intersection points from left to right, you see that, unlike the case for pure diffusion, they are regularly spaced (except for the first two time periods). That is, **the wave front of population advance moving across this space quickly reaches a constant velocity.** This constancy is very different from the dynamics when only diffusion occurs (Figure 2.5).

Figure 2.9 plots the number of cells in Figure 2.8 that reached more than 40 individuals as time goes on. This is repeated for a greater movement rate, $d = 0.2$.

These lines have steps because even though the distribution curve spreads out smoothly over time, the number of individuals in a cell has to exceed the threshold level

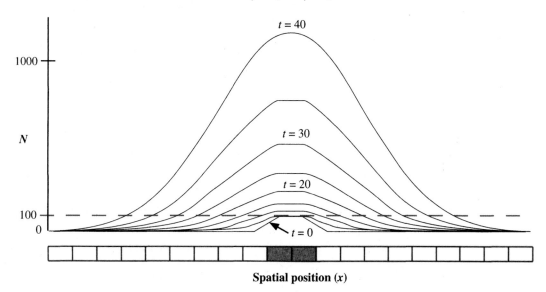

Figure 2.8
Diffusion with geometric growth. At $t = 0$, 100 particles are introduced into each of the two central cells. For all cells, $\lambda = 1.1$ and $d = 0.1$. The red dotted line at $N = 100$ serves as a marker to follow the population wave as it moves outward from the center. For purposes of simulation, the population in each cell first changes through its λ, then movements occur, and so on, for each time step.

Figure 2.9
The one-dimensional spread of a diffusing and geometrically growing population ($\lambda = 1.1$, so $r = 0.095$) for two different movement rates, d. The length of the occupied region ($N(x) > 40$) grows approximately linearly with time. Since the process is symmetrical in space, the occupied length advances in steps of 2 cells, one on each side. Note that doubling d did not double the slope, but increased it by a factor of $\sqrt{2}$.

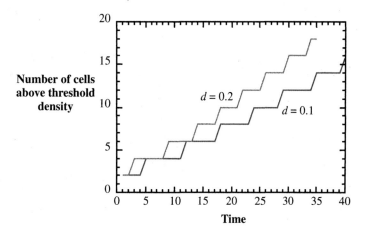

(set at 40 per cell) for that cell to be counted in the range. Since we have only 20 cells, this produces the steps. If we draw a line connecting the step centers, we see that the advance of the population is roughly linear with time in both cases. This result is also true for two dimensions. That is, assuming that each spatial location is identical in terms of the λ that it supports, **the radius of the population's occupied area eventually grows linearly with time.** Thus the area occupied grows at a rate proportional to the *square* of time. In order for the occupied area to grow 4 times as large, it takes only twice as much time. In order for the area to grow 9 times as large, it takes 3 times as long.

Intuition suggests that the area occupied by the population will grow at a faster rate if a greater fraction of individuals move each time step and if local populations have a higher growth rate λ. Intuition is correct; diffusive movements on the plane in combination with exponential growth yields the result that the area of the circle that is expected to contain all but a single individual has a radius^2 given by (Pielou 1969)

$$\text{Radius}^2(t) = 4Dt \ln(N(t)).$$

Since $\ln(N(t))$ is itself growing according to $\ln(N(t)) = rt + \ln(N_0)$, where r is the intrinsic growth rate,

$$\text{Radius}^2(t) = 4Dt[rt + \ln(N_0)]. \qquad (2.5)$$

As time goes on, the first term in the brackets will dwarf the second, so we can write for later times

$$\text{Radius}^2(t) = 4Dt\,[rt]$$

and

The asymptotic rate of spread of the circle's radius (distance/time) = $2\,[rD]^{0.5}$. $\quad$ (2.6)

Note that this asymptotic rate is constant over time (t does not appear on the right-hand side of Eq. (2.6)), and this rate does not depend on the initial numbers, nor the value of the threshold numbers used to define the enclosing circle. If the population is not increasing (i.e., if r = 0), we are back to pure diffusion and the only individuals are the

Figure 2.10
Five muskrats were introduced into Bohemia near Prague in 1905. (a) Contour lines for the spread of the muskrat (*Ondatra zibethica*) 1905–1927. After Elton (1958). (b) The total area occupied plotted by year. Note the straight-line relationship between the square root of the area and time since the muskrats were introduced near Prague. After Skellam (1951). These values were obtained by approximating the occupied area by a circle. Dividing the slope of this line by $\sqrt{\pi}$ gives the average increase of the radius in kilometers per year. This procedure yields an estimate of about 11.8 km/yr.

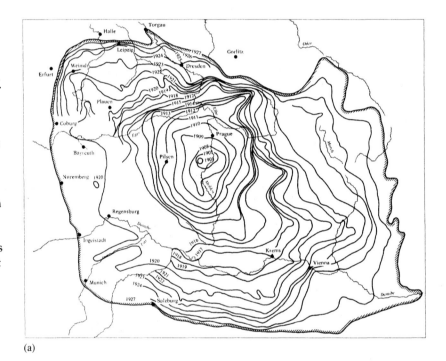

(a)

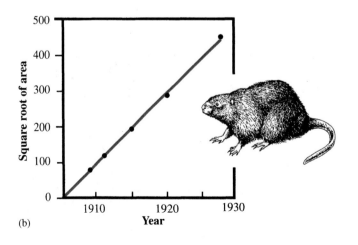

(b)

initial numbers N_0 that simply redistribute themselves over space. Going back to Eq. (2.5), the first term in brackets is now 0, so we have for the circle holding all but a single individual

$$\text{Radius}^2(t) = 4Dt \ln(N_0).$$

This expression shows in quantitative detail that for pure diffusion in two dimensions, the occupied *area* increases linearly with time (or the radius increases with the square root of time). The *rate* of increase in this radius is

$$\text{Rate of spread of the circle's radius (distance / time)} = \frac{2\sqrt{D \ln(N_0)}}{\sqrt{t}}.$$

This rate declines as time advances. Also we see that doubling D does not double the rate of advance, since this rate depends on $\sqrt{D}$. Finally, the rate of spread for diffusion alone depends on the initial numbers.

Introduced species provide an intriguing application of this model. Muskrats, a large rodent native to North America, provide one example. In the beginning of this century a landowner near Prague in Bohemia (now the Czech Republic) imported muskrats from North America to raise the animals for their pelts. Unfortunately, five animals escaped in 1905. As the population grew and individuals dispersed, the boundaries of the muskrat population increased over time (Figure 2.10). As predicted by the model, the occupied radius grew at roughly a constant rate (about 12 km/yr, averaged over all the compass directions). The student interested in this invasion will enjoy the more detailed analysis provided by Andow et al. (1990).

The spread of exotic Argentine ants (*Linepithema humilis*) in an old field near San Diego, California (Figure 2.11), provides another example—this time on a much finer spatial scale (Erickson 1971).

These ants disperse by colonies budding off new colonies adjacent to them, rather than aerial flights of new queens across unoccupied space. Consequently, the rate of spread averages only about 100 m/yr. Table 2.1 shows some other rates of spread for biological invasions, all of which are greater than the figure for Argentine ants. The spread of any species eventually must stop, since it will eventually occupy all the habitable area between physical or biotic barriers to further dispersal.

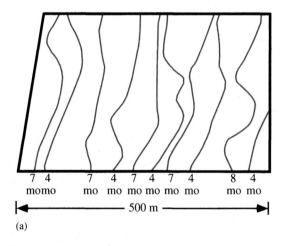

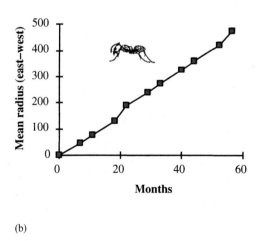

(a)   (b)

Figure 2.11
(a) The spread of Argentine ants across an old field. Each contour gives the approximate limit of the ants at various times. Time is expressed in months since the last census. (b) The positions in the east–west direction are averaged (across the north–south direction), and these averages are plotted as a function of elapsed time in months. Again we see an approximate linear increase in the occupied radius over time. Modified from Erickson (1971).

Table 2.1. **The Observed Rates of Range Expansion for Some Biological Invasions. from Grosholz (1996).**

Species	Latin name	Observed velocity of spread (km/yr)
Terrestrial species		
Weedy plant	*Impatiens glandulifera*	9.4–32.9
Gypsy moth	*Lymantria dispar*	9.6
Cabbage butterfly	*Pieris rapae*	14.7–170
Cereal leaf beetle	*Oulema melanopus*	26.5–89.5
Muskrat	*Ondatra zibethica*	0.9–25.4
Grey squirrel	*Sciurus carolinensis*	7.66
Collared dove	*Streptopelia decaocto*	43.7
European starling	*Sturnus vulgaris*	200
Plague bacterium (in human host)	*Yersinia pestis*	400
Marine Species		
Tunicate	*Botrylloides leachi*	16
Bryozoan	*Membranipora membranacea*	20
Crab	*Carcinus maenas*	55
Crab	*Hemigrapsus sanguineus*	12
Barnacle	*Elminius modestus*	30
Snail	*Littorina littorea*	34
Mussel	*Mytilus galloprovincialis*	115
Mussel	*Perna perna*	95

SPATIAL VARIATION

In the models we have developed so far, each point in space is treated as ecologically equivalent, yet in reality different positions on the earth's surface have different physical features, different weather, and different biotic conditions. A population's growth rate is likely to vary from one place to another. The regions of space where populations increase and areas where $\lambda < 1$ may form a patchwork quilt. Fragmentation of an animal's natural habitat by human activities such as deforestation and urban development creates a mosaic of small islandlike favorable patches of natural habitat isolated from one another by agricultural lands or urban/suburban regions.

Let's now explore some consequences of habitat patchiness by letting λ vary with spatial position, x. To keep the analysis simple, we assume as before that space is simply one dimensional. In these examples, we suppose that some cells in the interior of the space can support a $\lambda(x) > 1$, while peripheral cells have a $\lambda(x) < 1$. Each cell begins with just two individuals. As before, a fraction d of the local population moves to each of the two neighboring cells each time period. Two simulations are shown in Figure 2.12.

Within the large patch of 7 favorable cells in Figure 2.12(a), the population steadily grows, but it shrinks in the unfavorable peripheral cells where $\lambda < 1$. Note too that the rate of population increase (seen as the increment in N between time steps) does not seem to be accelerating as it did in Figure 2.10 when all cells supported $\lambda > 1$. In fact, for the example in Figure 2.12(a), the asymptotic rate of growth for the entire population summed over all 20 cells is very close to 1 (1.006), and even the central cell has a growth rate of only 1.0068, much less than the λ based only on local birth and death of 1.025. The favorable patch's growth rate is being dragged down by net movements of individuals into the unfavorable peripheral sites. We can think of the interior patch as a **source** population "discharging" a net flow of individuals out. The peripheral cells represent a **sink** population that receives a net flow of individuals. In this way sink populations are being subsidized by the large interior patch.

For the small favorable patch size of 4 cells in Figure 2.12(b), the population first increases for about the first 15 time periods, but then begins to shrink. As time goes on, the populations in both the favorable patch and the peripheral unfavorable cells decline

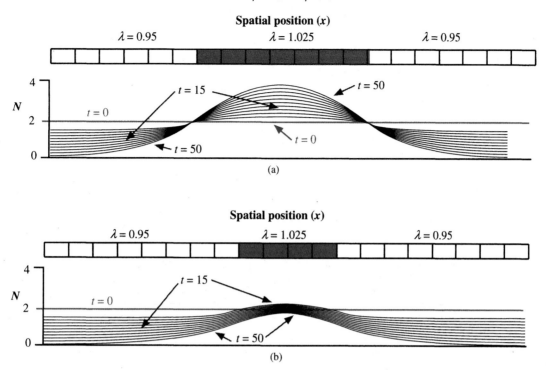

Figure 2.12
The intrinsic growth rate, $\lambda(x)$, varies across space, x. In the red region $\lambda > 1$, and in the white region $\lambda < 1$. Each cell begins with two individuals at $t = 0$ (the red horizontal line). In both cases $d = 0.2$; the only difference is the size of the region with $\lambda > 1$. The population density across space is drawn every 5 time steps.

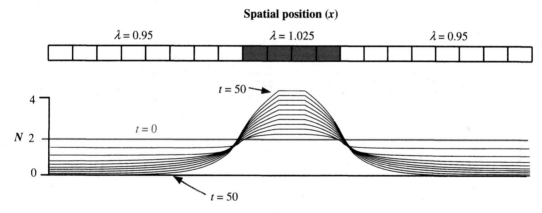

Figure 2.13
As in Figure 2.12(b), we have a patch of 4 cells with $\lambda(x) > 1$, surrounded by cells in which $\lambda(x) < 1$. The movement rate is reduced from $d = 0.2$ to $d = 0.05$. Now the population in the central cells continues to increase over time.

asymptotically to 0. The net loss of individuals moving out of the favorable patch is greater than the growth of the population within these cells. The favorable patch is a net source of births, but dispersal moves them to the peripheral cells, which are a dead-end sink. This effect can be reversed by decreasing the fraction of individuals that move each generation (Figure 2.13).

Figure 2.14 presents two favorable patches, both with $\lambda(x) > 1$. They are separated by a region of space where $\lambda(x) < 1$. The simulation shows that the population in each patch continues to increase over time, although this rate of increase is much less in the small patch than the larger one.

Now imagine that the large patch in Figure 2.14 is destroyed. We do this by decreasing the $\lambda(x)$ for these 8 cells to the background level of 0.95. The small patch is left unharmed, but a large source of immigrants to it has been eliminated by the destruction

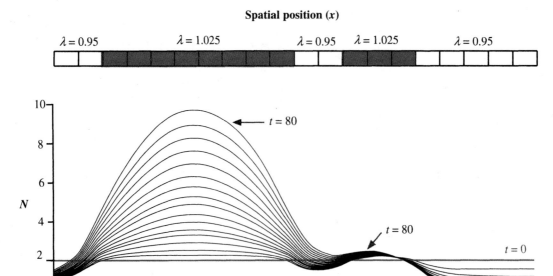

Figure 2.14
Two patches with $\lambda(x) > 1$ in a sea of cells where $\lambda(x) < 1$. The fraction of individuals in each cell moving to the left and to the right per time period is 0.1. $N(x)$ is shown every 5 time steps.

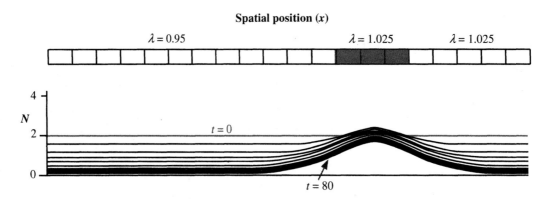

Figure 2.15
One small patch with $\lambda(x) > 1$ in a sea of cells where $\lambda(x) < 1$. Here $d = 0.1$. $N(x)$ is displayed every five time steps.

of the large patch. Even though the habitat in the small patch still supports a $\lambda > 1$, the population in the small patch, after an initial spurt of positive growth, declines asymptotically to 0 once its larger neighbor is destroyed (see Figure 2.15).

Thus a population occupying a habitat patch is prone to extinction from a combination of low growth rate, high dispersal rate, small patch size, and isolation from neighboring patches. The loss of individuals dispersing out of the patch can then be greater than the population growth rate per time period within the patch. **One important conclusion is that it might be difficult to preserve highly motile species with small conservation parks.**

This interaction between patch size and dispersal rate can also be developed using a simple analytical argument. Imagine a two-dimensional habitat patch within a defined region (Figure 2.16). The number of organisms inside the patch at time t is N_t and the density (D_t) is the number per unit area, or N_t/A. Birth and death of individuals within the patch create a local growth rate of λ. To consider the total change in numbers, however, we must also include the movement of individuals out of the patch. Once individuals leave, they die. We assume that the patch is so isolated by regions with $\lambda < 1$ that the immigration of individuals into the patch is negligible. The total emigration rate is

Figure 2.16
A habitat patch. Dispersers come only from a band of width w along the edge.

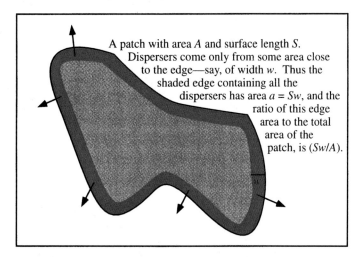

A patch with area A and surface length S. Dispersers come only from some area close to the edge—say, of width w. Thus the shaded edge containing all the dispersers has area $a = Sw$, and the ratio of this edge area to the total area of the patch, is (Sw/A).

proportional to the length of the edge of the patch (rather than its area), since only individuals at the edge are candidates for leaving the patch.

Thus numerical changes in the patch population can be approximated by the difference equation;

$$N_{t+1} = \lambda D_t A - efN_t, \tag{2.7}$$

where e describes the emigration rate for individuals in the edge area and f is the fraction of the edge area compared to the total patch area. We assume that the population density in the edge is equivalent to the density in the core area. As shown in Figure 2.16, this fraction f is given by Sw/A, where S is the surface length of the patch Substituting this expression for f in Eq. (2.7) yields

$$N_{t+1} = \lambda D_t A - e\frac{Sw}{A}N_t.$$

Dividing each term by A and collecting terms in D_t gives

$$D_{t+1} = D_t\left[\lambda - \frac{Sew}{A}\right].$$

Surface length S is proportional to the square root of area by some constant k. (For example, if the patch is a circle, $S = 2(\pi A)^{0.5}$ and $k = 2\sqrt{\pi}$; for other shapes, k will be a somewhat larger constant). Making the substitution $S = k\sqrt{A}$ yields

$$D_{t+1} = D_t\left[\lambda - \frac{kew\sqrt{A}}{A}\right]$$

and thus

$$D_{t+1} = D_t\left[\lambda - \frac{kew}{\sqrt{A}}\right].$$

If the population in the patch is to increase, $D_{t+1} > D_t$, which implies that the term in brackets must be greater than 1, which in turn implies that the following inequality be satisfied:

$$\lambda > \frac{kew}{\sqrt{A}} + 1.$$

This inequality is easier to fulfill as area increases, as the emigration rate e decreases and as λ increases. Note also that it is no longer sufficient for λ to simply exceed 1 if the patch's population is to increase (an observation we witnessed earlier in the simulations of Figure 2.12b).

Exercise: Take a look again at Figure 2.12(b). Here's a blow-up of the population change in just the 6 central cells:

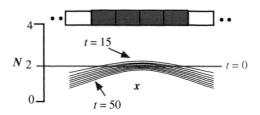

Since the growth rate $\lambda(x)$ is a constant over time in each cell x and is greater than 0 in the central 4 cells, why does the population size N first increase in these central cells but later reverse to become negative after $t = 15$? What happens to cause this reversal?

INDIVIDUAL VARIATION AND TEMPORAL VARIATION

A population's per capita growth rate, λ, incorporates each individual's probability of reproducing or dying per unit time. An organism's physiology, age, body size, behavior, prior experience, anatomy, and location all influence its probability of giving birth and of dying. Since these factors can vary over time and across individuals, the population growth rate for collections of individuals should, in fact, be a variable, not a constant as we have been assuming so far. The fate of a finite population, particularly a small population, is influenced by these factors, which contain chance elements. It is sometimes useful to distinguish **temporal** (or **environmental**) **variation** in the favorableness of the environment and **between-individual variation** (also called **demographic variability**) at the same moment of time.

Even without a complete understanding of the actual cause of this variability, we can still describe statistical properties of collections of individuals and predict, for example, the expected size of a population, whether the population might ultimately decline to extinction, and, if so, in what time frame. Shortly, we make a stab at some of these predictions. Some students may want to refer to Appendix 6 for definitions of the terms **expected value, mean,** and **variance.** Figure 2.17 introduces the logic and notation that we will follow.

For a population of N individuals, we figuratively draw λ_i out of a hat (the universe distribution) for each of the $i = 1$ to N individuals at each time period. For example, if individual i has a λ_i of 1.5, it will be replaced by 1.5 individuals in the next year. This could be accomplished by it surviving over the year plus having a 50% chance of giving birth to a single offspring, or it might die but have a 50% chance of giving birth to three offspring. We will not concern ourselves with these individual details, since we want to focus here on the populational consequences of variation in λ_i. The realized population growth rate in year t, $\lambda(t)$, is the arithmetic average of these λ_i values over the N individuals. To get the individual λ_i we must specify the distribution from which they are chosen. This is the **universe distribution,** shown as a cloud in Figure 2.17. Its shape (normal or uniform, etc.) and its mean and variance need to be specified.

We explore the case of pure temporal variability later by letting the universe distribution collapse to a single number (without any variance) each year, applied to all individuals, but this number varies from one year to the next, as specified by still another random distribution. We explore the case of pure between-individual variation by assuming that the universe distribution has some variance within a single year as shown in Figure 2.17, but the mean and variance remain constant over time.

When there is some between-individual variation, the "realized mean" λ for a finite population of N individuals will usually, by chance, not equal the mean of the universe distribution, μ_λ. For example, let's choose λ_i for each individual i by a toss of a coin for

Figure 2.17
To model individual variation requires that we find a plausible way to assign different values to different members of a population. Selecting these values from some continuous random variate is a reasonable approach. Here this random variate is called the universe distribution and is represented by a cloud, since it is never depleted by sampling and because its parameters may shift over time. Two types of variation affect $\lambda(t)$ and thus N_{t+1}: possible temporal variation in the universe distribution (i.e., if u_λ varies over time) and the chance sampling of individuals that goes into determining the realized values for λ_i that contribute to the mean, $\lambda(t)$, for the population at each time step.

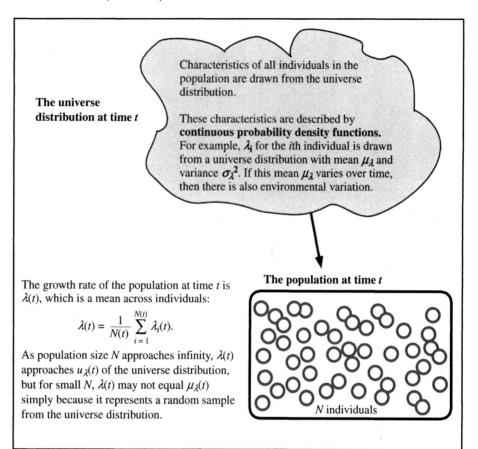

The universe distribution at time t

Characteristics of all individuals in the population are drawn from the universe distribution.

These characteristics are described by **continuous probability density functions.** For example, λ_i for the ith individual is drawn from a universe distribution with mean μ_λ and variance σ_λ^2. If this mean μ_λ varies over time, then there is also environmental variation.

The growth rate of the population at time t is $\lambda(t)$, which is a mean across individuals:

$$\lambda(t) = \frac{1}{N(t)} \sum_{i=1}^{N(t)} \lambda_i(t).$$

As population size N approaches infinity, $\lambda(t)$ approaches $u_\lambda(t)$ of the universe distribution, but for small N, $\lambda(t)$ may not equal $\mu_\lambda(t)$ simply because it represents a random sample from the universe distribution.

The population at time t

N individuals

a very small population of $N = 2$ individuals. The universe distribution λ_i is either 0 (for tails) or 1 (for heads) with probability 0.5 for each outcome. The mean of this random distribution, μ_λ, is 0.5, and the variance is $(0.5)^2 = 0.25$. With only two individuals, by chance, both might have $\lambda = 1$, and this event has the substantial probability of $(0.5)(0.5) = 0.25$. This particular **sample** of $N = 2$ has a realized mean λ of 1 even though the universe distribution has a mean of 0.5. This discrepancy arises because of sampling error: **simply by chance you are not likely to duplicate the true mean and variance in a sample.** As population size, N, increases, large divergences between the realized mean for a sample, λ, and the universe mean, μ_λ, become increasingly unlikely. For example, the probability of drawing a population of $N = 100$ with a realized sample mean of $\lambda = 1$, using the same coin toss, is extremely remote: 0.5^{100}.

To summarize, first we imagine an underlying universe distribution from which the probabilities for individuals are assigned. Second, we have for a finite number of individuals a sample realization of those probabilities, which forms the between-individual variation. Finally, we wish to combine these processes to explore the population consequences of those realizations. Both shifts in the universe distribution over time and different possible realizations for samples drawn from the same universe distribution produce population variation. We first consider the consequences of temporal variability alone, then between-individual variability alone, and finally put the two together.

Temporal Variability Alone

Here we assume that the variance in the universe distribution, σ_λ^2, is 0, so sampling error for between-individual differences can be ignored as a determinant of variability in λ. All individuals in the population behave identically, and the population mean $\lambda(t)$ at time t exactly equals the universe mean $\mu(t)$ at each time step, since the universe distribution is simply a point. Yet, as the notation suggests, this point varies over time.

Consider a simple comparison of two populations with the same initial size: in population A, the $\lambda(1)$ for year 1 is 1/year and $\lambda(2)$ is 2/year. In the second population, B, the λ in both years is 1.5. Which population will be the largest after two years? Applying Eq. (1.3a) to both populations we get

Population A: $N_2(\text{year 3}) = N_0\lambda(1)\lambda(2) = N_0(1)(2) = 2\,N_0$

and

Population B: $N_2(\text{year 3}) = N_0\lambda(1)\lambda(2) = N_0(1.5)(1.5) = 2.25N_0.$

Over this two year period, population B has grown more than A even though both populations have the same arithmetic average population growth rate of 1.5. Because population growth is inherently a geometric process, the arithmetic mean λ is the wrong average to consider for answering the question. Instead, **when λ's vary over time, the geometric mean λ serves as the appropriate measure of net population growth.**

The geometric mean gives the yearly λ, which, if reproduced year after year for n years without variation, would give the same final population size as does a series of varying $\lambda(t)$. The geometric mean of n numbers is the nth root of the product of those n numbers, or

$$\text{Geometric mean } \lambda = \sqrt[n]{\lambda(1)\lambda(2)\lambda(3)\cdots\lambda(n)} = \left[\prod_{i=1}^{n}\lambda(i)\right]^{1/n}. \qquad (2.8)$$

In the preceding two-year example, both populations have an arithmetic mean λ of 1.5, but for population A

$$\text{Geometric mean } \lambda = \sqrt{(1)(2)} = \sqrt{2} = 1.414$$

and for population B

$$\text{Geometric mean } \lambda = \sqrt{(1.5)^2} = 1.5.$$

In fact, the arithmetic mean is *always* greater than or equal to the geometric mean. To understand why this is the case, note that taking the natural log of Eq. (2.8) yields

$$\ln \text{ of the geometric mean } = \ln\left[\prod_{i=1}^{n}\lambda(i)\right]^{1/n}$$

$$= (1/n)\ln\big(\lambda(1)\lambda(2)\lambda(3)\cdots\lambda(n)\big)$$

$$= (1/n)\big[\ln\big(\lambda(1)\big) + \ln\big(\lambda(2)\big) + \ln\big(\lambda(3)\big) + \cdots + \ln\big(\lambda(n)\big)\big]$$

$$= (1/n)(n)(\textbf{arithmetic mean of } \ln(\lambda))$$

$$= \textbf{arithmetic mean of } \ln(\lambda). \qquad (2.9)$$

Now let's look at what the ln function does graphically, as illustrated in Figure 2.18. Note how equal intervals on the x axis in Figure 2.18 get "translated" into nonequal intervals on the y axis plotting $\ln\lambda$. In this way, higher values of λ are discounted in the mean for $\ln\lambda$ compared to the mean of λ. Consequently, the arithmetic mean λ (the mean of the values on the x axis) will be greater than the geometric mean λ. The geometric mean equals the arithmetic mean only when the numbers have no variance, as when the geometric mean, 2, of 2, 2, 2 is identical to the arithmetic mean. The greater the variance among the numbers, the greater is the difference between the arithmetic and geometric means.

Problem: Consider a hypothetical bird population: 50% of the time the population increases by 20% and the other 50% of the time the population decreases by 20%. Assuming that the two kinds of years are independent over time, what will happen to this population?

> **Solution:** Over a long sequence of n years: $n/2$ years will yield $\lambda = 1.2$, and $n/2$ years will yield $\lambda = 0.8$. The geometric mean then is
>
> $$\lambda = (0.8^{n/2}1.2^{n/2})^{1/n} = \sqrt{(0.8)}\sqrt{(1.2)} = 0.9798.$$
>
> This population will ultimately become extinct since the geometric mean λ is less than 1.0. Starting with an initial abundance of 400 birds, 90% of the population becomes extinct (<1 individual) before 400 years, by simulation.

The long-term behavior of a population, and particularly its persistence, are consequently influenced not only by the mean λ across years but also the temporal variance in these λ's. As the variance increases, the geometric mean λ decreases, all else being equal (this is proven in Chapter 7).

Figure 2.19 plots ln(population size) over time for 10 simulations in which λ is selected at random from a uniform distribution between 0.5 and 1.5 (the universe distribution). Thus λ is now a **continuous random variable,** meaning that its value each year is chosen at random from some continuous distribution. All real numbers between 0.5 and 1.5, are equally probable. When we select a particular $\lambda(t)$ for the year, it applies to every individual in the population (i.e., we assume no between-individual variability); thus in the terminology of Figure 2.17, $\lambda(t) = u_\lambda(t)$. The next year we select another $u_\lambda(t)$ and thus $\lambda(t)$ for all individuals, and so on. The **mean** of a uniform distribution is simply the midpoint of its range, (min + max)/2 and the variance is range^2/12 (see Box 2.1). The **range** here is $1.5 - 0.5 = 1$, so the mean λ is $(0.5 + 1.5)/2 = 1$ and the variance is 1/12, or about 0.08333.

Figure 2.18
The natural logarithm of a number (red curve) versus the number. The vertical lines show two values of λ (2 and 3) and how they map onto the ln λ scale. The arithmetic mean of these two numbers is 2.5, the midpoint between them. The mean of ln 2 and ln 3 is the midpoint between them and is also shown and compared with ln 2.5, the ln of the arithmetic mean. Because the slope of the ln function is always positive but declining (i.e., a negative second derivative), the ln(arithmetic mean) must always lie above the mean of the ln's of a series of different numbers.

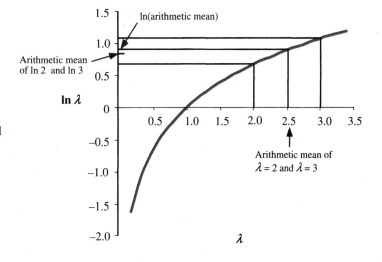

Figure 2.19
Geometric growth when λ is drawn randomly each time step from the uniform distribution [0.5 to 1.5]. Here the arithmetic mean $\lambda = 1$, but the populations tend to decline to extinction because the geometric mean $\lambda < 1$ (≈ 0.959).

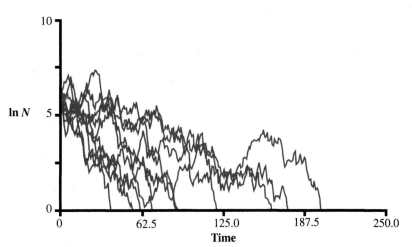

Box 2.1 (Advanced; Requires Calculus).   *The Mean of a Function of a Continuous Random Variable*

For the uniform distribution, it is possible to find an exact solution for the geometric mean to compare to the approximation formula, Eq. (2.12). Most of you are familiar with discrete random events like the flip of a coin or the roll of a die. The formula for the mean, or expected value, of a discrete random variable x is defined as

$$\text{Mean } x = \bar{x} = \sum_{i=\min}^{\max} p_i x_i .$$

Here p_x is the probability of each possible discrete outcome x. For a roll of a die there are 6 possible outcomes x, the numbers 1 through 6, and each x has equal probability $p_x = 1/6$. Thus the mean outcome is

$$\sum_{i=1}^{6} \frac{1}{6} x_i = \frac{1}{6}\sum_{i=1}^{6} x_i = \frac{1}{6}[1+2+3+4+5+6] = 3.5.$$

By extension, for a continuous random distribution x, the formula for the mean of a continuous random variable x with a **probability density distribution $p(x)$** is

$$\text{Mean } x = \int_{\text{lower limit}}^{\text{upper limit}} p(x)\,x\,dx.$$

The density function $p(x)$ for the uniform distribution between two numbers a and b looks like

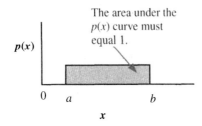

The area under the $p(x)$ curve is 1 since it represents all possible outcomes. Therefore the height of $p(x)$ can be determined since area $= 1 = $ (width)(height) $= (b - a)$(height). Thus the height is $1/(b - a)$.

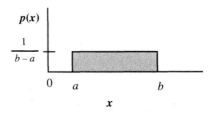

While the probability of getting any particular exact value of x is infinitesimally small, the density function $p(x)$ is defined such that for any interval of x—say, from m to n—the probability of getting any of the possible outcomes within that interval (m, n) is

$$\int_{m}^{n} p(x)\,dx = \text{probability of outcomes between } m \text{ and } n.$$

In graphical terms:

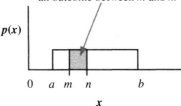

$p(x)$ is defined such that the area under $p(x)$ between m and n is the probability of gettting an outcome between m and n.

Now, with this understanding of what a probability density function is, we can return to the formula for the mean. If $p(x)$ is the density function for the uniform distribution, then the

Arithmetic mean of the uniform distribution from a to b

$$= \int_{a}^{b} \frac{1}{b-a}\,x\,dx = \frac{1}{b-a}\int_{a}^{b} x\,dx = \frac{1}{b-a}\left[\frac{x^2}{2}\right]_{a}^{b}$$

$$= \frac{1}{2(b-a)}[b^2 - a^2] = \frac{1}{2(b-a)}(b-a)(b+a).$$

In Figure 2.19 each population was initiated with a population size of 400. If the population falls below one individual, it is considered extinct. Consistent with a geometric mean $\lambda < 1$, all the populations out of 10 simulations went extinct before time 250.

What is the geometric mean λ for such a random process? By taking natural logs we transform this multiplicative process into a simpler additive process (see Eq. (2.9)):

$$\ln(\text{geometric mean}) = \text{arithmetic mean of } \ln \lambda.$$

But what is the arithmetic mean of $\ln(\lambda)$ when λ is a continuous random variable? Note that we want to find the arithmetic mean of the logs, *not* the log of the arithmetic mean.

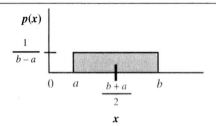

Mean of a uniform distribution = $(b + a)/2$ = the midpoint between a and b.

mean of $\ln(x)$ = $[1.5 \ln(1.5) - 1.5] - [0.5 \ln(0.5) - 0.5]$

$$= -0.892 + 0.847$$

$$= -0.0452.$$

By this analytical method, we find that the geometric mean = $\exp(\text{mean } \ln(x)) = \exp(-0.0452) = 0.956$, which is very close to the approximate solution, 0.959, that we found earlier.

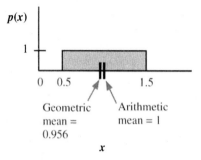

For this problem we want to know the mean, not of x, but of a function f of x, $f(x)$. Our particular concern is with $f(x) = \ln x$. Yet, in general, for any $f(x)$, the mean will depend on the probability density function $p(x)$. To convey this dependence we write $f(x)$ as $f(p(x))$. Then

Mean of $f(p(x)) = \int_{\text{lower limit}}^{\text{upper limit}} p(x) f(x) dx.$

The formula for the mean of $f(x)$ when x is a continuous random variable with probability density function $p(x)$

In our case $f(x)$ = the natural logarithm, $\ln(x)$, and $p(x)$ is the uniform distribution from a to b. Substituting these expressions for $p(x)$ and $f(x)$ into the formula, we get

$$\text{Mean of } f(x) = \ln x = \int_a^b \frac{1}{b-a} \ln(x) dx$$

the $p(x)$ function   the $f(x)$ function

With a trick from calculus, this integral evaluates as

$$\frac{1}{b-a} \left[x\ln(x) - x \right] \Big|_{x=a}^{x=b}$$

For the problem at hand, $b = 1.5$ and $a = 0.5$, so $1/(b-a) = 1$, and

The advantage of the approximate method is that finding an analytical formula giving the exact solution for the mean and variance of $f(x)$ may be impossible for a particular choice of $p(x)$ and $f(x)$. The simplicity of the uniform distribution and the $\ln(x)$ function allowed for an easy solution, but this ease is typically the exception rather than the rule. The Taylor's expansion in Eq. (2.12) is a good approximation when $\sigma^2 << \mu^2$. The means and variances for some random distributions are presented in Appendix 6.

Exercise: Show that the variance of a uniform distribution between a and b is $(b-a)^2/12$, by applying the formula for the variance of a continuous random variable x:

$$\text{Variance of } x = \int_{\text{lower limit}}^{\text{upper limit}} p(x) x^2 dx - (\text{mean } x)^2 .$$

Initial solution: Substitute into the variance formula the density function and mean of the uniform distribution:

$$\text{Variance of } x = \left[\frac{1}{b-a} \int_a^b x^2 dx \right] - \left[\frac{b+a}{2} \right]^2$$

These are not the same quantities, as emphasized in Figure 2.18. Using some tricks from calculus, we can apply a Taylor's expansion (see Chapter 5), to approximate both the mean and variance of any function of a random variable. If λ is a random variable with arithmetic mean μ and variance σ^2 and f is some function of the random variable λ, then the mean and variance of the function, $f(\lambda)$, are approximately

$$\text{Mean of } f(\lambda) \approx f(\mu) + (f''(\mu)/2)\sigma^2 \tag{2.10}$$

and

$$\text{Variance of } f(\lambda) \approx f'(\mu)^2 \sigma^2, \tag{2.11}$$

Figure 2.20
Geometric growth when λ is a random variable each time step. In this case, λ is drawn randomly from a uniform distribution [0.55 to 1.65], and the arithmetic mean and the geometric mean λ are both greater than 1. Compare to Figure 2.19.

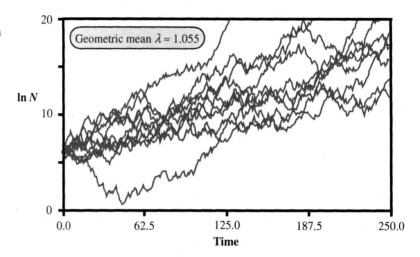

where f' is the first derivative of f with respect to λ and f'' is the second derivative. These derivatives are evaluated at the mean $\lambda = \mu$. Applying the relationship in Eq. (2.10) to the ln function, we reach

$$\ln(\text{geometric mean } \lambda) = \text{arithmetic mean of } \ln \lambda \approx \ln(\mu) - (\sigma^2/2\mu^2). \qquad (2.12)$$

The derivative $f'(\lambda)$ of $\ln \lambda$ is $1/\lambda$; the second derivative $f''(\lambda)$ is the derivative of $1/\lambda$, which is $-\lambda^{-2}$. We evaluate these derivatives at $\lambda = \mu$, substituting μ for λ, to get $-1/\mu^2$.

If we apply Eq. (2.12) to the uniform distribution from 0.5 to 1.5 for λ and substitute in $\mu = 1.0$ and $\sigma^2 = 1/12 = 0.08333$, we calculate $\ln(\text{geometric mean } \lambda) \approx [\ln(1) - (1/12)/(2(1^2))] = [0 - (0.08333/2) = -0.0417$. Finally, taking antilogs, the geometric mean λ is $\approx e^{-0.0417} \approx 0.959$. This value compares favorably with the exact geometric mean, 0.956, obtained using calculus (Box 2.1).

Populations with $\lambda(t)$ drawn randomly between 0.5 and 1.5 each year should usually trend toward 0 because the geometric mean is less than 1. This trend matches the results in Figure 2.19. Figure 2.20 shows the results of simulating 10 populations again beginning each with $N = 400$ individuals) when $\lambda(t)$ is drawn randomly from a uniform distribution with range 0.55 to 1.65. The arithmetic mean λ is now greater than 1, 1.1, but the range is also larger, 1.1. Hence the variance is $(1.1^2/12) = 0.10083$. The ln of the geometric mean (using Eq. 2.12) is approximately $\ln(1.1) - (0.10083/((2)(1.1)^2) = 0.0536$, and the geometric mean λ is therefore $\approx e^{0.0536} = 1.055$. Since the geometric mean is greater than 1 (in spite of the larger variance), the typical population should trend upward.

Not a single population in Figure 2.20 became extinct after 250 time steps, although one nearly did so before recovering. Having a geometric mean λ greater than 1 does not necessarily mean that each population is immune from extinction. Rather, it means that the average population will increase in size geometrically as time goes on. As the population size continues to increase, it "escapes" from the threat of extinction since it takes an increasingly longer succession of bad years to bring it back down. If there were an upper limit to population size (a subject we explore in depth in Chapter 5), then populations could not as readily escape the threat of extinction. The likelihood of extinction increases as the upper limit of population size decreases (Middletown et al. 1995).

Finally, one more example will illustrate the importance of the difference between geometric mean and arithmetic mean rates of population growth. For this example, the arithmetic mean is greater than 1 but the geometric mean is less than 1 (Figure 2.21). In this simulation not a single population out of 10 survives to time 200.

Figure 2.21
Geometric growth when λ each year is drawn randomly from a uniform distribution [0.1 and 2.1]. Compare to Figures 2.19 and 2.20. These populations tend to decline because the geometric mean λ is less than 1.

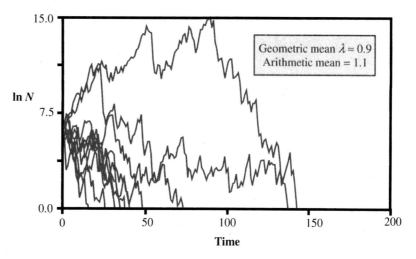

Problem: Given a uniform distribution for temporally varying λ, with a mean λ of 1.1, what range of this uniform distribution is necessary to make the geometric mean λ equal to 1.0?

Answer: Apply Eq. (2.11) for the variance of a uniform distribution and Eq. (2.12) to calculate the variance and mean after the ln transformation: The required range $\approx$ 1.664.

The initial population size also makes a difference in the average persistence time before a population becomes extinct. For example, if all the populations begin with 800 individuals instead of 400, the average time to extinction is much greater. Conversely, if the populations begin with a very small size (e.g., < 10), many might become extinct even though the geometric mean, λ, might be greater than 1. A complete development of these expected extinction rates is beyond the scope of this primer, but interested students can learn more from Pielou (1969), Nisbet and Gurney (1982), and Box 2.2.

Between-Individual Variability Alone

As we discussed earlier, another source of variability in population growth is due to chance events that affect whether an individual gives birth or dies at any single moment. Even two identical twins will not necessarily produce the same number of offspring. They may both be drawn from the very same universe distribution, but by chance one gets struck by lightning and thus has a low λ, while the other becomes rich and prosperous, leaves many children, and thus has a high λ.

If the initial size of a population is N individuals at time t, then on average at time $t + 1$, it will have size

$$\bar{N}_{t+1} = \mu_\lambda N_t, \qquad (2.13)$$

where the over bar represents expected values. Again μ_λ is the arithmetic mean of the universe distribution for λ_i and this distribution in not changing over time. For T time units into the future,

$$\bar{N}_{t+T} = \mu_\lambda^T N_t. \qquad (2.14)$$

The only factor causing variability in growth rate here is the chance sampling by random draws from the universe distribution for each individual. As you can see for the

Box 2.2 (Advanced)

In the preceding section, we were a bit casual about defining what is meant by the behavior of the "typical" population. In this box, we make a more precise probability statement. We certainly do not mean that the expected population size increases at a rate given by the geometric mean. In fact, the expected (or mean) population size after T years is

$$E(N_T) = N_0 \, \bar{\lambda}^T,$$

where $\bar{\lambda}$ is the *arithmetic mean* λ for those T years (Lewontin and Cohen 1969). How can this be? We were just convinced that the geometric mean λ—not the arithmetic average—tells us whether the population is likely to increase or increase. That too is correct. The discrepancy arises because, while the average population size is growing toward infinity as long as $\bar{\lambda} > 1$, the probability of the population becoming extinct after T years is governed by the geometric mean. How can the expected population size increase at the same time that the probability of extinction increases? This can happen because the probability distribution for population size at time T becomes increasingly skewed as time goes on—it becomes increasingly L-shaped. While most populations are close to zero, a thin tail of the probability distribution contains a tiny fraction of populations that are huge in size. Thus the mean population size $E(N_T)$ increases geometrically with time, despite the fact that nearly all populations are below this mean population size and, in fact, are growing smaller!

It is too difficult for us to calculate this highly skewed probability distribution for N at time T, but without too much effort we can calculate precisely the probability that after T years a population will be below its initial number, N_0. Such a population is on its way to extinction. We can show that this probability is given directly by the geometric mean λ and the elapsed time T. Let's write this probability as

$$\Pr(N_T < N_0).$$

Since the logarithm of a number is a smoothly increasing function of that number, we may also write

$$\Pr(N_T < N_0) = \Pr(\ln N_T < \ln N_0).$$

And since

$$\Pr(N_T < N_0) = \Pr\left(\sum_{i=1}^{T} \ln \lambda_i < 0 \right)$$

for the population size at time T to be below N_0, the sum of the logs of λ_i for those T years must be less than 0. Dividing both terms in the last inequality on the right by T, we get

$$\Pr(N_T < N_0) = \Pr(\overline{\ln\lambda})_T < 0).$$

The term on the right is the probability that the mean of ln λ_i over the T-year sequence is less than 0. From Eq. (2.9), we can also write this probability as

$$\Pr(N_T < N_0) = \Pr(\ln(\text{geometric mean } \lambda)_T) < 0).$$

We assume that ln λ_i from one year to the next are independently drawn from the same universe distribution with mean u_G and standard deviation σ_G. For any given string of T years, the realized $\overline{(\ln\lambda)}_T$ for those years is a sample mean from this universe distribution. From the Central Limit Theorem, this sample mean will be approximately normally distributed and have mean u_G and variance σ_G^2/T. We are interested in the probability of $\overline{(\ln\lambda)}_T$ being less than 0; this corresponds to a population that has decreased by year T. This probability is given by the integral of the normal distribution (N) (with mean u_{ln} and variance σ_G^2/T) from $-\infty$ to 0:

$$\Pr(N_t < N_0) = \int_{-\infty}^{0} N(x; u_G, \sigma_G / \sqrt{T}) \, dx.$$

If the geometric mean λ is less than 1, then u_G, which is the log transform, is less than 0. As time T goes on, the sample standard deviation decreases, since it is equal to $\sigma_G/\sqrt{T}$, and thus an increasingly smaller portion of the distribution crosses above 0 (Figure 2.22). By the same token, if μ_G is greater than 0, then as time goes toward infinity, the probability of extinction approaches 0 (Lewontin and Cohen 1969).

Figure 2.23 shows 100 replicate populations for the case of Fig 2.21; λ_i each year is a random variable uniformly distributed between 0.1 and 2.1. For this range the arithmetic mean λ is greater than 1 but the geometric mean is less than 1. Each population begins with only 2 individuals. The observed mean population size, 277.4, is close to the predicted size $(2)(1.1)^{50} = 234.7$, yet the vast majority of populations fell below a single individual, indicated by the horizontal line.

simulations in Figure 2.24, sampling error alone will lead to some deviation in the average growth rate of the population over time. This "temporal" variability is not, however, what we are calling environmental variability; this term is reserved for actual shifts in the universe distribution from year to year.

Note in Figure 2.24 that, early on when population sizes are low, the lines are jagged—populations sizes can actually dip, **despite the fact that the mean λ is 1.2.** But as time goes on and population sizes increase, the lines for each population smooth out and become nearly linear since there is less sampling error.

Figure 2.22
Each normal curve gives the probability density function for $\overline{(\ln\lambda)_T}$ (the mean of the $\ln \lambda_i$ after T years); $T = 1, 5,$ and 20. The ln of the geometric mean λ for these years is equal to this mean. In this example, $\ln \lambda_i$ for each year is drawn from a universe distribution with a negative mean: $u_G = -0.5$ and a standard deviation of 1. Hence the average λ is less than 1 ($e^{-0.5} = 0.6065$). The shaded regions represent the probability that $\overline{(\ln\lambda)_T}$ is greater than 0, which corresponds to a population that has increased in size after T years. As time goes on, this probability decreases. When the geometric mean λ is less than 1, the typical population declines to extinction as T increases.

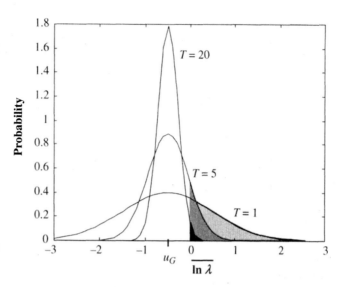

Figure 2.23
An example of a situation where the mean population size increases geometrically, but most populations decline to extinction. Here λ is drawn randomly from a uniform distribution [0.1 to 2.1]. The red lines show 100 replicate populations beginning from $N_0 = 2$. The black line marks a population size of one individual.

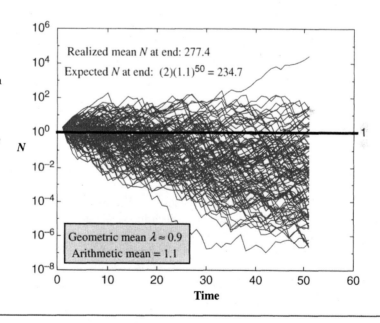

One last word of caution: even though the mean population is expected to grow with rate λ in a given time step, there is still some finite probability that a particular population will become extinct even if $\mu_\lambda > 1$, just as there is some finite probability that you might go home a winner from a Las Vegas casino even though the odds are stacked against you. A complete development of these probabilities is beyond the scope of this primer, but interested students can learn more from Pielou (1969) and Nisbet and Gurney (1982).

Figure 2.24
A simulation of 20 populations responding to the same demographic variability. For each individual in a population, its λ_i is chosen from a uniform distribution from [0.2 to 2.2]; thus the mean λ across individuals is 1.2. These λ_i are then summed to get the number of individuals at the next time step (which is then rounded to an integer). The initial population size is 10. From Eq. (2.14), we expect the average population size after 20 time steps to be $10(1.2)^{20} = 383$, shown by the arrow at the right, and this agrees well with the actual value averaged over these 20 replicates of 377.

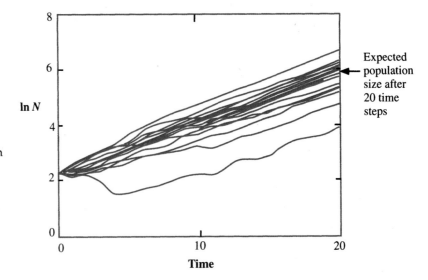

Figure 2.25
Combining temporal and individual variability. Each rectangle is the universe distribution for λ_i in a given year. The grand arithmetic mean of all the yearly means is $E(\mu_\lambda)$. The variance in the yearly means, σ_E^2, provides a measure of the temporal environmental variability.

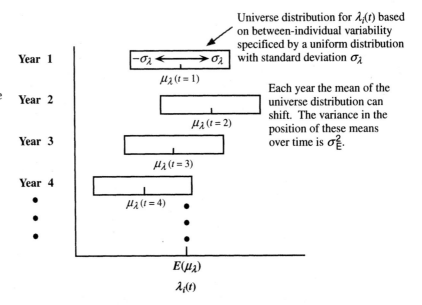

Combining Between-Individual and Temporal Variability (ADVANCED)

In this section we explore the interaction between temporal variability and individual variability. Each individual has a λ_i drawn from the universe distribution for that year, and the population $\lambda(t)$ is the mean of that sample. Even without any temporal variability in the universe distribution, the mean λ for a sample of individuals can differ by chance from year to year, just as it did in Figure 2.24. With temporal environmental variability, the universe distribution itself changes over time. To keep things simple, we use a uniform distribution for the universe distribution and simply shift the mean μ_λ randomly each year while keeping the range the same. In this way, the shape and variance, σ_λ^2, of the universe distribution do not change over time. To accomplish these temporal random shifts in the mean, we specified yet another uniform distribution in the section on "Temporal variability alone." Here we leave the details of this distribution vague, except to say that it has a variance, σ_E^2. This situation is illustrated in Figure 2.25.

Each year we can determine the population size for the next year by selecting λ_i from the universe distribution for each of the $N(t)$ individuals. When $N(t)$ is very large, the sample is very large, and $\lambda(t)$ will typically be similar to the mean of the universe distribution $\mu_\lambda(t)$ for that year t. Let's now consider the distribution of these sample means when $N(t)$ is not necessarily large.

The Central Limit Theorem states that the sum (call it S) of a number (N) of independent random variables (x_i) approaches a **normal** distribution as n increases, even if the individual variables being summed have nonnormal distributions and different means. Furthermore, the expected value of the sum, $E(S)$, equals the sum of the means of the individual distributions μ_i, or

$$E(S) = \sum_i^N \mu_i .$$

The variance of this sum, S, equals the sum of all the individual variances, or

$$\sigma^2(S) = \sum_{i=1}^N \sigma_i^2 .$$

Since we have N individuals, all chosen from the same distribution with variance σ^2, the variance of the sum is

$$\sigma^2(S) = N\sigma^2.$$

Now the rate of growth for a population of size N is $\lambda_N(t)$, which is a sample mean; it is just a sum of all the λ_i divided by a constant, the population size N. The distribution of $\lambda_N(t)$ for repeated samples of population size $N(t)$ will thus be approximately normal, with an expected value equal to the mean of the universe distribution $\mu_\lambda(t)$. Its variance can be found by invoking one additional rule for functions of random variables: if S is any random variable with a variance σ^2 and k is any constant, then the variance of kS is $k^2 \sigma^2$. In our case, the constant k is $1/N$ and $k^2 = 1/N^2$. Hence we have

$$\sigma_N^2 = \frac{N\sigma_\lambda^2}{N^2} = \frac{\sigma_\lambda^2}{N} , \qquad (2.15)$$

where σ_N^2 is the variance in λ_N for a population of size N (note that we dropped the t(s) only for notational simplicity). Since N is in the denominator of the right-hand side of Eq. (2.15), the larger the N, the smaller is the value of σ_N^2 (Figure 2.26). The square root of σ_N^2 is σ_N and in statistics is called the **standard error of the mean.**

Figures 2.24 and 2.25 may be combined, as in Figure 2.26, to show how the population growth rate $\lambda_N(t)$ may change over time (Figure 2.27). Note that for a single population only a single $\lambda_N(t)$ is selected from the red, bell-shaped curves at each time step.

The long-term growth rate of a population is determined by the geometric mean of these yearly sample means over time, which are related through a log transformation to the mean of $\ln(\lambda_N)$, as we developed earlier. Without any environmental variability, the only factor leading to variation in sample means across years is chance sampling effects. In that case, applying the approximation formula for the mean of the ln of a random variable (i.e., Eq. 2.12), we have

$$\ln \text{ (geometric mean)} = \text{mean of ln } \lambda_N \approx \ln(\mu_\lambda) - \sigma_N^2/2 \, \mu_\lambda^2 .$$

Since the mean λ across years and samples is $E(\mu_\lambda)$,

$$\text{Mean of } \ln(\lambda_N) \approx \ln(E(\mu_\lambda)) - \frac{\sigma_\lambda^2}{2N(t)\big[E(\mu_\lambda)\big]^2} .$$

It is difficult to evaluate this formula because $N(t)$ is changing over time in response to the value of λ at each time step. Nevertheless, for large N the second term is negligible, and therefore the mean of $\ln\lambda_N$ is only slightly less than $\ln(E(\mu_\lambda))$, which would be the ln of the population growth rate in an unchanging environment.

Environmental variability, the change in the mean of the universe distribution from year to year, provides an additional source of variation in $\ln\lambda_N$, which arises from σ_E^2 (Figure 2.25). The total variance in $\ln\lambda_N$ from sampling error *and* temporal environmental variability is just the sum $(\sigma_\lambda^2/N) + \sigma_E^2$ because variances of independent variables are additive (See Appendix 6). Again applying the approximation formula for the

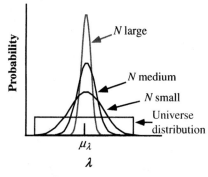

Figure 2.26
The probability plotted is that of obtaining λ with a sample of N individuals whose λ_i are drawn from the universe distribution. The larger the sample size N, the smaller is the standard error for the estimate of the mean in the universe distribution μ_λ. Regardless of the shape of the universe distribution, which in this case is rectangular, the density function for the mean of a sample from the universe is approximately a normal distribution with mean μ_λ.

Figure 2.27
Each year the universe distribution for λ_i shifts (the rectangles) and therefore so does the distribution for the mean, λ_N, for a population of size N sampled from this universe (in red). We chose a rectangular distribution for the universe distribution solely to make it conceptually clear and graphically distinct from the sample mean distribution, not because a uniform distribution is necessarily a more realistic assumption than, say, a normal distribution.

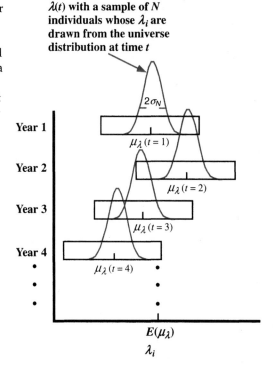

The probability of obtaining $\lambda(t)$ with a sample of N individuals whose λ_i are drawn from the universe distribution at time t

Figure 2.28
The geometric mean rate of growth declines with environmental variance and with decreases in population size; $E(\mu_\lambda) = 2$, and $\sigma_\lambda = 1$.

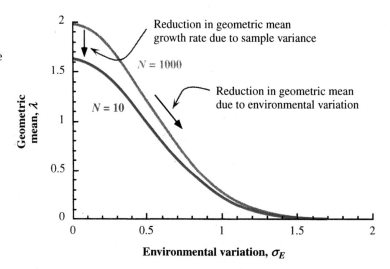

mean of the ln of a random variable (Eq. 2.12) and noting that the arithmetic mean λ across years and samples is the grand mean $E(\mu_\lambda)$, we finally reach

$$\ln(\text{geometric mean } \lambda_N) = \text{mean of } \ln(\lambda_N) \approx \ln(E(\mu_\lambda)) - \frac{\dfrac{\sigma_\lambda^2}{N(t)} + \sigma_E^2}{2[E(\mu_\lambda)]^2}. \quad (2.16)$$

The addition of environmental variability, σ_E^2, in the numerator always increases the value of the fraction on the right and thus decreases the geometric mean. This function is graphed in Figure 2.28 for some selected parameter values. The greater the temporal environmental variability specified by σ_E^2, the lower is the geometric growth rate of the average population. The sample variance also reduces the geometric mean—but to a significant degree only when N is small. The conclusion that we can draw is that small populations face a greater threat of chance extinction than do large populations from both demographic and temporal environmental variability. However, in large populations, the role of environmental variability is proportionately more important in determining their dynamic behavior.

For some types of models of population growth, it is possible to derive formulas for the probability of extinction after a fixed number of years (e.g., Pielou 1969, Nisbet and Gurney 1982, Goodman 1987, Lande 1993), but doing so is often not easy and we are often forced to rely on simulation models. However, under no situation does the addition of temporal variability alone stabilize a population by preventing it from ultimately climbing toward infinity or collapsing to zero. This stabilization requires some sort of density dependence in the growth rate such that λ tends to get smaller as N gets larger, a theme developed in Chapter 5.

PROBLEMS

1. Suppose that populations are governed by the equation

$$N(t + 1) = N(t) + \text{births} + \text{immigrants} - \text{deaths} - \text{emigrants}.$$

Consider two populations found in two adjacent patches (cells) with the following geometry.

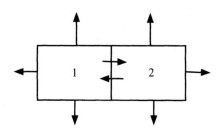

Individuals that leave the cells perish. Each population has the same fraction of emigrants leaving each side of the square per year. Call this emigration rate per individual per side of each cell e. Thus three sets of emigrants leave square 1 and perish and one set of emigrants goes to square 2. Assume that each population has the same per capita birth and death rates and the same initial population size. Further, both populations grow geometrically in discrete time. If the per capita birth rate is 0.8 and the death rate is 0.1, what emigration rate e can each population afford without eventually going extinct? How does this critical level of e change as the birth rate and death rate change? For a given birth rate b and death rate d, find the emigration rate where $N(t + 1) = N(t)$.

2. Per capita growth parameters for a population that exists in two adjacent patches are:

Patch	Birth rate	Death rate	Emigration rate
A	0.33	0.1	0.06 in each direction
B	0.2	0.1	0.04 in each direction

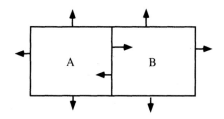

Assume that both populations grow exponentially and that $r = \text{births} - \text{deaths}$. Emigrants leaving A's right side all enter B as immigrants and vice versa (as shown by the arrows). If both populations start with equal numbers of individuals, will they persist? If patch A is eliminated, will patch B persist? If patch B is eliminated, will patch A persist?

3. Another formula for the diffusion coefficient D involves the mean dispersal radius of recaptured individuals after t time units (rather than the square of these distances as in Eq. 2.4). This is a measurable quantity in field ecology and can best be determined by putting small transmitters on animals to help find them. Let's call this distance measure $U(t)$, which is related to the diffusion coefficient D as $U(t)^2 = \pi D t$ (see, e.g., Shigesada and Kawasaki 1997, page 38). If the λ for muskrats is approximately 4/yr and the population has been expanding by 12 km/yr, what is $U(t)$ per year? (*Hint:* use the formula above and Eq. 2.6).

4. An exotic species of elf ant has recently been introduced to the fictional island of Qualam (area = 200,000 acres). The ant population grows geometrically with a λ of 1.5 and expands into new territory as it spreads, according to a diffusion model. The natives have calculated its occupancy of territory as

1999: Just introduced;

2006: Area of Qualam occupied = 400 acres.

At this rate of spread, how much area of Qualam is likely to be occupied by 2001? When will all of Qualam be colonized by elf ants? (What a Qualamity!).

5. A population of birds is influenced by its environment such that two-thirds of the time the population increases by 50% per year and the other one-third of the time the population decreases by 50% per year. What happens to the size of this population as time goes toward infinity?

6. (Advanced) What is the long-term expectation of N for a population growing geometrically when λ each year is a random variable with a *normal* distribution whose mean is 1.1 and whose variance is 2? (Assume no individual variation.)

7. Two individuals—one a male and the other a female—found a new population. This pair produces exactly 4 offspring that year; then both adults die. Offspring sex is determined by the flip of a coin with equal probability for males and females. What is the probability that this population will become extinct at the end of the first year because of a sex ratio imbalance?

3

Population Growth with Age or Stage Structure

As we demonstrated in Chapter 1, if a population has a constant per capita discrete growth rate λ, its population size can easily be projected into the future. Clearly, things become more complicated if different-aged individuals have different probabilities of dying or giving birth. How can we average or combine these separate age-specific birth and death rates to produce a growth projection for the entire population? Charles Darwin (1872) in the 6th edition of the *Origin of Species* wanted to illustrate the geometric process of population growth by calculating the number of elephants that could be produced beginning with a single pair.

> *The elephant is reckoned the slowest breeder of all known animals, and I have taken great pains to estimate its probable minimum rate of natural increase; it will be safest to assume that it begins breeding when thirty years old, and goes on breeding till ninety years old, bringing forth six young in the interval, and surviving till one hundred years old; if this be so, after a period of from 740–750 years there would be nearly nineteen million elephants alive, descended from the first pair.*

Actually this calculation went through some revisions in his book's various editions. It appears that Darwin wasn't completely sure how to combine the elephant's **vital rates** (the birth and death rates of each age) to get the correct λ for the entire population. At the end of this chapter we return to this elephant problem and see if Darwin's calculation was correct.

To perform these calculations today ecologists use "age-structured" population models. We now develop a model that conveniently keeps track of this age variation in birth and death rates and allows a ready prediction for the population growth rate λ and the resulting mix of age classes that is produced over time. **Demography** is the branch of population biology dealing with population projections for age-structured populations.

KEEPING TRACK OF THE SIZE OF EACH AGE GROUP

The *Constants* of the Equations

We use a discrete time difference equation approach. The first step is to chop time and therefore ages into some equal intervals and to number these intervals. As simple as this sounds, it often leads to confusion. The first age class (i.e., those individuals whose age is somewhere from birth to the moment of their first birthday) can be labeled as either the number 1 or the number 0 age class. People generally use the convention of saying that someone is 1 year old once they have already *passed* their first birthday but not reached their second. Thus by this system, if someone is 30 years old, then he or she is now in the 31st year of life. This is why when we talk of say the 20th century, we mean

Figure 3.1
Numbering time intervals and age classes.

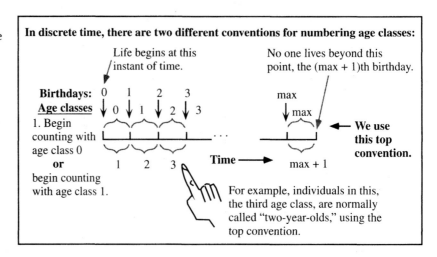

the years 1900–1999. The other convention is to begin numbering age classes as we number objects, beginning with the number 1. Since these two conventions simply differ by the integer 1, and since it's arbitrary how we number things, there is no problem as long as we're consistent. Here we use the convention of labeling the 0- to 1-year-olds as age class 0 (see Figure 3.1).

For brevity and to make time concrete, we mark time in units of years, but we could just as readily measure time in months, days, minutes, or decades. As smaller and smaller time units are adopted, the time intervals and the difference between actual age and age class number becomes smaller and smaller. For example, using years as time units, a two-year-old would mean anyone past the second birthday but not yet reaching the third—a span of an entire year. Using days as the time units, a 489-day-old individual encompasses a range of ages spanning only one day, but giving us many more age classes to deal with.

The parameters (or constants) controlling per capita births and deaths in the model are defined as follows. Let s_x be the survival rate, or probability, that an individual in age class x will survive from the beginning of the year to the beginning of the next year and thus reach age $x + 1$. By forming age classes by years and counting individuals only by age classes, we have lost more precise information on the abundance of ages at a finer resolution than years.

Let d_x = the corresponding death rate of age class x. Then

$$s_x = 1 - d_x.$$

The survivorship from age class 1 to the *beginning* of age class x is called l_x, so

$$l_x = s_1 s_2 \cdots s_{x-1}.$$

In other words, to survive to the beginning of age class 3, you must first survive from age 1 to age 2, and then survive from age 2 to age class 3. Since these probabilities are assumed to be independent, we can multiply them to get the survivorship to the beginning of age class 3. Note too that the product ends at $x - 1$ (not x), since l_x is defined to the *beginning* of age class x (not all the way through it). Also note that by this method the survival rate from birth to age 1, s_0, is not in the l_x product. The zero-year-olds in the present year, t, were born during the breeding season the previous year, $t - 1$; they are now just turning 1 year old. In this method of account keeping, newborn survival from birth to age 1 is incorporated into the birth rate function (see "net fecundity" below). The survivorship of age class 1 is defined to be 1.0. All subsequent ages then have fractional survivorship. For example, if $s_1 = 0.4$, and $s_2 = 0.2$, and $s_3 = 0$, we have: $l_1 = 1$, $l_2 = 0.4$, $l_3 = 0.08$, and l_4 (as well as the survivorship to any older ages) = 0.

Figure 3.2
The numbering scheme for ages and time periods.

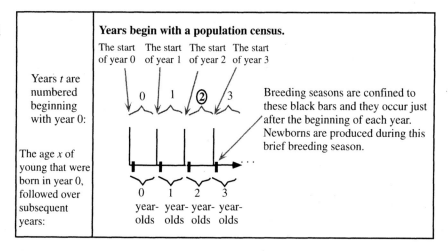

This expression for l_x can also be written in a shorthand way as

$$l_x = \prod_{i=1}^{x-1} s_i. \tag{3.1}$$

The Π symbol (capital "pi") in this context implies multiplication of a series of numbers subscripted i.

Let b_x be the **fecundity** of age x, which is the average number of female offspring born during the year to females of age x at the start of the year. Since only females give birth, ecologists usually keep track of population growth solely in terms of female numbers. For the purposes of tracking population change, it is convenient to let the year begin with the start of breeding season. If 2000 two-year-old females are present at the start of the breeding season and they produce a total of 200 female offspring during the breeding season, then $b_2 = 200/2000 = 0.1$.

The *Variables* of the Equations

The number of individuals in age class x at the beginning of year $t = n_x(t)$. It will usually be more convenient to begin the year, not with January 1, but with the beginning of the breeding season.

Setting Up the Equations for Discrete Population Growth

Consider a population that has a very short breeding season so that young are produced each year in nearly a single burst. The breeding pulse can be arbitrarily chosen at the time point that begins each year. For example, if the breeding pulse is the first week of April, each year begins with April 1. Also imagine that the population census is taken each year just *before* the yearly reproductive period. These assumptions are really not necessary, but they are helpful for visualizing the process. We could take the census at any time of the year and make some adjustments to the equations to be consistent. More on that later. Figure 3.2 illustrates these assumptions on the relative timing of breeding and census taking.

We denote consecutive years' *beginnings* as t, $t + 1$, $t + 2$, etc. Assuming that we have a population closed to immigration and emigration and that the environment stays the same so that survival and fecundity rates are constant from year to year, then the number of individuals in each age class at the start of this year (at time t) can be predicted based on the numbers counted at the start of last year (at time $t - 1$). We wish to develop the iteration equations that predict these changing numbers from one year to the next.

We begin with an expression that is correct for all individuals that were alive at the beginning of the previous year. All these individuals at the beginning of year *t* are produced by the **aging** and survival of individuals already alive at the start of last year (year *t* − 1). For example, the next figure illustrates this process for *x* = two-year-olds. They were one-year-olds last year at the start of year *t* = 0, but some died during the year.

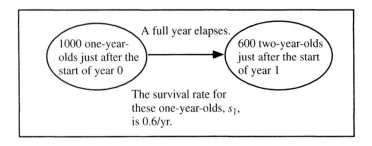

For the situation in this diagram,

$$n_2(t) = s_1 n_1(t - 1) = (0.6)(1000) = 600.$$

More generally, we can write for any age *x* > 1 at time *t*,

$$n_x(t) = s_{x-1} n_{x-1}(t - 1) \quad \text{for all ages } x > 1. \tag{3.2}$$

In words, Eq. (3.2) says: "The number of *x*-year-olds this year = the number of (*x* − 1)- year-olds last year that survived through the year" (for *x* = two-year-olds and older). This equation does not give us a formula for the number of one-year-olds at time *t*, since we did not count the number of zero-year-olds (newborns) in the previous year (year *t* − 1). Those newborns were born during the previous year's breeding pulse, which came after our census.

Figure 3.3
Calculating the number of offspring born in year *t* − 1 and surviving to year *t* (where they become one-year-olds), given the number of mothers of each age in year *t* − 1 and their fecundities.

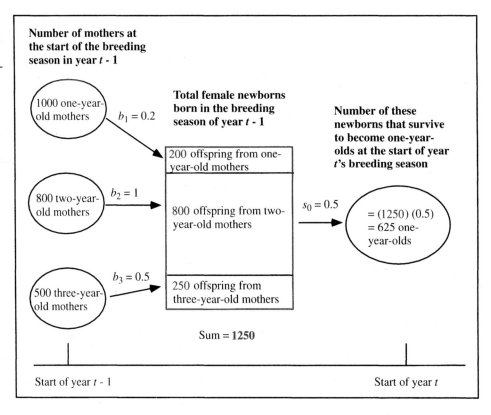

Exercise: Do you see why the subscripts on the right-hand side of Eq. (3.2) are $x - 1$, not x?

Exercise: Consider the following equation. Is it equivalent to Eq. (3.2)?

$$n_x(t + 1) = s_{x-1}n_{x-1}(t) \quad \text{for all ages } x > 1.$$

We now remedy this deficiency by developing an expression for the number of one-year-olds at the start of year t. Remember there are no zero-year-olds at the start of year t (the young of the year are born just *after* the birth pulse and the newborns from last year are having their first birthdays). Figure 3.3 presents an example of how births in one year end up as one-year-olds the next year.

More generally, this logic can be expressed with an **iteration** equation that gives the number of individuals of age 1 at the beginning of year t, based on the number of their mothers and their birth rates in year $t - 1$.

$$n_1(t) = \sum_{x=1}^{max} s_0 b_x n_x(t - 1).$$

This is the number of births produced by females of age x last year at the beginning of year $t - 1$.

This product is the number of those births that survive from year $t - 1$ to year t. Note that we assume that these newborns have the same survival rate from $t - 1$ to t, which is s_0, regardless of the age of their mother, x.

The $\sum$ means sum the surviving births from mothers of all ages, x. Note that the sum begins with age 1 (not 0) and goes to the oldest age (age max).

Since s_0 and b_x are both constants, their product is a constant. To simplify the notation, we can label this product $F_x = $ **net fecundity of mothers of age $x = s_0 b_x$.** F_x has units: surviving births/mother/year.

$$n_1(t) = \sum_{x=1}^{max} F_x n_x(t - 1)$$

$$= F_1 n_1(t - 1) + F_2 n_2(t - 1) + F_3 n_3(t - 1) \ldots + F_{max} n_{max}(t - 1). \tag{3.3}$$

The F's express the **net fecundity of mothers per year** in the sense that only a portion, s_0, of the newborns survive from birth to the beginning of the next year (they are sometimes denoted m_x in other texts). For example if the average two-year-old mother produces three offspring during a single breeding season and if, on average, half these offspring survive to reach their first birthdays, age 1, then $s_0 = 0.5$ and $F_2 = 1.5$.

Exercise: Why write the t's after the n's but not after the b's or F's?

Note that we don't have an iteration equation for the numbers of newborns, n_0. Equation (3.3) is for one-year-olds and Eq. (3.2) is for two-year-olds and older. In this method, newborns get counted only when they age and survive into age class 1. To see what's going on here, focus on the population at the time point of the third census tick, right under the "2" that is circled in Figure 3.2. The newborns of the previous year that are still alive at this point were born nearly a year ago and were not included in the last census at year 1 since they were not yet born. They will be counted for the first time in census 2. Because they are just about to have their first birthdays, they are counted as age 1 individuals. Thus the numbers of age class 0 (n_0) are not an explicit part of the accounting system in Eqs. (3.2) and (3.3).

We may write the Eqs. (3.2) and (3.3) as a single compact **matrix** equation that expresses all the same information and relationships. We illustrate this approach for a maximum age of 4:

$$
\begin{bmatrix} n_1 \\ n_2 \\ n_3 \\ n_4 \end{bmatrix}(t) = \begin{bmatrix} F_1 & F_2 & F_3 & F_4 \\ s_1 & 0 & 0 & 0 \\ 0 & s_2 & 0 & 0 \\ 0 & 0 & s_3 & 0 \end{bmatrix} \begin{bmatrix} n_1 \\ n_2 \\ n_3 \\ n_4 \end{bmatrix}(t-1). \tag{3.4}
$$

In words: "The population age vector at the start of year t equals the **population projection matrix** times the population **age vector** at the start of year $t-1$." The population projection matrix is also referred to as the **Leslie matrix.** Its name comes from the person who first introduced it to demography in 1945, P. H. Leslie, as a means of tracking population changes.

Note that the first entry in the population age vector is the one-year-olds. The 0 age class is not completely ignored since its s_0 term is contained in each of the net fecundity terms, F_x.

Finally, and completing the definitions:

$N(t)$ = the total number of females in the population at time t. $N(t)$ a scalar

$\qquad = n_1(t) + n_2(t) + n_3(t) + \cdots$.

$$
= \sum_{x=1}^{max} n_x(t). \tag{3.5}
$$

We can write the projection of all the age group numbers through time more compactly using matrices and vectors:

$$
\mathbf{n}(t+1) = \mathbf{L}\,\mathbf{n}(t). \tag{3.6}
$$

The bold face notation in Eq. (3.6) symbolizes the population age vector $\mathbf{n}$ and the Leslie matrix $\mathbf{L}$. As developed, the Leslie matrix is a *square matrix* (the number of rows equals the number of columns). Even though in this example there are five age categories (0, 1, 2, 3, 4), the zero-year-olds are not explicitly followed in this method and thus the Leslie matrix is a 4×4 matrix with 4 rows, 4 columns, and 16 elements. The first element in the population age vector is n_1 and the last element is n_4. (The convention in matrices is to index rows first and columns second. For example a 3×2 matrix has 3 rows and 2 columns. The (1, 3) element in the Leslie matrix refers to F_3).

To multiply a matrix times a vector, each row of the matrix is multiplied by the population size vector to get the elements in the new population size vector for the next year. Think of this as tipping the population vector on its side, multiplying each element in turn by the element in the same position in the Leslie matrix, and then summing all these products.

$$
F_1 n_1(t) + F_2 n_2(t) + F_3 n_3(t) + F_4 n_4(t) = n_1(t+1)
$$

$$
s_1 n_1(t) + 0 n_2(t) + 0 n_3(t) + 0 n_4(t) = n_2(t+1)
$$

$$
0 n_1(t) + s_2 n_2(t) + 0 n_3(t) + 0 n_4(t) = n_3(t+1)
$$

$$
0 n_1(t) + 0 n_2(t) + s_3 n_3(t) + 0 n_4(t) = n_4(t+1)
$$

At this point you may want to know why the Leslie matrix has the form that it does. Why do the survival terms appear in those positions one step down from the diagonal, and why are the F terms all aligned along the top row? The answer is that these are the places that these terms must appear, given the rules of matrix multiplication, so that our matrix equation Eq. (3.4), expresses exactly the arithmetic that we developed to get Eqs. (3.2) and (3.3). If you're confused at this point, review the rules of matrix multiplication in Appendix 2. Then perform the multiplication indicated by the right-hand side of Eq. (3.4) and see if you arrive at Eqs. (3.2) and (3.3).

So what do we have so far? We have a way of compactly writing in a single matrix equation the iterations that must be followed to project the age vector into the future 1 year (or some other time step) at a time. Our ultimate goal is to arrive at the ultimate growth rate, λ, for this population, so we still have a way to go. The next example illustrates this iterative process and provides some insight into the progression of growth rates over time, converging in the limit on λ.

Consider an animal with the following vital rates:

$$
\begin{array}{lll}
F_1 = 0 & s_1 = 0.5 & l_1 = 1 \\
F_2 = 0 & s_2 = 1.0 & l_2 = 0.5 \\
F_3 = 2 & s_3 = 0.5 & l_3 = 0.5 \\
F_4 = 4 & s_4 = 0 & l_4 = 0.25
\end{array}
$$

This animal reaches **sexual maturity** at age 3, and no individual survives beyond age 4; thus the oldest individuals in the population are four-year-olds. Now put this information together into the Leslie matrix and initial age vector:

$$
\mathbf{L} = \begin{bmatrix} 0 & 0 & 2 & 4 \\ 0.5 & 0 & 0 & 0 \\ 0 & 1 & 0 & 0 \\ 0 & 0 & 0.5 & 0 \end{bmatrix} \quad \text{and} \quad \mathbf{n}(t) = \begin{bmatrix} 100 \\ 0 \\ 0 \\ 0 \end{bmatrix} \tag{3.7}
$$

Begin at time t with a population of 100 one-year-olds and no other ages. At the start of the breeding season of $t + 1$, the population age vector is

$$
\mathbf{n}(t+1) = \begin{bmatrix} 0 \\ 50 \\ 0 \\ 0 \end{bmatrix}.
$$

Exercise: Do you see how this vector was calculated?

Verify that the next two time steps produce

$$
\mathbf{n}(t+2) = \begin{bmatrix} 0 \\ 0 \\ 50 \\ 0 \end{bmatrix} \quad \text{and} \quad \mathbf{n}(t+3) = \begin{bmatrix} 100 \\ 0 \\ 0 \\ 25 \end{bmatrix}.
$$

Box 3.1 points out a connection between this process and the one that we obtained in Chapter 1 for simple scalar geometric growth. A graph of this process is plotted in Figure 3.4. It shows the proportional size of each age class for many time steps (years) into the future; the sum of all four bars is 1. Above each chart is shown the total population size (summing all ages) and the lambda from year $t - 1$ to year t, given by $\lambda(t) = N(t)/N(t - 1)$.

Note that the age structure of the population initially changes radically from one year to the next. Also the total population size temporarily declines from year 1 to year 2 but then increases in year 3. In this example, the geometric rate λ, finally begins to home in on an asymptotic value of about 1.22. This yearly variation occurs even though the vital rates are constant from year to year. Note too that at about the same time the population size starts to grow at a constant geometric rate (around $t = 37$), the shape of the age structure also stays relatively constant from one year to the next. These are general features of age-structured population growth with constant vital rates.

Box 3.1

We have just shown how to iterate the equation

$$\mathbf{n}(t + 1) = \mathbf{L}\,\mathbf{n}(t)$$

to follow the growth of each age class from one time step to the next. For example, the double iteration to reach $\mathbf{n}(t + 2)$ is

$$\mathbf{n}(t + 2) = \mathbf{L}\,\mathbf{n}(t + 1)$$
$$= \mathbf{L}\,\mathbf{L}\,\mathbf{n}(t)$$
$$= \mathbf{L}^2\,\mathbf{n}(t).$$

This last expression says that we could also get the same answer for $\mathbf{n}(t + 2)$ by multiplying $\mathbf{L}$ times itself and then multiplying this product times the initial age vector $\mathbf{n}(t)$.

Appendix 2 shows how to multiply two matrices. By extension, for T time periods in the future, we could write

$$\mathbf{n}(t + T) = \mathbf{L}^T\,\mathbf{n}(t).$$

To simplify notation even further, let's set the initial time to $t = 0$ and call the initial population size at $t = 0$, $\mathbf{n}_0$. Then we can write

$$\mathbf{n}(T) = \mathbf{L}^T\,\mathbf{n}_0.$$

Note the similarity of this matrix equation for age-structured population growth to the scalar geometric growth equation in Box 1.1, (Eq. 1.3a), which was $N(T) = N_0\lambda^T$. We return to this similarity later in the chapter and look for a way of extracting λ from the matrix $\mathbf{L}$.

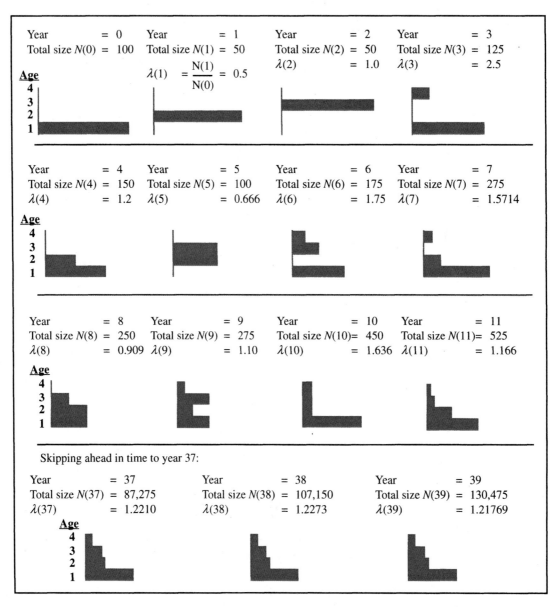

Figure 3.4
Yearly changes in the growth rate and age structure of a population with the Leslie matrix, Eq. (3.7), and initialized with 100 females of age 1 at year 0.

Figure 3.5
(a) Yearly changes in age structure for the Leslie matrix shown and 100 females of age 1 at time 0. (b) The results of Figure 3.5(a) plotted on a ln scale for the numbers of each age group.

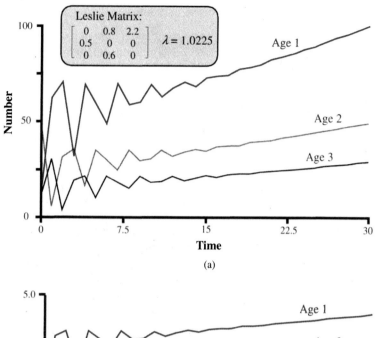

(a)

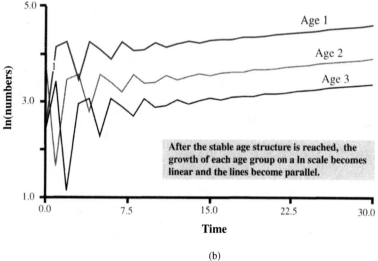

After the stable age structure is reached, the growth of each age group on a ln scale becomes linear and the lines become parallel.

(b)

Figure 3.6
Total population size, *N*, over time for the population in Figure 3.5. The slope of the ln *N* line near the end of this time series is 0.0221; $e^{0.0221} = 1.0225 = \lambda$.

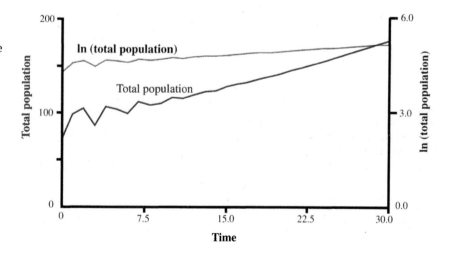

Another example, plotted a little differently, is shown in Figure 3.5(a). Plotting the same results on a ln scale gives Figure 3.5(b). Summing over all the ages to calculate total population size produces the results shown in Figure 3.6.

It turns out that several features of this example are, in fact, quite general. They can be summarized as follows.

Summary of Characteristics of Age-Structured Geometric Growth

1. Assuming that the birth and death parameters (i.e., the vital rates) remain constant (and this is a big assumption), populations eventually reach a stable age structure. (One complication: mathematically it's possible that, for some very special cases of the vital rates, the stable age structure will never be reached and that population numbers continually oscillate. However, the kinds of Leslie matrices that are capable of producing this result are generally biologically unreasonable, so we won't worry about them further.)

2. Before the stable age structure is reached, population growth from one year to the next can vary. Once the stable age structure is reached, populations grow geometrically with a constant discrete rate λ.

3. λ can be calculated solely from the F_x and s_x data, although its calculation may not be easy. For large Leslie matrices, it is usually easier to iterate the equations beginning with any arbitrary initial numbers, plot the resulting growth in ln N versus time, and then calculate the slope of the line after the stable age structure is reached. The slope of this line equals r and $\lambda = e^r$.

4. The greater the growth rate λ, the broader is the base of the age structure pyramid relative to its breadth at the top, the oldest ages. For $\lambda > 1$, at stable age structure the proportion of each age in the population must decrease with age. We derive this last point in Chapter 4.

SOME EXAMPLES OF SURVIVAL AND BIRTH FUNCTIONS

Pearl (1928) suggested that survivorship curves usually fall into the three categories or "types" displayed in Figure 3.7. Typically l_x is plotted on a logarithmic scale and the ages are scaled by the maximum age.

Type II has constant survival rates s_x across all ages; on a log scale this results in a straight line. In Type III, younger ages have higher mortality rates than older ages and the reverse is true for Type I. For many mammals, the situation is one that combines elements of Types I and III. For example, in Dall sheep, most mortality is suffered by the very young and the very old (Figure 3.8). This is expected since wolves are the chief source of mortality, and they hunt the most vulnerable prey (Deevey 1947). The general shape for the mortality schedule in some other species is shown in Figure 3.9.

Fecundity is by definition zero until sexual maturity. For many animals and plants whose body sizes continually increase throughout life (e.g., fish and lobsters), birth rates may continue to increase as their body sizes increase. However, other species, such as humans, annual plants, and rice weevils, show senescence or decay in fecundity at older ages (Figure 3.10).

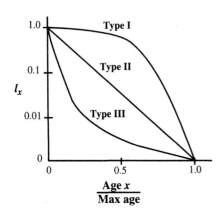

Figure 3.7
Different possible shapes for survivorship (l_x) curves. Note the log scale for l_x and the relative scale for age.

Figure 3.8
(a) Death rates, d_x, and survivorship, l_x, for a population of dall mountain sheep (*Ovis dalli*) on Mount McKinley in Alaska. (b) The same survivorship data plotted on a log scale. These data were constructed from a collection of sheep skulls (Deevey 1947).

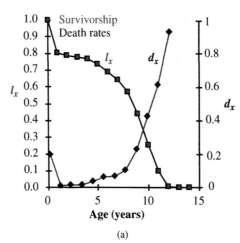

(a)

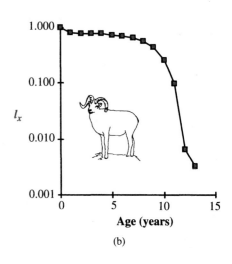

(b)

Figure 3.9
Death rate curves for (a) Costa Rican females, 1963 (Keyfitz and Flieger 1968), (b) rice weevils *Calandra oryzae* in the laboratory (Birch 1948), (c) red deer on the island of Rhum off Scotland (Lowe 1969), and (d) annual bluegrass (*Poa annua*) in fields (Law et al. 1977).

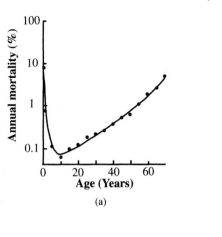

(a)

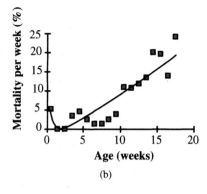

(b)

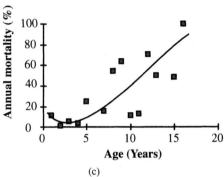

(c)

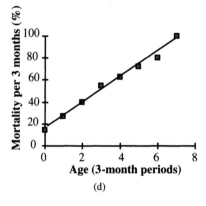

(d)

Figure 3.10
Fecundity versus age for (a) humans in the United States in 1965 (Keyfitz and Flieger 1968), (b) rice weevils *Calandra oryzae* in the laboratory (Birch 1948), (c) red deer on the island of Rhum off Scotland (Lowe 1969), and (d) annual bluegrass (*Poa annua*) in fields (Law et al. 1977).

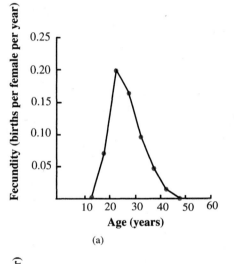

(a)

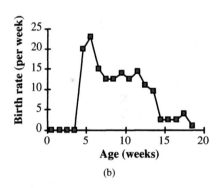

(b)

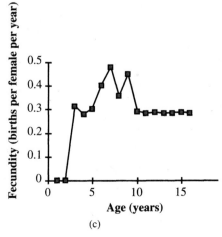

(c)

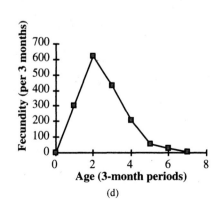

(d)

Figure 3.11
Changes in the numbers of young and adults from year to year. The *y* axis represents number of young and adults. The thin tic marks on the *x* axis indicate the beginning of each year. Young are produced in a birth pulse just after the start of each year (the thick tics). The solid black dots at the start of each year show the number of adults at the beginning of each year. The red hatched areas show the number of young, N_y. The total population, N, is the sum of adults, N_a, and young, N_y. The young mature into adults at their first birthday.

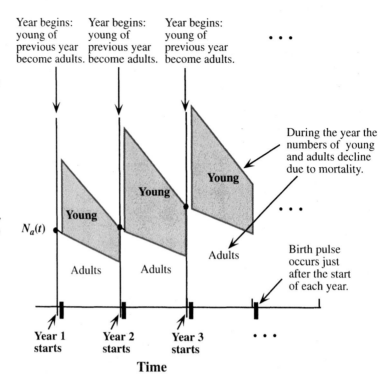

THE BIRTH PULSE RELATIVE TO THE BEGINNING OF THE YEAR

The date chosen to begin each year is arbitrary. Biological events serve as useful punctuation marks to divide time, rather than the solar cycle, which our calendars invoke. The purpose of this section is to show that the calculation of λ for a population is not affected by the timing of the birth pulse relative to the start time for each year, as long as we adjust our bookkeeping appropriately and equally space time intervals. To make this clear, imagine a very simple life history where all the adults, regardless of their age, have the same yearly survival and fecundity schedules. The young mature when they reach 1 year of age and potentially can have a different survival rate from the adults. Figure 3.11 illustrates the changes in population size of adults and juveniles from year to year. As for our earlier development, the birth pulse takes place immediately after the start of each year.

Everyone is an adult at census time. The young of the year have not yet been born and the young from last year have reached maturity and so now are adults. The solid black dots show the number of adults present at the beginning of each year.

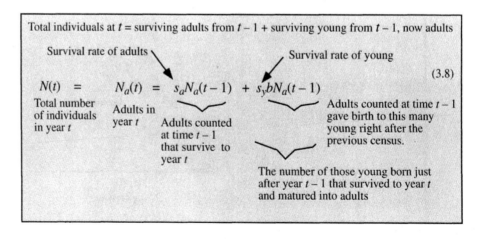

Figure 3.12
Changes in the number of young and adults from year to year. The thin tic marks on the *x* axis indicate the beginning of each year. Young are produced in a birth pulse just before the start of each year (the thick tics). The hatched area shows the number of young, N_y. The total population, N, is the sum of adults, N_a and young, N_y. The solid black dots show the number of adults at the time of census (the beginning of each year). The open black circles indicate the total number of individuals (young plus adults) at the time of census.

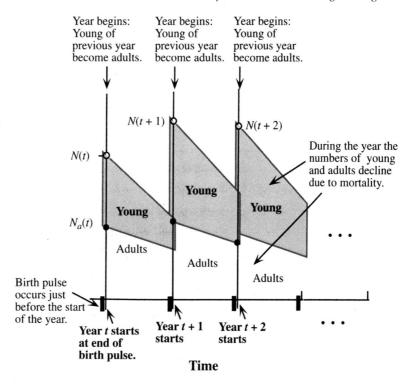

Rearranging

$$N_a(t) = s_a N_a(t-1) + s_y b N_a(t-1)$$

we get

$$N_a(t) = N_a(t-1)(s_a + s_y b).$$

For this life history schedule, where all adults have the same vital rates, we have arrived at a simple formula for growth of the adult population. It grows with a geometric growth rate of

$$\lambda = s_a + s_y b. \tag{3.9}$$

Since at a stable age structure all age groups grow at the same rate, the λ for adults is also the λ for young.

Would our calculation of λ be altered if the timing of the birth pulse and the start of the year (our census point) were reversed, with each year's start date being immediately after the birth pulse? Figure 3.12 illustrates this case.

Note that in Figure 3.12 not everyone is an adult on the first day of each year; the new young have just been born. The young produced in the previous year are having their first birthday, when they mature into adults, although a proportion, $1 - s_y$, have died during the time between the two censuses. Finally, some fraction s_a of adults alive at the beginning of the previous year have survived to the beginning of this year; the solid dots show the counts for adults at each census period, and the open circles indicate the total number of individuals at each census period. The number of immature young is given by the distance between the solid dots and the open circles. We can express the number of adults at the start of one year as a function of the number at the start of the previous year by using a difference equation:

$$\begin{bmatrix} \text{The number of} \\ \text{adults at time } t \end{bmatrix} = \begin{bmatrix} \text{The number of adults} \\ \text{from } t-1 \text{ that have} \\ \text{survived to } t \end{bmatrix} + \begin{bmatrix} \text{The number of new} \\ \text{adults from births} \\ \text{just before } t-1 \text{ that} \\ \text{are maturing at } t \end{bmatrix},$$

or

$$N_a(t) = s_a N_a(t-1) + N_a(t-1)bs_y. \tag{3.10}$$

Grouping terms in Eq. (3.10) gives

$$N_a(t) = N_a(t-1)(s_a + bs_y).$$

The adult population thus grows at a geometric growth rate of

$$\lambda = s_a + bs_y. \tag{3.11}$$

Since at the stable age structure all ages grow at the same rate, the λ for adults is equal to the ultimate λ for the young and for the entire population. Since Eq. (3.11) is identical to Eq. (3.9) we may conclude that the timing of the birth pulse relative to the census point does not affect the calculation of λ.

Next we explore this equivalence further using an example with a much more complicated life history. We also introduce an alternative method for forming the Leslie matrix and projecting population growth. Although the form of the Leslie matrix is different, it yields the same λ.

COHORT ANALYSIS AND AN ALTERNATIVE METHOD FOR FORMING THE LESLIE MATRIX

How do demographers estimate the vital rates? Two different approaches are depicted in Figure 3.13. The *segment-based* approach looks at the births and deaths of all individuals in the population during some slice of time; it requires the ability to know the ages of individuals. The *cohort analysis* follows all individuals born during a time

Figure 3.13

Each individual is represented by a diagonal line that begins at its birth date on the bottom horizontal line and ends as a dot or *x* when it dies. These time lines have slope 1 since, as time passes, individuals age at exactly the same rate. Where a diagonal line crosses a horizontal age line, an individual reaches a birthday and enters the next age class. Individuals that entered year 1991 alive are depicted by thick gray lines; if they died during year 1991, the lines end in *x*'s. The cohort born in year 1991 is represented by a shaded gray background and red lines. One individual was born in and died in year 1991 and is thus depicted with a red line ending in an *x*. This individual is included only in the cohort-based mortality rate calculation since it did not enter 1991 alive.

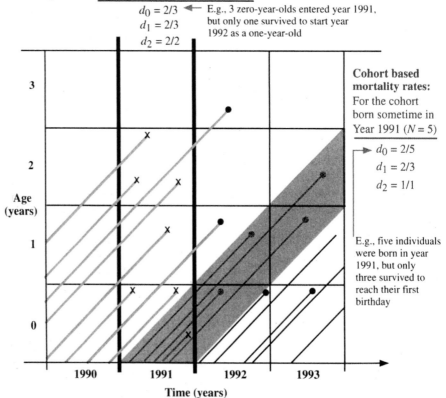

Segment-based mortality rates: For the total population alive at the begining of Year 1991 ($N = 8$)

$d_0 = 2/3$ ← E.g., 3 zero-year-olds entered year 1991, but only one survived to start year 1992 as a one-year-old
$d_1 = 2/3$
$d_2 = 2/2$

Cohort based mortality rates: For the cohort born sometime in Year 1991 ($N = 5$)

$d_0 = 2/5$
$d_1 = 2/3$
$d_2 = 1/1$

E.g., five individuals were born in year 1991, but only three survived to reach their first birthday period, determining the ages that they die and their birth rates as they age. Eventually, of course, all individuals in the cohort will die. Life tables constructed from the segment-based method are often called **vertical** and from the cohort-based method, **horizontal**. If the vital rates are constant from year to year, a stable age structure will eventually be reached. With large samples in the censuses, the death rate calculation for each age based on time segments yields the same death rates as that based on following cohorts over time.

It is possible to estimate death rates in a third way. Assuming that the population in Figure 3.13 has been at a stable age distribution throughout the time period and that $\lambda = 1$, from year 0 to year 3, we can determine the proportion of all these deaths by each age group. For example, the total number of deaths in Figure 3.13 is 16. The number of those that died at age 0 is 6; therefore the estimated death rate for newborns is $d_0 = 6/16$. The number of individuals left is $16 - 6 = 10$. Of these 10 individuals, 5 die as one-year-olds; therefore $d_1 = 5/10$. The number of individuals left is 5. Of these 5 individuals, 4 die as two-year-olds; therefore $d_2 = 4/5$. Finally $d_3 = 1/1$. This method works well when you can find and age dead individuals; it also is based on the assumption that $\lambda = 1$ and that the population has been in a stable age structure (The method can be modified if $\lambda \neq 1$; more on this in Chapter 4.) For example, the survivorship curve for dall mountain sheep shown in Figure 3.8 was determined in this manner. The skulls of dead animals could be aged based on distinguishing teeth and horn characteristics.

> **Conclusion:** Assuming that vital rates stay constant from year to year, once the stable-age structure is reached, the two methods will yield equivalent results for all cohorts and all year segments (also assuming large samples).

Unlike the situation depicted in Figure 3.13, in most species deaths can only be inferred when individuals disappear between two censuses. The following method for the calculation of survival rates, s_x, and survivorship, l_x, is based on a cohort analysis. Demographers traditionally use only females for analysis. The goal is to deduce the age-specific survivorship rates solely from counts of live individuals and their birth rates. Deaths are inferred.

Imagine that we obtained a perfect census of a hypothetical deer population during the breeding season in two consecutive years. To make the calculation even easier, assume that this population is in stable-age distribution and is not growing ($\lambda = 1.0$). Since the census takes place during the birth pulse, all newborns are counted. Furthermore, we imagine that the age of each mother can also be determined, although aging animals in the field usually is difficult. Table 3.1 shows what we counted.

Table 3.1 **Hypothetical Census of Female Deer**

Age x	Age class	Number time 1	Number time 2	Female newborns born to mothers of age x, time 1	Female newborns born to mothers of age x, time 2	Births/female of age x, times 1 and 2 = b_x
1-year-olds	1 to 2	600	600	0	0	0
2-year-olds	2 to 3	540	540	378	378	378/540 = 0.7
3-year-olds	3 to 4	486	486	379	379	379/486 = 0.78
4-year-olds	4 to 5	243	243	243	243	243/243 = 1.0
5-year-olds	5 to 6	0	0			
Total newborns ($x = 0$)				1000	1000	

Method 1. Each Year *t* Begins Just before the Birth Pulse

This is the technique developed at the beginning of this chapter. Age classes are subscripted beginning with age 0. Thus age class 1, the one-year-olds, represents the individuals *in* their second year of life between their first and second birthdays. The survivorship of newborns is $s_0 = 600/1000$. The life table is:

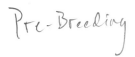

Age x	n_x	s_x	l_x	d_x	$F_x = s_0 b_x$
1	600	540/600 = 0.9	1.0	0.1	(0.6)(0) = 0
2	540	486/540 = 0.9	0.9	0.1	(0.6)(0.7) = 0.42
3	486	243/486 = 0.5	0.81	0.5	(0.6)(0.78) = 0.468
4	243	0/243 = 0	0.405	1.0	(0.6)(1.0) = 0.6

The F_x terms are formed by multiplying $s_0 = 0.6$ by each age-specific fecundity, b_x. Note also that the 0-to-1 age class is obscured by this approach; there is no place for it in the Leslie matrix or the age vector. The Leslie matrix and population vector are

$$\mathbf{L}\ (\textbf{before}) = \begin{bmatrix} 0 & 0.42 & 0.468 & 0.6 \\ 0.9 & 0 & 0 & 0 \\ 0 & 0.9 & 0 & 0 \\ 0 & 0 & 0.5 & 0 \end{bmatrix} \quad \text{and} \quad \mathbf{n} = \begin{bmatrix} 600 \\ 540 \\ 486 \\ 243 \end{bmatrix} = \begin{bmatrix} n_1 \\ n_2 \\ n_3 \\ n_4 \end{bmatrix}.$$

Because cohorts are indexed by age classes starting with zero, the first element of the age vector is n_1, which represents the number of individuals between their first and second birthdays.

Method 2. Each Year *t* Begins Just after the Birth Pulse

We desire an iteration formula that takes us from the numbers at the beginning of one year to the numbers at the beginning of the next. If the census begins just after the birth pulse, the newborns have just been born and no time has accumulated for their mortality. The number of newborns at the start of year t is $n_0(t)$. However, at this time the counts for the number of adult females that produced these young does not represent all the females that were counted 1 year earlier at $t-1$ and are now *entering* the next year class at time t. Some mothers have died in the year between these two censuses. The adults present to give birth at time t are those mothers that have survived from the census in year $t-1$ to the census in year t. This will become clearer as we work through an example.

We may write the equation for the number of individuals that are 0-years old at time t as $n_0(t)$,

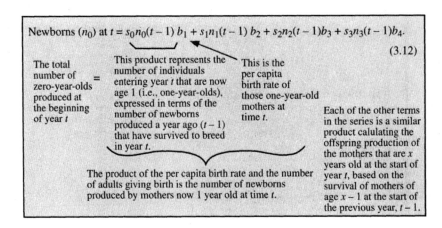

Then

$$n_0(t) = \sum_{x=0}^{\text{max}} s_x b_{x+1} n_x(t-1). \tag{3.13}$$

Here's the life table that does the calculations:

Post-Breeding

Age x	n_x	s_x	l_x	d_x	$F_x = b_{x+1}s_x$
0	1000	0.6	1	0.4	$(0)(0.6) = 0$
1	600	0.9	0.6	0.1	$(0.7)(0.9) = 0.63$
2	540	0.9	0.54	0.1	$(0.78)(0.9) = 0.702$
3	486	0.5	0.486	0.5	$(1.0)(0.5) = 0.5$
4	243	0	0.243	1.0	$(0)(0) = 0$

The Leslie matrix and population age vector are:

$$\mathbf{L}\ (\text{after}) = \begin{bmatrix} 0 & 0.63 & 0.702 & 0.5 & 0 \\ 0.6 & 0 & 0 & 0 & 0 \\ 0 & 0.9 & 0 & 0 & 0 \\ 0 & 0 & 0.9 & 0 & 0 \\ 0 & 0 & 0 & 0.5 & 0 \end{bmatrix} \quad \text{and} \quad \mathbf{n} = \begin{bmatrix} 100 \\ 600 \\ 540 \\ 486 \\ 243 \end{bmatrix} = \begin{bmatrix} n_0 \\ n_1 \\ n_2 \\ n_3 \\ n_4 \end{bmatrix}.$$

Note that in method 2 the Leslie matrix is 5 rows by 5 columns and in terms of the vital rates has the form

$$\begin{bmatrix} s_0 b_1 & s_1 b_2 & s_2 b_3 & s_3 b_4 & s_4 b_5 = 0 \\ s_0 & 0 & 0 & 0 & 0 \\ 0 & s_1 & 0 & 0 & 0 \\ 0 & 0 & s_2 & 0 & 0 \\ 0 & 0 & 0 & s_3 & 0 \end{bmatrix} \quad \text{and} \quad \begin{bmatrix} n_0 \\ n_1 \\ n_2 \\ n_3 \\ n_4 \end{bmatrix}(t)$$

The zero-year-olds are put in the first position of the age class vector; the one-year-olds, in position two; the two-year-olds, in position three, and so on. When you multiply this Leslie matrix by the population vector at time t to form the population vector at $t + 1$, you are accounting, for example, for the fact that two-year-olds at time t will be three-year-olds at time $t + 1$ when they give birth, and they will have a 10% chance of dying in the intervening year ($s_2 = 0.9$). Thus b_3 (not b_2) is in the third position of the first row of **L (after)** and multiplies n_2 (not n_3) in the age vector for time t.

In summary, the iteration of this Leslie matrix system,

$$\begin{bmatrix} n_1 \\ n_2 \\ n_3 \\ n_4 \end{bmatrix}(t+1) = \begin{bmatrix} s_0 b_1 & s_0 b_2 & s_0 b_3 & s_0 b_4 \\ s_1 & 0 & 0 & 0 \\ 0 & s_2 & 0 & 0 \\ 0 & 0 & s_3 & 0 \end{bmatrix} \begin{bmatrix} n_1 \\ n_2 \\ n_3 \\ n_4 \end{bmatrix}(t) \qquad (3.14)$$

yields the same population dynamics as the iteration of this Leslie matrix system,

$$\begin{bmatrix} n_0 \\ n_1 \\ n_2 \\ n_3 \\ n_4 \end{bmatrix}(t+1) = \begin{bmatrix} s_0 b_1 & s_1 b_2 & s_2 b_3 & s_3 b_4 & s_4 b_5 \\ s_0 & 0 & 0 & 0 & 0 \\ 0 & s_1 & 0 & 0 & 0 \\ 0 & 0 & s_2 & 0 & 0 \\ 0 & 0 & 0 & s_3 & 0 \end{bmatrix} \begin{bmatrix} n_0 \\ n_1 \\ n_2 \\ n_3 \\ n_4 \end{bmatrix}(t), \qquad (3.15)$$

except that Eq. (3.15) explicitly tracks newborn numbers, n_0, while Eq. (3.14) doesn't. Another important difference is that, since newborns are directly counted in Eq. (3.15), the survivorship of age x, l_x, is now calculated from birth to age x (instead of from age 1 to age x as for method 1; see Eq. 3.1.

Since the two methods give the same result, it is not too surprising that we can arbitrarily choose to begin the year at any time between consecutive birth pulses and still calculate an F_x that combines the correct proportion of both young and adult mortality so as to construct again a Leslie matrix with the same λ. For example, suppose that the birth pulse takes place at three months (or 0.25 year) after the beginning of the year when the census is taken. Then each F_x term needs to be weighted by 1/4 of the newborn survival rate and 3/4 of the adult survivorship from $x - 1$ to x. Thus

$F_x = 0.75 s_0 b_x + 0.25 s_x b_{x+1}$. Of course, if the census occurs outside the breeding season we won't be able to observe and count the number of offspring produced by mothers of different ages.

Moreover, since it is unlikely that all individuals of any organism actually give birth every year in one instant pulse, we can let the F's contain mortality terms based on the "average" individual's birth period relative to the census interval and still not be led astray. In other words, imagine that birth was continuous throughout the year. Now individuals in age group x may differ in age by as much as 364 days, or essentially a full year. We can still form a discrete model with years as indices and discrete vital rates in the Leslie matrix, but we must do some averaging to gather similarly aged individuals into 1-year age groups and then average the b_x and s_x within these age groups. (See Caswell (1989) for details on this type of averaging.)

It seems a nuisance to have two very different methods to perform the identical calculations—Why bother? Biologists generally base their choice of method on which formulation more conveniently suits the life history of the organism and the timing of the data collection. Someone studying seabirds typically finds an island where they breed. In this case, the census counts include all the newborns sitting side by side with their mothers. The observables are the birth rates, b_x, and the number of mothers that are breeding, so it is convenient to have a method that fits these data. On the other hand, someone studying polar bears would have trouble counting the number of young produced, since births takes place in ice caves away from view. In the summer when the adults emerge with their young, an observer can see how many cubs have survived but not necessarily how many were born. Here the investigator does not know the birth rate nor the survival rate of the newborns, but does know the product of these two terms, the net fecundity, F_x. Method 1 allows a Leslie matrix to be constructed based on such partial knowledge.

PROBLEMS

1. The flow diagram in Figure 3.14 represents a cohort of a closed population counted in 4 years, from birth in 1990 to 1993. What is s_0? Using the limited information that you have and method 1, fill in the following life table.

Age x	s_x	b_x	F_x
1			
2			
3			

As a cohort is followed over time, it produces offspring, as shown in Figure 3.15. These offspring then age and those that survive produce offspring; new cohorts are thus born each year.

In Figure 3.15, individuals in the same cohort are connected in an enclosed box. Reproduction is indicated by red arrows and survival by black arrows. Information on the fate of the new cohorts as well as the original 1990 cohort is provided. There are three separate estimates for the yearly survival of zero-year-olds, s_0:
(1) based on 1990 to 1991 (1000/2000),
(2) based on 1991 to 1992 (500/1000), and
(3) based on 1992 to 1993 (400/800).
In each the survival rate, s_0 is 0.5. Other vital rates may be calculated for different years. Look at the birth rate of one-year-olds, b_1 in the three years 1991, 1992, and 1993: $b_1(1991) = 1.0$ (1000/1000); $b_1(1992) = 1.0$ (500/500); but $b_1(1993) = 0.25$ (100/400). The year 1993 also was a relatively bad year for breeding for two-year-old mothers, since $b_2(1992) = 0.5$ (300/600) but $b_2(1993)$ was substantially less, 0.2 (50/250).

We can conclude that the vital rates are not staying constant from year to year in the hypothetical population represented by this example. This population therefore cannot be in a stable age structure and the geometric rate of growth is changing from year to year.

Figure 3.14

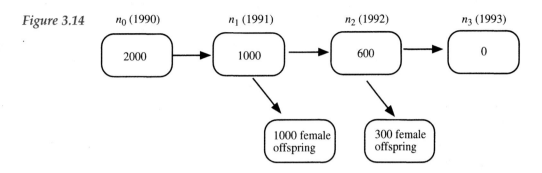

Figure 3.15
Demographic transitions over time from an initial cohort of 2000 newborns in year 1990. Subsequent cohorts of newborns are boxed in red.

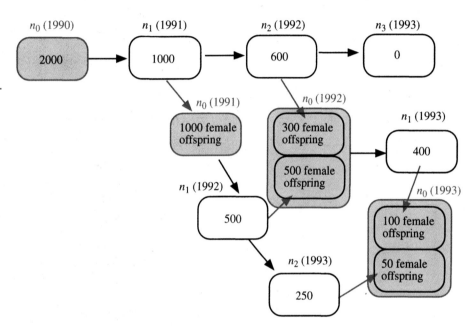

Nowadays, there are computer programs that can be applied to large data sets like that in Figure 3.15, containing several years data on the numbers of several age groups (Pollock et al. 1990, Lebreton et al. 1992). Statistical trends in the data can be determined to separate the effects of year and age on vital rates. Do the vital rates vary consistently from year to year, across several or all age groups? Do some age groups have statistically different vital rates than other age groups? Is the population size of different age groups growing at the same rates, indicating a stable age structure? Or do different years of data tend to produce different rates of change for the same age group?

2. You are an animal demographer studying bears. Bear cubs are born while their mothers are hibernating in hollow trees or caves. Consequently, they cannot be counted until they and their mothers emerge from hibernation in the spring. From data gathered previously you know that two- and three-year-old females (for the sake of simplicity let's say that these bears don't live past age 3) give birth to an average of 1.5 female cubs each year and that 30% of those cubs die during their first year of life. Annual survival is 80% in the next 2 years of life.

a. Complete the following life table, using method 1.

Age x	Survival, s_x	l_x	Fecundity b_x	Net fecundity F_x
1		1		
2				
3				

b. What is the 3×3 Leslie matrix for this population (first age is 1)?

c. You take a census of emerging bears in the spring and find the population age vector to be

40   20   20

Calculate the population age vector for next year and the year after.

d. What is the growth rate of the population from year 1 to year 2? From year 2 to year 3?

e. Is this population in stable age structure? How can you tell?

f. What is the 4×4 Leslie matrix for this population, based on the application of method 2 (first age is 0)?

3. Make a life table for the fictional shrewd shrew. Two consecutive breeding censuses produce the following data for it.

Age x	No. of females (year 1)	No. of females (year 2)	No. of female births of mothers of age x in year 1
1	1000	1110	0
2	600	700	600
3	540	540	1620

a. What is the number of newborns (age $x = 0$) in year 1?

b. It is not known whether this population is at a stable age distribution. Your life table for ages 0 through 3 should include survival rates, s_x, death rates, d_x, fecundity, b_x, and net fecundity, F_x, given the limited data that you have.

c. Construct a 3×3 Leslie matrix for this species (for ages 1 through 3). (*Hint:* You have enough information in this problem to construct the Leslie matrix by either method 1 or method 2; however, you are asked to apply method 1 to the life table).

d. Verify that your Leslie matrix is correct by seeing if it computes the year 2 numbers based on the year 1 numbers as an input. Assuming that the vital rates calculated above remain constant for one more year. What would the population size of each age group be in year 3?

e. Starting with the age distribution $n_1(0) = 20$, $n_2(0) = 4$, and $n_3(0) = 4$, compute the population vector for eight consecutive years. Round off so that you have only whole animals in the population vector. Make the following plots.

(1) Total population by time

(2) Ln (total population) by time

(3) Plot λ_t by time. Is λ_t becoming constant after 8 years?

(4) Plot the age distribution (the relative proportion of each class) at each time. For the last plot you should have three lines—one for the proportion of each of the three age classes. Is the age distribution stabilizing after 8 years?

LESLIE MATRIX FORMATION AND A NEW TRICK

Imagine a species that reaches reproductive maturity at age 1. The immature young are then all those individuals between birth and 1 year old. We label their survival rate $s_0 = s_y$, and all adults (age 1 and older) have the same yearly survival rate, s_a, regardless of their age. Imagine also that the per capita birth rate, b, per adult female does not vary with age.

If the beginning of each year is delimited such that the birth pulse occurs immediately *after* the year begins, and applying method 1, the Leslie matrix would look like this:

$$\mathbf{L} = \begin{bmatrix} s_y b & s_y b & s_y b & s_y b & \cdot & s_y b & s_y b \\ s_a & 0 & 0 & 0 & \cdot & 0 & 0 \\ 0 & s_a & 0 & 0 & \cdot & 0 & 0 \\ 0 & 0 & s_a & 0 & \cdot & 0 & 0 \\ \cdot & \cdot & \cdot & \cdot & \cdot & \cdot & \cdot \\ 0 & 0 & 0 & 0 & \cdot & s_a & 0 \end{bmatrix}. \tag{3.16}$$

Exercise: Do you see why the term in the second row and first column of the matrix **L** is s_a and not s_y?

Now, you're going to learn a very neat trick for handling large Leslie matrices. It is due to Lefkovitch (1965) and can be an immense time saver. Lefkovitch noticed that, as long as consecutive ages have basically the same vital rates, he could lump them into a single composite stage and in so doing get a reduced number of life **stages** (instead of ages). Moreover, he showed some rules for doing this lumping such that the calculation of the geometric growth rate λ would be conserved.

The usual situation is one in which, after a certain number of years, all the older individuals simply have the same s's and b's. Lefkovitch showed that for such situations you could truncate the age structure with a final category (e.g., called "adults") and that the Leslie matrix could be truncated with a slight modification such that λ was preserved. All that is needed is to simply place the common adult survival rate, s_a, in the bottom right-hand corner of this modified stage matrix. To see why this works consider Figure 3.16, which is based on the assumption that the vital rates are constant from year to year.

The numbered circles in Figure 3.16 represent age classes, the b's are fecundities (age-specific birth rates), and the s's are age-specific survival rates. For age class 3 and beyond, all the ages have the same vital rates, s_a and b_a. Thus we can short-circuit this aging and survival cascade by creating a composite "older adult (a)" **stage class** that lumps age 3 and beyond. We next shunt the surviving older adults each year back into the older adults stage class for the next year. This shunting represents a transition of some proportion of individuals, s_a, back into the same stage class that they came from: adult age class 3 and above. Therefore the correct position for it in the Leslie matrix lies in the extreme bottom right-hand corner of the matrix (i.e., on the diagonal in position 3, 3).

Using the Lefkovitch rule that follows the pattern of vital rates shown in Figure 3.16, we may replace the matrix and age vector

$$\begin{bmatrix} F_1 & F_2 & F_a & F_a & \cdot & \cdot \\ s_1 & 0 & 0 & 0 & \cdot & \cdot \\ 0 & s_2 & 0 & 0 & \cdot & \cdot \\ 0 & 0 & s_a & 0 & \cdot & \cdot \\ 0 & 0 & 0 & s_a & \cdot & \cdot \\ \cdot & \cdot & \cdot & \cdot & & \end{bmatrix} \quad \text{and} \quad \begin{bmatrix} n_1 \\ n_2 \\ n_3 \\ \cdot \\ \cdot \end{bmatrix}$$

Figure 3.16
Age category reduction. The older ages all have the same vital rates and are lumped together into an "old adults" stage.

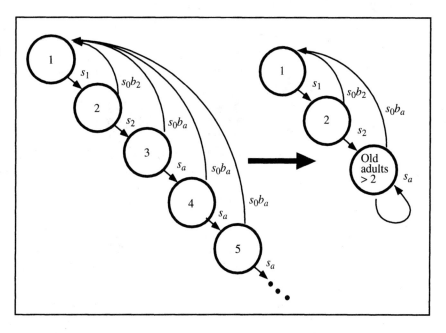

with

$$\begin{bmatrix} F_1 & F_2 & F_a \\ s_1 & 0 & 0 \\ 0 & s_2 & s_a \end{bmatrix} \quad \text{and} \quad \begin{bmatrix} n_1 \\ n_2 \\ n_a \end{bmatrix}.$$

The two Leslie matrices will have the same λ. Of course, some information is lost regarding the age composition for age 3 and beyond in the reduced system, since these ages are now lumped into the single stage *a*.

As soon as the *F* terms and the *s* terms no longer change with age, the Leslie matrix can be short-circuited in this manner. If you've done your homework, you'll appreciate how much simpler it is to work with these small matrices compared to very large ones.

As a second example, the large Leslie matrix of Eq. (3.16) can be reduced to a dynamically equivalent 2×2 Leslie matrix with the same λ. That is,

$$\mathbf{L} = \begin{bmatrix} s_y b & s_y b & s_y b & s_y b & \cdot & s_y b & s_y b \\ s_a & 0 & 0 & 0 & \cdot & 0 & 0 \\ 0 & s_a & 0 & 0 & \cdot & 0 & 0 \\ 0 & 0 & s_a & 0 & \cdot & 0 & 0 \\ \cdot & \cdot & \cdot & \cdot & \cdot & \cdot & \cdot \\ 0 & 0 & 0 & 0 & \cdot & s_a & 0 \end{bmatrix}$$

becomes

$$\mathbf{L} = \begin{bmatrix} s_y b & s_y b \\ s_a & s_a \end{bmatrix}. \tag{3.17}$$

This matrix can be used with the population stage vector

$$\mathbf{n} = \begin{bmatrix} n_1 \\ n_a \end{bmatrix} \quad \text{for all } x > 1 \tag{3.18}$$

to project population growth.

> **Exercise:** What are the 3×3 Lefkovitch matrix and the stage vector that are based on a birth pulse that begins immediately *before* the beginning of the census year?
>
> **Answer:**
>
> $$\begin{bmatrix} s_y b & s_a b & s_a b \\ s_y & 0 & 0 \\ 0 & s_a & s_a \end{bmatrix} \quad \text{and} \quad \begin{bmatrix} n_y \\ n_1 \\ n_a \end{bmatrix} \tag{3.19}$$

λ IS THE DOMINANT EIGENVALUE OF THE LESLIE MATRIX (ADVANCED)

The enigmatic title of this section should not scare you. In this discussion we explain what eigenvalues are, how they can be calculated, and what you can do with them. It turns out that they are very useful, not only in predicting population growth, but also predicting the stability of equilibria.

What is an eigenvalue? An **eigenvalue** (also called a characteristic exponent) is a number that encapsulates something about a matrix in a particular way. Let's return for a moment to the scalar geometric growth equation from Chapter 1 (no age structure here):

$$N(t + 1) = \lambda N(t).$$

Let growth begin at $t = 0$ with an initial population size of N_0. After an elapsed time of T periods, the population size will be

$$N(T) = N_0 \lambda^T.$$

When we have age-specific differences in the vital rates, in place of the scalar λ we have the Leslie matrix **L** that contains all the age-specific birth and death rates that go into determining λ. (Although we still don't know exactly what the analytical formula is for calculating λ—this is our goal.) Once at stable age structure, the analog of the geometric growth equation is

$$\mathbf{n}(T) = \mathbf{n}_0 \mathbf{L}^T, \tag{3.20}$$

where $\mathbf{n}(T)$ is the age vector at time T and $\mathbf{n_0}$ is the initial age vector. But what does it mean to take a matrix and raise it to the power T? If $T = 2$, multiply the matrix by itself. If $T = 3$, multiply it three times; **L L L**, and so on. Is there a way around all this arithmetic? Perhaps there's a shortcut. Let's pose the proposition that there might be some vector **x** such that the action of the matrix **L** on this vector **x** gives the same result as multiplying **x** by a simple scalar: Let's call this scalar λ, since it turns out to be the growth rate of the population. In other words, we propose

$$\mathbf{L}\,\mathbf{x} = \lambda\,\mathbf{x}, \tag{3.21}$$

$$(n \text{ by } n)\,(n \text{ by } 1) = (\text{scalar})(n \text{ by } 1)$$

where λ is a scalar, not a matrix. Equation (3.21) makes the bold proposition that the matrix multiplication **Lx** can be simplified into a multiplication of a scalar times a vector **x.** So if both a λ exists and an **x** has the property specified by Eq. (3.21), we could use λ (a scalar) in place of **L** (a matrix) and that would greatly simplify our ability to track the population with easy arithmetic, since the matrix in Eq. (3.20) would be replaced with

$$\boxed{N(T) = g\lambda^T \mathbf{x},} \tag{3.22}$$

where λ is a scalar, the vector **x** is a special vector with the property given by Eq. (3.21), and g is a scalar determined from the initial conditions N_0. Note that Eq. (3.22) says that, once **x** and λ are known, the population size at any future time T is just this vector **x** multiplied by a λ^T and then multiplied by a constant scalar g. But this implies that the special vector **x** must be the stable age vector, by definition. Thus from Eq. (3.22), we not only can solve for λ, but we can also find the stable age vector **x** in the process. Quite a feat! But how exactly do we do it?

Rearranging Eq. (3.21) gives

$$\mathbf{Lx} - \lambda\mathbf{x} = \mathbf{0},$$

where **0** is a vector of zeros. Collecting terms, we can also write

$$(\mathbf{L} - \lambda\mathbf{I})\mathbf{x} = \mathbf{0}, \tag{3.23}$$

where **I** is the *identity matrix;* **I** is a square matrix of all 0's except on the diagonal, where all the elements are simply 1. Here is a 3×3 identity matrix:

$$\mathbf{I} = \begin{bmatrix} 1 & 0 & 0 \\ 0 & 1 & 0 \\ 0 & 0 & 1 \end{bmatrix}.$$

Think of **I** as the matrix equivalent of the scalar value 1, in other words, **I** is defined such that $\mathbf{L\,I} = \mathbf{I\,L} = \mathbf{L}$ for any square matrix **L.** Of course, Eq. (3.23) has the trivial solution $\mathbf{x} = \mathbf{0}$, but this doesn't seem very interesting since it would give a trivial result to Eq. (3.22). It turns out that Eq. (3.23) also has a nonzero solution for **x**, if the matrix **L** – λ**I,** has a zero determinant (see Appendix 2 for *determinants*). The values of λ and

x that produce nonzero solutions to Eq. (3.23) are given by setting this determinant to 0. In other words, the solution of

$$\det (\mathbf{L} - \lambda\mathbf{I}) = 0 \qquad (3.24)$$

yields the value of λ that we are looking for (in fact, as we will see, *n* values). Equation (3.24) is called the **characteristic equation** of a square matrix **L**. As an example, suppose that we have the 2×2 Lefkovitch matrix **L**, Eq. (3.17), from the previous section,

$$\mathbf{L} = \begin{bmatrix} s_y b & s_y b \\ s_a & s_a \end{bmatrix}.$$

Then

$$(\mathbf{L} - \lambda\mathbf{I}) = \begin{bmatrix} s_y b - \lambda & s_y b \\ s_a & s_a - \lambda \end{bmatrix},$$

and the determinant of $\mathbf{L} - \lambda\mathbf{I}$ is $(s_y b - \lambda)(s_a - \lambda) - s_y s_a b$, which we set equal to 0, as in Eq. (3.24), to get the characteristic equation:

$$(s_y b - \lambda)(s_a - \lambda) - s_y s_a b = 0. \qquad (3.25)$$

We then use the characteristic equation to solve for λ:

$$\lambda^2 - \lambda s_a - \lambda s_y b + s_y b s_a - s_y b s_a = 0.$$

After canceling the two last terms of opposite sign on the left-hand side and collecting terms in the same power of λ, we arrive at

$$\lambda^2 - \lambda(s_a + s_y b) = 0.$$

This is a **quadratic** equation (see Appendix 1—Visualizing Equations), so we know that it has two roots (i.e., two solutions for λ). They are

$$\lambda_1 = (s_a + s_y b) \quad \text{and} \quad \lambda_2 = 0 \qquad (3.26)$$

Note that the first root is positive, is also a real number, and is exactly what we determined earlier to be the ultimate geometric growth rate of this population (Eq 3.9). This result is general: **the "dominant" eigenvalue of the Leslie matrix yields an age-structured population's ultimate geometric growth rate λ,** assuming that the vital rates remain constant over time. We use the word *ultimate* because we've already shown by example that a constant λ isn't reached until the population reaches its stable age structure.

Exercise: What is the dominant eigenvalue of this matrix?

$$\begin{bmatrix} 1 & 2 \\ 1 & 0 \end{bmatrix}$$

Solution: The characteristic equation is

$$\det \begin{bmatrix} 1-\lambda & 2 \\ 1 & 0-\lambda \end{bmatrix} = 0.$$

Thus

$$\lambda^2 - \lambda - 2 = 0,$$

$$(\lambda - 2)(\lambda + 1) = 0,$$

and the dominant λ is 2.

The stable age structure can also be calculated. Equation (3.23) suggests the answer. It's simply the **x** vector associated with the dominant λ. In other words, once you know λ, you plug it into Eq. (3.23) and solve for the associated eigenvector, which is the stable age vector **x.** For our example, **x** is obtained by solving

$$\begin{bmatrix} s_y b - (s_a + s_y b) & s_y b \\ s_a & s_a - (s_a + s_y b) \end{bmatrix} \begin{bmatrix} x_1 \\ x_a \end{bmatrix} = \begin{bmatrix} 0 \\ 0 \end{bmatrix}. \tag{3.27}$$

After doing the matrix multiplication in Eq. (3.27), we get two equations in two unknowns (x_1 and x_2), but the two equations are not independent (Eq. 3.23 guarantees this). We solve for a *relationship* between x_1 and x_2. Again after canceling terms in Eq. (3.27) we reach

$$\mathbf{L} - \lambda \mathbf{I} = \begin{bmatrix} -s_a & s_y b \\ s_a & -s_y b \end{bmatrix}. \tag{3.28}$$

Note that the determinant of this matrix is $s_a s_y b - s_a s_y b$ which equals zero, as it should since we have followed the prescription in Eq. (3.24).

Applying Eq. (3.23), we get

$$(\mathbf{L} - \lambda \mathbf{I})\mathbf{x} = \begin{bmatrix} -s_a & s_y b \\ s_a & -s_y b \end{bmatrix} \begin{bmatrix} x_1 \\ x_a \end{bmatrix} = \begin{bmatrix} 0 \\ 0 \end{bmatrix}.$$

Multiplying though we get two identical equations for the relationship between x_1 and x_2, which can be expressed as

$$x_1 = \frac{s_y b}{s_a} x_a.$$

If we arbitrarily set $x_1 = 1$, then $x_a = s_a / s_y b$. If we set $x_1 = 2$, then $x_a = 2 s_a / s_y b$, and so on. We have the general relationship that the vector **x** has the form

$$\mathbf{x} = g \begin{bmatrix} 1 \\ \dfrac{s_a}{s_y b} \end{bmatrix}, \tag{3.29}$$

where g can be any scalar. Typically the age distribution vector is expressed so that the abundance of each age class is given as a proportion of the total population (i.e., the n_x sum to 1. This means that g in Eq. (3.29) should be chosen as $1/(1 + s_a/s_y b)$. The stable age distribution is

$$\mathbf{x} = \frac{1}{1 + \dfrac{s_a}{s_y b}} \begin{bmatrix} 1 \\ \dfrac{s_a}{s_y b} \end{bmatrix}.$$

In conclusion, for this example, after the stable age structure is reached, there are $s_a/s_y b$ times as many adults as juveniles and the population grows at the discrete geometric rate $\lambda = s_a + s_y b$.

Problem: Suppose that newborns have survival rate 0.5, and ages 1 and older have survival rate 0.8. Reproductive maturity occurs at age 1 (the second year of life) and all adult females produce two female offspring per year. What is λ and what is the stable age distribution?

Solution:

$$\mathbf{L} = \begin{bmatrix} s_y b & s_y b \\ s_a & s_a \end{bmatrix} = \begin{bmatrix} (0.5)(2) & (0.5)(2) \\ 0.8 & 0.8 \end{bmatrix},$$

set

$$\det \begin{bmatrix} 1 - \lambda & 1 \\ 0.8 & 0.8 - \lambda \end{bmatrix} = 0,$$

$$\lambda^2 - 1.8\lambda = 0,$$

which has two solutions for λ:

$$\lambda = 0, \quad \text{and} \quad \lambda = 1.8.$$

From Eq. (3.29), the stable age structure is the eigenvector paired with $\lambda = 1.8$, or

$$\begin{bmatrix} 1-1.8 & 1 \\ 0.8 & 0.8-1.8 \end{bmatrix} \begin{bmatrix} n_1 \\ n_a \end{bmatrix} = \begin{bmatrix} 0 \\ 0 \end{bmatrix},$$

where n_a is the number of adults of age 2 and older. Then

$$\mathbf{n} = \frac{1}{1.8} \begin{bmatrix} 1 \\ 0.8 \end{bmatrix} = \begin{bmatrix} 0.556 \\ 0.444 \end{bmatrix}.$$

Problem: What is the dominant eigenvector for the matrix

$$\begin{bmatrix} 1 & 2 \\ 1 & 0 \end{bmatrix}?$$

Solution: We earlier determined that the dominant eigenvalue of this matrix is 2. So we have

$$\begin{bmatrix} 1-2 & 2 \\ 1 & 0-2 \end{bmatrix} \begin{bmatrix} x_1 \\ x_2 \end{bmatrix} = \begin{bmatrix} 0 \\ 0 \end{bmatrix} \quad \text{or} \quad \begin{bmatrix} -1 & 2 \\ 1 & -2 \end{bmatrix} \begin{bmatrix} x_1 \\ x_2 \end{bmatrix} = \begin{bmatrix} 0 \\ 0 \end{bmatrix}.$$

Solving for the relationship between x_1 and x_2 gives $x_1 = 2x_2$. If we set $x_2 = 1$, we get

$$\mathbf{x} = \begin{bmatrix} 2 \\ 1 \end{bmatrix}.$$

and normalizing by the total gives

$$\mathbf{x} = \begin{bmatrix} 2/3 \\ 1/3 \end{bmatrix}.$$

The computation of the eigenvalues of larger matrices can be very tedious. In general, however, there are n eigenvalues for an $n \times n$ matrix although they may not all be different numbers (i.e., some of the λ's may be duplicated).

These days it is easy to solve for λ using computer software like Matlab©, Maple©, Mathematica©, or MacMath©. One nice software application is Matlab©. Using it, all you have to do to get the eigenvalues of a matrix **A** is simply to type the matrix into the computer and give it a name—for example, A. Then type the command "eig(A)" and hit the carriage return. The software does the rest and all the eigenvalues quickly appear on the computer screen.

Here's an example of this process in Matlab© with a four-age-group Leslie matrix. The input typing is in boldface type; the computer's output is in regular font.

```
> A = [0,2,3,1;
.3, 0, 0, 0;
0, .4, 0, 0;
0, 0, .2, 0]

A =

     0      2.0000   3.0000   1.0000
  0.3000      0        0        0
     0      0.4000     0        0
     0        0      0.2000     0

»eig(A)
```

answer =

0.9933

–0.4585 + 0.3264i

–0.4585–0.3264i

–0.0763

That's all there is to it. A 4×4 matrix has four eigenvalues. In this example some of the eigenvalues are complex numbers, with a real part and an imaginary part containing $i = \sqrt{-1}$. The complex numbers always come in pairs called **complex conjugates,** which have a real part plus or minus the same imaginary part. In this example, the dominant eigenvalue, which is the ultimate geometric rate of growth, equals 0.9933. Note that this is the only eigenvalue that is both real and positive. Since λ is less than 1, the population growth according to the vital rates in matrix **A** will decline exponentially over time. The absolute values of the eigenvalues of **A** (remembering that, for complex numbers, $a + bi,$ the absolute value is defined as $(a^2 + b^2)^{0.5}$) are

$$0.9933 = \text{dominant eigenvalue}$$

$$0.5628$$

$$0.5628$$

$$0.0763$$

Leslie matrices are nonnegative matrices because all the terms in them are greater than or equal to zero. The **Perron–Frobenius theorem** derives some properties of the eigenvalues and eigenvectors of nonnegative matrices. If a matrix **A** is nonnegative (and it has a few other mathematical features that Leslie matrices will typically share), then one of **A**'s eigenvalues must be a real number (i.e., not a complex number with an imaginary part) and positive. There may be more than one positive real eigenvalue, but you can identify **the "dominant" eigenvalue as the one that is larger than all the others.** Also the dominant eigenvalue is the only one with a nonnegative eigenvector (i.e., all its numbers are real and nonnegative). For most biologically realistic conditions, there is only a single positive real eigenvalue, so there will be no doubt about which is the dominant one.

Why can we forget about all the other eigenvalues and eigenvectors? Why does the *ultimate* population behavior depend simply on the *dominant* one? To see this requires that we go back to the beginning—way back to Eq. (3.22). This solution is true for each and every eigenvalue. It's also true for the sum of all the solutions:

$$\mathbf{n}(t) = g_1\,\mathbf{x}_1\lambda_1^t + g_2\,\mathbf{x}_2\lambda_2^t + g_3\,\mathbf{x}_3\lambda_3^t + \cdots + g_{max}\,\mathbf{x}_{max}\lambda_{max}^t. \tag{3.30}$$

The g's are scalars set by the initial conditions—the initial population size of each age class—and the **x**'s are the different eigenvectors. This equation is often called the "general solution" of Eq. (3.20). Let's suppose that the dominant eigenvalue is $\lambda_1 = 2$ and that the next largest in absolute value is $\lambda_2 = 0.5$. At time $t = 1$ the ratio of the two is $(2/0.5) = 4$. In other words, λ_1 is contributing four times as much as λ_2 to the value of N at time 1. At time 2, the ratio is $(4/0.25) = 16$; at time 3, the ratio is $(8/0.125) = 64$. So as time goes on, the first term in the sum (with the dominant eigenvalue and eigenvector) has a greater proportional influence on $\mathbf{n}(t)$. After a while, you might as well lop off all those other terms that keep getting smaller and smaller relative to the total. The behavior of $\mathbf{n}(t)$ ultimately is reflected by the stable age structure $\mathbf{x}_1$, and the total population size N grows at rate λ_1.

The subject of the mathematical manipulation of complex numbers is beyond the scope of this primer. We want to keep things here as simple as possible. However, **there is a connection between imaginary numbers and oscillations.** Perhaps you recall this identity from trigonometry:

$$e^{iz} = \cos(z) + i\,\sin(z).$$

Another expression, when λ is a complex number, is

$$\lambda^n = r^n(\cos n\theta + i\,\sin n\theta), \tag{3.31}$$

where r is the absolute value of λ and θ is the angle formed by the complex number when represented as a vector in the plane of $x =$ real part and $y =$ imaginary part.

The important point for our purposes is simply that there's a deep mathematical connection between imaginary numbers and functions that **oscillate.** Because Eq. (3.30) shows that the growth of $\mathbf{n}(t)$ is controlled by the λ's raised to the power of time and because Eq. (3.31) shows that such functions can produce oscillations if λ is a complex number, you should not be too surprised to find out that $\mathbf{n}(t)$ can oscillate over time. Look again at Figure 3.5, which shows the growth of the population with three age classes. You can see how the numbers of each age jump up, then down, then up, and so on, before they settle down to grow smoothly at a constant geometric rate. That's because some of the **nondominant** eigenvalues of that Leslie matrix had imaginary parts. As time goes on, these terms with the imaginary parts are premultiplied by increasingly smaller numbers and the fluctuations damp out. Sometimes when two species interact, like a predator and its prey, oscillations like these may never dampen out—more on this in Chapter 12.[1]

MORE ON STAGE-STRUCTURED GROWTH

The Lefkovitch rule for absorbing terminal ages into a single stage can be generalized. For many organisms, their actual chronological age may be less important in influencing their vital rates than their body size or developmental stage, that is, whether it is an adult or a larvae—a seed, a seedling, a sapling, or an adult tree. Nearly all the techniques that we have developed for dealing with age-structured geometric growth may be co-opted to deal with **stage-structured** growth. However, since we ultimately need the change in numbers of each stage over time, we must know precisely the growth rate from one stage to another and this information must be incorporated into a transition matrix. An example illustrates the point.

Suppose that we have a forest tree species and divide its life history into four stages: seeds (1), seedlings (2), saplings (3), and adult trees (4), as shown in Figure 3.17.

The matrix $\mathbf{G}$ contains these transition terms:

$$\mathbf{G} = \begin{bmatrix} g_{11} & 0 & 0 & F_4 \\ g_{21} & g_{22} & 0 & 0 \\ 0 & g_{32} & g_{33} & 0 \\ 0 & 0 & g_{43} & g_{44} \end{bmatrix}.$$

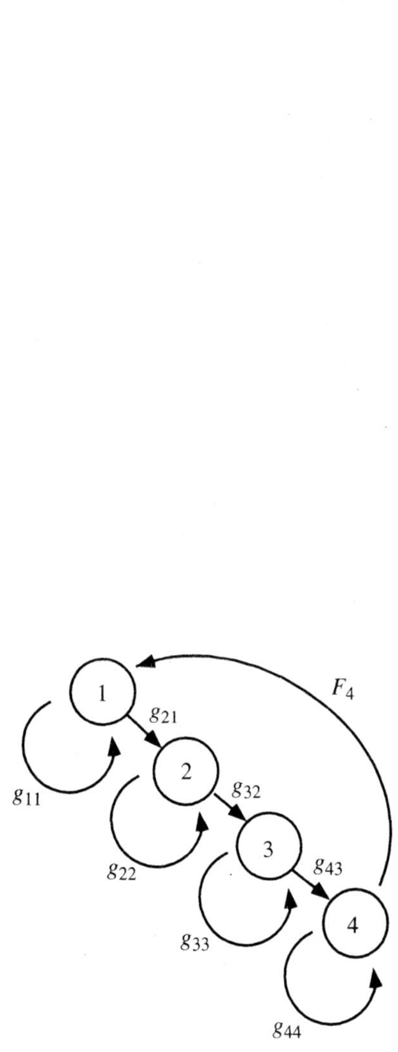

In contrast to a Leslie matrix, there are now terms along the diagonal of the transition matrix $\mathbf{G}$, which represent the probability that each stage survives but does not develop into some other stage. Moreover, the advancement from one stage to the next depends not simply on the passage of time, aging, and surviving, but also on growing or developing from one stage to the next. For example, some fraction of seeds (g_{11}) may remain dormant every year and not germinate. Consequently, we do not use the symbol s_i, which we used before for survival, but instead g, which involves both survival and growth or development.

It is best to think of these g terms as probabilities. For example, the probability of changing from a sapling (regardless of the absolute age of that sapling) into an adult tree per time step is g_{43}. It may even be the case for some plants that they can shrink in size from one year to the next. Thus we might have nonzero g_{ij} elements *above* the diagonal of the transition matrix. Regardless of these possible complexities, as long as the elements in the transition matrix are constants, the population will eventually reach a constant rate of geometric growth and a stable stage structure (with the exception of a

Figure 3.17
A transition scheme for a hypothetical tree. Seeds (stage 1) may remain dominate or may germinate into seedlings (stage 2). If seedlings grow and survive, they turn into saplings (stage 3). If saplings grow and survive, they turn into adult trees (stage 4), which can produce new seeds.

---

1. Recall that in Chapter 1 we compared single-species' geometric and exponential growth and reached the conclusion that $r = \ln \lambda$. With age-structured growth, we have now found that, for the difference equation approach taken here, some of the λ's may be complex numbers. Since the logarithm operation is undefined for imaginary numbers, it is impossible to convert Eq. (3.30) to a sum involving terms each with e raised to the power of $\ln \lambda$. However, since the dominant eigenvalue is strictly real, it is still fair to express the *ultimate* population growth interchangeably as $r = \ln \lambda$.

Box 3.2   Some Shortcuts for Finding λ, the Discrete Rate of Growth for a Leslie Matrix.

Reduction Rule 1: If all ages past some age have zero net fecundity, then the Leslie matrix has the form

$$\begin{bmatrix} F_0 & F_1 & F_2 & 0 & \cdot & \cdot & 0 & 0 \\ s_0 & 0 & 0 & 0 & \cdot & \cdot & 0 & 0 \\ 0 & s_1 & 0 & 0 & \cdot & \cdot & 0 & 0 \\ 0 & 0 & s_2 & 0 & \cdot & \cdot & 0 & 0 \\ 0 & 0 & 0 & s_3 & \cdot & \cdot & 0 & 0 \\ \cdot & \cdot & \cdot & \cdot & \cdot & \cdot & \cdot & \cdot \\ 0 & 0 & 0 & 0 & \cdot & \cdot & \cdot & 0 \end{bmatrix}.$$

The dominant λ of this matrix equals the dominant λ of the reduced matrix formed by deleting the columns with all 0's and the corresponding rows:

$$\begin{bmatrix} F_0 & F_1 & F_2 \\ s_0 & 0 & 0 \\ 0 & s_1 & 0 \end{bmatrix}.$$

Conclusion: Postreproductive ages can be ignored, and method 2 Leslie matrices may be reduced in size by truncating the last row and column.

Reduction Rule 2: If all ages beyond the first position in the Leslie matrix (i.e., either age 0 if the Leslie matrix is formed by method 2 or age 1 if the Leslie matrix is formed by method 1) have the same F_x and s_x, then the Leslie matrix has the form

$$\begin{bmatrix} F & F' & F' & F' & \cdot & \cdot & F' \\ s & 0 & 0 & 0 & \cdot & \cdot & 0 \\ 0 & s' & 0 & 0 & \cdot & \cdot & 0 \\ 0 & 0 & s' & 0 & \cdot & \cdot & 0 \\ \cdot & \cdot & \cdot & \cdot & \cdot & \cdot & \cdot \\ 0 & 0 & 0 & 0 & \cdot & \cdot & 0 \end{bmatrix}.$$

The dominant λ of this matrix will be identical to that of the reduced matrix, or

$$\begin{bmatrix} F & F' \\ s & s' \end{bmatrix}.$$

This last rule also means that the 3×3 Lefkovitch matrix produced by method 2 (i.e., the birth pulse occurs immediately before the beginning of the census; Eq. (3.19),

$$\mathbf{L} = \begin{bmatrix} s_y b & s_a b & s_a b \\ s_y & 0 & 0 \\ 0 & s_a & s_a \end{bmatrix},$$

can in turn be reduced to a 2×2 matrix with the same dominant eigenvalue, or

$$\mathbf{L} = \begin{bmatrix} s_y b & s_a b \\ s_y & s_a \end{bmatrix}. \tag{3.32}$$

Problem: Show that this last matrix, Eq. (3.32), has the same dominant eigenvalue as the method 1 analog,

$$\mathbf{L} = \begin{bmatrix} s_y b & s_y b \\ s_a & s_a \end{bmatrix},$$

whose eigenvalues were given in Eq. (3.26).

Answer:

$$\det \begin{bmatrix} s_y b - \lambda & s_a b \\ s_y & s_a - \lambda \end{bmatrix} = 0,$$

so

$$\lambda^2 - \lambda s_a - \lambda s_y b + s_y b s_a - s_y b s_a = 0.$$

After canceling the two last terms of opposite signs on the left-hand side and collecting terms in the same power of l,

$$\lambda^2 - \lambda (s_a + s_y b) = 0.$$

Therefore

$$\lambda_1 = (s_a + s_y b) \quad \text{and} \quad \lambda_2 = 0.$$

Note that these are the same eigenvalues that we got for the method 1 matrix in Eq. (3.26).

Problem: Show that the eigenvector associated with this dominant eigenvalue is

$$\mathbf{x} = \frac{1}{1+b} \begin{bmatrix} b \\ 1 \end{bmatrix}.$$

Similarly, the Lefkovitch matrix

$$\mathbf{L} = \begin{bmatrix} s_0 b_1 & s_a b_a & s_a b_a \\ s_0 & 0 & 0 \\ 0 & s_a & s_a \end{bmatrix}$$

has the same dominant eigenvalue as

$$\mathbf{L} = \begin{bmatrix} s_0 b_1 & s_a b_a \\ s_0 & s_a \end{bmatrix} \tag{3.33}$$

Figure 3.18
A simulation of the tree stage matrix G given by Eq. (3.34) over 12 consecutive years. The λ for this matrix is very close to 1.

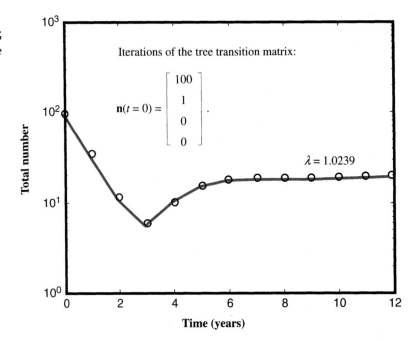

very few mathematical oddities). This rate of growth can be determined, as before, either by simulation or by finding the dominant eigenvalue, λ, of the transition matrix, **G**. The **stable stage distribution** is, as before, the eigenvector associated with this dominant eigenvalue.

The likelihood of the vital rates actually being constant needs to be considered. In age-structured growth, after 1 year, all surviving x-year-olds are now $x + 1$ years old. Aging is linear with time and not dependent on environmental factors, such as temperature or rainfall. Environmental factors may affect the vital rates, but they do not influence the advancement of time and chronological age. In stage-structured growth, the environment may effect both the transition rate from one stage to the next and the vital rates at each stage. In a warm year, individuals may grow and mature faster than in a cold year. Consequently, the assumption of the transition matrix maintaining constancy over time is less likely to be met. This distinction is important and is an important caution in the application of stage-structured models to real situations.

As an example, we may fill in the matrix **G** with some values that might be expected for a tree:

$$\mathbf{G} = \begin{bmatrix} 0.3 & 0 & 0 & 100 \\ 0.03 & 0.3 & 0 & 0 \\ 0 & 0.1 & 0.1 & 0 \\ 0 & 0 & 0.2 & 0.9 \end{bmatrix}. \tag{3.34}$$

The survival rates for seeds are very low, while adult survival is quite high. The population is initiated with 100 seeds and a single seedling and iterated over 12 years, as shown in Figure 3.18. The dominant eigenvalue for this matrix is $\lambda = 1.0239$, so the geometric growth rate is very close to 1; this asymptotic rate of growth is reached after only about 8 years. The corresponding changes in stage structure are shown in Figure 3.19. The stable age distribution has many times more seeds than adult trees. Note one other feature in this figure. The proportion of adult trees at stable stage structure, 0.0069, is greater than the proportion of saplings, 0.0043, the stage that proceeds it. Since all old trees come from young trees and since some young trees may die before they grow into adult trees, it seems paradoxical in a way that old trees can outnumber young trees in an increasing population. Indeed, as we prove in Chapter 4, for age-structured population growth, such an inverted age pyramid is impossible in an increasing population at stable age structure. However, with stage-structured population growth, not all stages necessarily have the same temporal duration. The young tree stage does not last as long as the adult tree stage (i.e. $g_{33} = 0.01 < g_{44} = 0.9$). Consequently, adult trees can outnumber young trees.

Figure 3.19
The stage structure for the first 8 years of the tree simulation shown in Figure 3.18. The stable stage distribution is (0.9495, 0.0393, 0.0043, 0.0069). The last two ages are so rare they can't be seen in some of the figures.

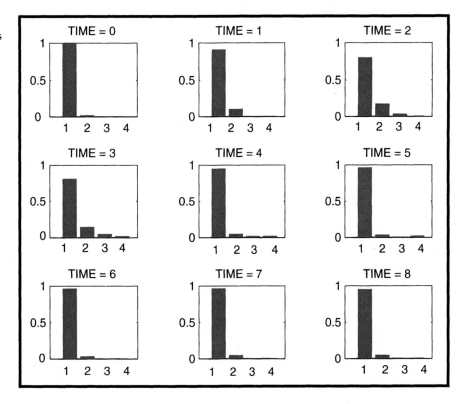

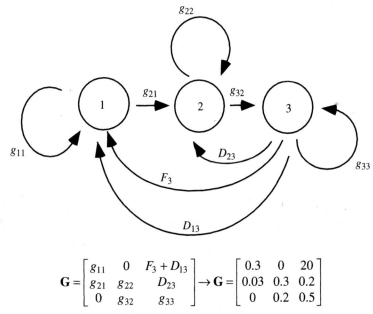

Problem: Consider the transition matrix **G** for a hypothetical sponge life history with three stage groups: small (1), medium (2), and large (3).

$$\mathbf{G} = \begin{bmatrix} g_{11} & 0 & F_3 + D_{13} \\ g_{21} & g_{22} & D_{23} \\ 0 & g_{32} & g_{33} \end{bmatrix} \rightarrow \mathbf{G} = \begin{bmatrix} 0.3 & 0 & 20 \\ 0.03 & 0.3 & 0.2 \\ 0 & 0.2 & 0.5 \end{bmatrix}$$

Large sponges (3) may divide into medium-sized sponges at rate D_{23} or into small sponges at rate D_{13}. Using the sample values for the **G** matrix, calculate the geometric growth rate of this population at the stable stage distribution. You may need some software for this.

Exercise: Suppose that seeds have a germination rate of 50% in their first year but 100% in their second year. How could you modify the stage matrix of Eq. (3.34) to account for this complication?

PROBLEMS

1. What is the answer to Darwin's elephant problem at the beginning of this chapter?

Answer:

We want to construct a Leslie matrix for elephants, given the information that Darwin had at his disposal. We can do so in 10-year intervals to get a 10×10 Leslie matrix. However, while we are told that a total of six offspring are produced from ages 30 though 90, we are not told how these births are distributed across these age classes. Also we are not given any information about the mortality schedule—only that the maximum age is 100. We make the generous assumptions that the young are distributed evenly across the adult age classes, 30–90, that the sex ratio at birth is 1:1, and that the survival rates are 1, for all but the last age class, which has survival rate 0. With these assumptions, we get

$$\begin{bmatrix} 0 & 0 & 0 & 0.5 & 0.5 & 0.5 & 0.5 & 0.5 & 0.5 & 0 \\ 1 & 0 & 0 & 0 & 0 & 0 & 0 & 0 & 0 & 0 \\ 0 & 1 & 0 & 0 & 0 & 0 & 0 & 0 & 0 & 0 \\ 0 & 0 & 1 & 0 & 0 & 0 & 0 & 0 & 0 & 0 \\ 0 & 0 & 0 & 1 & 0 & 0 & 0 & 0 & 0 & 0 \\ 0 & 0 & 0 & 0 & 1 & 0 & 0 & 0 & 0 & 0 \\ 0 & 0 & 0 & 0 & 0 & 1 & 0 & 0 & 0 & 0 \\ 0 & 0 & 0 & 0 & 0 & 0 & 1 & 0 & 0 & 0 \\ 0 & 0 & 0 & 0 & 0 & 0 & 0 & 1 & 0 & 0 \\ 0 & 0 & 0 & 0 & 0 & 0 & 0 & 0 & 1 & 0 \end{bmatrix}.$$

Using Matlab©, we get the dominant eigenvalue of this matrix: $\lambda = 1.1923$/decade. This means that in 75 decades a single pair of elephants (i.e., one female and one male) would produce

$$\ln N_t = \ln(1) + 75(\ln 1.1923)$$

$$= 0 + (75)(0.1759)$$

$$= 13.191.$$

Therefore the number of female elephants, N, at 750 years is $e^{13.191} = 535,692$, which implies a total of 1,071,384 elephants of both sexes. This is certainly a lot of elephants but considerably less than Darwin's estimate of "nearly nineteen million elephants" over the same time interval. Armbruster and Lande (1993) reviewed more recent work on the demography of African elephants and calculated a maximum population growth rate of about 30% per decade.

We could approximate the large 10×10 Leslie matrix that we used by assuming that for age classes 40–50 and beyond, all ages have the same fecundity of 0.5 and mortality rate of 0.1. While this makes elephants effectively immortal, the mean life span is still around 100 years (see Chapter 4). With these vital rates, we get the 4×4 Lefkovitch matrix

$$\begin{bmatrix} 0 & 0 & 0 & 0.5 \\ 1 & 0 & 0 & 0 \\ 0 & 1 & 0 & 0 \\ 0 & 0 & 1 & 0.9 \end{bmatrix}.$$

The dominant eigenvalue of this matrix is $\lambda = 1.1938$/decade, which is slightly larger than that based on the larger matrix.

2. The spotted owl has two subspecies: the northern spotted owl (primarily in Oregon, Washington, and British Columbia) and the California spotted owl, which ranges from northern to southern California. Both live in old growth forest. The owl matures at age 2 (i.e., in its third year of life). Survivorship in the first year of life has been estimated to be $s_0 = 0.3$; yearling subadults and adults of all ages have approximately the same survival rates of $s = 0.75$, and female fecundity is $b = 0.3$ (Noon and McKelvey 1996). Construct a Lefkovitch matrix that yields the correct λ and allows projections of the newborns and yearlings plus adults of all ages from year to year.

Answer:

Since this problem asks for projections of newborns, n_0, we must apply method 2 to form the Leslie matrix. The three age groups are the newborns, yearlings, and adults of all ages ($x = 2$ and beyond). Yet yearlings have the same survival rate as adults, s_a. This gives

$$\mathbf{L} = \begin{bmatrix} 0 & s_a b & s_a b \\ s_0 & 0 & 0 \\ 0 & s_a & s_a \end{bmatrix} \quad \text{and} \quad \mathbf{n} = \begin{bmatrix} n_0 \\ n_1 \\ n_a \end{bmatrix}.$$

What may seem odd about it is that, although reproductive maturity occurs in the third year of life, the first reproduction occurs in the second column of the matrix. Referring back to Eq. (3.15), the second element in the first row is $s_1 b_2$. The term b_2 is the fecundity of two-year-olds (the third age class). Hence we put the first reproductive bout of b in column 2 (not column 3). Also note that yearling subadults have the same survivorship as adults, $s_1 = s_a = 0.75$, so the number of two-year-old and older females giving birth must be discounted by adult survival, s_a. We also know from Box 3.2 that this matrix can be further reduced to one with the same dominant eigenvalue yet containing only two age groups: the zero-year-olds and one-year-olds and older; this matrix is

$$\mathbf{L} = \begin{bmatrix} 0 & s_a b \\ s_0 & s_a \end{bmatrix}.$$

It has the structure of Eq. (3.33). Now, filling in terms with their actual numerical values, we reach

$$\mathbf{L} = \begin{bmatrix} 0 & 0.225 \\ 0.3 & 0.75 \end{bmatrix} \quad \text{and} \quad \mathbf{n} = \begin{bmatrix} n_0 \\ n_{y+a} \end{bmatrix},$$

where $y + a$ refers to yearlings and all adults combined. The dominant eigenvalue is $\lambda = 0.831$. Populations of this subspecies appear to be shrinking by about 17% each year.

If Method 1 was applied to the data, the matrix system would be

$$\mathbf{L'} = \begin{bmatrix} 0 & 0.09 \\ 0.75 & 0.75 \end{bmatrix} \quad \text{and} \quad \mathbf{n} = \begin{bmatrix} n_1 \\ n_a \end{bmatrix}.$$

Exercise:

Verify that the matrix $\mathbf{L'}$ has the same eigenvalues as $\mathbf{L}$. Also verify that

$$\begin{bmatrix} 0 & 0.225 & 0.225 \\ 0.3 & 0 & 0 \\ 0 & 0.75 & 0.75 \end{bmatrix}$$

has the same *dominant* eigenvalue as $\mathbf{L'}$.

3. If you have access to a software program that calculates eigenvalues, show that the three Leslie matrices used to describe deer population growth in the section on Cohort Analysis,

$$\begin{bmatrix} 0 & 0.63 & 0.702 & 0.5 \\ 0.6 & 0 & 0 & 0 \\ 0 & 0.9 & 0 & 0 \\ 0 & 0 & 0.9 & 0 \end{bmatrix}, \quad \begin{bmatrix} 0 & 0.42 & 0.468 & 0.6 \\ 0.9 & 0 & 0 & 0 \\ 0 & 0.9 & 0 & 0 \\ 0 & 0 & 0.5 & 0 \end{bmatrix},$$

and

$$\begin{bmatrix} 0 & 0.63 & 0.702 & 0.5 & 0 \\ 0.6 & 0 & 0 & 0 & 0 \\ 0 & 0.9 & 0 & 0 & 0 \\ 0 & 0 & 0.9 & 0 & 0 \\ 0 & 0 & 0 & 0.5 & 0 \end{bmatrix}$$

all have the same dominant eigenvalue, 1.0.

4. What is the eventual geometric growth rate of a population with the following Lefkovitch modified Leslie matrix?

$$\begin{bmatrix} 0 & 2 \\ 0.5 & 0.8 \end{bmatrix}$$

Solution:

The eigenvalues of this matrix are found by setting its determinant equal to zero and solving for the λ's:

$$\det \begin{bmatrix} 0-\lambda & 2 \\ 0.5 & 0.8-\lambda \end{bmatrix} = 0.$$

This yields the characteristic equation

$$\lambda^2 - 0.8\lambda - 1 = 0. \qquad (a)$$

Recall that the formula for the solution of a quadratic equation,

$$ax^2 + bx + c = 0, \qquad (b)$$

with coefficients a, b, and c, is,

$$x = \frac{-b \pm \sqrt{b^2 - 4ac}}{2a}.$$

By comparing Eq. (b) with Eq. (a), we see that $a = 1$, $b = -0.8$, and $c = -1$. Thus the two eigenvalues are

$$\lambda = \frac{0.8 \pm \sqrt{0.64 + 4}}{2} = 1.477 \quad \text{and} \quad -0.677$$

and that the "dominant" λ is 1.477. This is the eventual geometric growth rate of this population.

5. Reese (1975) surveyed swans on the Chesapeake Bay during the fledging period (when the young birds are just about to fly). At this time of year the population can be counted according to four discernible age groups: fledglings, one-year-olds, two-year-olds, and adult birds (three-year-olds and older). The fledglings are about 4 months old at census, and an unknown amount of mortality has taken place since they were eggs. Female swans breed for the first time when they are 3. Nearly every adult (96%) breeds once each year, and the sex ratio of offspring is 1:1. The average number of fledglings seen per active nest is 3.1. The survival rate beyond fledging is high (0.9). Show that

$$\mathbf{L} = \begin{bmatrix} 0 & 0 & 1.3 & 1.3 \\ 0.9 & 0 & 0 & 0 \\ 0 & 0.9 & 0 & 0 \\ 0 & 0 & 0.9 & 0.9 \end{bmatrix}$$

can serve as a Leslie matrix for the female population for the stage categories that can be counted. This Leslie matrix has the same dominant eigenvalue as

$$\mathbf{L} = \begin{bmatrix} 0 & 0 & 1.3 \\ 0.9 & 0 & 0 \\ 0 & 0.9 & 0.9 \end{bmatrix}.$$

What is it?

4 *Demographic Relationships*

We have shown that the vital rates determining population growth can be compactly combined into a Leslie matrix. The Leslie matrix can be iterated to follow the growth of each age group over time. Moreover, the dominant eigenvalue of the Leslie matrix gives the ultimate geometric rate of growth of each age group (as well as that for the total population), and the eigenvector associated with this dominant eigenvalue gives the stable age composition. In this chapter we present another way to determine the stable age composition that doesn't require matrix math. It also has a nice feature: its development flows logically into a derivation of an important formula in demography, the **Lotka-Euler[1] equation.**

THE RELATIVE ABUNDANCE OF THE DIFFERENT AGE CLASSES

To develop the argument, focus on a simple situation with only three age classes and a population that is already at the stable age distribution. This population is growing at discrete rate λ, and thus each age group is growing at rate λ. Figure 4.1 shows the time course on a log scale for the size of each age class—hence the parallel straight lines. For purposes of illustration, let's focus on the situation at time 1992.

To simplify notation, we can renumber years so that $t = 1990$ is simply $t = 0$. Thus $1991 = 1$, $1992 = 2$, and so on. Time (i.e., year) is indicated by the number in parentheses, and age class is indicated by a subscript; e.g., $n_3(2)$ is the number of three-year-olds in year 2. Rearranging Eq. (4.1) to form

$$\frac{n_3(2)}{n_1(2)} = l_3 \lambda^{-2} \tag{4.2}$$

provides an expression for the ratio of the three-year-olds to the one-year-olds. By extension to the other ages and to other future times, the number of any age class x at time t relative to the number of one-year-olds at time t is

$$\frac{n_x(t)}{n_1(t)} = l_x \lambda^{-x+1}, \tag{4.3}$$

and the size of age class x relative to the number of newborns at time t is

$$\frac{n_x(t)}{n_0(t)} = l_x \lambda^{-x}. \tag{4.4}$$

We ultimately want an expression for the size class of x at t *relative to the total population size,* not just relative to the number of one-year-olds or newborns. To reach that

---

1. Alfred Lotka worked on mathematical descriptions of biological processes in the first part of this century. Leonhard Euler (pronouced "oiler") was a prolific mathematician of the 18th century.

Figure 4.1
The stable age distribution can be derived by calculating the number of one-year-olds that exist at $t = 1990$ in two different ways under the assumption that a stable age structure has been reached.

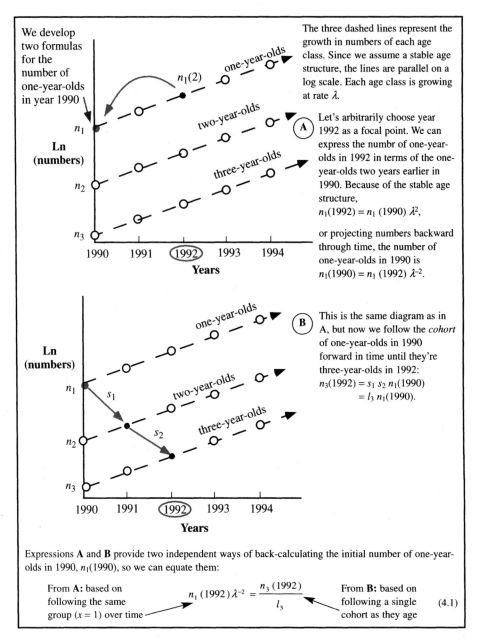

We develop two formulas for the number of one-year-olds in year 1990

The three dashed lines represent the growth in numbers of each age class. Since we assume a stable age structure, the lines are parallel on a log scale. Each age class is growing at rate λ.

A Let's arbitrarily choose year 1992 as a focal point. We can express the numbr of one-year-olds in 1992 in terms of the one-year-olds two years earlier in 1990. Because of the stable age structure,

$n_1(1992) = n_1(1990)\,\lambda^2$,

or projecting numbers backward through time, the number of one-year-olds in 1990 is

$n_1(1990) = n_1(1992)\,\lambda^{-2}$.

B This is the same diagram as in A, but now we follow the *cohort* of one-year-olds in 1990 forward in time until they're three-year-olds in 1992:

$n_3(1992) = s_1 s_2 n_1(1990)$
$\qquad = l_3 n_1(1990)$.

Expressions **A** and **B** provide two independent ways of back-calculating the initial number of one-year-olds in 1990, $n_1(1990)$, so we can equate them:

From **A**: based on following the same group ($x = 1$) over time

$$n_1(1992)\,\lambda^{-2} = \frac{n_3(1992)}{l_3}$$

From **B**: based on following a single cohort as they age   (4.1)

point we write expressions like Eq. (4.4) for each age group x and then sum across all the different ages:

$$\frac{n_0(t)}{n_0(t)} + \frac{n_1(t)}{n_0(t)} + \frac{n_2(t)}{n_0(t)} + \frac{n_3(t)}{n_0(t)} \cdots + \frac{n_{max}(t)}{n_0(t)} = \sum_{x=0}^{max} \lambda^{-x} l_x. \qquad (4.5)$$

The total population size including newborns at time t is $n_0(t) + n_1(t) + n_2(t) + \cdots + n_{max}(t) = N(t)$. Every term in Eq. (4.5) has the same denominator, and the numerators all sum to $N(t)$. Therefore we may rearrange Eq. (4.5) to yield an expression for the proportion of newborns in the population at time t, $n_0(t)/N(t)$:

$$c_0 = \frac{n_0(t)}{N(t)} = \frac{1}{\displaystyle\sum_{x=0}^{max} \lambda^{-x} l_x}. \qquad (4.6)$$

The proportion of year class 0 in the population is labeled c_0. When we multiply Eq. (4.4) by Eq. (4.6), a similar expression for all the other c_x terms pops out:

$$c_x = \frac{n_x(t)}{N(t)} = \frac{n_x(t)n_0(t)}{n_0(t)N(t)}$$

$$= \frac{\lambda^{-x}l_x}{\sum\limits_{x=0}^{max}\lambda^{-x}l_x}, \tag{4.7}$$

which is the relative proportion of each age group. Thus Eq. (4.7) represents the stable age distribution (also called the stable age *structure*).

You can get a more intuitive feeling for this formula by noting that the numerator is simply the number of females in a particular age class relative to the zero age class; this is the number born x time units earlier (λ^{-x}) times their survival rate to the present, l_x. The total population across all the ages is the sum of these terms in the denominator and converts this numerator into a fraction of the total.

If the population isn't growing (i.e., $\lambda = 1$), the stable age distribution reduces to

$$c_x = \text{proportion of females of age } x$$

$$= \frac{l_x}{\sum\limits_{x=1}^{max}l_x} \quad \text{for} \quad \lambda = 1. \tag{4.8}$$

In other words (assuming that the stable age structure has been reached), if $\lambda = 1$, you can essentially read the l_x curve to get the shape of the age structure pyramid. You might think at first glance that the numerator of Eq. (4.8) alone, l_x, would do the job. Why do we scale l_x by the sum of all the l_x in the denominator to determine the proportion of age x individuals in the population? Suppose that we have two stable populations and in both the survivorship to age 3 is 0.5. Imagine further that both populations have $\lambda = 1$, but in one population the expected total lifetime of a newborn female is 100 years and in the other it is only 5 years. Which population will have the highest proportion of age 3 individuals? Common sense tells us that the second population has many fewer old individuals. The relative proportion of three-year-olds is greater when subsequent ages are less common. Later in this chapter, we demonstrate that the denominator in Eq. (4.8) is roughly equal to the expected lifetime of a newborn. Thus (assuming that the population isn't growing) the relative abundance of each age is given by the survivorship to that age divided by the average lifetime from birth.

One interesting feature about the stable age distribution, Eq. (4.7), is that when a population is increasing at a constant rate (i.e., $\lambda > 1$), the relative abundance of any age x must be less than the relative abundance of the previous age $x - 1$. The ages diminish in relative abundance in a continuous (i.e., monotonic) way. This is because l_x cannot increase as age x increases. It's impossible to have a higher survivorship to age 4 than to age 3. Also, when $\lambda > 1$, λ^{-x}, which is $1/\lambda^x$, gets smaller and smaller as x gets larger and larger. Therefore $l_x\lambda^{-x}$ is the product of two pieces, each of which decreases with age x. Hence, when $\lambda > 1$, stable age pyramids with relatively less abundant successively older ages, like this, will result:

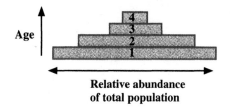

On the other hand, what if population size is *decreasing* at a constant rate $\lambda < 1$? Then l_x still must be a nonincreasing function of x. But now λ^{-x} will increase with x. (Suppose that $\lambda = 0.5$. Then $\lambda^2 = 0.25$, $\lambda^3 = 0.075$, and so on, and $1/\lambda^x$ gets larger as x gets larger.) Hence, since the two pieces of the age structure expression work in opposite directions as age increases, declining populations can have nonmonotonic shapes; that is, they can have intermediate bulges in their age pyramids, like this:

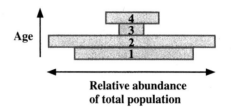

This kind of age structure would be impossible as a *stable* age structure for an increasing population. However, recall from Chapter 3, that for stage-structured population growth, not all life stages necessarily have the same temporal duration. A stage with a short half-life can be less numerous than later stages with longer half-lives, even at the stable stage structure of an increasing population. This was the case for the tree life history illustrated in Figure 3.17.

Problem: The number of females of each age in a hypothetical population at the beginning of 1996 are

$$n_1 = 1000,$$

$$n_2 = 500,$$

$$n_3 = 300,$$

and

$$n_x = 0 \quad \text{for } x > 3.$$

If this population has a stable age structure and is growing with $\lambda = 1.2$/year, what are the yearly survival rates s_1, s_2, and s_3?

Answer: Recall our basic iteration formula, Eq. (3.2), for changes in numbers (except for the first age group) from Chapter 3.

$$n_x(t) = s_{x-1}n_{x-1}(t-1) \quad \text{for all ages } x > 1. \tag{3.2}$$

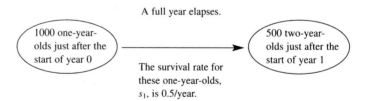

In our problem though we have a different situation. All the numbers for the age classes are given for the *same* year. To solve for the s_x terms, we desire an expression that relates the numbers in consecutive age groups within 1996 to their survival rates. We have also shown (see Figure 4.1) that, once the stable age structure is reached, each age group grows with rate λ; therefore age group $x - 1$ changes by a factor of λ/year. The problem statement supplies the value of λ/year, which is 1.2. Thus we can project the population for each age to 1997:

Age	1996	1997
1	1000	1200
2	500	600
3	300	360
4	0	0

The arrows now show how to calculate the survivorships by applying Eq. (3.2). The algebraic equivalent is

$$s_{x-1} = \frac{n_x(t)\lambda}{n_{x-1}(t)}.$$

Therefore

$$s_1 = \frac{n_2(1.2)}{n_1} = \frac{(500)(1.2)}{1000} = \frac{600}{1000} = 0.6;$$

$$s_2 = \frac{n_3(1.2)}{n_2} = \frac{(300)(1.2)}{500} = \frac{360}{500} = 0.72;$$

$$s_3 = \frac{n_4(1.2)}{n_3} = \frac{(0)(1.2)}{300} = \frac{0}{300} = 0.$$

Problem: What could you conclude from the data given in the preceding problem if the population was not in stable age structure and you were told only that in the previous year $\lambda_t = 1.2$ for the total population?

Problem: In Chapter 3, we worked primarily with Leslie matrices that ignored the newborns and began with age class 1. What is the formula for the stable age composition, using this convention?

Answer: To find the solution we pursue the same development as before but begin with Eq. (4.3), with age class 1 in the denominator, instead of (4.4), with age class 0 in the denominator. After we work through the same steps as before, we get the relative proportion of each age group

$$c_x = \frac{\lambda^{-x+1} l_x}{\sum_{x=1}^{max} \lambda^{-x+1} l_x},$$

which is the analog of Eq. (4.7), except that a factor of λ can now be factored out of both the numerator and denominator, giving

$$c_x = \frac{\lambda\lambda^{-x} l_x}{\sum_{x=1}^{max} \lambda\lambda^{-x} l_x},$$

which can then be canceled out to give an expression nearly identical to the one obtained before:

$$c_x = \frac{\lambda^{-x} l_x}{\sum_{x=1}^{max} \lambda^{-x} l_x},$$

the only difference being that the lower bound of the summation now is age class 1 instead of age class 0 in Eq. (4.7).

LOTKA–EULER EQUATION

Perhaps the most fundamental relationship in age-structured population growth is the **Lotka–Euler equation.** In Figure 4.1 we developed two different iteration expressions showing how the number of individuals in the xth age cohort changes through time. The first, **A,** is based on a constant rate of population growth, once the stable age structure is reached:

$$n_x(t) = \lambda n_x(t-1).$$

The second, **B,** is based on aging and survival and shows the advance of one age, $x - 1$, to the next age, x, through survival and the passage of one year of time:

$$n_x(t) = s_{x-1}n_{x-1}(t - 1).$$

These expressions hold for all but the very first age category, the births. What we ultimately want is something much stronger than these iteration equations; we desire an expression that allows us to solve for λ all at once, in terms of the vital rates, the s_x and b_x terms. In Chapter 3 we introduced such a method, using the Leslie matrix, but noted the numerical difficulty of solving for the eigenvalues of large matrices without a computer. We now develop an alternative technique.

By applying expression **B** we can express the number of four-year-olds, for example, at $t + 1$ in terms of the number of two-year-olds that existed *two* time steps earlier:

$$n_4(t) = s_3n_3(t - 1),$$

and since

$$n_3(t - 1) = s_2n_2(t - 2),$$

then

$$n_4(t) = s_3s_2n_2(t - 2). \tag{4.9}$$

Similarly, the number of two-year-olds can be expressed as the surviving one-year-olds, from an additional year back in time, or

$$n_2(t - 2) = s_1n_1(t - 3). \tag{4.10}$$

Substituting Eq. (4.10) into Eq. (4.9) gives

$$n_4(t) = s_3s_2s_1n_1(t - 3). \tag{4.11}$$

Iteration expression **A** provides a complementary approach based on population growth: for each year's passage, the population size of each age group increases by the same factor λ. So, projecting forward in time three time steps (from $t - 3$ to t), the population of n_4 (and in fact each age group) is λ^3 times bigger than it was at $t - 3$. In particular, the one-year-old's (i.e., the yearling's) population size has grown by λ^3, or

$$n_1(t) = \lambda^3 n_1(t - 3). \tag{4.12}$$

Projecting backward in time, we can rearrange Eq. (4.12) to solve for the number of yearlings that existed three years earlier, $n_1(t - 3)$, given the number yearlings this year, $n_1(t)$, or

$$n_1(t - 3) = n_1(t)\lambda^{-3}. \tag{4.13}$$

Substituting Eq. (4.13) for yearlings at $t - 4$ into Eq. (4.11) we get

$$n_4(t) = \underbrace{s_3s_2s_1}_{(a)}\underbrace{n_1(t)\lambda^{-3}}_{(b)}. \tag{4.14}$$

(b) This term represents the number of yearlings existing three years earlier than year t.

(a) This term is the survival of those yearlings from $t - 3$ to time t, at which time they become four-year-olds.

Recall the definition of survivorship when the year begins *after* the breeding pulse: $l_x = s_1s_2s_3 \cdots s_{x-1}$, and $l_1 = 1.0$. We use it to collapse the product of the yearly survival rates, s_x, in Eq. (4.14) into

$$n_4(t) = l_4n_1(t)\lambda^{-3}. \tag{4.15}$$

Now we write Eq. (4.15) more generally for any age x as

$$n_x(t) = l_x n_1(t) \lambda^{-x+1}. \tag{4.16}$$

The right-hand side of Eq. (4.16) still involves terms of $n_1(t)$, but we'd like to find an expression that gives λ as a function of the vital rates (not the variables).

We forge ahead by substituting into Eq. (4.16) an expression for the yearlings, $n_1(t)$. At this point there are two possible routes, and they account for some of the variation in the form of the Lotka–Euler equation between text books. One route follows the Leslie matrix iteration approach and uses the net fecundities, F_x, that relate $n_1(t + 1)$ to the number of all the mothers giving birth in the previous year, $n_x(t)$. With this approach, the growth equation for yearlings, borrowing from Eq. (3.3), is

$$n_1(t + 1) = s_0 n_1(t) b_1 + s_0 n_2(t) b_2 + s_0 n_3(t) b_3 + \cdots .$$

At stable age structure,

$$n_1(t + 1) = \lambda n_1(t) = F_1 n_1(t) + F_2 n_2(t) + \cdots + F_{max} n_{max}(t). \tag{4.17a}$$

Another route is to concentrate on the births that are produced *within* the breeding season of year t rather than transitions from t to $t + 1$, as in Eq. (3.3). This entails using the raw birth data from the life table. Then

$$n_0(t) = b_1 n_1(t) + b_2 n_2(t) + \cdots + b_{max} n_{max}(t). \tag{4.17b}$$

Note that both sides of Eq. (4.17b) are for the same year t, while in Eq. (4.17a), we went from year t on the right to year $t + 1$ on the left. Still a third way is to use the Fs but to adopt the alternative methodology for iterating Leslie matrices based on demarcating the start of each year immediately after the breeding season (instead of just before). In Chapter 3 we showed that with this technique the 0 age class is now an explicit part of the iteration scheme. The development based on this method is taken up in the next section.

We use Eq. (4.17a) to finish the derivation and then return to Eq. (4.17b) to develop alternative formulations. We substitute Eq. (4.16) into Eq. (4.17a) for each of the $n_x(t)$ terms:

$$n_1(t + 1) = F_1 l_1 n_1(t) \lambda^{-0} + F_2 l_2 n_1(t) \lambda^{-1} + F_3 l_3 n_1(t) \lambda^{-2} + \cdots . \tag{4.18}$$

Next, we divide each term of Eq. (4.18) by $n_1(t + 1) = \lambda n_1(t)$:

$$1 = \frac{F_1 l_1 n_1(t) \lambda^{-0}}{\lambda n_1(t)} + \frac{F_2 l_2 n_1(t) \lambda^{-1}}{\lambda n_1(t)} + \frac{F_3 l_3 n_1(t) \lambda^{-2}}{\lambda n_1(t)} + \cdots . \tag{4.19}$$

Simplifying Eq. (4.19), we reach the final result:

$$1 = F_1 l_1 \lambda^{-1} + F_2 l_2 \lambda^{-2} + F_3 l_3 \lambda^{-3} + \cdots ,$$

the Lotka–Euler equation 1 (denoted LE1), which can be more compactly expressed as

$$\boxed{1 = \sum_{x=1}^{max} F_x l_x \lambda^{-x}.} \tag{LE 1}$$

Equation (LE 1) is based on $F_x = s_0 b_x$ for ages 1 to max and survivorship of yearlings $l_1 = 1$. As you can see, this equation is a relationship between the parameters of the model and the constant 1.0. Unfortunately, we cannot usually rearrange Eq. (LE 1) to find a closed form solution for λ in terms of all the F_x and l_x data. Thus we are usually forced to find a solution based on trial-and-error solutions to Eq. (LE 1). As we showed in Chapter 3, we can also iterate the population growth over and over, or we can solve for the dominant eigenvalue of the Leslie matrix.

VARIATIONS ON THE LOTKA–EULER EQUATION (SUPPLEMENTAL)

When the breeding pulse takes place immediately before each year begins, the first age class tallied is the zero-year-olds, not the yearlings. We have to adjust that extra year in the terms of the Lotka–Euler equation, so it now looks like this:

$$1 = \sum_{x=0}^{max} F_x l_x \lambda^{-(x+1)}. \qquad \text{(LE 2)}$$

Equation (LE 2) is based on $F_x = s_x b_{x+1}$ for ages 0 to max with $l_0 = 1.0$ and $l_1 = s_0$. Also since n_0 is now the first term in the age vector, the total population size now includes newborns:

$$N(t) = n_0(t) + n_1(t) + n_2(t) + \cdots + n_{max}(t).$$

These alternative account-keeping systems are both valid but lead to slightly different versions of the Lotka–Euler equation. **If we are using the Lotka–Euler equation to solve for λ, it is important to know how the F_x and l_x terms were formed so that the correct equation can be applied.**

Finally, the Lotka–Euler equation may be cast in a slightly different and easier form, based on using Eq. (4.17b):

$$n_0(t) = b_1 n_1(t) + b_2 n_2(t) + \cdots + b_{max} n_{max}(t). \qquad \text{(4.17b)}$$

In this formulation, instead of using F_x in the Lotka–Euler equation, we use the raw fecundities, b_x. Next, we modify Eq. (4.16) so that the number of each age class at time t is expressed in terms of the number of newborns n_0 (rather than yearlings). Since the age interval from n_0 to n_x is one time step more than the interval from n_1 to n_x, we have to adjust the exponent of λ accordingly, or

$$n_x(t) = l_x n_0(t) \lambda^{-x}. \qquad \text{(4.16b)}$$

Substituting (4.17b) into Eq. (4.16b) for each of the $n_x(t)$ terms yields

$$n_0(t) = b_1 l_1 n_0(t) \lambda^{-1} + b_2 l_2 n_0(t) \lambda^{-2} + b_3 l_3 n_0(t) \lambda^{-3} + \cdots. \qquad \text{(4.18b)}$$

Finally, we divide both sides of Eq. (4.18b) by $n_0(t)$ to reach $1 = b_1 l_1 \lambda^{-1} + b_2 l_2 \lambda^{-2} + b_3 l_3 \lambda^{-3} + \cdots$. However, since b_0 is undefined in this discrete-time model, we will write this sum as beginning with age 0:

$$1 = \sum_{x=0}^{max} b_x l_x \lambda^{-x}. \qquad \text{(LE 3)}$$

This gives the same numerical result and helps remind us that the use of Eq. (LE 3) (like LE 2) requires that survivorship of newborns $l_0 = 1.0$ and $l_1 = s_0$. While there are three different ways to write the Lotka–Euler equation correctly, it's very easy to express it incorrectly. Even many textbooks contain mistakes. For example, note that the exponent in Eq. (LE 3) is $-x$, not $-(x + 1)$, as in Eq. (LE 1). Equation (LE 3) is the most straightforward equation to apply when you are using raw life table data, since it involves the b_x data, not the F_x data. On the other hand, the other two equations involving F_x lend themselves readily to matrix algebra and iteration and to data collected from natural populations where births might be difficult to observe. While all these variations on the same theme might give you the idea that this is all quite fluid, **remember that a Leslie Matrix cannot be constructed correctly using b_x data (rather than F_x data) along the top row of the matrix.** The following examples illustrate the use of the Lotka-Euler equation.

Let's use the life table data for a hypothetical population of deer from Chapter 3 (Table 3.1) to verify that both method 1 and method 2 for forming a Leslie matrix will yield a sum of 1 in the Lotka–Euler formula if the correct value of $\lambda = 1.0$ is inserted into that formula.

- **A breeding pulse and the year is marked to begin just before the breeding pulse:**

$$F_x = s_0 b_x$$

$$s_0 = 0.6$$

We use Eq. (LE 1).

Age x	s_x	l_x	b_x	$F_x = s_0 b_x$	$1.0^{-x} l_x F_x$
1	0.9	1	0	0	0
2	0.9	0.9	0.7	0.42	0.378 [= $(1/1^2)(0.9)(0.42)$]
3	0.5	0.81	0.78	0.468	0.3790
4	0	0.405	1.0	0.6	0.243
				Sum	1.0000

Note that the 0 age class is left out and that l_1 is normalized (divided by s_0) so that the surivorship of age 1 is $l_1 = 1$.

Note that in this method we use the F_x data, not the b_x data, and that the summing begins with age 1.

- **A breeding pulse and the year is defined to begin just after the breeding pulse:**

$$F_x = s_x b_{x+1}.$$

We use Eq. (LE 2).

Age x	s_x	l_x	b_x	$F_x = s_x b_{x+1}$	$1.0^{-(x+1)} l_x F_x$
0	0.6	1	0	0	0.000
1	0.9	0.6	0	0.63	0.378 [= $(1/1^2)(0.6)(0.63)$]
2	0.9	0.54	0.7	0.702	0.379
3	0.5	0.486	0.78	0.5	0.243
4	0	0.243	1.0	0.0	0.000
				Sum	1.000

Note that the 0 age class is included and assigned $l_0 = 1$.

Note that in this method we use the F_x data, not the b_x data, and that the summing begins with age 0.

- **The year begins at any time relative to the breeding season.** We use the b_x data and Eq. (LE 3).

Age x	s_x	l_x	b_x	$1.0^{-x} l_x b_x$
0	0.6	1	0	0.0
1	0.9	0.6	0	0.0 [= $(1/1^1)(0.6)(0)$]
2	0.9	0.54	0.7	0.378
3	0.5	0.486	0.78	0.379
4	0.0	0.243	1.0	0.243
			Sum	1.000

Note that the 0 age class is included and assigned $l_0 = 1$.

Note that in this method we use the b_x data and that the summing begins with age 0.

Exercise: Verify that the entries in the last column of each table in the examples are all correct.

Table 4.1   Trial-and-Error Solution for λ Using Eq. (LE 3)

	Vital Rates			$\lambda = 1.0$	$\lambda = 2.0$	$\lambda = 1.5$	$\lambda = 1.5296$
Age x	s_x	l_x	b_x	$1.0^{-x}l_xb_x$	$2.0^{-x}l_xb_x$	$1.5^{-x}l_xb_x$	$1.5296^{-x}l_xb_x$
0	0.5	1	0	0	0	0	0.0000
1	0.2	0.5	2	1	0.5	0.667	0.6538
2	0	0.1	8.1	0.81	0.25	0.444	0.3462
		Sum		1.81	0.75	1.111	1.0000
		Sum is		Too high	Too low	Too high	Exact
		λ is		Too low	Too high	Too low	Exact

For all three methods, λ must be solved by trial and error. However, with the first two methods, the vital rate data may be used to form the Leslie matrix as well. Then, instead of a trial-and-error solution for the Lotka–Euler equation, the dominant eigenvalue may be sought. Of course, for matrices bigger than 2×2, this usually requires some software.

The following life table contains the vital rates for a hypothetical pulse-breeding species:

Age x	s_x	l_x	b_x
0	0.5	1	0
1	0.2	0.5	2.0
2	0	0.1	8.1

In Table 4.1 we apply the b_x and l_x data to Eq. (LE 3) to show that $\lambda = 1.5296$. We first guess that λ is 1.0, then 2.0, then 1.5, and finally 1.5296.

Exercise: Only two of the following four Leslie-like matrix constructions of this life history will yield the correct projection of future population growth and a dominant eigenvalue of 1.5296. Which two are they?

$$\mathbf{a} = \begin{pmatrix} 0 & 2 & 8.1 \\ 0.5 & 0 & 0 \\ 0 & 0.2 & 0 \end{pmatrix}. \qquad \mathbf{b} = \begin{pmatrix} 1 & 1.62 & 0 \\ 0.5 & 0 & 0 \\ 0 & 0.2 & 0 \end{pmatrix}.$$

$$\mathbf{c} = \begin{pmatrix} 0 & 1.0 & 4.05 \\ 0.5 & 0 & 0 \\ 0 & 0.2 & 0 \end{pmatrix}. \qquad \mathbf{d} = \begin{pmatrix} 1 & 4.05 \\ 0.2 & 0 \end{pmatrix}.$$

Answer: Matrix **b** and **d** are the correct alternatives. Using the software Matlab©, we find that the dominant eigenvalue of both **b** and **d** is $\lambda = 1.5296$. Matrix **b** represents the application of method 2 to the vital rates, while matrix **d** results from the application of method 1. Matrix **a** is wrong because it has the raw bs across the top row, not Fs. Matrix **c** is based on inappropriate application of method 1's calculation of the F_x terms, which are then placed inappropriately in a matrix that explicitly tracks the newborns. The λ for matrix **a** is 1.2782, while for matrix **c**, $\lambda = 0.9601$; thus neither correctly predicts the true λ.

Exercise: Verify that Lotka–Euler's equation correctly applied to matrix **b** and **d** in Exercise 1 will also yield a sum of 1 using the F_x data.

Answer: First, matrix **b**:

Age x	s_x	l_x	b_x	F_x	$1.5296^{-(x+1)} l_xF_x$
0	0.5	1	0	1.0000	0.654
1	0.2	0.5	2	1.6200	0.346 [= (1/1.5296²)(0.5)(1.62)]
2	0	0.1	8.1	0.0000	0.000
				Sum	1.000

Now, matrix **d**:

Age x	s_x	l_x	b_x	F_x	$1.5296^{-x} l_x F_x$
1	0.2	1	2	1	0.654 [= $(1/1.5296^1)(1)(1)$]
2	0	0.2	8.1	4.05	0.346
				Sum	1.000

Note that the exponent of λ used in the Lotka–Euler equation for the two matrices is different ($x + 1$ versus x). Also, the two formulations begin with different ages: 0 for matrix **b** and 1 for matrix **d**.

Exercise: Given the following s_x and b_x data, use Eq. (LE 3) to compute λ by trial and error. As you proceed, plot the sum given by the right-hand side of Eq. (LE 3) versus your guesses for λ. Use this plot to guide your selection of choices for λ.

Age x	s_x	b_x
1	0.9	0
2	0.7	0
3	0.4	2
4	0.2	1
5	0	0

As we have shown, it can be difficult to calculate λ for age-structured populations. Before computers were as accessible as they are today, demographers searched for approximate formulas that allow less tedious calculations to predict trends in population growth. In the next section we show how this can be accomplished once a new term *generation time* is defined.

SOME MORE IDENTITIES AND CONNECTIONS FOR AGE-STRUCTURED POPULATIONS

In Chapter 1, without the complication of different vital rates for different-aged individuals, we had a simple relationship:

$$\lambda = 1 + R \quad \text{and} \quad e^r = \lambda \quad \text{or} \quad r = \ln \lambda.$$

With age structure, some connections between terms can also be made by defining a new term: the net reproductive rate, R_0, that bears a loose relationship to R:

$$R_0 = \sum_{x=0}^{max} l_x b_x. \tag{4.20}$$

One way to describe R_0 in words is: "the average number of female offspring produced per female over her lifetime." That is, if we follow a cohort of newborn females through their entire lives, the average number of (female) offspring each female produces is R_0. For example, an R_0 of 1.3 female offspring means that when a female is at the end of her lifetime and has finished reproducing, she will be replaced by 1.3 female offspring, on average.

Yet, there is another meaning that can be ascribed to R_0: "the per capita growth rate of the population per generation at stable age structure." This second definition implies that, barring any changes in age-specific birth and death rates and assuming a stable age distribution, an R_0 of 1.3 implies that the population will grow 30% per **generation.**

At first glance, these two definitions of R_0 do not seem to be equivalent. The first deals with a time unit of a lifetime and the second deals with generation times; the first with the number of offspring produced per female and the second with the growth rate of the population (which has units of per time period—e.g., per year). The trick is to define

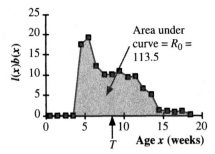

Mean age = 8.3 weeks = generation time, T.

Figure 4.2
A graphical illustration of generation time from data on rice weevils (Birch 1948). The calculation of T follows Eq. (4.22).

generation time in such a way that these two definitions are equivalent. Moreover, by defining generation time T more precisely, we can connect R_0 with λ as follows:

$$e^{rT} = \lambda^T \approx R_0,$$

$$\lambda \approx (R_0)^{1/T},$$

and

$$r \approx \ln(R_0)/T. \tag{4.21}$$

The approach we take is to imagine that all the R_0 babies born to these mothers over their lifetimes were all born to them at some age T instead. This gives us a correspondence with the case of discrete nonoverlapping generations. If the population grows by a factor of R_0 in approximately T years, its growth per year is $\lambda \approx (R_0)^{1/T}$. The trick is to now define generation time T so that this approximation is valid and the calculation of T is not too onerous.

There are several possible definitions for generation time. We desire one that is not sensitive to the rate that the population is growing (or shrinking). The most commonly used is to follow a cohort of females from their births, count up all the daughters that they produce during their lifetimes at each of their ages x, and average them to produce a mean age of reproduction:

$$T = \frac{\sum_{x=0}^{\max} l_x b_x x}{\sum_{x=0}^{\max} l_x b_x} = \frac{1}{R_0} \sum_{x=0}^{\max} l_x b_x x. \tag{4.22}$$

The numerator of T sums all the lengths of time x between the birth of each mother and the births of each of her own offspring. Mothers give birth only at age x if they survive to age x—hence the l_x terms. The denominator (R_0) divides this sum by the total number of births produced during the average mother's lifetime. The $l_x b_x$ curve in Figure 4.2 is based on data collected on rice weevils (*Calandra oryzae*) by Birch (1948). The generation time T is 8.3 weeks.

For these rice weevil data, the approximate formula for r can be compared to the exact solution. The approximate formula yields $r = \ln(113.5)/8.3 = 0.570$/week. The exact solution, using the Lotka–Euler equation, yields $r = 0.762$/week.

Zero Population Growth

For humans in the United States, the generation time is about 28 years, so if the population had an R_0 of 1.3, it would grow at about $r \approx \ln(R)/28 = \ln(1.3)/28 = 0.0094$ per year, or a little under 1% per year. Lately the R_0 of the U.S. population has been decreasing; in 1989 it was only 0.9069 (*Statistical Abstract of the United States* (1991)). This translates into a negative r of –0.0035. However, this does not imply that the U.S. population was shrinking, since this calculation leaves out a consideration of immigration and emigration. It also ignores the fact that, once the vital rates change, the new stable age structure will take some time to materialize. If $R_0 = 1$, then $\lambda = 1$ (or equivalently $r = 0$) **once the stable age structure is reached.** A population that is not growing is said to be at zero population growth (ZPG). For a growing population, like that of most developing countries, if R_0 suddenly reaches 1, perhaps through the implementation of birth control programs, it will take about one lifetime for the population to reach the stable age structure and thus ZPG where $\lambda = 1$. Some developing countries are growing at about 5% per year, or $\lambda = 1.05$/year. If birth control was put into place today to make R_0 sudently fall to 1, it would still take about 50–80 years before the population would stop growing. Rapidly growing populations build up much demographic momentum in their age structures, which are heavily weighted toward young individuals with much future reproductive potential. Keyfitz (1985) has calculated that populations in developing countries could increase by 60% during the lag period between $R_0 = 1$ and the gain of a stable age structure where $\lambda = 1$.

Problem: Show that a definition for R_0 equivalent to that of Eq. (4.20) is

$$R_0 = \sum_{x=1}^{max} l_x F_x ,$$

(4.23)

when l_1 is defined to be 1.0 and $F_x = s_0 b_x$.

The Rankings of R_0 and λ

In general, if $\lambda = 1$ *and* the population is at stable age structure, R_0 must also equal 1 from the Lotka–Euler equation:

$$1 = \sum_{x=1}^{max} F_x l_x \lambda^{-x}$$

Since the number 1 raised to any power still equals exactly 1, the λ^{-x} term vanishes from the Lotka–Euler equation, giving $1 = R_0$. Is the reverse also true? If $R_0 = 1$, must λ also equal 1 once the stable age structure is reached? You can see that this is also true by using the approximate formula for λ from the previous section: $\lambda^T = R_0$. Take the natural logarithm of both sides, which yields $T \ln \lambda = \ln R_0$. Now suppose that $R_0 = 1$: we have $T \ln(\lambda) = 0$. There are only two ways to make the last equality true—either $T = 0$ or $\ln \lambda = 0$. But if the generation time $T = 0$, the population doesn't even exist; consequently, the only sensible alternative is that $\ln(\lambda) = 0$, implying $\lambda = 1$. Thus we have this important result:

$\lambda = 1$ **if and only if** $R_0 = 1$ (**once the stable age structure is reached**).

Given this conclusion, we might naively guess that if a group of populations or species differed in their life histories (i.e., $l_x F_x$ schedules), and if we ranked those life histories according to the magnitude of their R_0's, then this ranking might be identical to the ranking that we would get for the same set of life histories ranked by their λ's.

But this is *not* true, even at stable age structure. It's quite possible that one life history has a lower R_0 than another but a higher λ. It's even possible for R_0 to be less than 1 but the λ of that same life history to be greater than 1 because its generation time is longer. Here R_0 is expressed per generation, but λ has units of per fixed time unit—for example, per year. Hence a generation represents multiple time units.

Exercise: Three survivorship curves are plotted in the figure shown. The l_x axis is plotted on a log scale. All three populations have the same maximum age of 10.

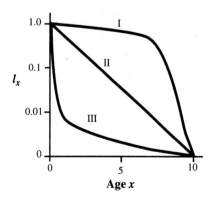

If the birth rates are equal for every age and across each population, which of the three survivorship curves yields the longest generation time?

Exercise: Here are two Leslie matrices with the same R_0 (the first age is 1 year old).

$$L_1 = \begin{bmatrix} 1 & 5 \\ 0.1 & 0 \end{bmatrix} \quad \text{and} \quad L_2 = \begin{bmatrix} 0 & 3 \\ 0.5 & 0 \end{bmatrix}.$$

What is R_0?
What is the λ for each?
What is the generation time for each?

Mean Lifetime of Individuals

How long can you expect to live? Another way of asking this question is: What is the mean age of individuals at death in a population? This question is of particular interest to insurance companies, since they hope to make a profit by basing the premium you pay for your policy on your expected lifetime. Imagine a cohort of individuals that has no mortality at all until age 10, and then everyone dies. The $l(x)$ curve looks like that in Figure 4.3.

The area under the $l(x)$ curve is 10, and all individuals have a lifetime of 10 years. Figure 4.4 shows another example, one where the $l(x)$ curve declines linearly with age. Now the average age of individuals in the population would intuitively seem to be one-half of 10, or 5, and in fact this is correct.

These examples suggest that the area under the $l(x)$ curve—that is, the integral of $l(x)$ from 0 to the maximum age—will give the average age of death (which equals the average lifetime or the *life expectancy* from birth). Box 4.1 gives a more formal proof of this relationship.

In Box 4.1 the assumption is that all the mortality takes place immediately at the beginning of the age interval (on the xth birthday). If deaths occur uniformly throughout the year, the life expectancy will be roughly a half year longer. Equation (c) can be adjusted for this difference as follows. In discrete time, the l_x schedule gives the proportion of individuals alive at the *beginning* of each time interval; l_{x+1} is the proportion alive at the *end* of the time interval. If mortality takes place evenly within the time interval, the average number alive from x to $x + 1$ is

$$L_x = \frac{l_x + l_{x+1}}{2}. \tag{4.24}$$

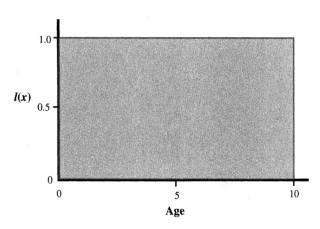

Figure 4.3
A hypothetical continuous survivorship curve in which no mortality occurs until age 10; then all individuals die.

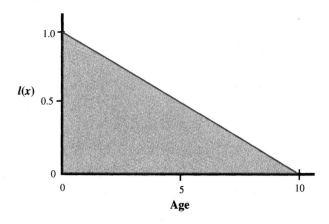

Figure 4.4
Another example of a possible continuous time survivorship curve with half the area under the $l(x)$ curve as that in Figure 4.3.

Box 4.1 *Life Expectancy Is the Area under the l_x Curve*

The life expectancy at birth can also be thought of as the mean age of death in the population. Using the definition of a mean age x, we get

$$\text{Life expectancy at birth} = \frac{\text{Mean age}}{\text{of death}} = \sum_{x=0}^{\max} x \; l_x \; d_x \qquad (a)$$

This is the proportion of individuals surviving to the beginning of age x.

This is the proportion of those that then die at age x.

The product of the two gives the probability of a newborn dying at exactly age x. Weighting this number by the age of death, x, and then summing over all ages gives the mean age of death (or the mean lifetime).

The trick now is to find a new expression for d_x in (a). The probability of death at age x is

$$d_x = 1 - s_x = 1 - \frac{l_{x+1}}{l_x} = \frac{l_x - l_{x+1}}{l_x}. \qquad (b)$$

Substituting Eq. (b) into Eq. (a) gives

Mean age of death

$$= \sum_0^{\max} x(l_x - l_{x+1}) = \sum_0^{\max} x l_x - \sum_0^{\max} x l_{x+1}$$

$$= 0 l_0 + 1 l_1 + 2 l_2 + 3 l_3 + \cdots + \max l_{\max}$$

$$\quad - 0 l_1 - 1 l_2 - 2 l_3 - 3 l_4 - \cdots - (\max - 1) l_{\max}$$

$$= l_1 + l_2 + l_3 + l_4 + \cdots.$$

$$= \sum_{x=1}^{\max} l_x. \qquad (c)$$

(Note that for discrete time the sum begins with age 1, since l_0 dropped out.)

The time lived since birth is the sum of the L_x values from birth to the last possible age class. Thus life expectancy at birth is

$$E_0 = \sum_{x=0}^{\max} L_x. \qquad (4.25)$$

Here is a simple example:

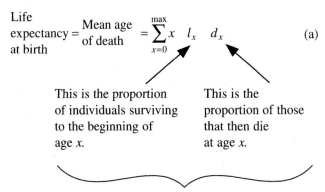

Age (years)	l_x	L_x
0	1.0	0.9
1	0.8	0.7
2	0.6	0.55
3	0.5	0.3
4	0.1	0.05
5	0	—
Σ	2.0	2.5

The mean lifetime or life expectancy from birth is the sum of the L_x column, or 2.5 years. If the l_x rather than the L_x are summed, we arrive at 2.0 years. Figure 4.5 illustrates the difference between summing the l_x and summing the L_x data.

In the continuous case, when time intervals are infinitesimally small, this discrepancy disappears and the analog of Eq. (4.25) is

$$E_0 = \int_0^\infty l(x)\,dx. \qquad (4.26)$$

A lifespan example:

$s_0 = 0.5$, $s_1 = 0.5$, and $s_2 = 0$. No one lives beyond age 2.

Here the shaded bars are the l_x data.

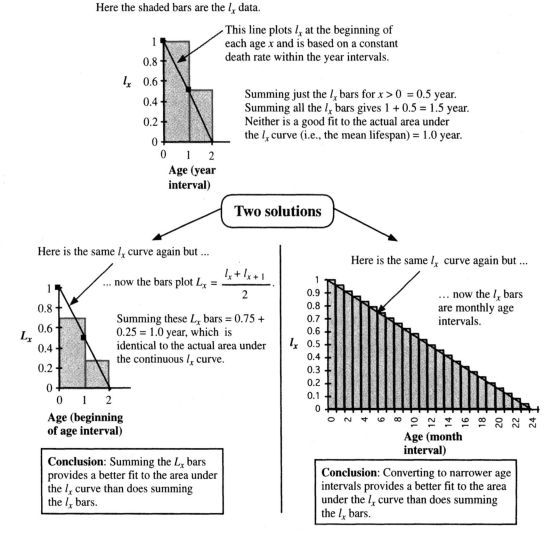

This line plots l_x at the beginning of each age x and is based on a constant death rate within the year intervals.

Summing just the l_x bars for $x > 0$ = 0.5 year.
Summing all the l_x bars gives $1 + 0.5 = 1.5$ year. Neither is a good fit to the actual area under the l_x curve (i.e., the mean lifespan) = 1.0 year.

Two solutions

Here is the same l_x curve again but ...

... now the bars plot $L_x = \dfrac{l_x + l_{x+1}}{2}$.

Summing these L_x bars = 0.75 + 0.25 = 1.0 year, which is identical to the actual area under the continuous l_x curve.

Age (beginning of age interval)

Conclusion: Summing the L_x bars provides a better fit to the area under the l_x curve than does summing the l_x bars.

Here is the same l_x curve again but ...

... now the l_x bars are monthly age intervals.

Age (month interval)

Conclusion: Converting to narrower age intervals provides a better fit to the area under the l_x curve than does summing the l_x bars.

Figure 4.5
The average lifetime of individuals is the area under a continuous survivorship curve l_x. In discrete time, the l_x data are binned into bars for each age. The area under the l_x curve can be approximated by summing L_x bars instead of the l_x bars or by binning the data into smaller age intervals.

If you are already at age y, what is your expected **future** lifetime? We sum the L_x from age y and beyond and then normalize that sum by the probability of surviving to time y. Thus the future life expectancy of an individual at the *beginning* of age class y is

$$E_y = \frac{\sum\limits_{x=y}^{max} L_x}{l_y}.$$

For this example, the future expected lifetime for someone who has just reached his or her third birthday is $0.35/0.5 = 0.7$ year, as calculated in Figure 4.6. Similarly, the future life expectancy of someone who is just about to have their fourth birthday is $0.05/0.1 = 0.5$ year.

For species with a continuous breeding season like that of humans, rather than pulse breeders, individuals that are "3 years old" may differ in age by as much as

Figure 4.6
An illustration of the calculation of future life expectancy, given the l_x information.

Age (years)	l_x	L_x
0	1.0	0.9
1	0.8	0.7
2	0.6	0.55
3	(0.5)	0.3
4	0.1	0.05
5	0	—
Σ		0.35

E_3 = Future life expectancy at third birthday

$$= \frac{0.35}{0.5} = 0.7 \text{ year.}$$

364 days. But if we assume that the death rate is constant within each year interval, the age of the average individual is halfway between the beginning of age y and the beginning of age $y + 1$ and we can calculate the

$$\text{Future life expectancy of the average individual, age } y = \frac{E_y + E_{y+1}}{2}.$$

Expected Lifetimes When Mortality Rates Are Constant across Age (ADVANCED)

Suppose that individuals in a population have a 0.2 probability of dying per year regardless of their age. What is the mean age of individuals at death? Intuitively, most people guess the correct answer: 1 divided by 0.2 = 5 years. However, it is hard for most people to justify their answer mathematically. Why does simply dividing the number 1 by the death rate yield the correct solution?

We provide two explanations: one based on calculus and the other based on discrete time probabilities. In the process of explaining the logic, we apply some new mathematical tools: the use of sums of **infinite series** and **imperfect integrals.**

Discrete Time

Let's follow a cohort of newborns over time and call the number of individuals reaching age x, N_x. At time $t = 0$, there are N_0 newborns. Without any loss of generality, and so that we can deal with proportions, we set $N_0 = 1$. Every year that passes, the surviving individuals get a year older, so there is a one-to-one correspondence between time and age for this cohort. If we call the constant discrete death rate per year D, the survival rate is $S = 1 - D$. Then

$$N_{x+1} = (1 - D)N_x = (1 - D) = S$$

and

$$N_{x+n} = S^n = l_x. \tag{4.27}$$

To die at age x, an individual must survive to age x and then die at age x. The number dying at age x is therefore

$$D_x = (1 - D)^{x-1}D. \tag{4.28}$$

The mean of a discrete function with possible outcomes x is

$$\text{Mean } x = \Sigma \, xp_x,$$

where the sum is evaluated across all possible outcomes x = min to max.

So to calculate the mean age of death we multiply the probability of an outcome of death at age x, p_x, by the value of x and then sum these products over all possible ages x. That is, the x values are the ages, and the p_x terms are the probabilities of individuals dying at a certain age x, which are the D_x terms given by Eq. (4.28). Substituting these probabilities into the definition of the mean gives

$$\text{Mean age of death} = \sum x(1 - D)^{x-1} D. \qquad (4.29)$$

This sum goes from age zero to age infinity. You may not readily know how to evaluate the sum of an infinite series of terms, so let's look at how to do it. For many infinite series, the sum simply grows to infinity. Here, however, the sum converges on a constant. Note that $1 - D$ is less than 1, so as individuals age and as x gets larger, each term $(1 - D)^{x-1}$ becomes progressively smaller and smaller, approaching zero. Here are three useful identities involving infinite summations to infinity when S is a constant <1:

$$\sum_{x=0}^{\infty} S^x = \frac{1}{1-S}, \qquad \text{(Infinite sum 1)}$$

$$\sum_{x=0}^{\infty} S^{(x-1)} x = \frac{1}{(1-S)^2}, \qquad \text{(Infinite sum 2)}$$

and

$$\sum_{x=0}^{\infty} S^x (x+1) = \frac{1}{(1-S)^2}. \qquad \text{(Infinite sum 3)}$$

We won't derive these expressions, but you can find them derived in many high school algebra textbooks.

Here's an example of Infinite sum 1: let $S = 0.4$. Then evaluating the right-hand side

$$\frac{1}{1 - 0.4} = \frac{1}{0.6} = 1.6666\ldots$$

The left-hand side of Infinite sum 1 says this is equal to

$$= 0.4^0 + 0.4^1 + 0.4^2 + 0.4^3 + \ldots$$

$$= 1 + 0.4 + 0.16 + 0.064 + \ldots$$

$$= 1.624 + \ldots.$$

So even by $x = 3$ we're getting close to 1.6666...; all the other terms, each growing progressively smaller, will ultimately sum to make this formula exact.

At the end of this chapter we use Infinite sum 1 in Problem 1. For our present problem, we substitute the right-hand side of Infinite sum 2 into Eq. (4.29):

$$\text{Mean age of death} = \sum x(1-D)^{x-1} D = D \sum x S^{x-1} = \frac{D}{(1-S)^2} = \frac{D}{D^2} = \frac{1}{D}. \qquad (4.30)$$

Thus the mean age of death is 1 divided by the discrete death rate D.

Calculus

There are a whole wealth of tricks in calculus for evaluating infinitesimal time series, one of which is integration. We start over with this problem but now use calculus to get an expression like Eq. (4.30) for the mean age of death based on continuous time. Our hope is that we will ultimately reach an easy integral to evaluate for the mean lifetime.

When we recast Eq. (4.27) into its continuous time form, the cohort numbers change over time t (and age x) due to one factor—death:

$$\frac{dN}{dt} = \delta N(t), \qquad (4.31)$$

where δ (pronounced "delta") is the *instantaneous* death rate.

Integrating Eq. (4.31) and noting the population size at time $t = 0$ as $N(0)$ gives

$$N(t) = N(0)e^{\delta t}.$$

Figure 4.7
The numbers surviving and dying after one year for a cohort with a constant instantaneous death rate δ.

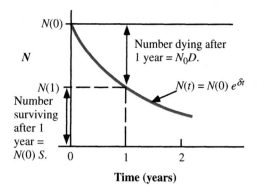

After 1 year, $N(1) = N(0)e^{\delta} = N_0(1 - D)$, where D is the discrete time death rate per year (Figure 4.7).

Since the discrete survival rate, $S = 1 - D$, we may also write for the numbers after 1 year:

$$N(0)e^{\delta} = N(0)S,$$

which implies that $S = e^{\delta}$, which in turn implies that

> **The instantaneous death rate δ equals the natural log of the finite survival rate S; or $\delta = \ln S$.**

Problem: A cage contains 400 fruit flies. If 50 of the original 400 flies die over the course of 1 day, what is the per capita instantaneous death rate δ?

Solution: $\delta = \ln((400 - 50)/400) = \ln(0.875) = -0.1335$/day (Note: not 50/400 = 0.125/day.)

Thus we may write Eq. (4.31) as

$$\frac{dN}{dt} = \ln(S)N(t). \tag{4.32}$$

Since S is less than or equal to 1, δ is less than or equal to 0. For example, if $D = 0.2$, then $S = 0.8$ and $\delta = \ln(0.8) = -0.2231$. The proportion of individuals surviving after 1 year is 80%. Note that instantaneous death rates are negative. They are also not bounded by 1. For example, if $S = 0.1$, then $\delta = \ln 0.1 = -2.302$. Instantaneous death rates may vary from 0 to $-\infty$.

With the passage of any unit of time, t, individuals also age an equivalent unit x, so Eq. (4.32) may be identically written in terms of age x:

$$\frac{dN(x)}{dx} = \delta N(x). \tag{4.33}$$

For continuous variables the definition of the mean is

$$\int_0^{\infty} xp(x)dx, \tag{4.34}$$

where $p(x)$ is the probability density of x. The $p(x)$ in our case is the instantaneous probability of dying at age x times $N(x)$, so

$$\text{Mean age of death} = \int_0^{\infty} x\delta N(x)dx. \tag{4.35}$$

Recalling the formula for continuous exponential growth, we can evaluate $N(x)$ as $N_0 e^{\delta t} = e^{\delta t}$, since we have defined N_0 to be 1. Substituting this into Eq. (4.35),

$$\text{Mean age of death} = \int_0^\infty x \delta e^{\delta x} \, dx. \tag{4.36}$$

Because the death rate, δ, is assumed to be a constant, we may take the first δ in Eq. (4.36) outside the integral:

$$\text{Mean age of death} = \delta \int_0^\infty x e^{\delta x} \, dx. \tag{4.37}$$

You can look up the evaluation of this integral in a table of integrals available in most mathematical handbooks. You will find that Eq. (4.37) evaluates as

$$\delta \left(\frac{e^{\delta x}}{\delta^2} [\delta x + 1] \right) \Bigg|_0^\infty$$

Evaluating integrals at infinity is referred to as taking "improper integrals" and yields finite answers only when the integral converges as x approaches infinity; that is, each term must get smaller and smaller as x gets larger. In this case, we're in luck: because δ is a negative number, as x goes to infinity, $e^{\delta x}$ approaches 0. The integral therefore converges to

$$\delta \left[0 - \left(\frac{1}{\delta^2} \right) \right] = \frac{-1}{\delta} = \text{Mean age of death}. \tag{4.38}$$

As an example, if $D = 0.2$, then $\delta = -0.2231$ and the average age at death (or the average lifespan from birth) in continuous time is 4.482 years. On the other hand, recalling our argument in the section on discrete time Eq. 4.30, if the instantaneous death rate were computed yearly instead of instantly, the rate of death per year would be D and the mean age of death in years would become $1/0.2 = 5$ years. A frequency distribution of the different lifetimes of individuals in the cohort is displayed in Figure 4.8.

The *modal* lifetime is 1 year; that is, more individuals die at age 1 then at any other age. Note that the modal lifetime is much shorter than the mean lifetime of 5 years. The mode and mean are usually different when the probability distribution is asymmetric, as in Figure 4.8.

A summary table of some age-structured demographic relationships in presented in Box 4.2.

Figure 4.8
The proportion of individuals who die at different ages when the yearly death rate is $D = 0.2$ for all ages.

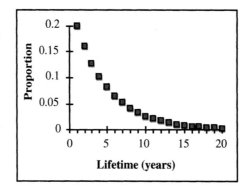

Box 4.2 *Summary of Demographic Relationships for Age-Structured Population Growth*

Discrete Time	**Continuous Time**

Properties of an individual or a cohort

$l_x = s_0 s_1 s_2 \cdots s_{x-1}$ (assumes that 0 age class is explicitly tracked and that $l_0 = 1.0$)

Discrete (finite) survival rate $= s_x = \dfrac{l_{x+1}}{l_x}$.

Discrete (finite) death rate $= d_x = 1 - s_x$

The average number of female offspring produced per female over her lifetime $= R_0 = \sum_{x=0}^{max} l_x b_x \quad (l_0 = 1.0)$.

Generation time $=$ mean age of reproduction $= T = \dfrac{1}{R_0} \sum_{x=0}^{max} l_x b_x x$.

Life expectancy at birth $= E_0 = \sum_{x=0}^{max} L_x$

where $L_x = \dfrac{l_x + l_{x+1}}{2} \quad (l_0 = 1.0)$.

Survivorship as a function of age $x = l(x)$; $l(0) = 1.0$

Instantaneous survival rate $= s(x)$.

Instantaneous death rate $=$

$$\delta(x) = \frac{\partial l(x)}{l(x) \partial x} = \frac{\partial \ln(l(x))}{\partial x} \quad \begin{array}{l} \delta(x) < 0 \\ \delta(x) = \ln(s(x)). \end{array}$$

$$R(0) = \int_0^\infty b(x) l(x)\, dx$$

$$T = \frac{1}{R_0} \int_0^\infty x b(x) l(x)\, dx$$

$$E(0) = \int_0^\infty l(x)\, dx$$

Properties of a population at its stable age structure

The proportion of each age group relative to first age class $= \dfrac{n_x}{n_0} = l_x \lambda^{-x}$.

The relative proportion of each age group to total N $\quad c_x = \dfrac{\lambda^{-x} l_x}{\displaystyle\sum_{x=0}^{max} \lambda^{-x} l_x}$.

Lotka–Euler equation to determine λ:

$$1 = \sum_{x=0}^{max} b_x l_x \lambda^{-x}, \ l_0 = 1.0 \quad \text{or} \quad 1 = \sum_{x=1}^{max} F_x l_x \lambda^{-x}, \ l_1 = 1.0.$$

Total population size at time T:

$$N_T = N_0 \lambda^T.$$

$$\frac{n(x)}{n(0)} = l(x) e^{-rx}$$

$$c(x) = \frac{l(x) e^{-rx}}{\displaystyle\int_0^\infty l(x) e^{-rx}\, dx}$$

$$1 = \int_0^\infty e^{-rx} b(x) l(x)\, dx, \quad l(0) = 1$$

$$N(T) = N(0) e^{rT}$$

PROBLEMS

1. Many vertebrate life histories can be approximated as follows. The animals reach reproductive maturity at age A (in years). Adults have constant yearly fecundity of b daughters and constant yearly mortality rate D for adults (ages A and greater; yearly adult survival rates are $D = 1 - D$). The death rate of juveniles is different from that of adults but is again constant across age; thus the proportion of offspring that survive to maturity is simply a function of the age at maturity, so finite juvenile survival rates may be written as $S_j(A)$. What is R_0 for this life history?

Answer:

$$R_0 = \sum_{x=0}^{\infty} l_x b_x$$

However, since none of the ages prior to A reproduce ($b_x = 0$), they can be ignored in the calculation. So

$$R_0 = \sum_{x=A}^{\infty} S_j(A) S^{x-A} b.$$

Bringing the constants outside the summation gives

$$R_0 = S_j(A) b \sum_{x=A}^{\infty} S^{x-A}.$$

If we define a new variable—the number of years beyond maturity, $y = x - A$—the last summation can be written as

$$\sum_{x=A}^{\infty} S^{x-A} = \sum_{y=x-A=0}^{\infty} S^y.$$

The advantage of this shift in variables is that it allows us to use infinite sum 1 to evaluate this summation, since now the summation starts at zero and goes to infinity; the summation evaluates as $1/(1 - S) = 1/D$, so

$$R_0 = \frac{S_j(A) b}{D}. \tag{4.39}$$

2. Find λ, R_0, and the generation time, T, for a life history where reproductive maturity occurs at 1 year, $S_0 = 0.6$/year, and each adult age has survivorship $S = 0.8$/year and fecundity = 2.

Answer:

$\lambda = S + S_0 b = 0.8 + (0.6)(2) = 2$/year = 200% per year.

$R_0 = (0.6)(2)/0.2 = 6$ offspring per generation.

$r \approx \ln(R_0)/\text{generation time}(T)$, so $T \approx \ln(R_0)/\ln(\lambda) = \ln(6)/\ln(2)$

= 2.58 years.

3. Given the yearly survival rates $s_0 = 0.6$, $s_1 = 0.7$, $s_2 = 0.8$, $s_3 = 0.3$, and $s_4 = 0$, what is the expected probability of living to age 4 when an individual is already at age 1?

Hint: First form the survivorships l_x, then convert these to L_x by applying Eq. (4.24) and finally apply the formula

$$E_y = \frac{\sum_{x=y}^{max} L_x}{l_y}$$

for age $y = 1$.

4. The numbers of females of each age in a hypothetical population at the beginning of 1996 are

$$n_1 = 1000,$$
$$n_2 = 1200,$$
$$n_3 = 900,$$

and

$$n_x = 0 \quad \text{for } x > 3.$$

If this population has a stable age structure and is growing with $\lambda = 0.8$/year, what are the yearly survival rates s_1, s_2, and s_3?

5. Figure 4.9 shows a survivorship curve, and Figure 4.10 shows these survivorship's plotted on a log scale.

Does survival rate s_x vary with age x?

What is s_x for each age?

Write an equation for $\ln(l_x)$ that is valid for all ages x given by these data.

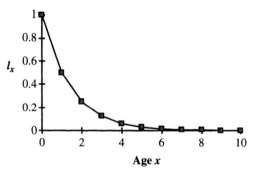

Figure 4.9

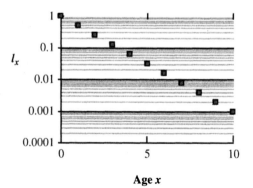

Figure 4.10

6. You wander through a cemetery looking at gravestones. Each gravestone gives the date of birth and the date of death of the person buried there. Based on this information, you determine the age distribution of the dead in the cemetery in 10-year intervals. Using these data you wish to construct a survivorship curve for the population. Assume that the gravestones are accurate, that the people buried in the cemetery are a nonbiased sample of the deaths, and that the population has been stationary ($\lambda = 1$). Complete the following table for l_x, the survivorship for the age class, and d_x, the death rate per 10-year intervals).

Age class	Number of dead	l_x	d_x
0 to 10	250	1.0	
10 to 20	150		
20 to 30	40		
30 to 40	20		
40 to 50	50		
50 to 60	90		
60 to 70	170		
70 to 80	150		
80 to 90	80		
>90	0		

Suppose that the population has been growing at a rate of 10% per year. Now, how would you calculate the death rates?

5 *Density Dependent Population Growth*

Using the analogy of Newton's first law of motion (a body in motion tends to stay in motion), we can ask: What forces might act like friction to prevent populations from growing ever larger or declining inevitably to extinction? What forces might allow populations to eventually come "to rest" at some intermediate level of abundance? As we have shown, temporal and spatial variation alone cannot explain when and where numbers stabilize. Several more deterministic biological factors, however, can act to produce a negative feedback between population size and population growth rate.

1. Available food may decrease as more individuals have to share the same food base.
2. The foraging success of the average individual may decrease because other foragers scare prey or interfere with each other's ability to find and successfully capture prey.
3. Available living space may become filled (this applies particularly to sessile species like plants, barnacles, or even mobile territorial animals if they have a fixed lower limit to how small their territories can become).
4. Aggression may increase, leading to detrimental effects on birth and survival rates.
5. Enemies of a species may numerically increase, leading to eventual declines in birth and survival rates.
6. As a prey species increases, its enemies will have more learning opportunities with the prey and thus perfect their hunting and handling abilities with the result that a greater proportion of prey are eaten or parasitized.
7. Emigration rates may increase with population density as individuals search for less occupied territories.
8. Fewer individuals may move into an overcrowded population (immigration rates decline).

The ecological literature is filled with examples of each of these mechanisms, all of which lead to qualitative decreases in population growth rate. The consequence is that the rate of population expansion will often slow down as population size increases. In the next section we examine this tendency in detail.

A GRAPHICAL MODEL FOR DENSITY DEPENDENCE

Density dependent growth is contrasted with exponential growth in Figures 5.1 and 5.2.

In Figure 5.2 the per capita growth rate declines with increasing population size, finally reaching zero at some relatively high population size where the population size becomes asymptotic. At this asymptotic population size, the population growth rate must equal zero since the population size is no longer changing. This in turn implies that for a closed population, birth rates must equal death rates at this asymptotic population size. Symbolically, then, the per capita birth rate and death rate are functions of N and at the equilibrium population size $b(N) - d(N) = 0$.

103

Figure 5.1
For exponential growth, population size, *N*, increases exponentially over time. The per capita rate of growth does not decline with population size.

Figure 5.2
With density dependent growth the population size reaches an equilibrium. The per capita rate of growth declines as population size increases.

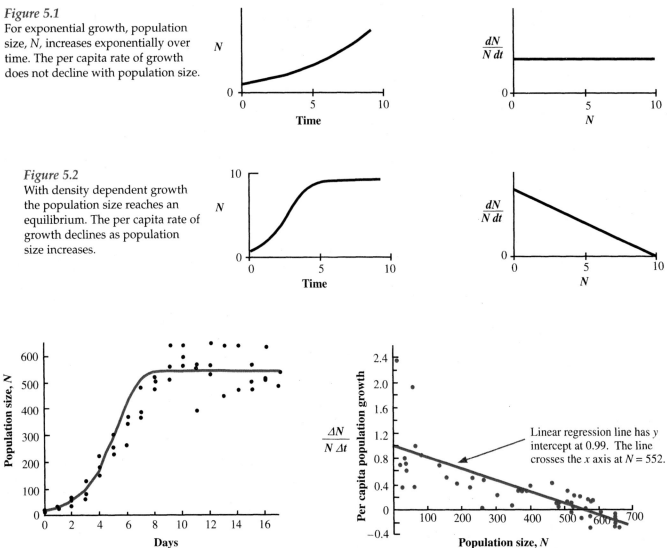

Figure 5.3
(a) The growth of *Paramecium aurelia* in test tubes containing Osterhaut culture medium with bacteria as food (after Gause 1934, as plotted by Roughgarden 1971). Population size is number per 0.5 ml. (b) The same data replotted to show the relationship between per capita growth rate and population size.

The population growth of the protozoan *Paramecium* in test tubes provides a match to this expectation as illustrated in Figure 5.3 (data from Gause 1934). Under the conditions of the experiment, the population stopped growing when there were about 552 individuals per 0.5 ml. The time points show some scatter, which is caused both by the difficulty in accurately measuring population size (only a subsample of the population is counted) and by environmental variation over time and between replicate test tubes.

Species colonizing new areas often exhibit a decrease in growth rate as their population size increases, as shown in Figure 5.4.

Of course, there is no strong reason why the decline in the per capita growth rate should necessarily be linear as population size increases. It could conceivably decline like any of the alternatives shown in Figure 5.5. All three populations have the same maximal per capita growth rate at very low *N*, and all three curves decline to zero per capita growth rate at about the same density but the shape of the decline differs. Above this density the per capita growth rate becomes negative as the average individual's likelihood of death at these high densities exceeds its birth rate.

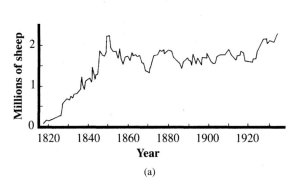

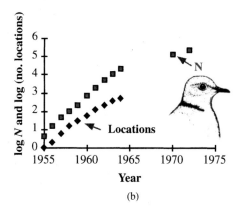

(a)

(b)

Figure 5.4
Two populations whose rate of growth slows as population size increases. (a) Sheep after being introduced into Tasmania (Davidson 1938). The variation in numbers around the asymptote is partly due to errors in estimating the yearly sheep population and partly due to variation in numbers due to climatic differences between years. (b) Collared doves (*Streptopelia decaocto*) after expanding their range into Great Britain (Hengeveld 1989). Total population size, *N*, and the number of locations with doves present increased over time but at a diminishing rate.

Figure 5.5
Some possible modes of decline in the per capita growth rate with population density.

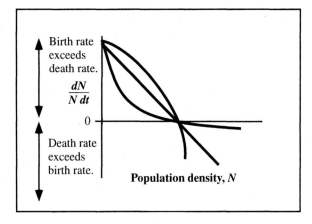

The reason that the per capita growth rate declines in a closed population could be because birth rates decrease with density or because death rates increase with density or some combination of both, as depicted in Figure 5.6.

The per capita birth rate equals the per capita death rate at the population density *N**. This state of affairs means that *N** is an **equilibrium** population size, since if births equal deaths, the population size will not change over time. Note too in Figure 5.6 that both the birth and death rate axes are scaled identically from 0 to 1. More typically the range for birth rates may be greater than that for death rates. Discrete death rates cannot exceed 1 (a single mother can die at most once), but a mother can give birth to several young per time period. If the axes were scaled differently, then the *N** at which births just equal deaths would, of course, not be the place where the two curves intersect, but it still would be the density at which birth rates equal death rates.

Note also that, when the population is below *N**—for example, at *N*low—the birth rate exceeds the death rate, and thus the population density will increase in the next generation as illustrated in Figure 5.7. On the other hand, when the density is greater than *N**—for example, at *N*high—the death rate exceeds the birth rate, and thus the population size will decrease in the next generation. In this way the population density will move toward *N** from sizes above or below it.

Consequently *N** is not only an equilibrium point but it is a stable equilibrium point. The concept of stability is different from the concept of equilibrium. Equilibrium

Figure 5.6
The per capita birth rate decreases with density while the per capita death rate increases in this example of density dependent growth. The density where these two rates are equal is an equilibrium, *N**.

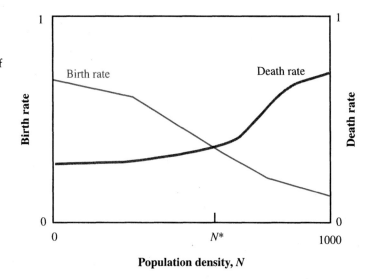

Figure 5.7
A hypothetical population experiencing density dependence in both per capita birth rates and death rates; *N** is the equilibrium population size. Perturbations of numbers above or below *N** lead to changes in the population growth rate that return the population to *N**.

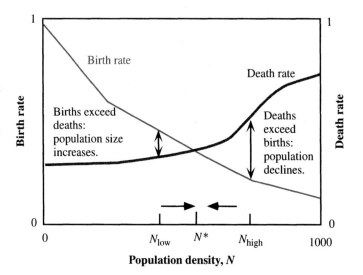

means an absence of change. **Stability** means that the system tends to return to its equilibrium state following perturbations away from that state. Consider the thought experiment suggested by Figure 5.7. Suppose that we have a population at size *N** and that we perturb the population size by the one-time addition of some additional individuals. Now we step aside and see what happens. At this new higher density (above *N**) the death rate exceeds the birth rate, leading to a population decline. In this way the population size or density tends to return to *N** following the perturbation. Similarly a perturbation in the opposite direction, by removing individuals from an equilibrium population, would result in a new lower density. At this lower density the birth rate exceeds the death rate and the population will climb back up to *N** over time. Thus *N** is a **stable equilibrium point.**

The diagrams in Figure 5.8 illustrate the concept of the stability of an equilibrium point, using an analogy with gravity. A ball placed on a landscape is subject to the force of gravity, which acts on the ball's mass. Of course, with population dynamics it is not the force of gravity operating on the system but the force of population growth, the inherent tendency of life to reproduce, and the influence of population density acting on birth and death rates. Consider a population at equilibrium at *N**. It is then perturbed by a one-time "nudge" (i.e., by the addition or subtraction of some additional individuals). Does population size tend to return to equilibrium following the perturbation (indicating a stable equilibrium), or does it grow or shrink even more (indicating an unstable

Figure 5.8
A gravity analogy to illustrate the concept of stability.

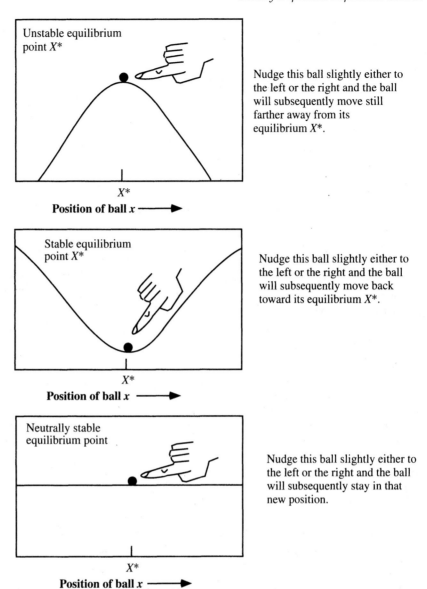

equilibrium)? The simple graphical model presented yields a stable equilibrium point by this criterion. At the end of this chapter we develop some mathematical tools for the evaluation of stability. For now, however, let's explore the surprising behavior of one general model with density-dependence, the logistic equation.

THE LOGISTIC EQUATION: A PARTICULAR MODEL OF DENSITY DEPENDENCE

Perhaps the simplest expression that produces a stable equilibrium population size is the logistic equation. It is presented in both continuous and discrete forms in Box 5.1. We explain the discrete logistic and Ricker diagrams a bit later; for now, let's begin with the continuous logistic equation,

$$\frac{dN}{dt} = rN\left(\frac{K-N}{K}\right).$$

Note that without the term in parentheses, the population growth rate is exponential with intrinsic rate r. Also, if $N = K$, the term in parentheses becomes zero so population growth becomes zero when the population size hits K, regardless of the initial

Box 5.1 Models of Density Dependent Growth: Logistic Equations

Difference Equations, Discrete Time

General form:

$$N_{t+1} = F(N_t).$$

The particular form for the discrete logistic equation:

$$N_{t+1} = N_t\left(1 + R\left(1 - \frac{N_t}{K}\right)\right). \tag{a}$$

To solve, the object is to try to find an expression for population size, N_{t+T}, after an arbitrary number T of time steps into the future. Unfortunately, unlike the case with exponential growth (Box 1.1), no neat closed form solution exists and the easiest method is to follow the dynamics over time by plotting **Ricker Diagrams:**

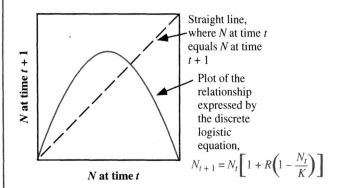

N at time *t* + 1

N at time *t*

Straight line, where *N* at time *t* equals *N* at time *t* + 1

Plot of the relationship expressed by the discrete logistic equation,

$$N_{t+1} = N_t\left[1 + R\left(1 - \frac{N_t}{K}\right)\right]$$

Differential Equations, Continuous Time

General form:

$$\frac{dN}{dt} = f(N).$$

The particular form for the continuous logistic equation:

$$\frac{dN}{dt} = rN\left(\frac{K - N}{K}\right). \tag{b}$$

To solve a differential equation we find an expression for $N(t)$ (not dN/dt) as a function of t. To find it we integrate the differential equation and plug in the initial conditions $N(0) = N_0$ and $t(0) = 0$. After several steps, the solution is

$$N(t) = \frac{K}{1 + \left(\dfrac{K - N_0}{N_0}\right)e^{-rt}}. \tag{c}$$

For exponential growth it was possible to make the substitution $\lambda = e^r$ so that the discrete and continuous models yielded identical results. **There is no such simple substitution for the two logistic equations (a) and (b). Although they look very similar they can produce qualitatively different population dynamics.**

population size. The population growth rate is also zero when $N = 0$. In words, the logistic equation says

$$\begin{bmatrix}\text{The rate of} \\ \text{increase of the} \\ \text{population}\end{bmatrix} = \begin{bmatrix}\text{the maximum rate of} \\ \text{population growth} \\ \text{per capita, } r\end{bmatrix}\begin{bmatrix}\text{the number of} \\ \text{individuals, } N\end{bmatrix}\begin{bmatrix}\text{the unutilized} \\ \text{opportunity for} \\ \text{population growth}\end{bmatrix}.$$

A typical time course of logistic growth is illustrated in Figure 5.9.

Figure 5.10 illustrates the approach to K from three different initial N_0.

Now let us explore the logistic equation algebraically. Multiplying through the parentheses of the logistic equation and simplifying gives

$$\frac{dN}{dt} = rN - \frac{rN^2}{K}. \tag{5.1}$$

Note that the expression on the right of the equals sign contains both a term in N and a term in N^2; thus a plot of dN/dt versus N must look like a parabola (see Appendix 1, Visualizing Equations). An example of population growth, dN/dt, plotted versus population size, N, for the case where the carrying capacity K is 200 is shown in Figure 5.11 for three different values of r.

Figure 5.9
Time course of a population growing
according to the continuous logistic
equation.

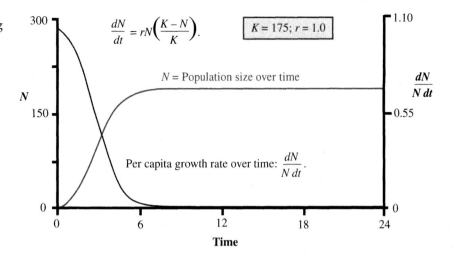

Figure 5.10
Logistic with three different initial
population sizes. The time course of
population size based on the logistic
equation, with initial density, N, of
150, 50, or 10.

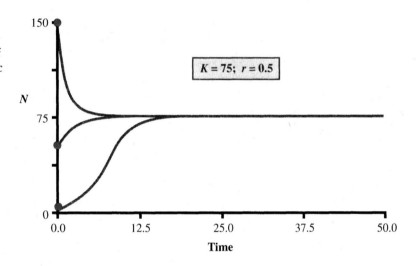

Figure 5.11
Logistic growth with different values
of r. The relationship between
population growth rate, dN/dt, and
population size, N, for three different
values of the intrinsic growth rate, r.
K = 200 for all three values.

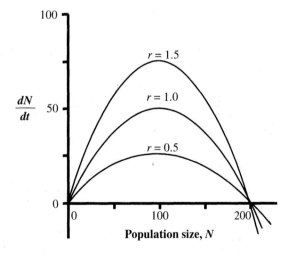

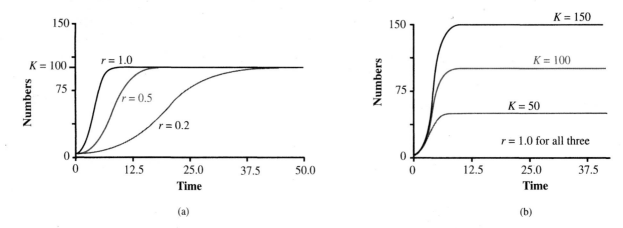

(a)

(b)

Figure 5.12
The effect of the parameters *r* and *K* on population growth in the continuous logistic equation. (a) Same *K*'s but different *r*'s. (b) Same *r*'s but different *K*'s.

The hump-shaped curve for *dN/dt* makes intuitive sense. To see this easily we rearrange the logistic equation:

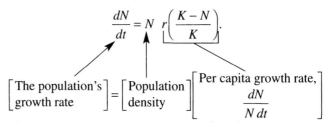

$$\frac{dN}{dt} = N \; r\left(\frac{K-N}{K}\right).$$

$$\begin{bmatrix} \text{The population's} \\ \text{growth rate} \end{bmatrix} = \begin{bmatrix} \text{Population} \\ \text{density} \end{bmatrix} \begin{bmatrix} \text{Per capita growth rate,} \\ \dfrac{dN}{N\,dt} \end{bmatrix}$$

Thus we have

$$\frac{dN}{dt} = N \frac{dN}{N\,dt}.$$

When the population size, *N*, is very low (near 0), the per capita growth rate, *dN/N dt* is high because there is little competition, but few individuals are present to reproduce so the growth rate of the total population is low (*dN/dt = N(dN/N dt)*). At the other extreme, when the population is very high (near *K*), there are many individuals but per capita growth rates are low or even negative, so again the population grows slowly. When population levels are intermediate, both per capita rates and population size are moderate, and their product is the greatest; thus the population grows faster.

In Figure 5.12, the total population size grows with an S-shaped curve, beginning from low N_0 and ultimately plateauing at *K*. The larger the intrinsic growth rate, *r*, the faster is the approach to this *K*.

We can also write the per capita growth rate as

$$\frac{dN}{N\,dt} = r - \frac{rN}{K}. \tag{5.2}$$

Per capita growth rate is perhaps the most intuitive measure we can think of, since people experience life as individuals not as populations. Per capita growth expresses an individual's quality of life—it is the difference between the average individual's birth and death probabilities. For the logistic equation, the birth and death rates themselves are not explicitly defined—only the difference between them is. This difference is *dN/N dt* and follows the equation of a straight line, Eq. (5.2), as a function of *N*. The growth rate declines linearly from a maximum of *r* (the *y* intercept) when the population size is zero to zero when the population size is at **carrying capacity, K.** Increases in population size *N* beyond *K* lead to negative per capita growth rates (the death rate exceeds the birth rate). The slope of the line is *–r/K*, as shown in Figure 5.13.

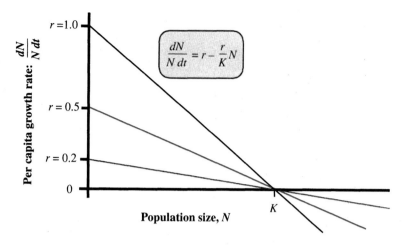

Figure 5.13
In the logistic equation, per capita population growth declines linearly with population size—illustrated here for three values of *r*.

Exercise: What are the values of *r* and *K* for the populations of *Paramecium aurelia* shown in Figure 5.3?

Behavior of the Discrete Logistic Equation

The discrete logistic equation gives N_{t+1} as a function of N_t, given the parameters R and K. We can derive the discrete logistic equation,

$$N_{t+1} = N_t\left(1 + R\left(1 - \frac{N_t}{K}\right)\right),$$

from the continuous logistic equation by substituting $\Delta N/\Delta t$ for dN/dt. When shifting to discrete time we should also change the symbol r to something else. Let's use R, as we did in Chapter 1, to remind us that we are no longer dealing with an instantaneous rate of population change but a discrete rate. Thus we write

$$\frac{\Delta N}{\Delta t} = RN_t \frac{(K - N_t)}{K}$$

$$\Delta N = N_{t+1} - N_t = RN_t \frac{(K - N_t)}{K} \Delta t,$$

and since $\Delta t = (t + 1) - t = 1$,

$$N_{t+1} = N_t + RN_t \frac{(K - N_t)}{K}.$$

After collecting terms we get

$$N_{t+1} = N_t\left(1 + R\left(1 - \frac{N_t}{K}\right)\right).$$

Note that the discrete logistic equation has a term in N_t and a term in N_t^2; thus this equation also describes a parabola (see Appendix 1, Visualizing Equations). When $N_t = 0$ and $N_t = K$, then $N_{t+1} = N_t$, which is the condition for an equilibrium point.

Exercise: Use the discrete logistic equation to determine the values of N_t, such that $N_{t+1} = 0$.

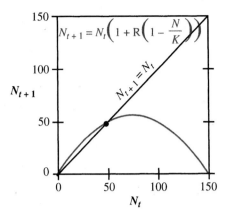

$$N_{t+1} = N_t\left(1 + R\left(1 - \frac{N}{K}\right)\right)$$

Figure 5.14
The discrete logistic equation is plotted in red, and the line $N_{t+1} = N_t$ is plotted in black. Where the two lines intersect, the population is at equilibrium since $N_{t+1} = N_t$.

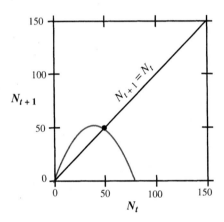

Figure 5.15
An example of a discrete logistic equation with a higher value of R but the same K as in Figure 5.14.

We take a graphical approach to understanding the dynamics of the discrete logistic equation in Figure 5.14. The parabola shown is simply the discrete logistic equation for $R = 0.5$ and $K = 50$. The line $N_{t+1} = N_t$ is simply a construction line—a line that helps to identify the condition of equality between the two axes. By definition, when $N_{t+1} = N_t$, the population is at equilibrium. There are two values of N that represent equilibria in Figure 5.14: $N = 0$ and $N = 50$. These two N's are easily seen as the only points where the construction line intersects the population growth parabola.

An example of a discrete logistic equation with a higher value of R but the same K ($R = 1.6$ and $K = 50$) is shown in Figure 5.15. The nonzero population equilibrium is still 50. The growth curve is shifted to the left compared to the one in Figure 5.14, and the parabola is shorter and somewhat steeper. This change in shape now causes the equilibrium point, K, to fall on the right-hand side of the peak of the parabola instead of the left, as it did in Figure 5.14. Now, let's trace the growth of a population following this equation. We arbitrarily begin with an initial density at time 0 of 20 individuals. We follow the approach taken in Chapter 1, Figure 1.7, except that now the population growth is logistic not geometric, as depicted in Figure 5.16.

First look at Figure 5.16(a). From the initial population size of $N = 20$, the population increases to about $N = 39$ at the next time step. By moving a line horizontally across to the construction line, we can begin the next iteration, which results in a population size of about 53, overshooting the carrying capacity of $N = 50$. The numbers along the vertical axis in Figure 5.16(b) indicate the various time steps (starting at 0); a running chart of population size over the first few time periods results. Note how the numbers line up in the two parts of the figure.

This diagram is known as a **Ricker diagram** (Ricker 1952); it is a convenient way to graphically follow the growth of a population in discrete time steps. The diagonal construction line (slope = 45°) allows us to simply convert the last N_{t+1} to the starting density, N_t, for the next iteration. In this example, the population initially overshot its carrying capacity at time step 2 and then it undershot the carrying capacity at time step 3. As time goes on these oscillations damp out and population size zeros in on its equilibrium of $N = 50$.

We increase R still more in Figure 5.17. The Ricker diagram and the corresponding time series plot are shown.

The dynamics in Figure 5.17 illustrate a stable two-point **limit cycle.** The cycles do not converge to the equilibrium point, K, which apparently is now an *unstable* equilibrium point. Instead the population oscillates between the same two levels of N—one above K, the other below.

In summary, as R increases, the discrete logistic equation parabola of N_{t+1} versus N_t becomes steeper. This in turn implies that the slope of the parabola is steeper at the equilibrium point, K. If steep enough, trajectories that begin near this equilibrium point wind outward instead of inward to the equilibrium point. In the next section, we provide a more formal description of this process so that we can precisely define "steep enough."

The two-point limit cycle obtained with $R = 2.1$ is a *stable cycle,* since alternative initial conditions converge on this same cycle, though not necessarily in phase, as the time series in Figure 5.18 shows.

For ectothermic animals, increasing ambient temperature increases their metabolic rate and growth rate, shortening generation times and thus increasing R. This can have the result in some laboratory populations of shifting population dynamics from a stable equilibrium point to apparently stable oscillations. Two examples are shown in Figure 5.19.

Increasing R still more in this model produces increasingly larger amplitudes for the two-point cycle, but something qualitatively different happens when R exceeds 2.449. An example of the time series for $R = 2.5$ ($K = 100$) is shown in Figure 5.20. The asymptotic behavior of this population is a four-point limit cycle. It takes a little while, but eventually the population oscillates continually over the same four values of N. We labeled these values points 1, 2, 3, and 4.

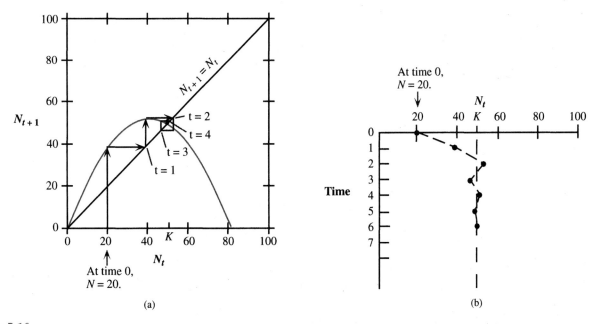

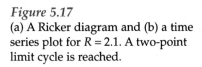

(a)   (b)

Figure 5.16
(a) The curve for the discrete logistic equation and an initial population size of 20. The black lines and arrows depict how the population grows over successive time steps. (b) Transcription of these population sizes to a time-step plot.

Figure 5.17
(a) A Ricker diagram and (b) a time series plot for $R = 2.1$. A two-point limit cycle is reached.

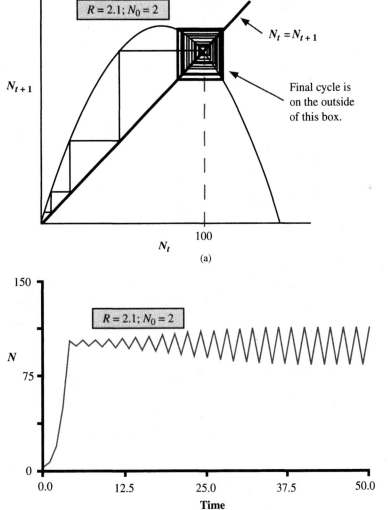

Figure 5.18
Three different initial conditions
($N_0 = 10, 20$, and 30). All three
populations converge on the same
two-point limit cycle but not
necessarily in phase with one another.
The dashed population initiated at
$N_0 = 30$ converges to a cycle that is out
of phase with the other two
populations.

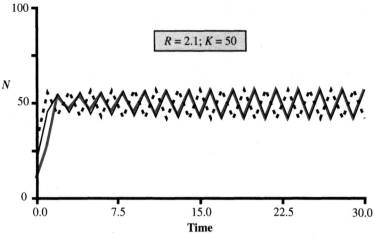

Figure 5.19
(a) Laboratory populations of *Daphnia
magna* growing at two different
temperatures (from Pratt 1943). The
population at 25° C has a higher
growth rate and oscillates. (b)
Laboratory populations of rotifers
growing at three different
temperatures. As the temperature
increases, so do the oscillations (from
Halbach 1979).

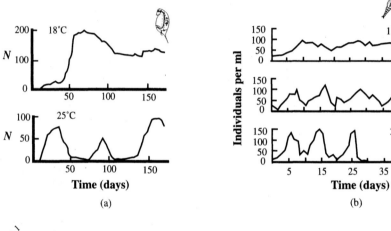

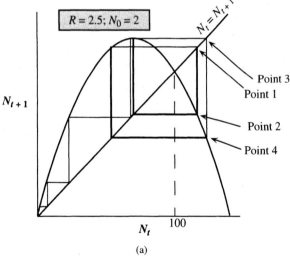

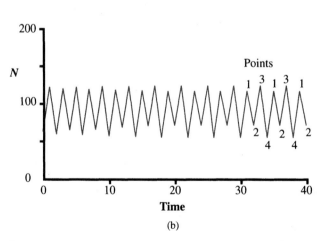

Figure 5.20
(a) A Ricker diagram and (b) time course plot illustrating a four-point limit cycle for $R = 2.5$.

Again, as R increases still more, the amplitude of these cycles increases until a
qualitatively different pattern emerges when $R > 2.544$, as illustrated in Figure 5.21.
This population converges on an eight-point limit cycle. At first glance, because the values of points 1 and 5 look similar, as do the values of points 2 and 6 and points 4 and
8, the cycle almost appears to be a four-point cycle. The range of R values that produces

Figure 5.21
A time course plot for a discrete logistic equation, illustrating an eight-point limit cycle for $R = 2.55$ and $K = 50$.

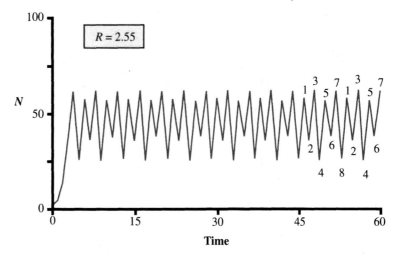

Figure 5.22
A time course plot of the discrete logistic equation for $R = 2.85$.

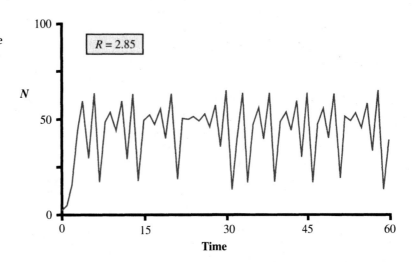

an eight-point cycle is even smaller than the range for a four-point cycle and, as before, is independent of K.

In summary, as R increases from 0 upward, the population behaves as a period-doubling scheme (May 1976):

$R < 2$	Equilibrium point is locally stable.
$R = 2.0$	Equilibrium point becomes unstable; a cycle of period 2 is born.
$R = 2.449...$	A cycle of period 4 is born.
$R = 2.544...$	A cycle of period 8 is born.
$R = 2.564...$	A cycle of period 16 is born.
$R = 2.5687...$	A cycle of period 32 is born.

Note that successive period doublings come faster and faster. Actually they are following a geometric series and converging on $R = 2.57$, in the limit of an infinite period. But what happens when R is larger than 2.57? Then the population growth becomes **chaotic** (May 1976). An example of chaotic dynamics under the discrete logistic equation is shown in Figure 5.22.

A Ricker diagram for another population in the chaotic region is shown in Figure 5.23. This population's behavior is highly irregular; it never seems to settle into exactly the same temporal pattern, and its behavior appears to be erratic—almost random. But remember that the population is following an exactly deterministic behavior specified precisely from one time step to the next by the discrete logistic equation. In the case of Figure 5.22, this equation has $R = 2.85$ and $K = 50$. While chaotic, the behavior is not random since the population is strongly density dependent; when the population's size

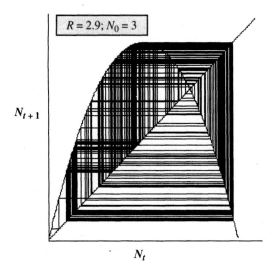

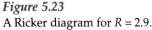

Figure 5.23
A Ricker diagram for $R = 2.9$.

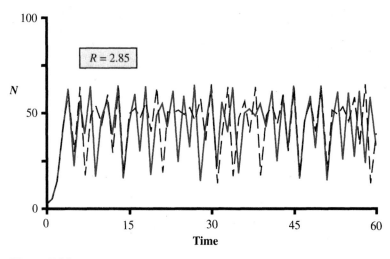

Figure 5.24
Time series for two populations that start at nearly identical points but quickly diverge.

is high, it tends to decrease at the next time step, and, when it is low, it tends to increase. One hallmark of **deterministic chaos** is that the dynamics show **sensitive dependence to initial conditions.** Think of throwing a die—we consider the face that comes up a random event because we cannot specify exactly all the variables that influence its roll and initial trajectory. All these difficult to specify initial conditions interact with the sharp 90° angles on the die to make the roll unpredictable. The slightest difference in the toss can produce an entirely different face up. Similarly, for chaotic population behavior, sensitivity to initial conditions means that the time course of the population appears unpredictable and that slightly different initial conditions produce increasingly divergent time series. The time series for two populations initiated at $N_0 = 1.0$ and 1.1 are shown in Figure 5.24. While the trajectories for the two populations are nearly identical over about the first five time steps, they rapidly diverge after that. In fact, as time goes on the probability of the two populations sharing the same size at the same time becomes vanishingly small.

It is this sensitivity to initial conditions that is the signature of deterministic chaos. Much current research is devoted to analyzing time series to assess the degree of density dependence and sensitivity to initial conditions. To quantify this degree of sensitivity, theoreticians have introduced a measure called the **Liapunov exponent.** Here's the intuition behind its definition. Imagine an initial population at size N_0 and consider a nearby point at $N_0 + \Delta_0$, where Δ_0 is very small. After n time steps we measure the sizes of the two populations, one starting at N_0 and the other at $N_0 + \Delta_0$. Let $|\Delta_n|$ be the absolute value of their difference. We do this again and again for different time periods, n, and then fit this data to an exponential model with elapsed time n as the independent variable:

$$|\Delta_n| = |\Delta_0| e^{n\lambda}. \tag{5.3}$$

This model has one parameter, the **Liapunov exponent,** λ. Note use of the same symbol that was used earlier for both the discrete growth rate and eigenvalues (Chapter 3). This is because there is a functional relationship between these concepts. If λ is negative, then populations are converging on the same temporal pattern as time goes by (either a stable equilibrium point or a limit cycle), while if λ is positive the deviation between populations initialized at slightly different numbers is growing over time: this is the signature of deterministic chaos. There are more computationally useful formulas for calculating Liapunov exponents directly from time series data, but they are less intuitive than Eq. (5.3); in general they require a computer. Figure 5.25 shows the relationship between the Liapunov exponent and the magnitude of R in the discrete logistic equation (Olsen and Degn 1985).

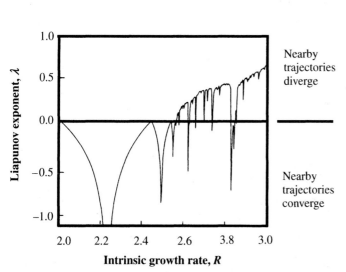

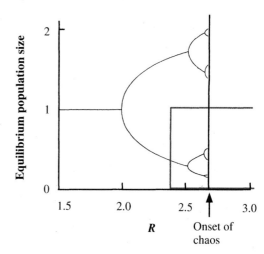

Figure 5.25
The Liapunov exponent, λ, as a function of R for the discrete logistic equation. When the Liapunov exponent is positive, then nearby trajectories diverge over time—the signature of deterministic chaos.

Figure 5.26
Equilibrium population size for the discrete logistic equation as a function of R; K is set to 1. The first bifurcation (to a two-point limit cycle) occurs at $R = 2.0$. The curve traces the magnitude of the two-cycle points as R increases. Each of these points, in turn, bifurcates to produce a four-point cycle at $R = 2.449$. A blowup of the red box is shown in Figure 5.27(a).

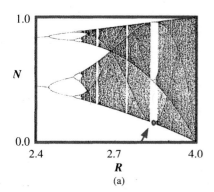

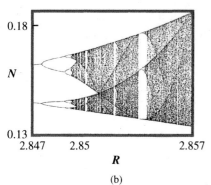

Figure 5.27
Blowups of equilibrium population size as a function of R. (a) The red box region of Figure 5.26. The very small region circled in red in (a).

Note that the first time λ becomes positive is at 2.57. For R greater than 2.57, the Liapunov exponent increases with sudden dips and spurts below 0. These dips are associated with regions of periodic cycles. The large dip at $R = 2.83$ is a window of R where cycles of period 3 emerge and later disappear.

Figures 5.26 and 5.27 show the relationship between the ultimate population size and the magnitude of the growth parameter R.

In Figure 5.27, even for R's beyond the onset of chaos, we see an unexpected mixture of intervals where chaos is interrupted with orderly cycles, mostly with odd numbers (e.g., cycles of period 3 at $R = 2.83$). When we blow up a portion of Figure 5.27(a), in the bottom part of the period 3 cycle region in the vicinity of $R = 2.83$, Figure 5.27(b), we see that the bifurcation diagram appears to reappear in miniature. The pattern of behavior is **fractal,** in the sense that the same geometric pattern is reproduced at successively larger or smaller scales.

The important lesson from this is that the discrete logistic equation, simple as it is, has a rich and surprising variety of behaviors compared to the continuous logistic equation, which was quite boring in comparison. **In discrete time a population has an implicit lag time of one time unit in its response to N. No such lag time exists with differential equations.**

Which model is the most realistic? The answer depends on the population being modeled. The fundamental difference between the two models is the inherent time lag in the discrete logistic equation caused by the form of the difference equations. The population responds in jerks. Imagine if your stereo had a short in the electronics such that there was a time lag between the time you adjusted the volume knob and the time that the volume in the speakers actually changed. If the sound was too low to hear, you would turn the volume up—but nothing would happen immediately, so you might keep turning the volume up more. After some time, the adjusted volume would finally click in, but now the volume would be higher than you had intended, so you would turn the volume down—again the time lag would prevent an immediate response between your adjustment of the knob and the volume coming out of the speaker, so you might overadjust before the new volume emerged from the speaker: this time too low. In this way, the volume would oscillate as you overadjust, then underadjust. The faster you turn the volume knob, the greater will be the amplitude of the volume cycle (higher highs and lower lows) emerging from the speaker—just as the greater the *R*, the larger are the population cycles. Ultimately, since humans are intelligent, you would catch on to the problem and adjust the knob slowly. But animal populations are sometimes kept from compensating for time lags by enforced features of their environment (e.g., strict seasonality) and the constraints set by the lengths of their generation times.

Consider the case of lemmings living in the tundra zone of the far north. Populations of these rodents cycle with an approximate 3-year period. What is the actual mechanism behind the density dependence and the creation of a time lag? Experiments have shown that the cycles are not apparently caused by delayed responses to food exploitation nor by a delayed response by predators to these lemming numbers—two obvious choices. Instead, current views suggest the following scenario.

The winter is actually a pretty favorable time for lemmings. They make tunnels underneath the snow and feed on frozen shoots and lichens. The snow insulates them from the cold and helps protect them from predators, so females continue breeding. When the snow melts in the summer, lemmings move on the surface and their numbers during an increase year will be large—but aggression now becomes important since individuals encounter each other more frequently. This aggression leads to much scarring, young don't mature quickly, and females eventually stop breeding. Next year when the snow melts—strangely, even though lemming numbers are now lower— lemmings still "act" like they are crowded. It's not known if this response is due to some long-term hormonal change in the surviving individuals or due to short-term genetic changes in the composition of the population. In any event, the cycle starts over again in the third year (Krebs 1988).

The Continuous Time-Lagged Logistic Equation

In later chapters on predation, we show that population cycles can also be generated by the interaction of predators and their prey. The numerical increase in predator numbers as prey increase leads to increased predator birth, but this takes some finite time to effect. Consequently oscillations frequently emerge in both theoretical and empirical predator–prey interactions. The important role of time lags in causing population cycles can be shown in another way. We can start with the continuous (differential equation) density dependent model and insert a time lag *T*. There are several ways to do this, however. Hutchinson (1948) introduced a model where the time lag occurred only in the density dependent term:

$$\frac{dN}{dt} = rN\left(\frac{K - N(t - T)}{K} \right). \tag{5.4}$$

How might this time lag arise? Let's think of this species as a herbivore. The vegetation may take τ time units to regrow to full size after being consumed. Even if the limitation to herbivore population growth is proportional to present food levels, these plant levels are related to not just the present number of grazers but also to those that lived (and consumed food) between now and τ time steps earlier. It is often then possible to approximate this integral of consumption by the numbers $N(t - T)$ at some specific earlier time *T* (where $0 < T < \tau$) (May 1973).

This model behaves like the continuous logistic equation when *r* (or *T*) is low, but it is more like the discrete logistic equation when *r* (or *T*) is high. However, instead of giving the sawtoothed behavior, it yields smooth population cycles. Figure 5.28 shows the behavior of this model for three different values of the time lag *T* and the same *r*. When the lag time is 2.8 (or nearly three times *r*), the population overshoots *K* so much that it comes crashing down to extinction.

This behavior is not unlike the population dynamics of reindeer introduced to some of the Alaskan Pribiloff Islands as illustrated in Figure 5.29. The reindeer lack predators on these islands, so their population growth responds primarily to food availability. Their food is lichens, short willows, and other plants in the tundra. The climate is generally cold with a short growing season. Thus these plants grow very slowly. The lag

Figure 5.28
An illustration of the time-lagged continuous logistic equation for three different values of the time lag, *T.* For the purposes of this simulation, this population becomes extinct if its numbers fall below a single individual, as is the case for a time lag of 2.8. Here *r* = 1.0 and *K* = 50.

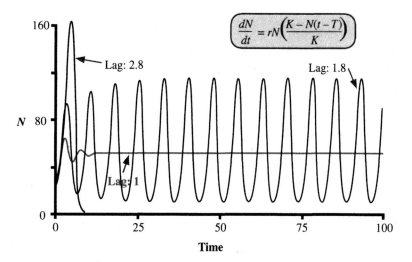

$$\frac{dN}{dt} = rN\left(\frac{K - N(t - T)}{K}\right)$$

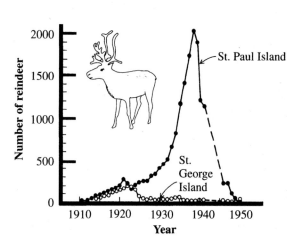

Figure 5.29
Introduced reindeer populations on two small islands in the Alaskan Pribiloff Islands. After Scheffer (1951).

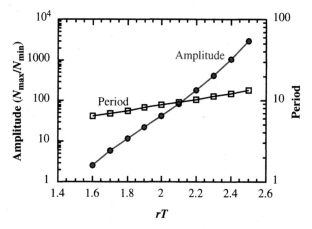

Figure 5.30
The amplitude and period of cycles for the time delayed continuous logistic as a function of the product *rT,* where *r* is the intrinsic growth rate and *T* is the time lag. Note the log scale on both vertical axes.

time between plant consumption and subsequent plant recruitment is large, consistent with a large lag time. Why these two reindeer populations behaved so differently, however, is not known. The model would predict that either the reindeer on St. George Island has a lower *R* for some reason or that the plants there had a faster recruitment rate, reducing the lag time.

The same progression of dynamic behaviors seen in Figure 5.28 can be reproduced by keeping the lag time, *T,* constant and instead increasing the intrinsic growth rate, *r.* This is because the behavior of this model is controlled only by the product *rT.* When $rT < \pi/2$, the carrying capacity is stable, but when $rT > \pi/2$, it is unstable and a stable limit cycle emerges (Wangersky and Cunningham 1957). As *rT* increases, the amplitude of the population cycle grows, as shown in Figure 5.30. The period also grows but much more slowly. If N_{max}/N_{min} exceeds *K,* then the population will crash to extinction before completion of one full cycle.

Lemming population cycles in the arctic north are nicely described by a time delay logistic equation, as shown in Figure 5.31.

Voles are small rodents in the same family as lemmings, and they again display roughly 3-year cycles at northern latitudes (Figure 5.32). Hornfeldt (1994) found a strong correlation between their growth rate and their numbers delayed about 9 months.

Figure 5.31
The black curve shows the density of lemmings (*Dicrostonyx groenlandicus*) in the Churchill area of northern Manitoba, Canada (number of individuals per hectare). The red dashed curve is the time delay logistic equation with $r = 3.333$/year and $T = 0.72$ year. After May (1981) and based on data of Shelford (1943).

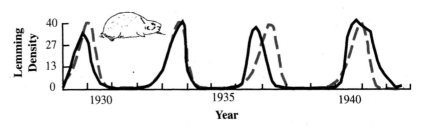

Figure 5.32
Approximately 3-year cycles in three species of voles in Sweden. Relative vole abundance is determined by trapping success (the number of voles trapped per 100 trap nights). The black bars on the *x* axis show the winter months. Trapping occurred immediately before and after winter. After Hornfeldt (1994).

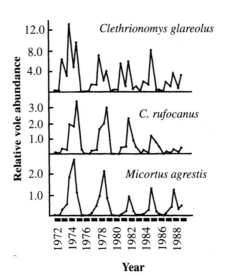

Figure 5.33
Blowfly numbers are shown as the black line. These are fit with the time delay continuous logistic equation, with $rT = 2.1$, shown as the red line. After Nicholson (1958) and May (1975).

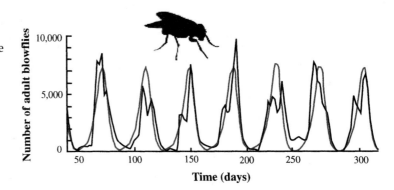

Another classic example of delayed density dependence comes from the experiments of the Australian population ecologist Nicholson, using laboratory cages of sheep blowflies, which can be a serious pest of sheep in Australia. Nicholson (1958) arranged a two-tiered cage to rear the blowfly adults separate from the eggs and larvae. In one experiment, adults were given an unlimited supply of liver, but larvae were provided with only 50 g of liver per day. In this experiment, the adult blowfly population exhibits regular fluctuations, as illustrated in Figure 5.33. Presumably the density dependence felt by the larvae—with a limited food supply—leads to a reduced survival to adulthood, but since each adult has unlimited food and can lay hundreds of eggs, the population tends to overshoot and oscillate. These dynamics can be fit quite well by the time delay logistic equation.

A clever test of the time delay hypothesis is to eliminate experimentally the lag time in the density dependent response by making the deleterious effects of competition at high fly densities felt more immediately by the adults. Nicholson (1958) was able to do this by restricting food to adults (only 1 g of liver per day) as well as to larvae. Figure 5.34 shows the results.

Figure 5.34
Fluctuations in laboratory populations of blowflies when the amount of food available for adults is limited to 1 g per day at the point of the dashed line (from Nicholson 1958).

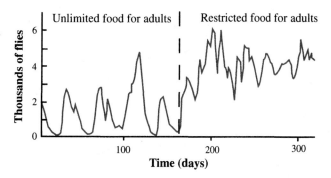

Figure 5.35
Roughly periodic oscillations in the density of *Paramecium aurelia* (per 15 cc) and the yeast *Sacchoaromyces exiguus* (per 0.001 cc). Lag time is longer with *Paramecium* feeding on yeast instead of bacteria. The *r* of *Paramecium* is ≈ 1/day; the *r* of yeast is ≈ 4/day; the r of bacteria is ≈ 20/day. Compare these dynamics to those in Figure 5.3.

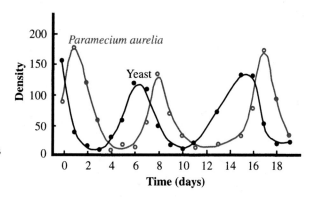

Regular cycles were eliminated with the restriction of food for adults, and the age structure became more even. Note also the surprising increase in the average number of adults over time. For unknown reasons, less food for adults led to higher average adult numbers. This last component is unexplained by our model since the equilibrium point, *K*, which is in the center of the cycles does not vary with changes in *r* and *T*. Clearly, there is more going on here than the time delay logistic can describe.

Still another example is provided by Gause's work with *Paramecium aurelia*, shown previously in Figure 5.3. This logisticlike curve was produced by *Paramecium* feeding on bacteria in laboratory test tubes. The *r* for these *Paramecium* is about 1/day compared to a much higher *r* of about 20/day for their bacteria prey. If we substitute a food organism that has a slower intrinsic growth rate, closer to that of the *Paramecium* themselves, we can introduce a lag time into this interaction. The results of such an experiment conducted by Gause (1936) are shown in Figure 5.35 and confirm the theoretical expectation of cycles. In addition to the reduction in prey, *r*, the switch to yeast also induces a lag time because yeast excretes toxic waste products that accumulate in the medium, ultimately reducing survival and reproduction in both *Paramecium* and yeast.

The behavior of time-lagged density dependent models can become still more complicated than the simple sinelike function shown in Figure 5.28, as we now demonstrate. Another way of deriving a biological meaning for the lag time is as a crude approximation to the maturation time for young in an age-structured population. Suppose that the birth function is logistic and that young require T_1 time units to mature into adults. Then we would have

$$\frac{dN}{dt} = rN(t - T_1)\left(\frac{K - N(t - T_1)}{K}\right),$$

where *N* now is just the adult population of herbivores. Note that the lag time occurs in both *N* terms. In addition, suppose that there is also a density independent adult death rate *d* that is instantaneous (i.e., not lagged), which gives,

$$\frac{dN}{dt} = rN(t - T_1)\left(\frac{K - N(t - T_1)}{K}\right) - dN(t).$$

Figure 5.36
Two illustrations of the double time-lagged logistic equation. (a) The parameters yield "shoulders" on the front edge of each cycle. (b) The dynamics are much more complicated.

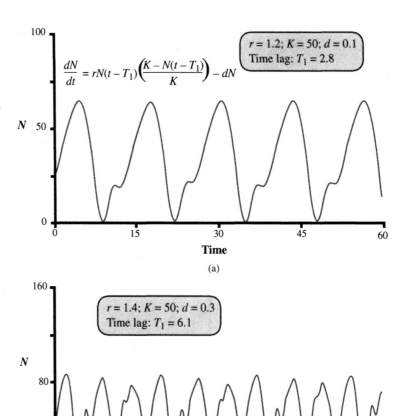

$$\frac{dN}{dt} = rN(t - T_1)\left(\frac{K - N(t - T_1)}{K}\right) - dN$$

$r = 1.2; K = 50; d = 0.1$
Time lag: $T_1 = 2.8$

(a)

$r = 1.4; K = 50; d = 0.3$
Time lag: $T_1 = 6.1$

(b)

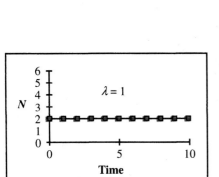

Figure 5.37
Geometric growth when $\lambda = 1$. The initial population size is 2.

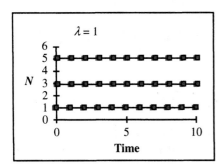

Figure 5.38
Geometric growth for three values of N_0 when $\lambda = 1$.

An example of the dynamics of this equation, the double time-lagged logistic equation is shown in Figure 5.36.

More complicated differential delay equations for density dependence like these can even produce chaotic behavior (May 1981, Nisbet 1997).

Stability Analysis of Discrete Density Dependent Population Growth

For unlimited geometric population growth, the qualitative behavior of a population falls into three categories according to the magnitude of λ (or the sign of R).

1. If $\lambda = 1.0$ ($R = 0$), then a population will remain through time at its initial size (Figure 5.37).

When $\lambda = 1.0$, there is an equilibrium, but it is determined solely by the initial conditions, N_0. In category 1 the initial population size, N_0, is 2.0, and the population remains at $N = 2$ over time. Whatever the value of N_0, that is where the population remains (Figure 5.38).

2. If $\lambda > 1.0$ ($R > 0$), then the population grows geometrically to infinity (Figure 5.39).
3. If $\lambda < 1.0$ ($R < 0$), then the population declines asymptotically to 0 (Figure 5.40).

Only if $\lambda < 1.0$ can the geometric model yield an equilibrium point ($N^* = 0$) that is stable. The zero point is a stable equilibrium point since trajectories from different nonzero population sizes all converge at 0, as in Figure 5.41 for $\lambda = 0.8$. This zero equilibrium point is often called the *trivial equilibrium,* since it's not very interesting. Obviously we

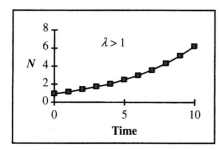

Figure 5.39
Geometric growth when $\lambda > 1$.

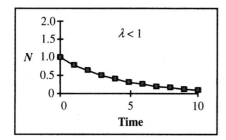

Figure 5.40
Geometric growth when $\lambda < 1$.

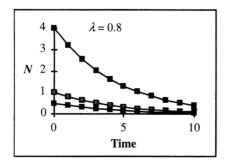

Figure 5.41
Geometric growth for three values of N_0 when $\lambda < 1$.

can't get a population to exist from zero individuals. However, this concept of exponential decline toward an equilibrium will prove useful later when we evaluate the stability of nontrivial equilibrium points (i.e., $N^* \neq 0$).

For both discrete or continuous time models of density dependent growth, the population has an equilibrium carrying capacity, K, which represents a nontrivial equilibrium point, $N^* = K$. This point can be reached from a wide variety of different initial conditions and for many different R's. However, in discrete time models for low R's the equilibrium, K, is stable, while for higher R's the same equilibrium point becomes unstable, becoming effectively unreachable. In this section, we develop the mathematics to evaluate the stability of an equilibrium point. We keep things as general as possible so that the conclusions apply not just to the discrete logistic equation but also to many possible density dependent population growth equations. Consequently, the details of the particular growth equation will be left unspecified; let's call it F and let F be a function of N_t. Then

$$N_{t+1} = F(N_t). \tag{5.5}$$

The function F converts (or maps) the present value of N, which is N_t, to a new value of N one time step later, N_{t+1}. The discrete logistic equation is one example of a possible $F(N_t)$ function, which involves N_t, N_t^2 and two parameters, R and K. We could easily imagine much more complex functional forms for F that might include higher-order or even transcendental terms involving N_t and many parameters. This is all permissible for what follows. However, we are not allowing F to be a function of still earlier population densities such as N_{t-1} or N_{t-2}, and so on. In this sense the population dynamics have no memory of former size but respond solely to present size and the parameters of growth.

To evaluate the stability of an equilibrium point we perform the mathematical equivalent of nudging the ball at equilibrium (see Figure 5.8). After the nudge, or perturbation, does the ball tend to move back to the original equilibrium position (the stable case), or does it tend to move even farther away in the next time step (the unstable case)? Figure 5.42 conveys this same idea in terms of population size.

A **local stability analysis** performs the mathematical equivalent of nudging the ball. Figure 5.43 shows how it operates.

Figure 5.44 shows a blowup of a portion of Figure 5.43 in the vicinity of the equilibrium point, N^*. At time t, the population density is perturbed very slightly away from N^* to a new point $N^* + n_t$. The number of individuals added to the population by this perturbation is n (or the number subtracted, if n is negative), so

$$N_t = N^* + n_t. \tag{5.6}$$

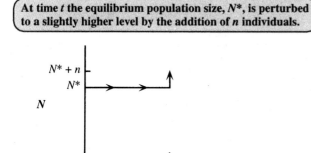

At time *t* the equilibrium population size, *N**, is perturbed to a slightly higher level by the addition of *n* individuals.

(a)

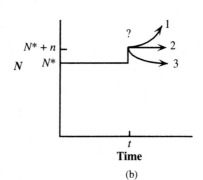

Possible dynamics following this perturbation:

1. Numbers may tend to move farther away from N^*; therefore N^* is <u>locally unstable</u>.

2. Numbers may stay exactly at new position $N^* + n$; therefore N^* is <u>neutrally stable</u>.

3. Numbers may tend to come back to N^*; therefore N^* is <u>locally stable</u>.

(b)

Figure 5.42
(a) The force of population growth—the nature of the density dependence—determines (b) behavior following a perturbation in numbers.

Figure 5.43
A local stability analysis around the point N^*.

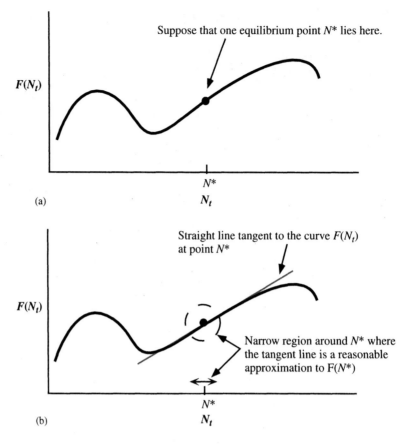

Figure 5.44
A blowup of a portion of Figure 5.43.

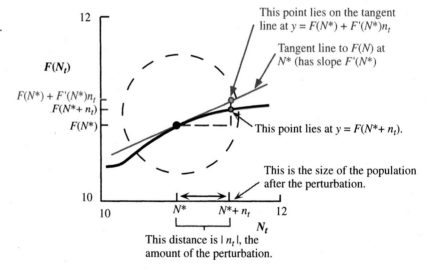

To proceed, we want to understand how this small difference in population size changes in subsequent time steps. Does it grow larger or smaller? This is equivalent to asking if the ball rolls back to equilibrium or runs off in Figure 5.8.

We now rearrange Eq. (5.6) to produce an expression for just the perturbation component, n_t:

$$n_t = N_t - N^*. \tag{5.7}$$

Based on our general growth equation Eq. (5.5), the population will change in the next time step as

$$N_{t+1} = F(N_t),$$

and since following the perturbation $N_t = N^* + n_t$, we have

$$N_{t+1} = F(N^* + n_t). \tag{5.8}$$

Rewriting Eq. (5.7) in terms of n_{t+1} instead of n_t yields

$$n_{t+1} = N_{t+1} - N^*. \tag{5.9}$$

Substituting Eq. (5.8) into Eq. (5.9) gives

$$n_{t+1} = F(N^* + n_t) - N^*. \tag{5.10}$$

This is the height on the y axis at $F(N^*)$ and, at equilibrium, $F(N^*) = N^*$.

This is the height on the y axis after the perturbation to $N_t = N^* + n_t$.

The difference represents the part of $y = F(N_t)$ due to the perturbation amount, n_t.

The motivation for Eq. (5.10) is important. We have reduced our original dynamical equation, Eq. (5.5), which specified changes in N over time, to a new related model, Eq. (5.10), that predicts changes in just the perturbation component, n, which, after all, is the component we are primarily concerned with when evaluating stability. Does n shrink or grow larger over time?

Since, as we've shown, F may be a complicated function, Eq. (5.10) may not be easy to evaluate. To make further headway we'd like to simplify the F function of Eq. (5.10). To do so we approximate F with a simple linear function. This linearization is shown graphically in Figures 5.43 and on Figure 5.44 as the tangent line to $F(N)$ at point N^*. (Don't confuse the tangent line in Figure 5.44 at N^* with the construction line $N_{t+1} = N_t$ in the Ricker diagrams—they are very different lines.)

How can we justify this simplification by which we convert a curved function F to a straight line? As you can see in Figures 5.43 and 5.44, in a narrow region around N^* the tangent line to the curve $F(N_t)$ at N^* is a reasonable approximation to the actual curve $F(N_t)$ around N^*, although it deviates wildly when n is large and thus $N^* + n$ is far from N^*. Since the perturbation n is a small quantity (just a nudge), this limitation is acceptable. As long as the region is narrow enough, any function can be approximated by a straight line. For wider regions, however, we could find better approximating functions than a straight line. A simple way is to decompose $F(N_t)$ into a polynomial. The more higher-order terms included in the polynomial, the greater is the similarity to the $F(N_t)$ function being approximated. A straight line (the tangent line) is this polynomial with all the terms except the first two (the slope and the y intercept) lopped off.

What are these additional terms? We want the polynomial to follow all the bends and curves of the real function F. Consequently, these additional terms are successive derivatives of F evaluated at the equilibrium point in question, in our case $F(N_t)$ evaluated at N^*. The inclusion of more terms corresponding to these successively higher derivatives would, in the limit (as the number of terms grows very large), match $F(N_t)$ everywhere, not just narrowly around N^*. (This assumes that $F(N_t)$ is a smooth continuous function so that these derivatives exist and that the derivatives are continuous.) The approximation of a function by a polynomial, which is the sum of terms containing successively higher derivatives of the original function, is called a **Taylor's expansion**. The general formula for a Taylor's expansion evaluated around a point x^* is

$$F(x) \approx F(x^*) + \frac{F'(x^*)(x - x^*)}{1!} + \frac{F''(x^*)(x - x^*)^2}{2!} + \frac{F'''(x^*)(x - x^*)^3}{3!} + \cdots, \tag{5.11}$$

where F' means the first derivative of F with respect to x; F'' indicates the second derivative, and so on. Note that, for small perturbations where $x - x^*$ is much less than 1.0, $(x - x^*)^2$ is smaller then $(x - x^*)$ and $(x - x^*)^3$ is smaller yet. In other words, the higher-order terms of $(x - x^*)$ add successively smaller corrections; this provides the justification for ignoring them if the perturbation is small.

Now let's return to the simplest case where all the higher-order terms in Eq. (5.11) are ignored. Using Eq. (5.11), we write Eq. (5.10) as

$$\begin{aligned} n_{t+1} &= F(N^* + n_t) - N^* \\ &\approx F(N^*) + F'(N^*)n_t - N^*, \end{aligned} \tag{5.12}$$

and, since by definition for an equilibrium, $F(N^*) = N^*$, Eq. (5.12) becomes

$$n_{t+1} \approx F'(N^*)n_t. \tag{5.13}$$

The value of $F'(N^*)$ is simply the slope of the tangent line in Figure 5.44 at the equilibrium point N^*. If we assume that n_t is positive (i.e., we did an addition perturbation), then according to Eq. (5.13) the magnitude of the distance from equilibrium after one time step (i.e., n_{t+1}) will be smaller at $t + 1$ than it was at time t (i.e., n_t), as long as the slope $F'(N^*)$ is less than 1. But since $F'(N^*)$ is simply the slope of the tangent line, the condition for the stability of the equilibrium point reduces to a condition on the slope of $F(N)$ at point N^*; it must be less than 1. On the other hand, if this slope is greater than 1, then the deviation will increase at $t + 1$ (i.e., now $n_{t+1} > n_t$). More generally, since the perturbations may be either additions (n_t is positive) or subtractions (n_t is negative), we can write the condition for stability, using the absolute value of the slope, as

$$\left| F'(N^*) \right| < 1. \tag{5.14}$$

Equation (5.14) is the central result: In words: **For a single species density dependent equilibrium point to be stable, the line tangent to $F(N)$ at this equilibrium point must have a slope whose *absolute value* is less than 1.**

Now you see why increases in R destabilized the equilibrium K in the discrete logistic equation. Increases in R increased the steepness of the parabola, $F(N_t)$, and this steepened the slope (made it more negative than –1) at point $N^* = K$. In Box 5.2, we solve for the value of R that just destabilizes $N^* = K$. In Box 5.3 we perform a stability analysis for an equilibrium point of a continuous time differential equation.

Because of the assumption of linearizing the dynamics, which is only a good approximation to the real population dynamics in a narrow vicinity around an equilibrium point, this method for evaluating stability is strictly valid only for relatively small

---

Box 5.2   *Solving for the Value of R That Just Destabilizes the Equilibrium Point*

The discrete logistic equation is

$$N_{t+1} = F(N_t) = N_t\left(1 + R\left(1 - \frac{N_t}{K}\right)\right)$$

$$= N_t + RN_t - \frac{RN_t^2}{K}.$$

The slope of this equation at K is the first derivative with respect to N_t,

$$F'(N_t) = 1 + R - \left(\frac{2RN_t}{K}\right).$$

We evaluate this derivative at point $N_t = K$:

$$1 + R - \left(\frac{2RK}{K}\right) = 1 - R.$$

For stability of K this derivative must have absolute value < 1, and since R is a positive number, $\left| 1 - R \right| < 1$, means that $R < 2$ is necessary for stability of the equilibrium point K.

Box 5.3   Stability Analysis for the Continuous Logistic Equation

In Box 5.2 we performed a local stability analysis on an equilibrium point of a difference equation in one variable: $N_{t+1} = F(N_t)$. Now we repeat this process but consider a differential equation $dN/dt = f(N)$. We have already shown that, for the continuous time logistic equation, the equilibrium point is always stable, regardless of the value of r, but here we bring this observation into line with the analytical methods that we just developed for a difference equation. Again let's consider a situation where $N(t)$ changes over time and eventually reaches a stable equilibrium point K. Assume that we begin at some initial population size N_0, and, after an elapsed time period Δt, we again measure the population size to produce the red arrow shown in the following diagram.

after time Δt

Population size, N

If N starts far below K, then its initial growth rate will be rapid; if N starts very close to K, then its growth rate will be slower, or

Population size, N

The tail of each red vector is at the beginning N, and the tip of arrowhead is the population size after some one time period. Thus the length of each vector equals the magnitude of growth from point N_0, and the direction of the vector indicates the sign of that growth rate. Now by flipping these vectors up on their tails, and assuming that Δt is a fairly small time interval, we have a direct measure of the magnitude and sign of dN/dt at each of the starting population sizes:

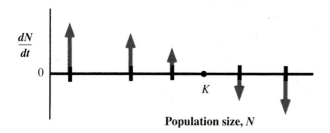

Population size, N

By repeating this process from many different additional starting points, we could determine the rate of growth from other N and, by interpolation, fill in the whole $f(N)$ curve:

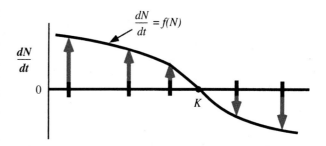

Thus the $f(N)$ function can be experimentally derived by such a procedure. The preceding diagram clearly shows that small perturbations in N below the equilibrium point K (e.g., $K - n$, where n is the perturbation amount) lead to increases in population size (upward pointing arrows). Similarly, small perturbations in N above K ($K + n$) lead to negative dN/dt and thus decreases in population size. Thus the equilibrium population size, K, tends to be restored, implying that K is a locally stable equilibrium point. Contrast this situation with the following:

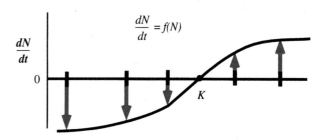

Now, K plus a small perturbation n, leads to a positive growth rate and therefore to an increasing population size; the population size escapes from K after the perturbation. In this diagram K is an unstable equilibrium point. Comparing the last two figures you can see that the condition for stability is simply that the slope of $f(N)$ evaluated at the equilibrium point, K, be negative. For the logistic equation this yields

$$f(N) = \frac{dN}{dt} = rN\left(\frac{K-N}{K}\right) = rN - \frac{rN^2}{K}$$

and

$$\text{Slope of } f(N) = \frac{\partial f(N)}{\partial N} = r - \frac{2rN}{K}.$$

At $N = K$,

$$\text{Slope} = r - \frac{2rK}{K} = r - 2r = -r,$$

as illustrated in the following diagram.

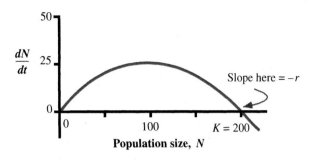

We conclude that the equilibrium point $N^* = K$ in the continuous logistic equation is stable as long as r is positive. From this same argument you can see that the equilibrium point $N^* = 0$ is not locally stable since the slope of $f(N)$ at $N = 0$ is positive.

In summary, for stability we have (a) for the discrete logistic equation, $|\text{slope at } N^*| < 1$; and (b) for the continuous logistic equation, slope at equilibrium point $N^* < 0$.

Discrete Logistic Equation

N^* will be locally stable when the slope of $F(N)$ at N^* falls in this region:

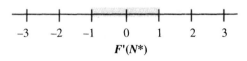

Continuous Logistic Equation

N^* will be locally stable when the slope of $f(N)$ at N^* falls in this region:

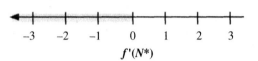

perturbations, n, from equilibrium. Consequently, this method is referred to as an evaluation of **local stability.** An equilibrium point that is locally stable (i.e., stable to small perturbations) might still be unstable to much larger perturbations. A metaphoric example of this is illustrated Figure 5.45. Local stability does not necessarily imply **global stability,** although if an equilibrium point is globally stable it surely must be locally stable. We explore an example of a modified density dependent model with two locally stable equilibria in the next section.

Also, although we have developed a method to evaluate the stability or instability of an equilibrium point, we have *not* developed an analytical method to determine the dynamics of the population if that equilibrium point is unstable. Will the population decline to extinction? Will it display a limit cycle? If so, what period and amplitude will the cycle have? Will chaos emerge? While mathematical methods do exist to determine the presence or absence of these more complicated dynamical behaviors for one-, two-, and sometimes even three-species systems, they are limited and general analytical predictions are much more difficult and often impossible to make for larger numbers of interacting species. In practice, then, we are often left to simulate the dynamics on a computer.

Modifications to the Logistic Equation

An Allee Effect

The logistic equations are not the only way that density dependence might be manifested. The techniques that you have just learned are applicable to other possible density dependent functions. In particular, the *linear* decline in per capita growth rate with increasing density is not biologically general. Most species do not show such a linear decline (Pomerantz et al. 1980, Fowler 1988). In some cases there is a continuous decline, but it is curved, not straight (see Figure 5.5). There is a simple fix to this problem: an additional parameter may be added to the logistic model to produce the so-called theta (θ) model. By adjusting the value of θ, you can create convex and concave per capita growth curves of infinite variety. This model is left as an exercise at the end of this chapter (Problem 3).

More fundamentally, some animal and plant species do not show a monotonic decline in per capita growth with increasing density. Instead per capita growth rates may *increase* with population density at low population levels and decline only at much

Figure 5.45
Gravity metaphors for local and global stability. (a) The equilibrium point X^* is (b) locally stable but (c) not globally stable.

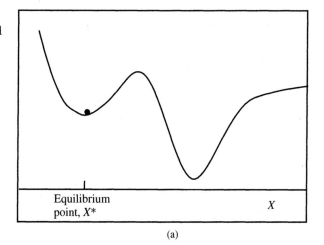

Equilibrium
point, X^*

X

(a)

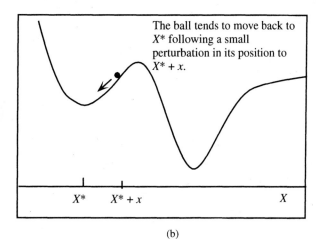

The ball tends to move back to X^* following a small perturbation in its position to $X^* + x$.

X^* $X^* + x$ X

(b)

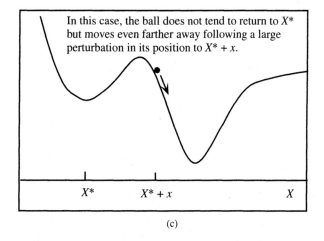

In this case, the ball does not tend to return to X^* but moves even farther away following a large perturbation in its position to $X^* + x$.

X^* $X^* + x$ X

(c)

higher population levels. One biological reason for the ascending part of the curve at low density is that species may require conspecifics for protection from predators or from climatic extremes; other species may forage more effectively in groups than alone. In sexual species, individuals may have a difficult time finding mates at low densities, so mating rates increase with population density. The result of these types of effects is that positive per capita growth rates might not even be possible until the population reaches some threshold size and per capita growth rates then increase with population density—at least up to a point. Only when population densities are far above this size might the negative effects of crowding become evident. This population response is often referred to as an **Allee effect** (Allee 1931). Figure 5.46 incorporates these

Figure 5.46
An Allee effect added to the discrete logistic equation. (a) This population displays an **Allee effect** at low densities. Note that three equilibrium points exist, at $N = 0$, N_1^*, and N_2^*. (b) Conclusion: 0 and N_2^* are both locally stable, but neither is globally stable.

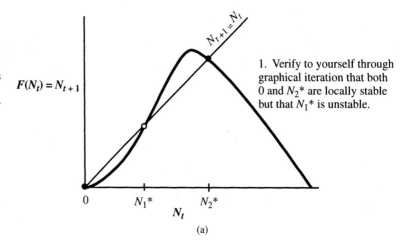

$F(N_t) = N_{t+1}$

$N_{t+1} = N_t$

1. Verify to yourself through graphical iteration that both 0 and N_2^* are locally stable but that N_1^* is unstable.

N_t

(a)

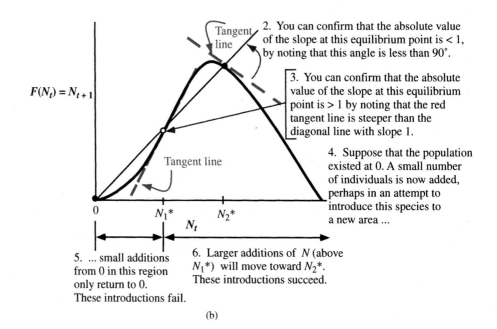

$F(N_t) = N_{t+1}$

Tangent line

2. You can confirm that the absolute value of the slope at this equilibrium point is < 1, by noting that this angle is less than 90°.

3. You can confirm that the absolute value of the slope at this equilibrium point is > 1 by noting that the red tangent line is steeper than the diagonal line with slope 1.

4. Suppose that the population existed at 0. A small number of individuals is now added, perhaps in an attempt to introduce this species to a new area ...

Tangent line

N_t

5. ... small additions from 0 in this region only return to 0. These introductions fail.

6. Larger additions of N (above N_1^*) will move toward N_2^*. These introductions succeed.

(b)

considerations in a discrete time model and explores the stability consequences. This is accomplished by skewing the $N_{t+1} = F(N_t)$ curve so that, for low levels of N_t, N_{t+1} is even lower than N_t (and thus below the diagonal line). The population cannot begin to increase unless $N_t > N_1^*$.

Problem: Consider a modified logistic differential equation that incorporates an Allee effect in the following way. The *per capita* growth rate begins at $-r$ when $N = 0$ and then rises linearly to a value of $+r$ at the point where N equals M. From $N = M$ onward, however, the per capita growth rate declines with a slope equal but opposite in sign to the first part of the curve. Below M, increases in density result in higher individual birth rates, while above M, crowding results in progressively lower birth rates with N. Figure 5.47 shows this curve for the values of $r = 0.5$ and $M = 15$.

Express these per capita growths in differential equations.

Answer:

$$\frac{dN}{N\,dt} = -r + \frac{2rN}{M} \text{ (for } 0 \le N < M) \text{ and } \frac{dN}{N\,dt} = r - \frac{2r(N-M)}{M} \text{ (for } N \ge M).$$

Problem: Solve for the carrying capacity, K, in Figure 5.47.

Answer: The upward portion of the curve has slope $= 2r/M$, and the downward portion has slope $-2r/M$. The population stops growing when the per capita growth is 0, which can be determined by solving $0 = 0.5 - (2)(0.5)(N-15)/15$. The solution for N is $N = 22.5 = K$.

Problem: Using the parameters $r = 0.5$ and $M = 15$, plot the population growth curve dN/dt.

Answer: We multiply the per capita equations by N to get the curve plotted in Figure 5.48.

The Ricker Logistic Equation

Another unrealistic feature of the discrete logistic equation is that $N(t+1)$ can become negative when $N_t \gg K$. In our simulations, we stopped this from happening by enforcing a rule: if $N(t+1) < 0$, then $N(t+1) = 0$. A different and more continuous approach is followed in the Ricker logistic equation (Ricker 1952):

$$N_{t+1} = N_t \exp\left(R\left(1 - \frac{N_t}{K} \right) \right). \tag{5.15}$$

Note that, when $N_t = K$, the exponent in parentheses becomes 0 and thus $N_{t+1} = N_t$, yielding an equilibrium point at $N^* = K$, as before for the discrete logistic equation. For $N_t > K$, the exponent is negative, but e raised to a negative number is still positive. As N_t approaches infinity, N_{t+1} approaches zero. Two examples of Ricker curves with different R's and sample trajectories are shown in Figure 5.49. Figure 5.49(a) shows the case of a stable equilibrium point; in Figure 5.49(b) R is larger, and the dynamics are complex cycles.

The Ricker logistic equation has dynamics that go through the same sequence of behaviors that we have already explored for the discrete logistic equation. The exact values of R that mark the transitions from one dynamical phase to another are similar but not always exact. For example, the onset of chaos with the Ricker logistic equation is at $R > 2.692$, rather than at $R > 2.570$ as for the logistic equation.

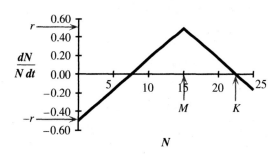

Figure 5.47

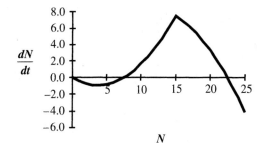

Figure 5.48
Population growth rate as a function of population size for the per capita growth shown in Figure 5.47.

$$F(N_t) = N_{t+1} = N_t \, \exp\left(R\left(1 - \frac{N_t}{K}\right)\right).$$

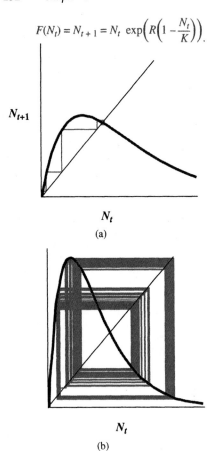

N_{t+1}

N_t

(a)

N_t

(b)

Figure 5.49
Ricker diagrams based on the Ricker equation: (a) $R = 1.5$ and (b) $R = 2.7$.

Exercise: For the Ricker logistic equation, at what value of R will the equilibrium point at K become unstable?

Solution: From the stability analysis of the previous section, we know that this will occur when the absolute value of the slope of $F(K) > 1.0$. For the Ricker logistic equation,

$$F(N_t) = N_{t+1} = N_t \, \exp\left(R\left(1 - \frac{N_t}{K}\right)\right).$$

We take the first derivative of $F(N)$ to find the slope of $F(N)$:

$$F' = N_t \, \exp\left(R\left(1 - \frac{N_t}{K}\right)\right)\left(\frac{-R}{K}\right) + \exp\left(R\left(1 - \frac{N_t}{K}\right)\right)$$

$$= \exp\left(R\left(1 - \frac{N_t}{K}\right)\right)\left(1 - \frac{RN_t}{K}\right).$$

At point $N_t = K$, this reduces to

$$F' = 1 - R.$$

The slope F' will become steeper than -1 when $R > 2$. This result turns out to be identical to that for the discrete logistic equation.

Density Dependence in Stage-Structured Models. Annual bluegrass, *Poa annua*, is a widespread weed and a good colonizer of open habitats. Figure 5.50(a) presents the fecundity (seed production) rates, and Figure 5.50(b) presents the survival rates for different age groups over a typical year. Law (1975) found that some of these vital rates were density dependent. He divided the life history into five stages (seeds, seedlings, young adults, A_1, medium adults, A_2, and old adults, A_3, with the transitions illustrated in Figure 5.51. Further, Law found that seedling survival rates, s_{seedling}, and seed production rates of adults B_i decreased with increasing density, N. Thus he was able to form the following transition matrix:

$$\begin{bmatrix} \text{Seed} \\ \text{Seedlings} \\ \text{Adults 1} \\ \text{Adults 2} \\ \text{Adults 3} \end{bmatrix}_{t+1} = \begin{bmatrix} 0.2 & 0 & B_1(N) & B_2(N) & B_3(N) \\ 0.05 & 0 & 0 & 0 & 0 \\ 0 & s_{\text{seedling}}(N) & 0 & 0 & 0 \\ 0 & 0 & 0.75 & 0 & 0 \\ 0 & 0 & 0 & 0.75 & 0 \end{bmatrix} \begin{bmatrix} \text{Seeds} \\ \text{Seedlings} \\ \text{Adults 1} \\ \text{Adults 2} \\ \text{Adults 3} \end{bmatrix}_t, \quad (5.16)$$

where the density dependent terms, determined from his field data, were

$$s_{\text{seedling}} = 1 - 0.25 \, \exp{(0.0005N)}$$

$$B_1(N) = B_3(N) = 100 \, \exp{(-0.0001N)}$$

$$B_2(N) = 200 \, \exp{(-0.0001N)}$$

and population density, N, is the sum of seedlings plus all adults.

Note that, in the transition matrix, 20% of the seeds remain dormant in the seed bank every year (the term in the first row, first column), while 5% of them germinate into seedlings (second row, first column). When $N = 0$, seedling survival rate is at a maximum of $1 - 0.25 = 0.75$ (since $e^0 = 1$), and adult fecundity is at a maximum of 100 in young and old adults and 200 in medium adults.

By iterating the transition matrix, we can trace the expected course of population growth from year to year, as depicted in Figure 5.52. This simulation is based on the density dependence as measured in the field.

The population initially fluctuates but eventually reaches an asymptotic density. On the other hand, as we demonstrated earlier with density dependence in discrete time,

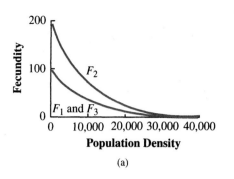
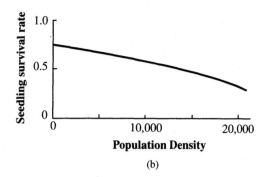

(a)   (b)

Figure 5.50
Density dependence in (a) adult seed production and (b) seedling survival in *Poa annua* (from Law 1975).

Figure 5.51
Stage transitions in the annual grass,
Poa annua.

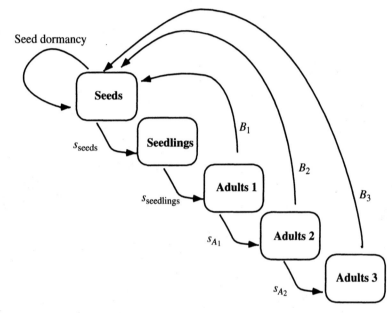

Figure 5.52
Time series for the grass *Poa annua*,
produced by iterating the transition
matrix of Eq. (5.16).

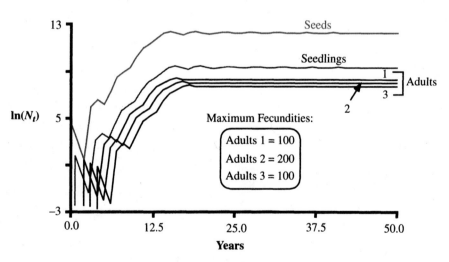

increases in the age-specific fecundities can produce unstable equilibrium points, oscillations, and chaos. Figure 5.53 shows a time series based on some elevated seed production rates. This simulation examines the effect of increasing the maximum seed production rates.

Each age group in this population settles into a five-point limit cycle during which each of the adult age classes crashes to zero at each cycle. Further increases in maximum fecundity can lead to even more erratic population behavior.

Figure 5.53
Time series for the grass *Poa annua*, produced by iterating the transition matrix with some hypothetical elevated maximum fecundities. Adults 1 and 2 fluctuate, as does Adult 3, but were left out of the figure for clarity. This pattern is a five-point limit cycle.

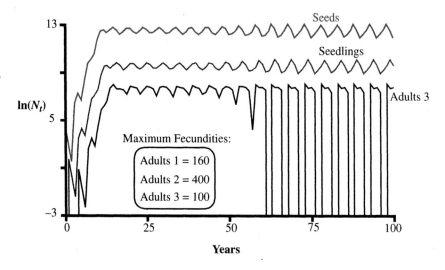

Figure 5.54
The population density over two-dimensional space (x, y) for a spreading population with a carrying capacity (in density) of K.

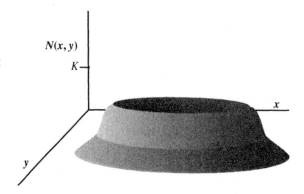

Rate of Spatial Spread with Density Dependent Growth

In Chapter 2, we added random walk movements of individuals across space to the local population growth model of geometric growth. We imagined a grided space (discrete space), with individuals moving each time step only to adjacent cells. This discrete time and discrete space model can be modified to continuous time and continuous space, that is, diffusion. We reached two important results earlier.

1. **The wave front of population advance quickly reaches a constant velocity.**
2. **The radius of the population's occupied area grows with the square root of time.**

In particular, for diffusion, this radius increases at the rate

$$\text{Radius} = 2(\ln(\lambda)D)^{0.5}, \tag{5.17}$$

where D is the diffusion coefficient and $\ln \lambda = r$ is the rate of exponential growth.

What happens if density dependent growth occurs in each cell instead of exponential growth? Not much—Eq. (5.17) still holds, but only approximately. The density dependence kicks in as population density builds up in a local area. But, by the time that happens, some individuals have already immigrated to adjacent vacant areas. Consequently, it is primarily the rate of growth, λ, not K, that dominates the rate of spatial spread. The spreading population looks like that shown in Figure 5.54, where the rate of advance in each direction is given by Eq. (5.17).

However, if the density dependence, had an Allee effect, then on the leading edge of the front, where the population is at low density, the individuals there would be at a disadvantage. This would retard the rate of population spread and lead to a "clumpiness" of populations at the edge of the wave of population advance.

PROBLEMS

1. Prove mathematically that the peak of the dN/dt versus N plot occurs at $K/2$ for the continuous logistic equation.

2. Under each example on the left enter the number(s) of the most appropriate answer(s) from the right.

a. Example(s) of a neutrally stable equilibrium point _____

b. Example(s) of a locally and globally stable equilibrium point _____.

c. Example(s) of an unstable equilibrium point _____.

d. System(s) that lacks any stable equilibrium points _____

e. Example(s) of a smooth limit cycle _____

f. System(s) that lacks any stable asymptotic behavior (points or cycles)_____

i. Initial density for exponential growth with $r = 0$

ii. Geometric growth and $\lambda > 1$

iii. K in the discrete logistic equation when $R = 2.2$.

iv. Time delay in the continuous logistic equation, $rT = 2$

v. $N = 0$ for geometric growth and $\lambda < 1$

vi. K in the continuous logistic equation

3. Consider the three-parameter density dependent model,

$$\frac{dN}{dt} = rN\left(1 - \left(\frac{N}{K}\right)^{\theta}\right).$$

It is called the "theta logistic" equation because of the introduction of the new parameter, θ. For $\theta = 0.5$, 1, and 2 draw the following plots (let $r = 1$ and $K = 1$).

a. dN/dt versus N

b. $dN/N\,dt$ versus N

c. Under what biological circumstances might θ be less than 1? Greater than 1?

4. In one set of Gause's (1934) experiments with *Paramecium* growing in his "one-loop medium," he produced the following results, which we have plotted in terms of per capita growth as a function of population size. The relationship deviates substantially from the linear relationship expected for the discrete logistic equation.

Day	N	Per capita growth rate
0	2	2.000
1	6	3.000
2	24	2.125
3	75	1.427
4	182	0.451
5	264	0.205
6	318	0.173
7	373	0.062
8	396	0.119
9	443	0.025
10	454	−0.075
11	420	0.043
12	438	0.123
13	492	−0.049
14	468	−0.145
15	400	0.180
16	472	

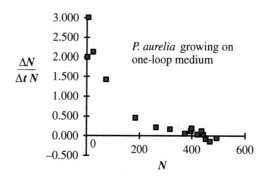

These data, however, can be fit quite nicely by the theta logistic equation. Estimate the values of r, K, and θ that provide reasonable fits.

5. In the discrete logistic equation, as R increases the growth curve becomes steeper and the equilibrium K eventually becomes unstable. What happens to the shape of the growth curve and the dynamic stability of the population as K instead of r gets larger and larger?

6. Consider the following Ricker diagram.

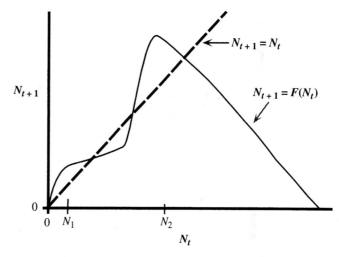

a. Draw small dots on all the equilibrium points and label each equilibrium point with a letter (a, b, etc.).

b. For each equilibrium point, state whether it is stable or unstable.

c. Trace the population growth trajectory, beginning at density N_1.

d. Trace the population growth trajectory, beginning at density N_2.

7. A discrete logistic equation for the population growth rate of some hypothetical lizards can be written as

$$N_{t+1} = (R+1)N_t - \left(\frac{RN_t^2}{100}\right).$$

a. What is the equilibrium population size? _____

b. If $R = 2.7$, is this equilibrium stable? _____.

c. If $R = 1.9$, is this equilibrium stable? _____.

8. For the discrete time logistic equation, what two different values of population size N_t at time t will produce exactly the value 0 for the population size one time step later? Let the growth rate parameter $R = 1$.

9. A species grows according to the discrete time logistic equation. A plot of N_{t+1}/N_t versus N_t for several populations of this species at different densities N_t reveals a linear relationship described by $N_{t+1}/N_t = 3.04 - 0.01N_t$. What are the values of R and K for this species? Is the equilibrium point, K, stable?

10. Another species grows according to the Ricker logistic equation. A plot of N_{t+1}/N_t versus N_t for several populations of this species at different densities N_t reveals a linear relationship described by $\ln(N_{t+1}/N_t) = 1 - 0.01N_t$. What are the values of R and K for this species?

11. Another discrete density dependent model that is sometimes applied to insect populations is $F(N_t) = \lambda N_t(1 + aN_t)^{-B}$, with three parameters, λ, a, and B. A graph of this function for one set of parameter values is the following.

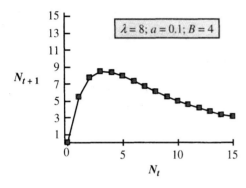

$\lambda = 8; a = 0.1; B = 4$

Is $N = 0$ a stable equilibrium for this model? What is the solution, in terms of the parameters of the model, for the nontrivial equilibrium point N^*?

12. The bobcat population in Wyoming was studied by Crowe (1975). Bobcats reach sexual maturity at age 1. The average litter size for adult females is $B = 2.8$ kittens/litter and the sex ratio is 1:1. All females breed once each year. The survival rate of kittens to year 1 is s_0 and is strongly density dependent. The survival rate of adults, which includes natural mortality and deaths due to trapping, is $s_a = 0.67$. This information can be combined into a Leslie matrix (census taken just before reproduction):

$$\mathbf{L} = \begin{bmatrix} s_0 B & s_0 B \\ s_1 & s_1 \end{bmatrix}$$

Find s_0 so that the population is stationary (i.e., $\lambda = 1.0$). If trapping is curtailed, adult survival jumps to 0.98. Now what s_0 is necessary for the population to be stationary?

13. A population grows according the equation

$$\frac{dN}{dt} = rN(t - T)\left(\frac{K - N(t-T)}{K}\right),$$

where $T = 5$ years, $r = 1$/year, and $K = 1000$. In 1990 the population size N is 30. What will be the population size 2000 years later?

14. Krebs (1986) reported the following data on the yearly discrete growth rate ($\lambda_t = N_{t+1}/N_t$) of snowshoe hare populations at different natural densities in the Yukon.

No. per acre	N_{t+1}/N_t	No. per acre	N_{t+1}/N_t
10	0.1	370	0.25
10	0.6	290	0.9
3	2.5	280	0.2
2	4	255	1.2
6	3.8	70	2.8
2	6.2	109	1.1
2	6.1	110	0.1
7	8.2	115	0.3
8	8.9	120	0.1
15	0.2	150	1.8
30	1.3	148	2.1
40	1.5	200	3.1
30	3.1	220	0.7
60	0.2	250	0.5
60	2.9		

We wish to determine if these data are consistent with a population that overshoots its carrying capacity, cycles, or experiences chaos—or, alternatively, a population that should have a stable equilibrium point. Make a determination by fitting these data to the discrete logistic equation.

Answer:

We plot these yearly growth rates on the y axis against population density and determine the regression line $\lambda = 3.34 - 0.011N$ (correlation coefficient, $r = 0.476$). Based on the form of the discrete logistic equation we may now make some parameter estimates:

$$\frac{N_{t+1}}{N_t} = 1 + R\left(1 - \frac{N_t}{K}\right) = 1 + R - \frac{RN_t}{K}.$$

Therefore the y intercept = $1 + R = 3.34$, which implies that $R = 3.34 - 1 = 2.34$. The slope = $-R/K = -0.011$; therefore $K = 2.34/0.011 = 213$. The plot of the equation is shown in the following diagram.

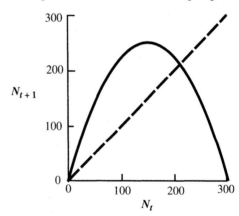

This high R implies population cycles. But a question remains: Is the discrete logistic equation the correct model for this species? Would we reach a different conclusion if we fit these data to a different model?

15. We've been using population numbers and population density interchangeably because we've been assuming a closed population in some circumscribed area; as numbers increase, so must density. However, it would be possible to decouple density and numbers, at least experimentally, by changing the physical dimensions of the area holding the population. Suppose that we did such an experiment, say, with flour beetles, in which in one set of vials we hold density the same, but vary absolute numbers and thus the sizes of the vials and the amounts of flour that they hold. In another set of experiments we keep absolute numbers the same but vary density by changing vial size. Will density dependence and population size dependence necessarily be equivalent? What biological functions might depend more on absolute numbers than density? What biological properties might depend more on density than absolute numbers?

6 Population Regulation, Limiting Factors, and Temporal Variability

We have shown how crowding can lead to density dependent population growth—population regulation is a result. The acid test for density dependence is a negative relationship between a population's per capita growth rate and its density. But this is only the beginning of an explanation. An important endeavor in ecology is understanding when and why density dependence occurs, since this helps explain the natural regulation of numbers. We ask:

- Does density dependence act through changes in birth rates, death rates, or both?
- What ages, sexes, and genotypes are involved?
- If density dependence occurs, what is the mechanism behind its operation?
- Is density dependence continuous or does it act only at certain times of the year or only in certain years?
- How do regulating (density dependent) environmental factors interact with density independent factors to shape overall population dynamics?

THE INTERACTION OF DENSITY DEPENDENT AND DENSITY INDEPENDENT FACTORS

As a way of thinking about the last question, let's return to the continuous logistic equation and imagine that, in addition to density dependence, some superimposed density independent mortality also exists, yielding

$$\frac{dN}{dt} = rN\left(\frac{K-N}{K}\right) - DN. \tag{6.1}$$

Here D is the per capita rate of this density independent mortality. Figure 6.1 depicts Eq. (6.1) graphically for three different levels of D.

This same relationship may be explored from a different angle by plotting per capita growth rates, as shown in Figure 6.2.

> **Exercise:** Use Eq. (6.1) to find the equilibrium population size, N^*, when $r = 1$, $D = 0.2$, and $K = 100$. What is N^* if r is doubled? What is N^* if D is doubled? What is N^* if K is doubled?

More generally, however, populations live in temporally variable environments; the relationship between density and growth rate is not as exact and thus not as predictable as the straight-line relationships depicted in Figures 6.1 and 6.2. We can think of this temporal variability as leading to a probability cloud rather than to an exact line or curve. This notion is illustrated in Figure 6.3, as so-called **density vague** population dynamics (Strong 1986).

Figure 6.1
The continuous logistic equation (in red) with three different levels of density independent mortality, *D*, superimposed. Where a *D* line intersects a logistic curve, the net $dN/dt = 0$. The corresponding value of *N* is thus an equilibrium, as shown by the dashed lines. Population turnover, dN/dt, is higher for higher values of *D*.

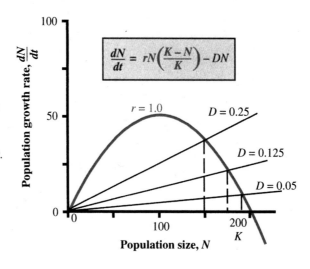

$$\frac{dN}{dt} = rN\left(\frac{K-N}{K}\right) - DN$$

Figure 6.2
As in Figure 6.1, but per capita population growth is now plotted on the *y* axis.

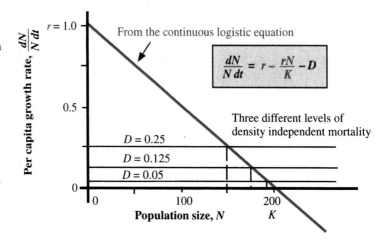

From the continuous logistic equation

$$\frac{dN}{N\,dt} = r - \frac{rN}{K} - D$$

Three different levels of density independent mortality

Figure 6.3
Density vague population dynamics. As in Figure 6.2, but now the scatter indicates probabilistic distributions for both the density dependent and density independent components of per capita population growth.

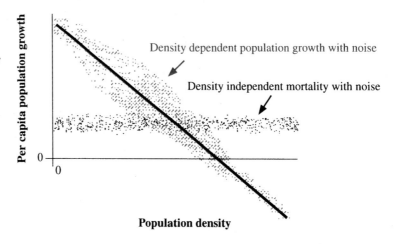

Density dependent population growth with noise

Density independent mortality with noise

A time series from a population responding to these probabilistic forces might at first glance look like a random walk, but the "signal" of the density dependence could be extracted from the population's time series with the proper statistical analysis. A few decades ago, a big debate raged in ecology on the relative importance of density dependent and density independent population growth for real populations. Tanner

Figure 6.4
A test for density dependence.

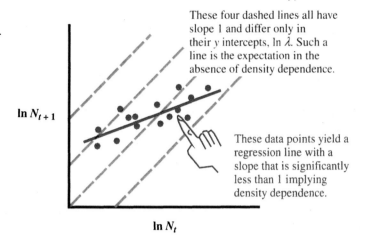

These four dashed lines all have slope 1 and differ only in their *y* intercepts, ln λ. Such a line is the expectation in the absence of density dependence.

ln N_{t+1}

These data points yield a regression line with a slope that is significantly less than 1 implying density dependence.

ln N_t

(1966) analyzed many time series for wild animal populations. One test that he applied used the following logic: if populations are generally growing in a density independent manner but λ is simply a random variable subject to temporal variation, then in discrete time

$$N_{t+1} = \lambda_t N_t. \tag{6.2}$$

Taking the natural log of both sides gives

$$\ln(N_{t+1}) = \ln(\lambda_t) + \ln(N_t). \tag{6.3}$$

Plotting $\ln(N_{t+1})$ versus $\ln(N_t)$ from Eq. (6.3) produces a linear relationship with a slope of 1 and a *y* intercept of ln λ_t. With real data there will be some scatter as λ changes from year to year and because, after all, we only get estimates of population size and they contain some measurement error. Consequently, as shown in Figure 6.4, if growth is simply exponential, the regression line going through the points ln N_{t+1} versus ln N_t, has a slope of 1 (the family of dashed lines) and a *y* intercept of $E(\ln \lambda_t) = E(r_t)$, where $E(\ )$ indicates the average value. On the other hand, if the average intrinsic growth rate, *r*, decreases with increasing population density, then the slope of the relationship in Eq. (6.3) should be less than 1.

Tanner (1966) performed such an analysis on 70 time series that he collated for natural animal populations. He found that for 63 out of 70 populations the slope based on linear regressions was significantly less than 1. This strongly suggests that much of the "balance of nature" that we see around us is due to the action of density dependent factors. However, there can be problems with this approach, as we will now illustrate. As an example of the application of this technique, we generated several time series on a computer, using the model

$$N_{t+1} = N_t \left(1 + \frac{R}{K}(K - N_t) + VRnd_t \right) \tag{6.4}$$

Like Eq. (6.1), Eq. (6.4) is just the discrete logistic equation with one extra term, $VRnd_t$, which is density independent. Unlike in Eq. (6.1), however, this term is a **random variable.** A random number, *Rnd*, is drawn each time period, *t*. In the simulations *Rnd* is drawn from a uniform distribution between –0.5 and +0.5. This random number is then multiplied by a constant *V* to produce a per capita density independent change to the population, where *V* represents the magnifier of this environmental noise level. When $V = 0$, temporal noise is completely absent. As *V* becomes larger, the density independent term becomes progressively large compared to the density dependent logistic term. Figure 6.5 shows how the noise multiplier, *V*, affects the growth of a population. For each case, $R = 0.2$ and $K = 500$.

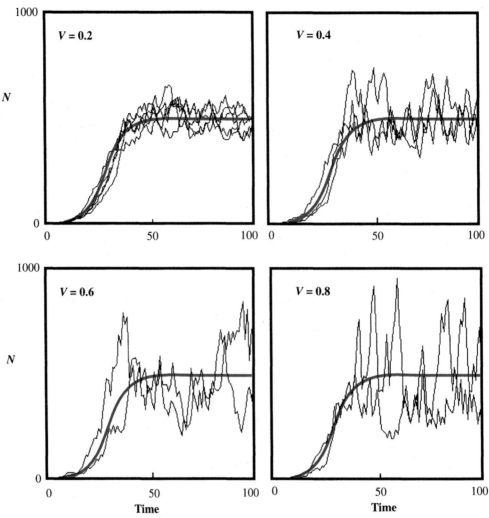

Figure 6.5
Time series of simulated populations based on Eq. (6.4). Each plot shows two or more simulations (in black) and a deterministic population, $V = 0$, (in red).

DETECTING DENSITY DEPENDENCE

Let's now explore some consequences of this type of random variation for detecting density dependence. Figure 6.6 shows a computer-generated time series based on $V = 0.5$, $R = 0.5$, and $K = 200$.

Figure 6.7 shows the corresponding ln–ln plot of densities from the time series plotted in Figure 6.6. Clearly this regression line has a slope less than 1. However, several problems exist with this technique for the detection of density dependence. For one thing, a linear regression assumes that the various points are independent. Yet, the x coordinate for one point becomes the y coordinate for the next point in the time series. The result is that accepting the hypothesis that the slope is less than 1 becomes too likely, since the real number of independent points is actually less than the actual number of points. More important, the mean and variance of N_t on the x axis will be nearly identical to the mean and variance of N_{t+1} on the y axis; after all, they represent exactly the same values, except for the first and last points. When this is the case, the slope of the regression line equals the correlation coefficient between the two variables. Now here's the rub: any measurement error in x and y will lead to a correlation coefficient that is less than 1 (the points don't all fall exactly on the regression line). This means that the slope of the regression may be less than 1 even if density dependence is lacking. The more measurement error in population size, the more likely it is that the slope will be less than 1, simply as a statistical artifact unrelated to density dependence.

Figure 6.6
A computer-generated population time series based on Eq. (6.4).

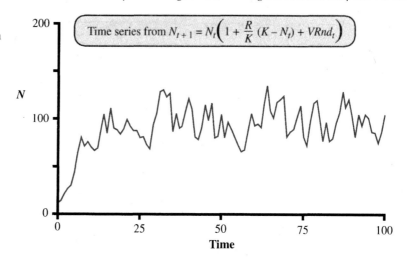

$$\text{Time series from } N_{t+1} = N_t\left(1 + \frac{R}{K}(K - N_t) + VRnd_t\right)$$

Figure 6.7
The time series in Figure 6.6 plotted on a ln–ln scale.

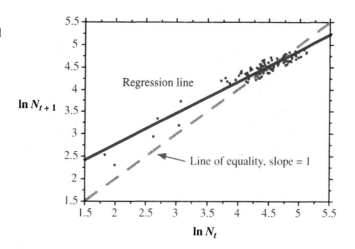

Regression line

Line of equality, slope = 1

Figure 6.8
A chaotic time series based on Eq. (6.4). Now environmental variability is absent since $V = 0$, but R = 2.8 is in the chaotic region.

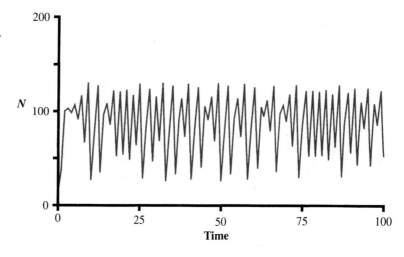

Finally, the expected relationship between N_{t+1} and N_t for most models of density dependence is a nonlinear one. For the discrete logistic equation, recall that the expected relationship between N_{t+1} and N_t is a parabola. Thus the expected relationship between $\ln N_{t+1}$ and $\ln N_t$ is also a humped relationship. Figure 6.8 presents a time series based on the chaotic region of the continuous logistic equation without any noise (Eq. 6.4, for $V = 0$ and $R = 2.8$).

Figure 6.9
The ln–ln plot of the time series shown
in Figure 6.8.

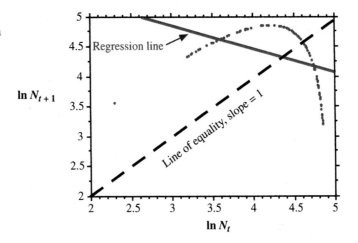

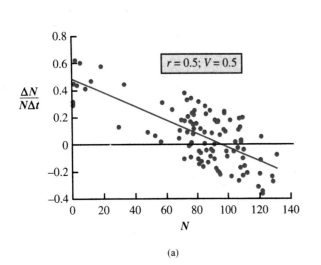

(a)

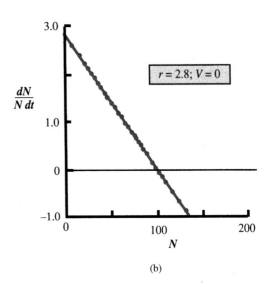

(b)

Figure 6.10
Per capita growth plotted against population size, *N*. (a) Plot of the continuous logistic in the chaotic region with noise
compared to (b) plot of logistic in chaotic region with no noise; all points fall exactly on the line.

The ln–ln density plot is shown in Figure 6.9. Fitting a straight line through these
points, which clearly describe a curve, is highly misleading. One way to avoid this
problem is to consider per capita growth rates instead of population growth rates. For
the continuous logistic equation, the expected relationship between $\Delta N_t/(N_t \, \Delta t)$ and N_t
is a straight line with a negative slope. If growth is density dependent, the slope is
expected to be negative, but the slope should be positive if an Allee effect occurs.

Thus a second test for density dependence is whether $\Delta N_t/N_t \, \Delta t$ decreases with
increases in N_t. The population growth rate and the per capita growth rates plotted as a
function of *N* will have a deterministic relationship for the logistic equation without
noise, even in the chaotic region, but will have much scatter if noise is present. Figure
6.10 compares these two in terms of per capita growth.

Figure 6.11
A time series from the chaotic region with added random noise. The population became extinct at time 65.

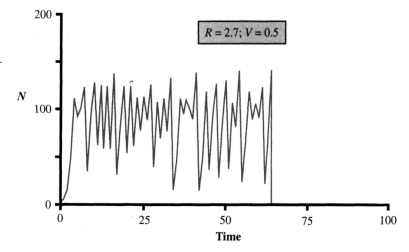

Figure 6.12
Plot of the discrete logistic equation with high *R* in the chaotic region and moderate environmental noise (*V* = 0.5).

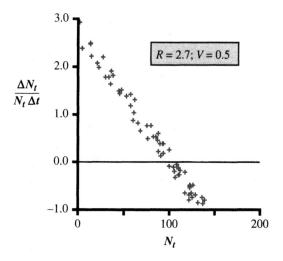

Exercise: Show that a test equivalent to test 2 for small Δt is
Does $\Delta \ln N_t$ decrease with N_t?

Where it becomes confusing is when a population has a large *R* and therefore emits complicated dynamics (e.g., chaos) but also experiences temporal noise (*V* > 0). Figure 6.11 illustrates this case.

Note that this population went extinct at time 65; this is a common feature of noisy populations with high *R*. The time series also shows the sawtooth pattern. If noise alone were driving population change, we would not expect to see this tendency for low populations to rebound and high populations to shrink. The per capita plot is shown in Figure 6.12. The development of new techniques to detect density dependence in population time series continues to be an active area of ecological research (Dennis and Taper 1994, Ives 1995).

Table 6.1 **Discrete Logistic Equation with Noise**

Growth rate, *R*	Noise level, *V*	Carrying capacity, *K*	Number of extinctions out of 200 model populations
0.1	1.5	70	116
0.1	1.5	300	60
0.1	1.5	700	47
R	*V*	*K*	**Extinctions**
0.1	1.0	700	1
0.1	1.5	700	60
0.1	2.0	700	173
R	*V*	*K*	**Extinctions**
0.05	1.0	70	28
0.1	1.0	70	6
0.5	1.0	70	0
1.0	1.0	70	5
1.5	1.0	70	75

THE INTERACTION OF POPULATION REGULATION AND ENVIRONMENTAL NOISE

The degree that population sizes fluctuate in nature is a reflection of the extent of population regulation and the amount of environmental noise affecting the birth and death rates. The more population size fluctuates, the more likely it is that the population will eventually become extinct. Returning to the discrete logistic equation with noise, changes in *R, K,* and *V* each can affect the likelihood of extinction. Table 6.1 is based on simulations wherein each population begins with 50 individuals and then grows for 50 time steps according to Eq. (6.4). For these simulations, the population becomes extinct if it drops to two individuals. The number of extinctions out of 200 initial populations is shown in the last column.

Table 6.1 shows three different comparisons. The top block shows that the likelihood of extinction diminishes with increasing carrying capacity, *K*. The second block shows that increases in the noise factor, *V*, increase the rate of extinction. Finally, in the third block, the effect of varying the growth rate, *R*, is explored. Here the results are more complicated. When *R* is very small, the population cannot recover quickly enough from these very bad years. When the next bad year hits, the population is still small and is now knocked into extinction. However, for very high *R*, the populations tend to overshoot carrying capacity, leading to subsequent population declines. If these density dependent declines happen to correspond to bad years, the population receives a double whack. The lowest extinction rates occur at intermediate levels of *R*. Interestingly, in laboratory cultures of *Tribolium* beetles propagated as discrete generations on limited food, the average persistence time is lower in strains with higher values of *R* (Wade 1980). These populations seem to be in the *R* > 1 portion of the last section of Table 6.1.

Exercise: Another possible way to add temporal variability would be to let *r* be a random variable in the continuous time logistic equation. How would the results of such a model differ from the behavior of the populations shown in Figure 6.5?

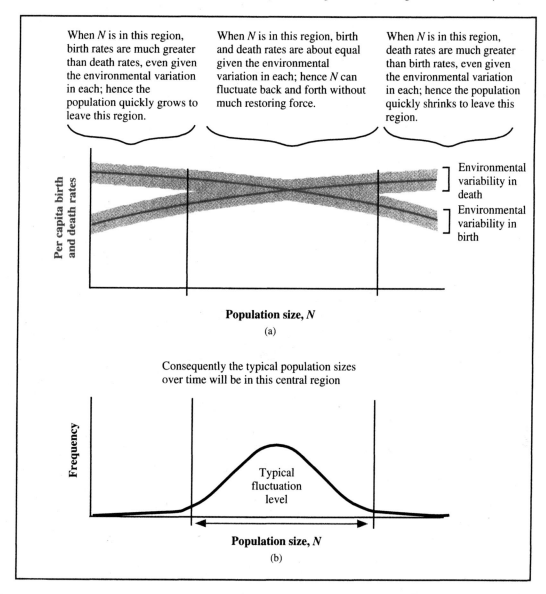

When *N* is in this region, birth rates are much greater than death rates, even given the environmental variation in each; hence the population quickly grows to leave this region.

When *N* is in this region, birth and death rates are about equal given the environmental variation in each; hence *N* can fluctuate back and forth without much restoring force.

When *N* is in this region, death rates are much greater than birth rates, even given the environmental variation in each; hence the population quickly shrinks to leave this region.

Per capita birth and death rates

Environmental variability in death

Environmental variability in birth

Population size, *N*

(a)

Consequently the typical population sizes over time will be in this central region

Frequency

Typical fluctuation level

Population size, *N*

(b)

Figure 6.13
(a) At very low population sizes, the population has a high capacity for positive growth; therefore it quickly grows beyond these low sizes. Similarly, at very high population sizes, the average death rate far exceeds the average birth rate, so the population quickly declines. At intermediate population sizes the noise level in birth and death rates is such that the two rates are about equal. (b) Consequently, the population tends to drift around in this size range.

One problem with the discrete logistic equation is that it adds noise to the population dynamics by adding a random component to the *population's* per capita growth rate each time period, independent of the magnitude of the birth and death rates. It is perhaps more sensible to think of single individuals and how their likelihood of giving birth or dying depends jointly on the population size and on the vagaries of the environment that they experience. A simple way to visualize the interaction of these two effects is depicted in Figure 6.13.

As environmental variation increases or population regulating forces decrease (i.e., the slope of the birth and death rate function becomes less steep), the variance in observed population size will increase (Figure 6.14).

Figure 6.14
The effect of changes in density
regulation and environmental
variability on the typical range
of population size fluctuations
(see Figure 6.13).

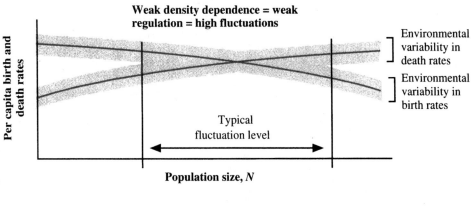

**Weak density dependence = weak
regulation = high fluctuations**

Environmental
variability in
death rates

Environmental
variability in
birth rates

Typical
fluctuation level

Per capita birth and death rates

Population size, *N*

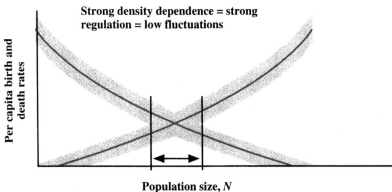

**Strong density dependence = strong
regulation = low fluctuations**

Per capita birth and
death rates

Population size, *N*

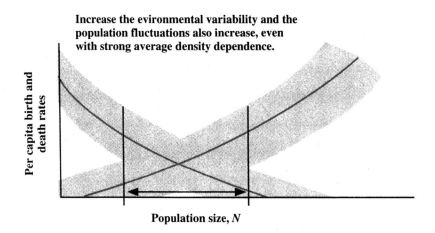

**Increase the evironmental variability and the
population fluctuations also increase, even
with strong average density dependence.**

Per capita birth and
death rates

Population size, *N*

These examples illustrate the joint role that density dependent factors and environmental variation play in determining the distribution and fluctuations in population size over time (Pimm and Redfearn 1988). A mechanical metaphor for this interaction is illustrated in Figure 6.15.

POPULATION REGULATION
AND POPULATION LIMITATION

You should not confuse limiting factors with regulating factors in the determination of population dynamics. They need not be the same. Limiting factors are environmental factors that act to check population size or geographic distribution. The limiting factor may be a nutrient or some essential resource. If you add more of this type of "enhancing" limiting factor, the population will expand. Other limiting factors may be "sup-

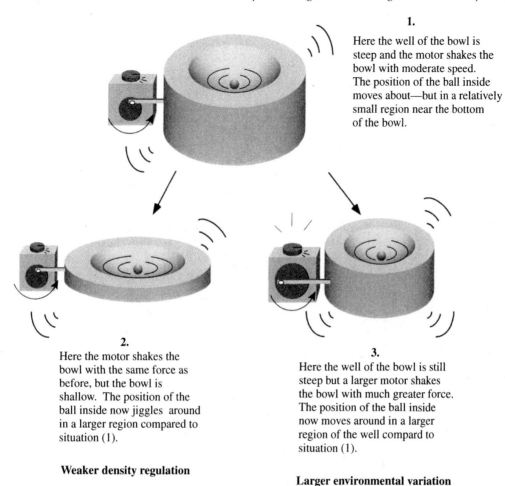

1.
Here the well of the bowl is
steep and the motor shakes the
bowl with moderate speed.
The position of the ball inside
moves about—but in a relatively
small region near the bottom
of the bowl.

2.
Here the motor shakes the
bowl with the same force as
before, but the bowl is
shallow. The position of the
ball inside now jiggles around
in a larger region compared to
situation (1).

Weaker density regulation

3.
Here the well of the bowl is still
steep but a larger motor shakes
the bowl with much greater force.
The position of the ball inside
now moves around in a larger
region of the well compard to
situation (1).

Larger environmental variation

Figure 6.15
A motor is connected to a bowl through a mechanical arm. As the gear of the motor turns, the arm shakes the bowl and a small ball inside the bowl is jiggled around. The fluctuations in the position of the ball in this bowl are analogous to the fluctuations in the size of a population. The degree of fluctuation in the ball's position (i.e., the extent of population fluctuation) is directly proportional to (1) the amount of jiggling of the bowl by the motor (the extent of environmental variation forcing the population growth and (2) the shallowness of the bowl's sides (the lack of density dependence).

pressers." If you *reduce* the levels of limiting parasites or predators, then the population will expand. Thus the acid test for a limiting factor is whether, by artificially adjusting its levels, the population adjusts in size in the appropriate direction.

Regulation refers specifically to density-dependence; a regulating factor is one that causes the per capita birth or survival rate to decrease as the population size increases. Limiting factors may or may not act in a density dependent way. For example, the availability of blood meals for a tick will limit its population growth, but since ticks do not normally deplete the numbers of their prey or the blood availability for other ticks, blood supply will probably not be a regulating factor. Regulation of a consumer population may come about through the consumption of resources (the trophic level below the consumer), interference with other consumers (the same trophic level as the consumer), or through the actions of parasites and predators (trophic levels above the consumer). Figures 6.16 and 6.17 illustrate this distinction, under the assumption that separate sources of mortality are additive.

Figure 6.16
A hypothetical example explaining the difference between population regulating and limiting factors. The two sources of mortality are assumed to be additive.

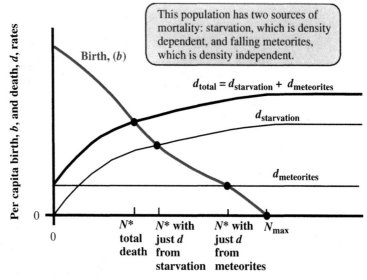

Population size would increase in the absence of meteorites from N^*_{total} to N^*_{starv}. Population size would increase in the absence of food limitation from N^*_{total} to N^*_{meteor}. Population size would increase in the absence of borth mortaility sources to N_{max}.

Conclusion   Both food and meteorite showers are limiting population size. However, only food is a **regulating** factor since increases in population size result in increases in $d_{starvation}$. In this example, the most limiting factor, food, is also the regulating factor:

After removing meteorites, the population change is

After removing the food limitation, the population change is

$$N^*_{starvation} - N^*_{total}.$$ $<$ $$N^*_{meteorites} - N^*_{total}.$$

Figure 6.17
Another example of regulating and limiting factors. Both sources of mortality, d_1 and d_2, are limiting, but only d_1 is regulating in the vicinity of the equilibrium population size, N^*, for total death rates. If the environment changed to eliminate the source of mortality d_1, the population would grow from N^* for total death rates to N^* for d_2 alone. At this new higher density, d_2 would be weakly regulating. The two sources of mortality are assumed to be additive.

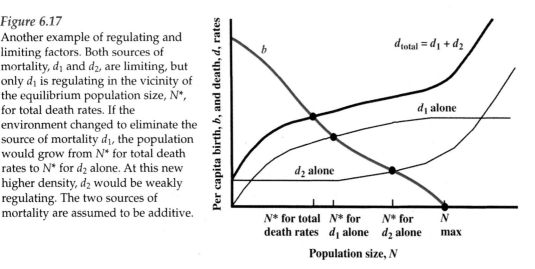

Exercise: A common student lament to this argument goes something like this: I thought that if two probabilities were independent they could be multiplied. Yet, here you are saying that, if different sources of mortality are independent, then they can be *added*. What gives?

Answer: An individual must survive each of n independent mortality factors if it is to live to age x. If it succumbs to any one factor, then it obviously dies before age x. Thus, in continuous time, we may write the survivorship to age x as the product of these separate survival probabilities:

$$l(x) = l_1(x)l_2(x)l_3(x) \cdots l_n(x) \qquad (6.5)$$

where $1, 2 \cdots n$ are n independent sources of mortality. For example, if the probability of dying by starvation before reaching age x is 0.2, the probability of dying from a predator is 0.5, and the probability of dying from any other cause is 0.5, then $l(x)$, the probability of surviving from birth to age x, equals $(1 - 0.2)(1 - 0.5)(1 - 0.5) = 0.2$. This assumes that causes of death are mutually exclusive. The instantaneous mortality rate, $\delta(x)$, at any age x can be deduced from the slope of the $l(x)$ curve at that age x. The greater the deflection of $l(x)$ at age x, the greater is the mortality rate at age x. More precisely and since $\delta(x)$ is a per capita measure, we may write (also see Chapter 4, Box 4.2)

$$\delta(x) = \frac{\partial l(x)}{l(x)\partial x} = \frac{\partial \ln(l(x))}{\partial x} \quad \text{for } \delta(x) < 0.$$

It follows, by taking logarithms and differentiating Eq. (6.5), that the total instantaneous death rate is the sum of the separate death rates from each factor, or

$$\delta(x) = \delta_1(x) + \delta_2(x) + \delta_3(x) + \cdots + \delta_n(x).$$

Conclusion: The additivity of the instantaneous death rates follows from the multiplicativity of the survivorships in Eq. (6.5).

Where things can get confusing, however, is when different sources of mortality are not additive. For example, if a predator takes only sick and postreproductive individuals, which are destined to die soon anyway, then the mortality caused by the predator only substitutes for mortality that would occur anyway. In this situation, the total mortality is less than the sum of the separate components—disease and predators. Such a predator may not be limiting or regulating, as shown in Figure 6.18. Such mortality is called **compensatory mortality.**

At the other extreme, different limiting factors can interact so that the combined effect of both together is greater than the sum of either factor alone. For example, one predator species might take prey exclusively from one type of microhabitat, while another predator might forage exclusively in another type of microhabitat. In the presence of either predator alone, the prey may avoid predators by moving to the alternative microhabitat where they are safe from predators. However, when both predators are present, there is no place to hide; the prey are caught between the proverbial rock and a hard place. Predation rates for both predators then go way up. Such mortality effects are **supra-additive.** Another example is provided by the work of Krebs et al. (1995) who studied the impact of food and predators on snowshoe hare cycles in the Yukon. Various hypotheses have been offered to explain the regular 10 cycles in this species and several others in the boreal forest, including nutrient limitation, predator-driven cycles, and food-driven cycles. Krebs et al. tried to prevent the hares from crashing by several different treatments. They added fertilizer to some plots to enhance plant growth. To other plots, they directly augmented hare food by using commercial rabbit chow. Hares in still other plots were protected from predators by electric fences (to keep out large mammals) and monofilament line draped over the top to exclude raptors. These two last treatments were combined in some replicates to look at the interaction of food addition and predator exclosure. The results demonstrated supra-additivity and are shown in Figure 6.19.

In community ecology, when the effects of different species on the per capita growth rate of a focal species are nonadditive due to either compensatory effects or

Figure 6.18
Nonadditive sources of population size, *N* mortality. This population has two sources of mortality: disease, which is density dependent; and predation, which is density independent. Predators kill only diseased individuals, so the total mortality is equivalent to mortality from disease alone. Population size would increase in the absence of disease from N^*_{total} to $N^*_{predation\ alone}$. Population size would not increase in the absence of predation. Conclusion: Predation here is neither limiting nor regulating.

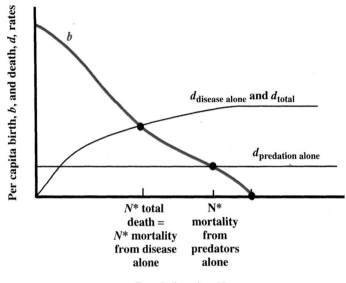

Figure 6.19
Hare survival rates under different experimental treatments. If predator removal and food addition were additive in terms of impacts to hare mortality, the combination of the two should equal the line indicated. The result that survival rate is higher than this line in the predator removal plus food addition treatment indicates a supra-additivity of these two effects.

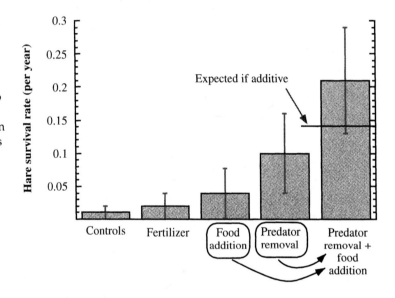

supra-additivity, we call them **higher-order interactions.** In such cases, the dynamic behavior of some focal species cannot be deduced simply as the sum of the separate interaction strengths with other species in a pairwise fashion because the effect of species A on B, might itself depend on the abundance of a third species, C. (More on this in Chapter 15.)

Resources are defined as items essential for the survival and reproduction of individuals and which are consumed or occupied by individuals in such a way that they are depleted; that is, they become unavailable to other individuals. Food, available nest sites, suitable living space, and water are all resources. Temperature and humidity are not resources because they are not "removed" from the habitat by the activities of individuals. These climatic features, however, are often limiting factors for poikilothermic animals and plants. In very seasonal habitats, the number of warm days strongly influences the development and reproductive rates of such organisms and thereby is one of the most important factors determining their population sizes at the end of the growing season. However, even if reproductive rates are sharply affected by temperature, temperature is not "removed" from availability to others because heat is "used" by an indi-

Figure 6.20
In this hypothetical example, birth rates are density dependent with some stochastic variation but death rates are not. Warm climates have higher birth rates than cold climates for all population densities. Consequently, warm climates have higher average population densities than cold climates.

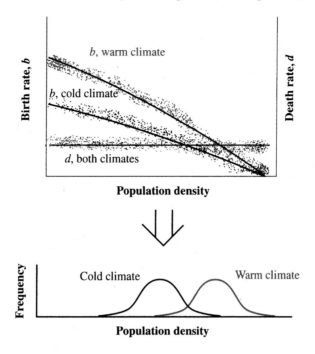

vidual. On the other hand, if the habitat is a mosaic of warm and cold spots created by the topography, (e.g., warm south-facing slopes and cooler north-facing slopes in the northern hemisphere), then warm living spaces might be both limiting and regulating the population size, as illustrated in Figure 6.20. In this case, though, suitable living space as determined by warm temperature is the true regulating factor, not heat per se.

Light can be a limiting factor for some understory plants. In one sense photons of light are "consumed" by plants in the process of photosynthesis. However, the infinite flow of photons prevents them from being consumed in any real sense because they are instantaneously replaced. Light, however, can be physically preempted by the activities of an individual; a tall plant shades a smaller plant growing beneath it. In this way light is locally "removed" from the physical space in a plant's shadow. An overall increase in ambient light levels might allow the shaded smaller plant to now receive enough light to grow, just as an overall increase in ambient temperature might allow organisms to grow on the cool north-facing slopes. This means that light is limiting. However, this does not mean that light per se is a regulating factor.

Distinguishing Cycles from Random Fluctuations

Ecologists would not only like to detect density dependence if it occurs, but they would also like to be able to distinguish regular population cycles from random and periodic ups and downs. Cycles that simply follow the yearly flow of the seasons do not present any great conceptual mystery. The climate is simply forcing numerical change on an annual basis. However, cycles with periods that span several years offer intriguing questions. They suggest that it is not the physical environment alone that is creating the cycle but something about the interaction of the population dynamics of different species, with different generation times that might be producing the periodicity. We have shown that very regular cycles like those of lemmings are quite obviously different from a random walk, but if regular cycles are also subject to environmental noise and estimation errors, it can be a challenge to identify and describe them. An **autocorrelation analysis** results in a **correlogram.** The correlogram is constructed by plotting the correlation between a population's size at time *t* and its size lagged some interval in the future. For each time lag, we again calculate the correlation coefficient, as performed in Figure 6.21.

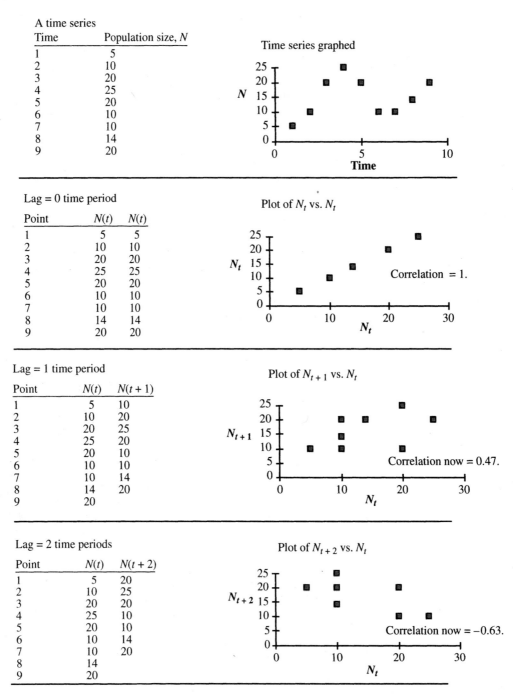

Figure 6.21
An illustration of autocorrelation analysis. Each plot shows the autocorrelation for a different lag time. Each resulting correlation is then plotted as a function of lag time in the next figure.

The various correlation coefficients are next plotted as a function of the time lag, as shown in Figure 6.22, to produce the correlogram. To interpret the correlogram, remember that if a population cycles, say, with a 6-year period, then a strong positive correlation coefficient, *r*, exists for a time lag of 6 years. For example, if the population size were in a trough in 1990, it would also be in a trough in 1996. If it were at a peak in 1990 then it would be at another peak in 1996. Hence the population sizes, with time lag 6, would be positively correlated to one another (this is the "auto" part of autocorrelation). Similarly, this population would show a positive correlation for time lags of 12 years and 18 years and, in fact, any integer multiple of 6 years. On the other hand,

Figure 6.22
Correlation coefficients from
Figure 6.21 plotted as a function of the
lag times to produce the correlogram.

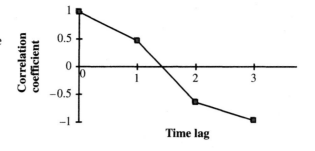

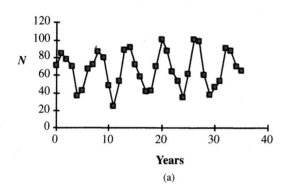

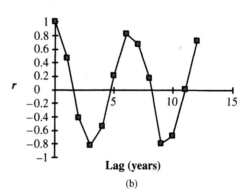

Figure 6.23
(a) A time series for a population that cycles about every 6 years. (b) The corresponding correlogram for lags from 1 to 12 years.

there would be a strong negative correlation coefficient at 3 years. If the population were at a peak in 1990, it would be in a trough in 1993. Similarly there would be another strong negative correlation at 8 years, 14 years, etc. This example of a 6-year cycle is shown in stylized form in Figure 6.23.

As the time lag becomes longer, the peak positive and negative correlations will gradually diminish since more random noise separates very distant peaks than it does adjacent ones. You can see a hint of this in Figure 6.23. As the noise level increases, the cycles in the correlation coefficient, r, damp out more quickly with increasing time lags.

Figure 6.24 shows an example from nature: the number of lynx pelts acquired by the Hudson Bay Company of Canada, for the years 1821 to 1934. Note that the cycles for lynx have a period of about 10 years and that they barely damp out, suggesting that there is some driving force keeping the cycles going. We discuss the possible mechanisms producing these cycles in Chapter 12.

To further understand the correlogram, we can subject the chaotic logistic time series in Figure 6.8 to a correlogram. The result is shown in Figure 6.25. The most pronounced tendency here is for population size to be negatively correlated with itself 1 year later. For other time lags, the autocorrelations are mostly small and tend to alternate up and down successively. The negative correlation for a time lag of one period follows from the sawtooth shape of the time series; it is due to the tendency of this population to overshoot carrying capacity and then to crash one time step later to a level far below carrying capacity, only to overshoot again the next time step.

When the population does not have such a tendency to overshoot, as in the case previously illustrated in Figure 6.6 where $R = 0.5$, then the population dynamics are dominated more by random noise. In this case, the most pronounced tendency is for the population to be positively correlated with itself 1 year later, as depicted in Figure 6.26 (the gray curve). Longer lags do not show much autocorrelation, and they cycle from positive to negative. The autocorrelation for the noise itself (the red curve) shows no significant correlation for any lag size.

Figure 6.24
(a) Time series for Canadian lynx trapped by the Hudson Bay Company in Canada from 1821 to 1934 and (b) The correlogram for Canadian lynx, which plots the correlation coefficient *r* as a function of the lag time.

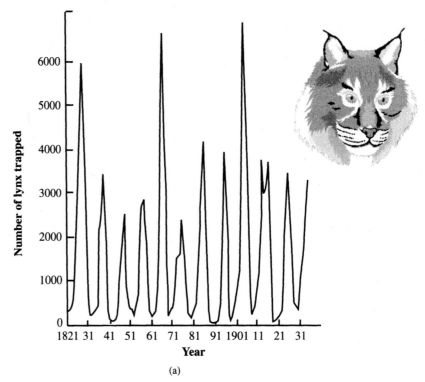

(a)

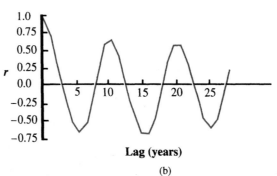

(b)

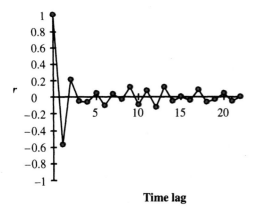

Figure 6.25
The correlogram for the chaotic time series shown in Figure 6.8.

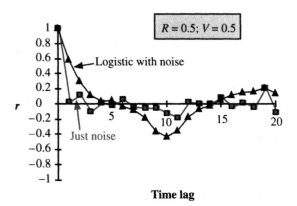

Figure 6.26
Correlograms based on a simulated population growing according to Eq. (6.4) is compared to the time series for only the random number component for this simulation. The noise itself does not have periodicity, but the density dependent growth of the population "filters" this noise into a signal for population dynamics that shows greater autocorrelation for small lags

PROBLEMS

1. Consider the discrete time growth equation

$$X_{t+1} = X_t \left(1 + \frac{\beta}{\omega}(\omega - X_t) + \alpha R_t \right),$$

where X represents population size, t represents time, and R_t represents a random number with a range of -1 to $+1$.

a. What are the variables in this equation and what are the parameters?

b. In words describe the biological meanings of ω, α, and β.

c. Rearrange the equation to express the per capita growth rate of the population as a function of population size.

2. Consider the following diagram, which shows the relationship between birth rates and death caused by two different factors as a function of population density.

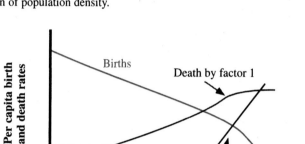

a. Place a tic on the x axis where the population density will have a stable equilibrium (assume continuous growth and that both sources of death act simultaneously and are independent and additive).

b. At this equilibrium, is factor 1 (i) a limiting factor, (ii) a regulating factor, or (iii) both?

c. At this equilibrium, is factor 2 (i) a limiting factor, (ii) a regulating factor, or (iii) both?

d. If you could eliminate one of the mortality factors, which one's elimination would produce the largest increase in population density?

3. Consider the following statement: In density regulated populations subject to environmental variability, increases in the maximal per capita growth rate will necessarily decrease average extinction times. Do you agree or disagree? Why?

7

Life History Trade-Offs

A giraffe's long neck is an adaptation, allowing it to successfully feed on foliage high in trees. The salt-concentrating kidneys of a kangaroo rat are an adaptation to a water-poor environment. The timing and extent of reproduction spread throughout an organism's lifetime is also adaptively attuned to its environment. One obvious life history feature with enormous variation across species is the number of offspring produced per breeding episode. Some species of insects and fish produce thousands and millions of young per year per adult, while other organisms like elephants or humans produce only a handful of young over their entire lifetimes. What environmental features might make natural selection go in such seemingly opposite directions in terms of reproductive effort?

LIFE HISTORY STRATEGIES

The production of offspring represents a gene's link to the future. One way to think of offspring is as lottery tickets—not all will be winning tickets, but the more tickets you buy, the greater are your chances of winning. All else being equal, this creates an advantage for genotypes associated with high birth rates. It creates a situation where natural selection might favor a more "wasteful" life history than seems necessary. If all your neighbors are buying lottery tickets, unless you buy *more* tickets, your *relative* chance of winning goes down. This creates a race to buy ever more tickets than your competitors. But, as applied to life history evolution, this is an oversimplification to make a point. Not all lottery tickets are equal in nature, and this gives natural selection room to operate in some alternative directions. Moreover, genes have another link to the future, at least for a while: the survival of individuals bearing those genes. If there is a trade-off among genotypes in the magnitude of their birth rates and death rates (i.e., between the number of offspring that can be produced and the individual body size or quality of these offspring), then the highest birth rate will not necessarily produce the highest fitness. Bigger, higher quality offspring might have a competitive advantage over smaller, lower quality offspring. Also a higher reproductive effort may come at the expense of that individual's ability to survive to reproduce again. Under what circumstances is the "salmon" or "mosquito" strategy of many small young favored and when is the "primate" or "elephant" strategy of a few well-cared-for young favored?

We explore this question in various steps. To set the stage, first look at Box 7.1, which shows a series of alterations in life history schedules and the value of the intrinsic growth rate, r, for each. Note how delaying the age of reproductive maturity has a huge effect on decreasing r (part a), regardless of the survivorship schedule. Even if the total number of births over the lifetime increases to compensate for this delay, r will generally be lower than a life history strategy with an early birth rate even if the total lifetime births is much lower (part b). Part (c) presents a situation where births occur only at a single age and the number of births increases geometrically as the age of reproductive maturity is delayed. Only for this case can delayed reproductive maturity

157

Box 7.1 *The Effect of Different Life History Schedules on the Intrinsic Growth Rate, r.*

A triangular birth schedule

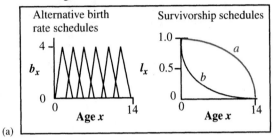

(a)

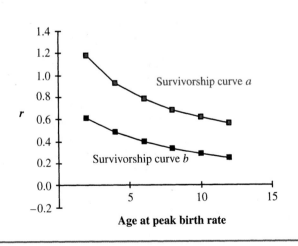

A different triangular birth schedule

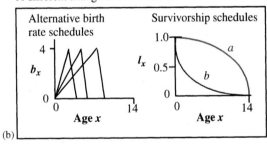

(b)

A single birth pulse at only one age, x; b_x increases with the square of age x.

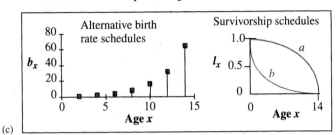

(c)

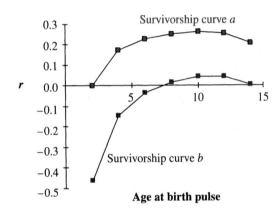

Two *Drosophila serrata* populations have approximately the same r of 0.16 despite their very different $l_x b_x$ schedules (from Lewontin 1965).

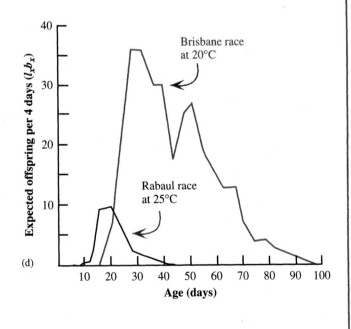

(d)

be compensated for by increasing the birth rate, and then only up to a point (about age 10). Finally, part (d) is an example of two populations of the Australian fruit fly *Drosophila serrata* reared in the laboratory under two different temperatures. Despite a roughly tenfold higher R_0 for the Brisbane race compared to Rabaul flies, the r's of these two life histories are both about 0.16 (Lewontin 1965).

OPTIMAL CLUTCH SIZE

We begin by exploring a model to determine the optimum clutch size—the number of eggs a bird lays in its nest. The arguments used, however, will extend to any situation where there is a trade-off between offspring number and offspring survival. In animals with parental care it is natural to expect a trade-off between the number of young that parents produce and the survival of the young under the parents' care. Often clutch sizes are greater in years of more abundant food, suggesting that energy or nutrient availability sometimes limits the number of young that are produced. Is this because females simply cannot find enough energy in lean times to form eggs? Energy alone does not seem to limit egg production, since if eggs are experimentally removed from nests, the mother generally has no trouble replacing them by laying new eggs. Further, energy budget calculations show that birds with posthatching care of young, expend much more energy feeding and rearing young to independence then they do in simply producing the eggs (King 1973, Case 1978). Thus, if energetic constraints influence clutch size, the rearing stages, not egg formation, should be the bottleneck. With more mouths to feed in a nest, the risk of starvation for all the young increases if the parents can't keep up with the feeding demands. More young means more foraging trips for the parents to find food and bring it back to the nest. This means that the nest is potentially left unattended and exposed to predators. This risk forces the father and mother to take turns on the nest or to subdivide the nesting and food gathering labors. However, even with one parent always guarding the nest, each time the other parent returns to the nest with food, there is the danger that it will be seen by observant predators. This is why birds, in fear of revealing their nest location, often move indirectly to their nest through thick brush or canopy, trying to dodge any following eyes.

We thus expect lower juvenile survival with more young in the nest and more mouths to feed, requiring more and longer feeding trips by the parents. Let's explore this trade-off as it affects optimum clutch size. First, we define **breeding success** as

$$\begin{bmatrix} \text{Breeding success} \\ \text{per year (surviving} \\ \text{young reared per} \\ \text{year)} \end{bmatrix} = \begin{bmatrix} \text{number of breeding} \\ \text{attempts (nests per} \\ \text{year)} \end{bmatrix} \begin{bmatrix} \text{average number} \\ \text{of eggs per nest} \\ \text{(young/nest)} \end{bmatrix} \begin{bmatrix} \text{survival of young} \\ \text{(probability of} \\ \text{each young} \\ \text{surviving)} \end{bmatrix}.$$

The last term, young survival, is often a decreasing function of clutch size for the reasons already given. An example based on a study of European starlings (Lack 1948) is shown in Figure 7.1. The number of breeding attempts per year is determined by

Figure 7.1
Survival rate (per individual) of young European starlings declines in larger clutch sizes. This is a study of birds that were banded in the nest. *Relative survival* is the number of birds that were recovered, divided by the number banded. This provides only a relative measure of chick survival since many more banded birds actually survived than were recovered. Over 3500 nests were involved in this study (from Lack 1948).

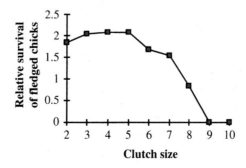

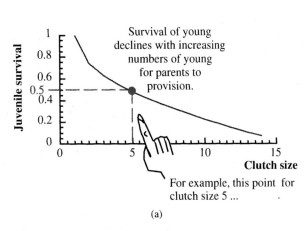

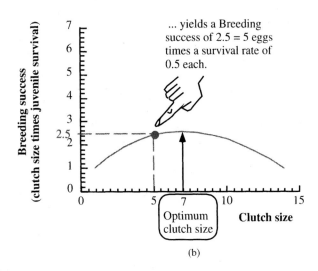

Figure 7.2
Lack's model for optimal clutch size. (a) Survival rate of young in the nest as a function of clutch size. (b) Conversion of this relationship to one for breeding success versus clutch size. Note that the declining curve in (a) yields a humped-shaped relationship between breeding success and clutch size. For this model, the optimum clutch size is that which produces the maximum breeding success.

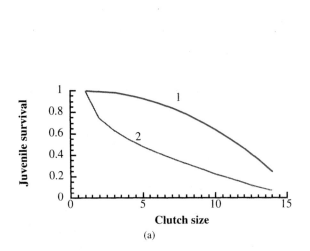

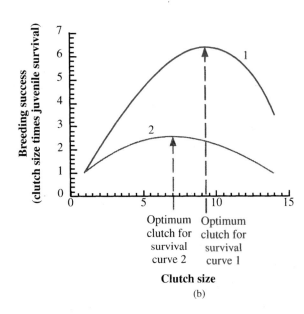

Figure 7.3
(a) The two different survival curves produce (b) two different breeding success curves. Conclusion: All else being equal, lower juvenile survival leads to a lower optimum clutch size.

environmental constraints interacting with the particular life history of the species; some birds have slower growth rates than others and cannot fit in more than one breeding season a year at high latitudes with short breeding seasons. For now, we assume that for a particular population this is simply a constant, 1. Then we can take the survival rate versus clutch size relationship and translate it into a breeding success versus clutch size relationship by multiplying, as in Figure 7.2. For this example, the optimum clutch size is seven eggs.

We can explore the effect of different environmental conditions on optimum clutch size by varying the survival versus clutch size curve. Figure 7.3 shows how breeding success is affected by the strength of the trade-off between clutch size and juvenile sur-

Figure 7.4
Clutch size as a function of latitude for two families of birds: (a) finches, *Emberiza,* and (b) flycatchers, *Tyrannidae* (from Cody 1971).

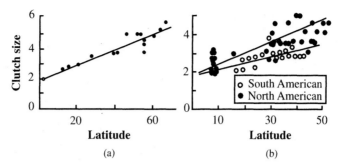

vival. Based on this trade-off, an intermediate clutch size often is the most productive in terms of breeding success. Also the steeper the trade-off, the larger is the optimum clutch size.

David Lack (1947) argued that natural selection should favor genotypes that have the highest breeding success. To make this argument one about evolution, we must first assume that the phenotypic expression of different clutch sizes within a species has a genetic component such that differences in clutch size are heritable. Second, we must assume that the trade-off between clutch size and juvenile survivorship is manifest between genotypes and is not simply due to phenotypic or environmental variability alone. Evidence that clutch size involves genetic factors comes from several sources. Generally, birds that lay greater than the mean clutch size one year, also tend to be higher egg layers in subsequent years (Boyce and Perrins 1987). More telling, artificial selection with chickens has resulted in dramatic increases in egg laying rates (Romanoff and Romanoff 1949). Genetic studies have frequently documented heritiabilities in the range of 20–40% for wild bird populations (van Noordwijk et al. 1981, Mousseau and Roff 1987). That is to say, 20–40% of the variance in clutch size between individuals in a population can be attributed to additive genetic factors. Thus there seems to be a heritable component to clutch size in many birds.

With this understanding of Lack's model, we're ready to examine some interesting empirical patterns in clutch size (Cody 1966, 1971; Lack 1968).

1. The observed mean clutch size of a population is often that which yields the greatest productivity of young, as predicted by this simple model. This optimum clutch size is well below the physiological maximum since, if eggs are experimentally removed, the mothers usually can continue to lay additional eggs to replace those taken by the investigator. Price (1998) summarized 53 experiments of another type: egg additions. He found that in 28 of the 53, nestling survival was lower in experimentally enlarged broods and in no study was it better—so the basic premise of Lack's model seems to hold.

   Does the most frequent clutch correspond to the most productive? Yodzis (1989) synthesized several field studies. He found 10 bird species for which the most productive clutch and the mean clutch were approximately equal. However, he also found that in 11 species the mean clutch was substantially lower than the predicted optimum. In only one species was the mean clutch greater than the clutch with the highest breeding success. In general then, **many birds seem to be laying less eggs than the clutch with the highest breeding success.** Why? (We pursue this further in a moment.)

2. Several bird families show enormous variation in typical clutch size. Among closely related species, lower latitude species generally have smaller clutches, as illustrated in Figure 7.4.

   In some species with wide geographic ranges, like the European robin, this same trend is visible intraspecifically as depicted in Figure 7.5. This trend also makes sense in terms of Lack's model in tropical communities, which have a large variety of nest and egg predators. Demographic studies of birds generally show higher nest predation rates in the tropics (Ricklefs 1969, Cody 1971), as indicated

Figure 7.5
Average clutch size of the robin (*Erithacus rubecula*) in various countries (from Lack 1954).

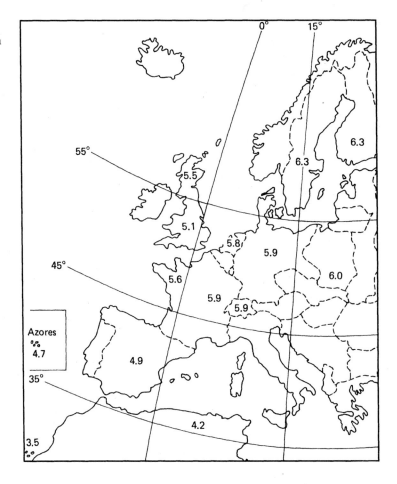

Table 7.1. **The percent survival of eggs in nests of altricial passerine species according to nest type and latitude. Sample sizes are in parentheses.**

Nest type	Temperate	Humid tropics
Open or domed	46.1 (17)	30.9 (9)
In holes	76.4 (7)	43.6 (16)

in Table 7.1. According to Lack's model, this will favor the evolution of smaller clutches, as shown in Figure 7.6.

3. Hole nesting birds (like woodpeckers, bluebirds, starlings, and several parrots) have larger clutches than open nesting birds. Again this pattern is consistent with Lack's model, since the young of hole nesting species generally suffer less predation than birds with more exposed nests. According to Lack's model (Figure 7.6) this would favor larger clutches. Also with this added protection, hole nesters prolong their nesting period, as indicated in Table 7.2.

4. Island species often have lower clutches than closely related species (or the same species) on the mainland. This last pattern holds mostly for temperate zone island–mainland comparisons (Cody 1971). Islands generally lack many predators—consequently, based on the within-year breeding success model, we would expect the opposite association, namely, higher clutches associated with lower juvenile mortality expected on the islands (see Figure 7.3). On the surface this trend seems to be a contradiction to Lack's model based on maximal breeding success.

In short, Lack's breeding success model, while partially successful, is not perfect. Why? One reason is that it leaves out some additional trade-offs influencing optimal reproductive effort. The most important factor, developed in detail in the next two sections of this chapter, is an additional trade-off between present reproductive expenditure, of which clutch size is one component, and **adult** survival rate. Unless they are annuals, adults can potentially breed more than once. The overall fitness of a genotype depends on its lifetime reproductive success, not simply that in a single year. If a large reproductive investment in one reproductive bout leads to poor survival for the parents, then genotypes with that adult strategy might not be as fit as those that hold back some reproductive effort to gain higher survival, increasing their opportunity for future reproduction. In a moment, we will develop a model that incorporates this additional trade-

Figure 7.6
Overall lower survival rates, as might be the case in the tropics, lead to lower predicted optimum clutch sizes according to Lack's model. In this example, we assume that survival rate declines with clutch size at the same rate in both situations. A lower clutch size will also be favored if the survival decline is steeper in the tropics which might be the case if more young in the nest attract more predators. In that case, one additional egg in the tropics attracts a greater increment of predators than an additional egg in the temperate zone, leading to a steeper survival decline.

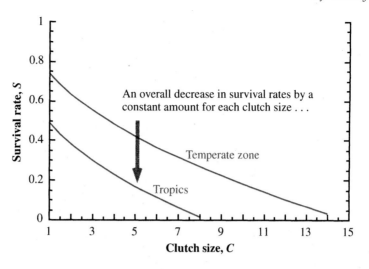

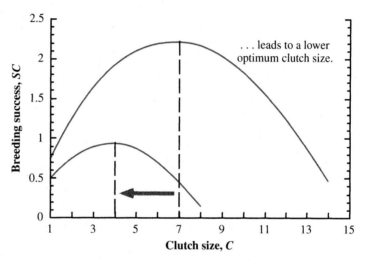

Table 7.2. **Nesting data for some European passerine birds (from Lack 1948).**

Nest location	Nest predation	Incubation and nesting period (days)	Clutch size	Number of species
In holes	Low	31.1	6.9	18
Roofed nests	↓	29.9	6.8	7
In a niche		28.4	5.5	5
Open	High	26.3	5.1	54

off. It can partially explain why birds should hedge their bets, producing *less* eggs than predicted based solely on the trade-off between clutch size and juvenile mortality. Moreover, the island–mainland differences in clutch size will become more understandable when we incorporate this additional life history trade-off.

WHY BE ITEROPAROUS?

Iteroparous life histories are those with repeated bouts of reproduction throughout life. Adults survive to reproduce again. The alternative is **semelparity** where adults die after their first reproductive experience. Annuals are organisms with a semelparous life history of a 1-year period. Biennials have a semelparous life history with a 2-year period.

In other species, like certain salmon, agave, and bamboo, the period may be several years or even decades. What environments favor one or the other of these alternatives? To answer this question we must evaluate the relative fitness of different life history strategies. For exponentially growing populations, the growth rate of a particular genotype, g, is the intrinsic rate of increase, r_g (or, equivalently, λ_g for discrete growth), of a homogeneous population of that genotype. For age-structured populations, the r_g of a particular genotype is also a measure of its Darwinian fitness, as long as selection is not too strong, mating is random, and frequency dependent interactions do not determine fitness (Charlesworth 1994). This means it will often (but not always) be possible to evaluate natural selection as it affects the evolution of life histories, in terms of the Malthusian parameter, r_g, or in discrete form, λ_g. **Those genotypes whose life histories have the highest r will increase in frequency relative to those with a lower r.** As we indicated in Chapter 4, the net reproductive rate, R_0, does not necessarily go hand in hand with r. Because R_0 is expressed on a per generation basis, R_0 could be high for a particular life history but the associated generation time might be high too, leading to a low r. The Malthusian parameter, r (or in discrete time, λ) has units of per year (or some other fixed time period). Nor does R_0 entirely capture the fact that early reproduction is much more important than late reproduction in contributing to population growth (see Box 7.1). Early reproduction is more important because, when young are produced earlier, they can get a head start at reproducing themselves and likewise with their young, and so on.

Cole's Paradox

Lamont Cole (1954) observed that an annual will have equivalent fitness (in terms of r) as a perennial that lives forever (i.e., no mortality), if the annual simply produces one more young per year than the perennial. To Cole it seemed paradoxical that perennials should even exist, since improvements in adult survival seemingly could be so easily compensated for by improvements in the early birth rate.

After we develop Cole's result, we'll explain how this paradox can be resolved. Let's return to the simple life history explored in Chapter 3 where $B =$ births b_x is the same for all adult ages x, $s_0 =$ survival of newborns to age 1, and $s_a =$ adult survival, which is constant for all ages 1 and beyond. Young reach sexual maturity at year 1 (i.e., in their second year of life). The total number of adults in year t is N_t and is given by the difference equation

$$N_{t+1} = (s_0 B + s_a)N_t$$

$$= \lambda N_t$$

$$\lambda = s_0 B + s_a = F + s_a.$$

For a perennial with no adult mortality ($s_a = 1$) with, for example $s_0 B = F = 10$/year, $\lambda = 10 + 1 = 11$/year. A population comprising only this life history genotype would increase elevenfold each year. But this is the same λ that would be reached for an annual life history if $F = 11$ and adults didn't survive at all, $s_a = 0$, or

$$\lambda = 11 + 0 = 11.$$

In terms of the ultimate rate of population growth, λ, this means that the following two Leslie matrices have the same λ:

$$\mathbf{L}_A = [11] \quad \text{and} \quad \mathbf{L}_P = \begin{bmatrix} 10 & 10 & 10 & \dots & 10 \\ 1 & 0 & 0 & \dots & 0 \\ 0 & 1 & 0 & \dots & 0 \\ 0 & 0 & 1 & \dots & 0 \\ \cdot & \cdot & \cdot & \cdot & \cdot \end{bmatrix},$$

where the dots at the bottom of the second matrix imply that it extends to infinity. This is Cole's paradox. Is his conclusion modified if perennials can't live forever but have some finite probability of dying?

Suppose that $F = 10$ and $s_a = 0.2$; then $\lambda = 10.2$. This is the same λ that would result from an annual life history where $F = 10.2$ and $s_a = 0$. So Cole's paradox persists.

Why then should perennials even exist in nature if a little extra birth rate can compensate so easily for low adult survivorship and the loss of opportunities to reproduce in the future?

Cole left out a factor in his argument. Again let's consider the same simple life history. For annuals (subscript the birth and λ terms with A for annual): $\lambda_A = s_0 B_A + s_a = s_0 B_A$, since adults have zero survivorship. For perennials (subscripted with capital P): $\lambda_P = s_0 B_P + s_a$. If the annual life history is to have an equivalent or greater fitness than the perennial life history, then $\lambda_A \geq \lambda_P$, or

$$s_0 B_A \geq s_0 B_P + s_a.$$

Rearranging this last expression to solve for the value of B_P such that λ_A will just equal or exceed λ_P, we find that

$$B_A \geq B_P + \frac{s_a}{s_0}.$$

Upon further rearrangement,

$$B_A - B_P \geq \frac{s_a}{s_0}. \tag{7.1}$$

Equation (7.1) says that the reproductive advantage of annuals (on the left) must exceed the ratio of adult to juvenile survival (on the right) if the annual strategy is to be favored. When this survival ratio is high, the adoption of an annual life history strategy may demand relatively large increases in birth rate relative to the perennial birth rates to yield equivalent fitness. Cole's paradox arose from the hidden and not very general assumption that $s_a/s_0 = 1$ (Charnov and Schaffer 1973). For example, if $s_a = 0.8$ and $s_0 = 0.001$, then $s_a/s_0 = 800$; the annual would have to produce 800 more young than the perennial to have equivalent fitness.

With this understanding of the circumstances that favor an iteroparous life history, we can determine how the trade-off between reproductive effort (as it affects fecundity, F), and adult survival rates will influence the optimal reproductive effort at each age.

OPTIMIZATION OF REPRODUCTIVE EFFORT

In the previous section we showed that the relative degree of mortality imposed on different ages by the environment can influence the relative advantage of different reproductive strategies. We now explore this notion in more detail by developing a model that predicts the optimal reproductive effort based on the mortality schedule imposed by the environment and the trade-off between reproduction and adult survival (Gadgil and Bossert 1970).

Reproductive effort is defined as the proportion of assimilated energy devoted to reproduction in a breeding season. An organism's energy gained can be partitioned into various categories of energy expenditure, as illustrated in Figure 7.7. All else being equal, the greater the proportion of energy that goes into reproduction, the less energy is left for storage, maintenance, growth, and defense. In some organisms energy may not be as important as some other limited nutrient (like nitrogen or water) in determining the amount of reproduction that occurs. Whatever this limiting nutrient is, though, its allocation, like that of energy, will create a trade-off between reproductive output and survival.

In Lack's model for the adaptive significance of clutch size, we found that the trade-off between offspring number and offspring quality, as it affects juvenile survival, sometimes results in an intermediate clutch size being optimal (most productive). Another trade-off that organisms face is that between reproductive effort and *adult* survival. Consider Figure 7.8 which plots fecundity as a function of reproductive effort, θ, the proportion of assimilated energy devoted to reproduction per time period.

Figure 7.7
The partitioning of ingested energy.

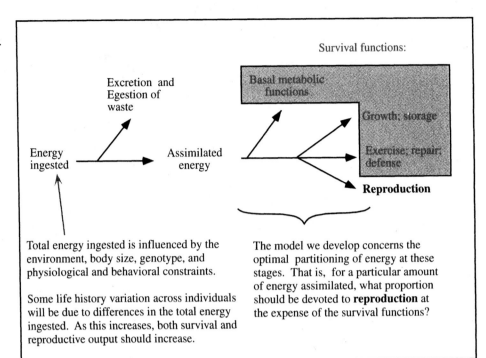

Survival functions:

Basal metabolic functions

Growth; storage

Exercise; repair; defense

Reproduction

Excretion and Egestion of waste

Energy ingested → Assimilated energy

Total energy ingested is influenced by the environment, body size, genotype, and physiological and behavioral constraints.

Some life history variation across individuals will be due to differences in the total energy ingested. As this increases, both survival and reproductive output should increase.

The model we develop concerns the optimal partitioning of energy at these stages. That is, for a particular amount of energy assimilated, what proportion should be devoted to **reproduction** at the expense of the survival functions?

Figure 7.8
Fecundity increases with reproductive effort but the exact shape of the relationship will depend upon several ecological factors.

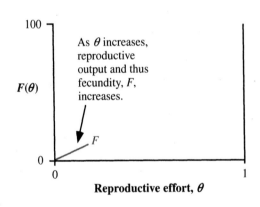

As θ increases, reproductive output and thus fecundity, F, increases.

$F(\theta)$

100

0

0   1

Reproductive effort, θ

F

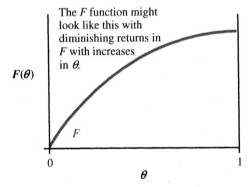

The F function might look like this with diminishing returns in F with increases in θ.

$F(\theta)$

F

0   1

θ

The F function includes some newborn survival. s_0 might decrease with the number of young produced because the parents cannot keep up with demands, because local predators locate larger clutches more readily, or (in plants) because pollinators become saturated.

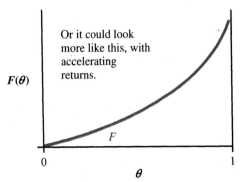

Or it could look more like this, with accelerating returns.

$F(\theta)$

F

0   1

θ

s_0 could increase with the number of young produced in other circumstances, e.g., due to thermoregulatory advantages, or because local predators become swamped or (in plants) because fruit production might increase exponentially with the size of floral displays.

Whatever the details, we expect F to generally increase with increases in reproductive effort.

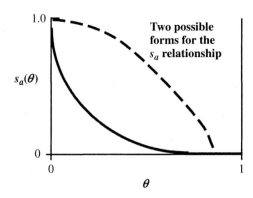

Figure 7.9
Adult survival rate is expected to decline with increasing reproductive effort.

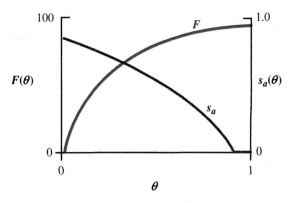

Figure 7.10
A possible relationship between net fecundity, F, and adult survival, s_a, with reproductive effort, θ. Note how these axes are scaled. Why isn't s_a necessarily equal to 1.0 when θ is 0? Why can't s_a be greater than 1? Why can F be greater than 1? Why is s_a exactly 0 when θ is 1? Why can't θ be greater than 1?

Figure 7.11
In addition to θ, the environment and other external or even internal constraints (like body size) also influence realized values of F and s_a.

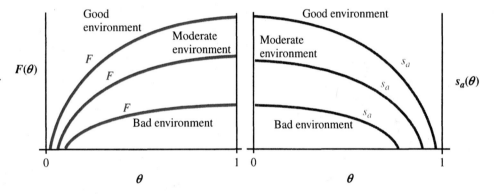

Similarly adult survival, s_a should decline with θ, although the functional form of this relationship may be unknown, as depicted in Figure 7.9.

We next combine the F and s_a functions on the same graph. One possible configuration is shown in Figure 7.10.

Suppose that different values of θ represent different genotypes in a population. The genotypes are ordered on the x axis from those with smallest θ on the left to largest θ on the right. Genotypes with high reproductive effort, θ, have high reproductive effort and high fecundity, F, but low adult survivorship, s_a. Genotypes with small θ have a low reproductive effort but high adult survival. The question is: What is the optimal level of θ?

Before we answer this question, let's reconsider two assumptions of the model. One assumption is that energy acquisition rates are constant across gentotypes and that only the allocation of this energy to various fitness-related components varies. Thus with this model we are asking: What is the optimum θ *for a given environment*? That is, the environment is fixed, and the optimal genotype for that environment is then found. Recalling Figure 7.7, we know reproductive output may be increased in two ways: increase the total energy assimilated or increase the proportion of assimilated energy devoted to reproduction. Figure 7.11 provides a graphical comparison. We know, however, that both acquisition and allocation may both vary (Price 1998).

For example, Galápagos ground finches respond to unusually high levels of rainfall by producing more clutches of eggs per year, as illustrated in Figure 7.12, as well as breeding in their first year of life. If both factors are simultaneously varied, the expected trade-off between reproductive output and survival may not be apparent. Consider an analogy. The income of a household is fixed. Money spent on a car leaves less available to make payments for a bigger home. Based on this trade-off alone, we might

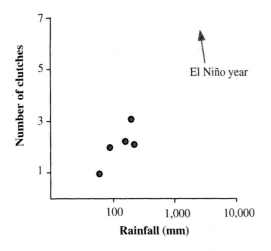

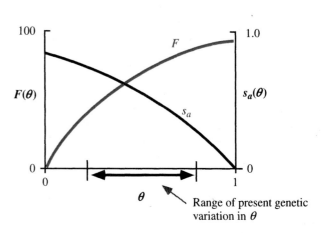

Figure 7.12
The mean number of clutches per pair of Galápagos finches (*G. conirostris*) increases with yearly rainfall (after Grant and Grant 1987).

Figure 7.13
Genetic variability in reproductive effort may be constrained.

Figure 7.14
Different shapes for the $F(\theta)$ and $s(\theta)$ curves lead to different optimal reproductive effort. Convex curves can produce an optimal reproductive effort, θ, which is intermediate—not maximal at 1 or minimal at 0, but something in-between. This is impossible for concave curves.

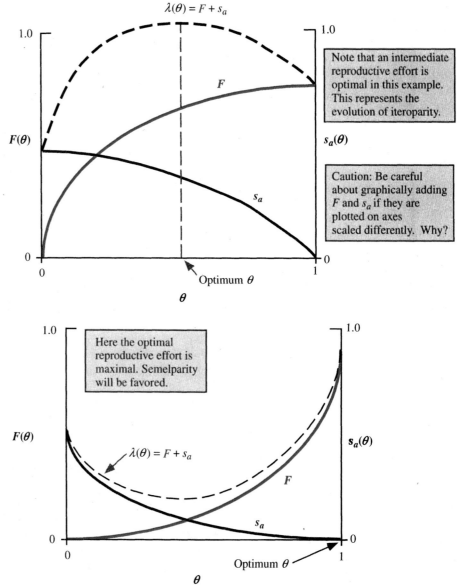

Note that an intermediate reproductive effort is optimal in this example. This represents the evolution of iteroparity.

Caution: Be careful about graphically adding F and s_a if they are plotted on axes scaled differently. Why?

Here the optimal reproductive effort is maximal. Semelparity will be favored.

expect to see cheap inexpensive cars in neighborhoods of expensive homes. Personal experience tells us that this is not the case because, although individual household incomes are fixed, these incomes vary greatly across neighborhoods. Rich families can buy bigger cars *and* bigger homes.

A second complication is that in real populations the genetic variation for all levels of θ may not be present; the optimal θ may be impossible for natural selection to achieve. In particular, the extreme low and high θ might be absent, as shown in Figure 7.13. This could represent a problem for evolution because natural selection might be stopped at suboptimal end-places without genetic variation to act on. We take on faith that, given enough time, mutational input will ultimately provide the genetic variation to allow the population to reach the predicted optimum.

Now we're ready to find the optimum θ. This will be the θ genotype with the highest fitness, λ. In Chapter 4 we found that, for most life histories, the calculation of λ can be a daunting task, but for the one outlined here it is especially easy. For this special case $\lambda(\theta) = F(\theta) + s_a(\theta)$. So to evaluate fitness, we simply add the $F(\theta)$ curve to the $s_a(\theta)$ curve to get the $\lambda(\theta)$ curve as a function of θ. Then we identify the θ that corresponds to the highest λ. To illustrate, Figure 7.14 presents an example where the F and s_a axes are scaled identically so that visually adding them will be easy. Note that, for an intermediate θ to be optimal rather than an extreme θ (i.e., 0 or 1), convex F and s_a functions are required.

Once we know how to find the optimal reproductive effort, we can determine the likely effect of environmental changes in different vital rates on the optimum life history. Figure 7.15(a) shows how the optimal θ is changed by uniformly halving the juvenile survival rate s_0 for all θ. On the other hand, decreasing adult survival rates leads to an increase in the optimal reproductive effort, as shown in Figure 7.15(b).

Exercise: Repeat the analysis in Figure 7.15(a) and (b), but use concave curves instead of convex curves for F and s_a.

Thus changes in either juvenile or adult survival rates change the optimum reproductive effort but in opposite directions. The varying direction of these effects makes intuitive sense. For the moment put yourself in the position of a bird trying to make the right decision about how many young to produce. If the young have a low probability of survival, but as an adult your own survival rate is high, then why expend a lot of energy to produce young? Most will die anyway, and your own chance of living to reproduce another day will be diminished for little potential gain. On the other hand, if chances were high that you were going to die tomorrow, why hold back any reproductive effort today? In this way, the ratio of s_0/s_a generally predicts the optimal reproductive effort. The greater this ratio, the greater is the optimum reproductive effort (Box 7.2 and Figure 7.16). Recall that iteroparity was expected to evolve when the ratio s_a/s_0 was high. This is consistent with our present result. The evolution of high reproductive effort and semelparity is favored, all else being equal, when the ratio of s_0/s_a is high, but the model also suggests another requirement for iteroparity: convex curves. Semelparity clearly can evolve with concave curves (as in Figure 7.14b), but it also can evolve for convex curves, as shown in Figure 7.17.

Now let's return to the problem of optimal clutch size. Simply looking at breeding success makes it hard to explain why so many bird species had a mean clutch size lower than the most productive size. **This lower clutch size is consistent with an additional trade-off between reproductive effort and adult survival.** This provides parents with an additional fitness penalty compared to the Lack model, which simply considered survival of offspring. Consequently, the number of eggs must be reduced compared to situations where only juvenile survival is affected by increasing clutch size. In 5 of 14 bird studies, where it was looked for, parental survival rate declined with artificially increased clutch size. In 8 of 14 cases, increased clutch size led to a lowering of future reproductive performance of the parents (Stearns 1992).

Figure 7.15
(a) Here are two birth functions (red): (1) has twice the magnitude of (2) because of twice the juvenile survival for all θ. A single adult survival curve, s_a (gray) applies to both. The corresponding fitness curves, $\lambda = F + s_a$, are superimposed (black dashed). (b) Here are two adult survival rates in gray: (1) has twice the magnitude of (2) for all θ. A single net fecundity curve, $F(\theta)$, applies to both (red). The corresponding fitness curves, $\lambda = F + s_a$, are superimposed (black dashed).

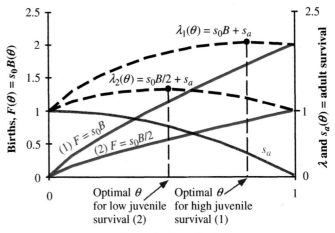

(a)

> **Conclusion: All else being equal, <u>decreasing</u> juvenile survival favors a <u>lower</u> optimal reproductive effort.**

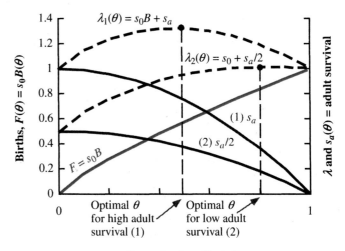

(b)

> **Conclusion: All else being equal, <u>decreasing</u> adult survival favors a <u>higher</u> optimal reproductive effort.**

The other exception to Lack's model that we encountered earlier was a pattern for lower clutch sizes in many island populations compared to their mainland counterparts. This is a pattern that is also seen in the litter sizes of insular mammals, and it has been shown to have a genetic basis in several birds and some shrews on Corsica (Fons et al. 1997). Islands lack many predators, so juvenile survival is often higher on islands than in comparable habitats on the mainland (George 1987, Flack 1976, Higuchi 1976). But this lack of predators also means that adult survival rates will be higher as well on islands (e.g., Atwood et al. 1990). Moreover, with less predators and competing species, densities of the species that are present tend to be higher than on the more species-rich mainland (e.g., MacArthur et al. 1972, Cox and Ricklefs 1977, George 1987). This means that islands generally will have higher densities of adults and that each adult will live longer because of the absence of predators and competitors. Consequently, it may be harder for young of the year to find vacant territories to occupy. For these reasons, we might expect that, while both s_0 and s_a are potentially higher on islands than in similar habitats on the mainland, the ratio of the two, s_0/s_a, will often be relatively lower on islands because of so little turnover among the territory-holding adult population. For example, the mean

Box 7.2

Imagine convex birth and survival functions that yield an intermediate optimal reproductive effort, as shown earlier in Figure 7.15. One simple mathematical form that gives such convex shapes is

$$F(\theta) = b\theta^{0.5} \tag{a}$$

and

$$s_a(\theta) = a(1 - \theta^2), \tag{b}$$

where $a \leq 1$, $s_a \leq 1$, and $\theta \leq 1$. When reproductive effort, θ, is 0, net fecundity, F, is 0 and the adult survival rate, s_a, is at its maximum value, a. As θ increases to 1, F increases to its maximum of b while s_a decreases to 0. The larger the parameter b is, the larger F becomes. The parameter a tunes the adult survival rates: increasing a increases adult survival rates. Thus the environment influences the upper and lower bounds of fecundity and survival, but in addition these vital rates are further constrained by the trade-offs inherent in survival versus reproduction. Figure 7.16 illustrates the model for two choices of the parameters a and b.

How does the optimal θ for this model change with changes in parameters a and b that control the juvenile survival rate s_0 (through its effects on F) and the adult survival rate, s_a? The λ for this simple life history is

$$\lambda(\theta) = F(\theta) + s_a(\theta).$$

Substituting the expressions from Eqs. (a) and (b) for F and s_a, we get

$$\lambda(\theta) = b\theta^{0.5} + a(1 - \theta^2). \tag{c}$$

To find the optimal reproductive effort, θ, we need to find the θ that maximizes $\lambda(\theta)$. If this is an intermediate maximum (rather than one at the extremes, $\theta = 0$ or 1), the slope of $\lambda(\theta)$ will be 0 at the peak. For this case, we must find the value of θ where $\partial\lambda/\partial\theta = 0$. We therefore take the derivative of Eq. (c) with respect to θ,

$$\frac{\partial\lambda}{\partial\theta} = \frac{b}{2}\theta^{-0.5} - 2a\theta, \tag{d}$$

and solve for the value(s) of θ that make this derivative 0. We label these special values $\hat{\theta}$. Thus

$$\frac{\partial\lambda}{\partial\theta} = \frac{b}{2}\hat{\theta}^{-0.5} - 2a\hat{\theta} = 0. \tag{e}$$

One solution of Eq. (e) is $\hat{\theta} = 0$, but this does not always correspond to a maximum for λ. Another solution is found by dividing both sides of Eq. (e) by $\theta^{-0.5}$, giving

$$\frac{\partial\lambda}{\partial\theta} = \frac{b}{2} - 2a\hat{\theta}^{3/2} = 0.$$

Solving for $\hat{\theta}$ gives

$$\hat{\theta} = \left(\frac{b}{4a}\right)^{2/3} \quad \text{for } \hat{\theta} < 1.0. \tag{f}$$

If $\hat{\theta} > 1.0$ by Eq. (f), then the optimal reproductive effort $\hat{\theta}$ is maximal at 1.0.

Since b is directly proportional to s_0 and the parameter a is directly proportional to adult survival rate, s_a, we can express Eq. (f) as saying that

$$\hat{\theta} \text{ is proportional to } \left(\frac{s_0}{s_a}\right)^{2/3}. \tag{g}$$

The choice of this particular model structure is a bit arbitrary; we could get convex curves by substituting any exponent less than 1 in Eq. (a) and any exponent greater than 1 in Eq. (b). The qualitative result is unaltered (as long as the curves remain convex).

Conclusion: Given the assumptions of this model, optimum reproductive effort is directly proportional to the ratio of juvenile to adult survival rates raised to some positive power, Eq. (g). By this model, the absolute value of either of the survival rates is not as important as their ratio in predicting optimal reproductive effort.

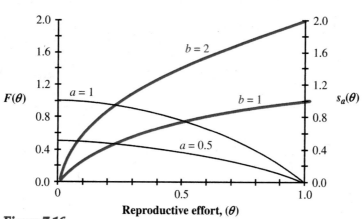

Figure 7.16

Figure 7.17
Semelparity can evolve with convex curves for $F(\theta)$ and $s(\theta)$. Maximal reproductive effort, θ, is optimal in this example. This represents the evolution of semelparity even though the curves are convex.

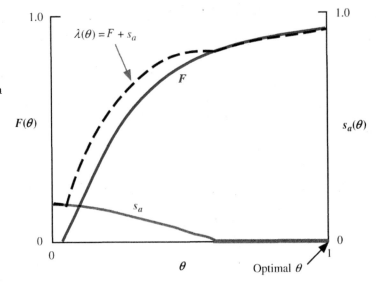

Table 7.3 **Life history traits in guppies (*Poecilia reticulata*) are shaped by the impact of large fish predators that eat adult guppies (Reznick 1982, Reznick and Endler 1982, and Reznick et al. 1997). These differences between natural populations are genetic and even appear under uniform conditions in laboratory-reared guppies. They can also be duplicated in relatively short periods of time by transplanting populations from one environment to the other.**

Life history trait	Low-predation populations	High-predation populations	High-predation populations 11 years after being placed in a low-predation environment
Sexual maturity	Later	Earlier	Later
Fecundity	Lower	Higher	Lower
Frequency of litters	Lower	Higher	Lower
Offspring size	Larger	Smaller	Larger

clutch size of scrub jays on Santa Cruz Island (3.71) is significantly smaller than the mean clutch size of scrub jays on the southern California mainland (4.34) (Atwood 1980). Adult Santa Cruz Island jays have a higher annual survival rate—0.935 compared to 0.833 on the mainland (Atwood et al. 1990)—yet the survival rate of nestlings and fledglings in the two places are not very different. Thus in this case the lower insular clutch size is associated with a lower ratio of s_0/s_a, as predicted. More demographic data contrasting sister island–mainland taxa are needed to determine if this result is general.

What about the tropical–temperate comparison? If the tropics also has lower adult survival rates, as well as lower survival rates for young, then it is not clear that lower clutch size would be favored in the tropics. Empirically, however, field studies find that, if anything, adult survival rates are greater in many tropical passerine species than in temperate species (Johnston et al. 1997), providing additional support for this theory that predicts the evolution of reproductive effort based on age-specific mortality patterns.

An experimental test of these predictions is provided by the work of David Reznick and colleagues with guppies in streams in Trinidad. Both in nature and when reared in uniform conditions in the laboratory, guppies in upstream habitats differ in several life history traits from downstream populations. The upstream populations lack several species of fish predators on the adult guppies. Guppies in these low-predation habitats have genetically diverged to attain maturity at a later age, have lower fecundity, have larger offspring size, and lay fewer litters per year. Reznick et al. (1997) found that guppies transplanted from high-predation habitats to low-predation habitats evolved similar life histories to the resident populations and thus diverged significantly from their high-predation ancestors after only 11 years as summarized in Table 7.3.

TEMPORAL VARIABILITY AND OPTIMAL REPRODUCTIVE EFFORT

Another possible reason why birds might lay smaller clutches than the most productive clutch in terms of breeding success, involves the effect of temporal variability on the vital rates. Temporal variability in juvenile mortality from year to year, even if it does not affect the average yearly survival rate, favors a lower optimal reproductive effort. The easiest way to see this is to imagine just two kinds of years: *good years* and *bad years,* which are equally frequent and occur randomly and independently over time (Schaffer 1974). Juvenile survival is augmented by some degree **e** in these two kinds of years such that

$$\text{in good years, } s_0 = s_0 + e$$

and

$$\text{in bad years, } s_0 = s_0 - e.$$

Since good and bad years are equally frequent, on average, after 2 years, the net rate of increase of the population will be

$$\lambda_g \lambda_b = [B(s_0 + e) + s_a][B(s_0 - e) + s_a]$$
$$= s_0^2 B^2 + 2s_a B s_0 + s_a^2 - \boxed{B^2 e^2}. \tag{7.2}$$

Yet, if the environment were constant (**e** = 0), the net rate of increase would be

$$\lambda^2 = s_0^2 B^2 + 2s_a B s_0 + s_a^2, \tag{7.3}$$

which is a greater quantity than that in Eq. (7.2) since Eq. (7.3) is identical to just the portion of Eq. (7.2) without the positive term surrounded by the box in Eq. (7.2), which is not subtracted in Eq. (7.3).

Hence **fluctuations in juvenile survival tend to decrease the "effective" juvenile survival rate, which in turn favors a lower optimal reproductive effort.**

Exercise: Show that the converse applies to fluctuations in adult mortality rates. That is, fluctuations in adult survival, s_a, cause $\lambda_g \lambda_b < \lambda^2$ by a term e^2. This, in turn, favors a greater optimal reproductive effort because the ratio s_0/s_a becomes higher than in the case of no fluctuations.

In summary, for this model where good years and bad years differ by some additive difference in juvenile or adult survival and assuming that the arithmetic means for adult and juvenile survival are constant,

$$\text{Optimum reproductive effort is proportional to } \frac{\text{variability in adult survival}}{\text{variability in juvenile survival}}.$$

This latter result appears fragile, however, in terms of some assumptions of the model. In particular Hastings and Caswell (1979) show that if good years and bad years differ by a multiplicative (rather than additive) factor, then fluctuations have the same effect on optimal reproductive effort, regardless of whether they effect juvenile or adult survival.

Another way to understand the role of environmental fluctuations in reducing fitness and potentially shifting optimum life history strategies is to recall that, if fitness fluctuates from year to year, the geometric mean fitness is the appropriate fitness to consider to determine ultimate population growth rates. In Chapter 2 (Eq. 2.9), we found that the geometric mean of a random variable like λ is proportional to its arithmetic mean, μ, but it is inversely proportional to its variance, σ^2, or

$$\ln G = \text{natural log of the geometric mean, } \lambda,$$

$$= \text{arithmetic mean of } \ln \lambda$$

$$\approx \ln(\mu) - \left(\frac{\sigma^2}{2\mu^2} \right).$$

To convert $\ln G$ to the geometric mean, G, we only need to exponentiate both sides. An additional approximation can also be used under some circumstances. When the absolute value of x is much less than 1, $e^x \approx 1 - x$. Hence (Kendall and Stuart 1969, Price and Liou 1989):

$$\text{Geometric mean, } G \approx \exp^{\ln(G)} = \exp\left(\ln(\mu) - \left(\frac{\sigma^2}{2\mu^2} \right) \right)$$

$$= \mu \exp\left(-\left(\frac{\sigma^2}{2\mu^2} \right) \right)$$

$$\approx \mu \left(1 - \left(\frac{\sigma^2}{2\mu^2} \right) \right)$$

$$= \mu - \frac{\sigma^2}{2\mu},$$

or, in words,

$$\text{Geometric mean, } G \approx \text{arithmetic mean} - \left(\frac{\text{variance}}{2(\text{arithmetic mean})} \right). \qquad (7.4)$$

An example is illustrative. Suppose that fitness λ take on the value 1 in a bad year and 4 in a good year. Further suppose that good years and bad years are equally frequent; then

$$\text{Arithmetic mean fitness is } (1 + 4)/2 = 2.5$$

and

$$\text{Geometric mean fitness is } [(1)(4)]^{0.5} = 2.$$

The variance, σ^2, of any variable $x =$ the mean value of x^2 minus the mean^2. So for our case we have $\sigma^2 = [0.5(1) + 0.5(16)] - 2.5^2 = 2.25$. Applying the approximation formula for the geometric mean in Eq. (7.4) we get $2.5 - (2.25/5) = 2.05$.

Thus the true geometric mean in this example is 2, and the approximation formula yields 2.05. When the ratio of the standard deviation to the mean, σ/μ is smaller, the approximation formula of Eq. (7.4) will be much closer to the true geometric mean.

Refer again to Eq. (7.4) and recall that fitness for simple life histories with exponential growth is fitness $= \lambda = Bs_0 + s_a$. If there is variability in any of the vital rates, especially if the arithmetic mean survival stays the same, fitness will decrease. Imagine that s_a is a random variable from year to year, with mean $E(s_a)$, variance σ_a^2, and the other vital rates constant. The geometric mean fitness from Eq. (7.4) is

$$G \approx Bs_0 + E(s_a) - \left(\frac{\sigma_a^2}{2(Bs_0 + E(s_a))} \right) = \bar{\lambda} - \frac{\sigma_a^2}{2\bar{\lambda}}.$$

Now we can form the ratio

$$\frac{\text{Geometric mean with temporal variation in } s_a}{\text{Geometric variation without temporal variation in } s_a} = \frac{\bar{\lambda}}{\bar{\lambda} - \dfrac{\sigma_a^2}{2\bar{\lambda}}} = 1 - \frac{\sigma_a^2}{2}.$$

This ratio shows the *relative* change in fitness for different life history strategies if they experience temporal variation in s_a. Note that the mean λ drops out of the equation and that we are left with a term only in the variance, σ_a^2. Since higher variances are usually associated with higher mean values, the relative fitness of a high adult survival strategy will often be reduced relatively more by temporal variation than that of a low adult survival strategy. This potentially favors species or genotypes with a higher reproductive effort but lower s_a. For example, at the extreme, the annual strategy will not be affected at all by variance in adult survival rates since $\lambda = Bs_0$.

On the other hand, when there is variability in juvenile survival or in birth rates only, geometric mean fitness is again reduced:

$$\frac{\text{Geometric mean with temporal variation is } s_B}{\text{Geometric variation without temporal variation in } s_B} = 1 - \frac{\sigma_B^2}{2}.$$

Now the relative fitness of the high B strategy (or high juvenile survival strategy) is reduced relatively more than that of the high s_a but low B (or low juvenile survival) strategy.

ANOTHER GRAPHICAL TECHNIQUE FOR EVALUATING OPTIMAL LIFE HISTORIES

Richard Levins (1968) suggested another way of graphically representing trade-offs between different components of overall fitness. It is called a *fitness set analysis*. The trade-off between F and s_a can be regraphed into a fitness set, as explained in Figures 7.18 and 7.19.

Figure 7.18
Optimal reproductive effort through fitness set analysis.

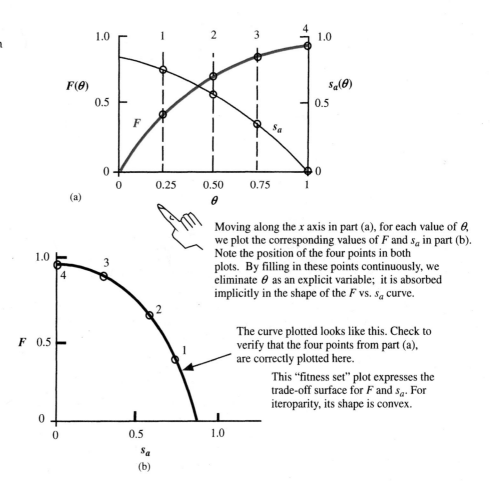

Moving along the *x* axis in part (a), for each value of θ, we plot the corresponding values of F and s_a in part (b). Note the position of the four points in both plots. By filling in these points continuously, we eliminate θ as an explicit variable; it is absorbed implicitly in the shape of the F vs. s_a curve.

The curve plotted looks like this. Check to verify that the four points from part (a), are correctly plotted here.

This "fitness set" plot expresses the trade-off surface for F and s_a. For iteroparity, its shape is convex.

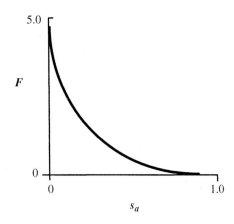

Figure 7.19
If the $F(\theta)$ and $s_a(\theta)$ curves were concave instead, the fitness set curve would also be concave, as shown here.

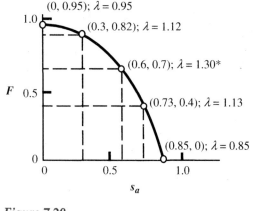

Figure 7.20
Five points are identified on the trade-off surface, and the fitness of each life history is calculated as $\lambda = F + s_a$. Of these five points, point (0.6, 0.7) has the highest fitness, but does it have the maximal fitness on the entire trade-off surface?

Figure 7.21
A fitness set analysis to predict optimal fecundity and adult survival. Each of the four lines—out of an infinite number of possible lines—has slope –1 but a different value for the y intercept, λ.

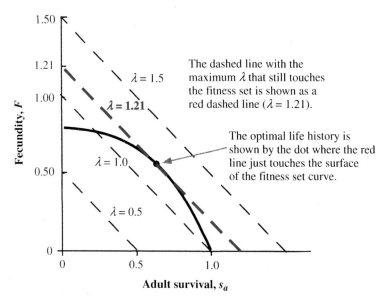

The optimal life history schedule can be identified from these diagrams. Since $\lambda = F + s_a$, we can assign to each point on the trade-off surface a fitness by simply summing the y coordinate (F) and the x coordinate (s_a), as shown in Figure 7.20.

Of course, it takes a lot of arithmetic gymnastics to calculate λ in this way for all points on the fitness set, but fortunately there is a graphical trick that provides a quicker way to identify the optimal life history. It works like this: since $\lambda = F + s_a$, lines of equal λ can be plotted in this F–s_a space, as shown in Figure 7.21). Straight lines have the general equation $y = mx + b$, where m is the slope and b is the y intercept. In our plots, net fecundity, F, is plotted on the y axis and s_a is plotted on the x axis. We therefore rearrange the fitness equation to match this graphical construction, which yields $F = \lambda - s_a$. When we choose different values of λ, a family of straight lines emerges, each with slope of –1 but differing in their y intercepts of λ.

The optimal life history for the circumstances depicted in Figure 7.21 has an adult survival rate, s_a, of about 0.63 and an effective birth rate, F, of about 0.58. This life his-

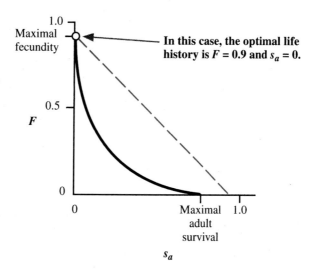

Figure 7.22
For this example of semelparity, the optimal life history is at maximal *F,* in this case *F* = 0.9, and s_a = 0. This describes the life history of an annual.

In this case, the optimal life history is $F = 0.9$ and $s_a = 0$.

tory yields a λ of about 1.21. The optimal life history shown in Figure 7.22 is that of maximal *F* and zero s_a —an annual strategy.

One advantage of disposing with reproductive effort as an explicit variable is based on practical considerations. Often it is difficult to measure reproductive effort precisely since it requires a complete understanding of energy input and partitioning. However, in principle, we can still use this technique to predict the optimal fecundity, *F,* and the corresponding adult survival rate, s_a, if we know the shape of the trade-off surface between *F* and s_a, even if we don't measure reproductive effort itself. However, it is important to ensure that the trade-off surface is based on genotypic variation in reproductive effort rather than just phenotypic variation associated with environmental differences.

Exercise: You can also use calculus to solve for the optimal life history. Suppose that the fitness trade-off set is given by the relationship $F = 4 - s_a^2$. Solve for the optimal life history again, assuming that $\lambda = F + s_a$.

Egg Size and Parental Care

Reproductive effort is a measure of the proportion of assimilated energy devoted to reproduction for a particular age group. The theory just discussed suggests that this proportion should evolve to greater levels in environments where adult survival rates are low and where juvenile survival rates are potentially high and temporally variable.

Another question is how reproductive effort should be packaged each year: Producing more offspring will increase fitness, but producing larger offspring with higher survival rates will also increase fitness, all else being equal. Given a fixed level of energy, producing more offspring will therefore translate into producing smaller offspring. Given a certain amount of energy devoted to reproduction in a breeding season, the organism could make many small eggs or a few large ones that sum to the same total energy investment. The solution to this problem will again depend on quantitative aspects of how fitness is increased by changes in egg number compared to changes in egg size. In the plant genus *Plantago,* some species are annuals while others are perennials (Primack 1978). The annuals typically produce a smaller number of larger seeds compared to the perennials, which produce a larger number of smaller seeds. An adult perennial *Plantago* is large with relatively deep roots. Any of its seeds that germinate in its shadow will experience strong competition with its parent for light, nutrients, and water. The best strategy here will be to produce many small seeds, which through dispersal might find a chance to grow elsewhere. On the other hand, annual plants are small with limited vegetative growth; they die back each year. Seeds produced by annuals can recruit into the same site that their now dead parent occupied if they can only survive through the winter. Annuals therefore should invest in larger seeds with more stored reserves to get through the winter and to give them a headstart for germination and growth in the spring.

In some species parental care of young is an option but one with a potentially large energetic cost (Case 1978). Should the parent expend all its energy making many eggs and have none left to nourish the brood, or should it produce fewer eggs and expend some additional time and energy caring for the eggs and the young after they hatch? In birds, with parental care, we have argued that a trade-off often exists between juvenile survival, s_0, and the number of eggs in the nest. Yet, even in species with little or no parental care, such a trade-off might still be expected. Larger eggs will produce larger hatchlings, which may have higher survival rates than smaller hatchlings. The same amount of reproductive effort could be devoted to producing many small eggs or a fewer number of larger eggs that produce larger young with higher survival rates.

Environments that allow higher juvenile survival rates, all else being equal, favor the evolution of higher clutch sizes at the expense of eggs of lower quality (e.g., smaller size). Joshi et al. (1996) believe that this principle helps explain cultural (rather than evolutionary) patterns in humans. As underdeveloped countries develop and industrialize, they experience a **demographic transition.** Death rates typically drop because of improved health and sanitation conditions, which now decrease the transmission rates of several infectious diseases like cholera and typhus. Also malnutrition is less of a problem. This leads to an increase in population growth rate unless birth rates also decline. Birth rates typically do decline, but the social factors that encourage families to decrease the number of offspring are less well understood. In developing countries, agrarian families must work hard physically to produce enough food to feed the household. The larger the family size, the more hands there are to help in this work, and large families are favored. As a nation industrializes, new technical jobs are available that require more sophisticated skills and training than simple physical labor. The subsequent decrease in birth rates and family size, according to this hypothesis, stems from more cultural emphasis on offspring "quality" (necessitating more costly education) than shear offspring number.

PROBLEMS

1. The flow diagram shown represents a cohort of 1000 newborn females in a population followed in time through 4 years.

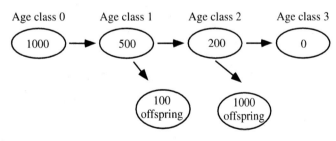

Age class 0   Age class 1   Age class 2   Age class 3

1000 → 500 → 200 → 0

100 offspring   1000 offspring

a. Fill in the following table.

Age class, x	s_x	l_x	b_x	$F_x(= s_0 b_x)$
1		1		
2				
3				

b. Construct a 2×2 Leslie matrix for this population, beginning with age class 1.

c. Assuming that the vital rates remain constant, will this population ultimately increase or decrease?

d. Graphically present a theoretically reasonable relationship between birth rate and adult survivorship (as the y axes) versus reproductive effort (as the x axis) for age class 2. Wherever possible give the numerical values at which the survival and birth curves intersect the axes.

2. For bird species A, the survivorship of young to fledging, S, falls with increasing clutch size, C, according to the relationship $S = 1.0 - 0.1C$. Species B follows the relationship $S = 0.5 - 0.1C$.

a. What is A's optimal clutch size, according to Lack's hypothesis?

b. What is B's optimal clutch size, according to Lack's hypothesis? If you had to guess,

c. which species might nest in holes and which might nest in more exposed conditions?

d. which species might be tropical and which might be temperate?

3. Two hypothetical seal species produce their young on the same isolated island. Seal species A forages far offshore for fish during the nursing period. Adults of species B put on much fat earlier in the year and don't need to feed at all during the period when they are nursing their young. When the young of both species are left alone, they are in danger of being trampled by other seals. Otherwise, the two seal species are very similar in habit and body size. Use life history theory to predict answers to the following questions.

a. Which of the two seal species probably has the highest litter size?

b. Which of the two seal species probably has the highest juvenile growth rate?

c. Which of the two seal species probably reaches reproductive maturity earliest?

d. Which of the two seal species probably is the most precocial (physically mature) at birth?

4. Given the following tables for populations A and B, which of the θ values is the optimal reproductive effort for each? F is fecundity, s is adult survival rate, and θ is a measure of reproductive effort that varies from 0 to 1. Describe the life history strategy for each population.

Population A			**Population B**		
θ	$s(\theta)$	$F(\theta)$	θ	$s(\theta)$	$F(\theta)$
0	0.70	0.0	0	1.00	0.00
0.2	0.62	0.30	0.2	0.60	0.02
0.4	0.56	0.56	0.4	0.45	0.05
0.6	0.47	0.73	0.6	0.18	0.17
0.8	0.25	0.84	0.8	0.07	0.60
1.0	0.00	0.90	1.0	0.00	1.90

5. A field study of breeding in a particular bird species yielded the following data on frequency of different sized clutches and the resulting survival of the young.

Clutch size	No. of such clutches	Average per capita survival of young from each clutch
1	10	0.6
2	100	0.4
3	50	0.3
4	4	0.1

a. What is the optimal clutch size, according to Lack's theory?

b. Is the optimal clutch size the most frequently seen clutch?

c. If the answer to (b) is "no," what biological factors might account for the discrepancy? In other words, why might the birds not behave as Lack predicted?

6. Some butterflies lay their eggs only on pipevine plants. In one species, the adult females prefer to oviposit in sunny areas even though the survival (in the field) of the eggs and larvae is lower in sunny areas then in shady areas. Assuming that these results are correct, why might such adult behavior have evolved?

7. Fruit flies, *Drosophila*, which breed in cactus (cactophilic) tend to be very host-specific. That is, a given species will lay its eggs in only one or two different cacti species in nature. On the other hand, mycophagous *Drosophila*, which lay their eggs in mushrooms, are host generalists. A given individual of a given species will accept many different host mushroom species. What selective factors might

account for these differences in host specificity between these two *Drosophila* groups?

8. Atlantic Salmon (*Salmo salar*) spawn in freshwater rivers and streams. Juveniles migrate downstream to the ocean. After a year or more, they then return to spawn. Some Atlantic salmon then die after spawning while others survive to return to the sea and then migrate upstream to spawn again. Schaffer and Elson (1975) found that salmon spawning in long, fast rivers are older, having spent more years at sea than salmon spawning in shorter, slower rivers. Also salmon stocks that have a higher growth rate at sea tend to spawn at later ages. Commercial fishing, which disproportionately takes older and larger fish, seems to have selected in recent times for an early age of first return. Relate these trends to the theory of optimal reproductive effort.

9. Suppose that we have a smooth juvenile survival versus clutch size curve, *s*, like that shown. How would the optimal clutch size according to Lack be changed if survival rates were halved to *s*/2 at all clutch sizes?

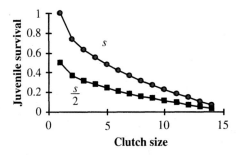

10. It may be argued that a more realistic shape for the fecundity function, $F(\theta)$, in animals is probably sigmoidal since, until some threshold amount of energy is devoted to reproduction, not even a single young can be produced, and at very high levels of reproductive effort, the number of offspring produced might be limited by physical space in the body cavity. The following diagram combines a sigmoidal $F(\theta)$ curve with a concave $s_a(\theta)$ curve to show two local maxima in fitness separated by a minimum. Note that there are now two peaks in fitness: at $\theta = 0$ and at $\theta = 0.74$. How might natural selection respond to this two-peaked fitness gradient?

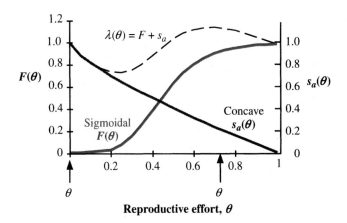

Reproductive effort, θ

Reproductive Value and the Evolutionary Theory of Aging

s individuals age, they grow in body size and gain valuable experience, which generally improves their success. This means that the trade-off relationship between survival and reproduction may be different for different ages—unlike the way we depicted the situation in Chapter 7. For example, a reproductive effort of, say, 50%, might result in the successful production of only one offspring for a yearling adult with a consequent halving of the survival rate compared to no reproduction at all. On the other hand, older, larger, and more experienced adults expressing the same reproductive effort of 50% might be able to produce three offspring and incur a survival decline of only 20% relative to not reproducing at all. With these complexities, how can we find the optimal reproductive effort, θ, for each age? We cannot simply successively optimize the reproductive effort at each age as if these events were independent. True, reproductive effort at age x cannot affect births or survival at earlier ages (they are now history), but it can most assuredly influence these vital rates at still later ages of life. An excessively high reproductive effort at an early age could decrease survival, body size, and potential reproduction at later ages and thus lead to a lower lifetime, λ. In this chapter we describe how to evaluate the relative importance of each age to future population growth and in the process discover some insights on aging.

THE REPRODUCTIVE VALUE OF DIFFERENT AGES

Several practical questions in ecology and wildlife management deal with the relative importance of different age groups to population growth and persistence. Should we restrict hunting in deer to certain ages to minimize the impact on the hunted population? If we wish to limit the population growth of an injurious pest insect, which ages would be the most effective to kill? If different disease pathogens strike some ages of their host more than others, which diseases will have the biggest impact on the growth rate of the host or its likelihood of eventual population extinction? If we want to introduce an endangered species into a new location, which age groups will lead to the largest expected population at some specified future time?

As an example of this last question, let's work through a simulation. Here is a Leslie matrix (ages: 1–4):

$$\mathbf{L} = \begin{bmatrix} 0 & 2 & 0.5 & 0.1 \\ 0.5 & 0 & 0 & 0 \\ 0 & 0.6 & 0 & 0 \\ 0 & 0 & 0.5 & 0 \end{bmatrix}. \tag{8.1}$$

Figure 8.1 shows the time course of total population size based on iterating population growth, beginning with 10 individuals of just age class 1; then 10 individuals of just age class 2, and so on. As time goes by, each simulation reaches the same stable age structure well before year 50, and, as expected, each simulation also converges on the same rate of population growth ($\lambda = 1.0736$) since the vital rates are constant. However, in

Figure 8.1
A graphical explanation of reproductive value of age x based on the iteration of the Leslie matrix in Eq. (8.1). All simulations begin with exactly 10 individuals (ln 10 = 2.303) from only one of the four age classes. If age x has a reproductive value of 2, then a population initialized with an innoculum of only age x individuals will be twice as large as that begun with an innoculum of only age 1 individuals, once the stable age structure has been reached.

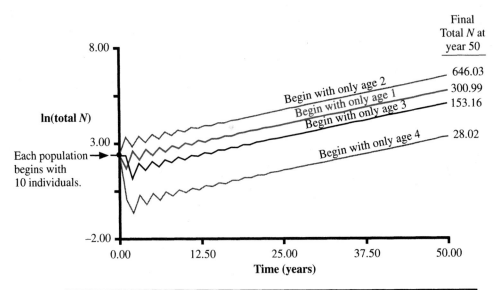

Begin with just age	Final Population size	Ratio of final population size to final size beginning with just age 1
1	300.99	1.00
2	646.03	2.146
3	153.16	0.509
4	28.02	0.093

After the stable age structure is reached, these ratios remain constant over time. These ratios are the **reproductive values** of each age.

spite of the fact that all the populations begin with the same total population size, 10 individuals, they do not converge to the same population size by year 50. After the stable age structure is reached, the lines of ln N become parallel as time goes by, implying that the population sizes from these different initializations remain different by the same constant factor. The endpoint of 50 years and the initial population size of 10 in this example are completely arbitrary and do not affect the resulting ratios calculated in Figure 8.1.

These ratios represent the relative reproductive values of different age groups to future population growth. In this example, two-year-olds have over twice the reproductive value of one-year-olds, and four-year-olds have less than one-tenth the reproductive value of one-year-olds.

Is there a formula that we could use to calculate these reproductive values, V_x based on the vital rates without having to rely on simulation? We begin with some guesses that will turn out to be incomplete and naive. We first guess that

$$\text{Reproductive value, } V_x, \text{ of age } x = F_x.$$

If F_x were all there was to reproductive value, V_x, then, if two different ages (say, ages 3 and 4) have the same F_x, they should have the same V_x. But it takes one more year of life to reach age 4 than age 3, and during this time some death, d_3N_3, occurs, so there should be fewer aged 4 years than 3 years, and this alone would lower their contribution to overall future offspring production. Clearly, F_x does not capture all of what we're searching for, and survivorship must be included in the formula for V_x.

What about setting $V_x = l_xF_x$? Again, let's explore the consequences of this guess. Suppose that we have two ages now with the same l_xF_x (say, $l_xF_x = 3.0$ for both ages 3 and 4). In a growing population the babies born to 3-year-old mothers start "compounding interest" sooner than the babies born a year later to 4-year-old mothers. On the other hand, the 3-year-old mothers will be 4-year-old mothers next year if they survive, but the 4-year-old mothers will never be 3-year-old mothers again (unless they go through some time warp). Somehow we need to reward this relative "head start" for early reproduction in our calculation of V_3 and V_4. Since the advantage of the early head start will be greater the greater the population growth rate λ is, we can conclude that our eventual formula for V_x must somehow be proportional to λ and inversely proportional to x.

Sir Ronald Fisher (1958) wrestled with this difficult problem and was able to make the connection. As long as the population is in stable age structure, the future production of offspring by individuals of age x, is equivalent to the number of offspring produced this year by individuals that are x or older. Reproductive value is this number divided by the number of individuals that are x this year. That is,

$$\text{Reproductive value at age } x, V(x) = \frac{\begin{array}{c}\text{No. of female offspring produced this}\\ \text{year by mothers age } x \text{ or older}\end{array}}{\text{No. of mothers that are age } x \text{ this year}} \qquad (8.2)$$

Now we flesh out the numerator and the denominator of Eq. (8.2). If the total population size at time T is $N(T)$, the

$$\text{Numerator} = \sum_{t=x}^{\text{max}} b_t c_t \ N(T) = \sum_{t=x}^{\text{max}} b_t \left(\frac{l_t \lambda^{-t}}{\text{sum}} \right) N(T),$$

where c_t is the proportion of individuals of age class t compared to all other age classes. This proportion can be expressed as the absolute number of age class t, c_t', divided by the total of all ages (sum). In Chapter 4 we found that at stable age structure $c_t' = \lambda^{-t} l_t$. Moreover the denominator of Eq. (8.2) is

$$\text{Denominator} = \frac{\lambda^{-t} l_t}{\text{sum}} N(T),$$

so in total we get,

$$\begin{array}{c}\text{Reproductive value}\\ \text{of age } x\end{array} = \frac{\displaystyle\sum_{t=x}^{\text{max}} b_t \left(\dfrac{l_t \lambda^{-t}}{\text{sum}} \right) N(T)}{\dfrac{l_x \lambda^{-x}}{\text{sum}} N(T)} = \frac{\displaystyle\sum_{t=x}^{\text{max}} \lambda^{-t} l_t b_t}{l_x \lambda^{-x}},$$

or, more compactly

$$\boxed{V_x = \frac{\lambda^x}{l_x} \sum_{t=x}^{\text{max}} \lambda^{-t} l_t b_t \quad \begin{array}{l} \text{for } x = 0 \text{ to max}, l_0 = 1 \\[4pt] \text{and } V_0 = 1 \end{array}} \qquad (\text{RV 1})$$

An equivalent expression that uses the Leslie matrix fecundity data, F_x, instead of the raw birth, b_x, data is

$$\boxed{V_x = \frac{\lambda^x}{l_x} \sum_{t=x}^{\text{max}} \lambda^{-(t+1)} l_t F_t \quad \begin{array}{l} \text{for } x = 1 \text{ to max}, l_1 = 1, F_x = s_0 b_x \\[4pt] \text{and } V_1 = 1 \end{array}} \qquad (\text{RV 2})$$

Note the change in the exponent of λ within the summation in Eq. (RV 2) compared to Eq. (RV 1) and the difference in the beginning age that is assigned a survivorship of one. The difference between these expressions arises in the same way that we found different formulations of the Lotka–Euler equation based on derivations involving b_x versus forming the F_x (Chapter 4).

> **Exercise:** Apply this last formula RV 2 to the Leslie matrix shown in Eq. (8.1). Calculate V_x for each age and verify that you reach the same values as those produced by simulation and given in the last column of the table at the bottom of Figure 8.1.

Figure 8.2
Curves of reproductive values for
human populations in three countries
(from Keyfitz 1985).

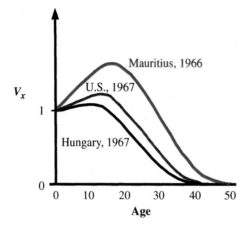

Figure 8.3
Reproductive values of rice weevils
(from data in Birch 1948).

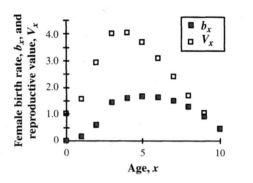

Figure 8.4
Age-related trends in birth rate b_x and
reproductive value V_x for
sparrowhawks. The birth rates are
smoothed estimates based on female
young produced per nest per adult
female; this includes nests that
produced no young. The reproductive
value estimates in this case assumed a
stationary population, $\lambda = 1$ (From
Newton and Rothery 1997).

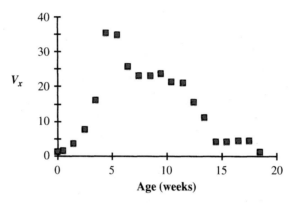

The reproductive value in three human populations is shown in Figure 8.2. Human
life histories are typical of many other animal life histories in that V_x increases with age
up to reproductive maturity and then subsequently declines. The differences in height
of the three curves parallels the differences in λ for the three countries in those years:
Mauritius, 1966 > United States, 1967 > Hungary, 1967.

In Chapter 3, we presented data on the death rates and fecundities of rice weevils
(Figures 2.9 and 2.10). From those data we may calculate the reproductive values for each
age, as shown in Figure 8.3. Rice weevils also show a gradual increase in reproductive
value up to the age of sexual maturity at age 4 weeks and then a decline thereafter. Note
the higher scale of the y axis for weevils compared to that for humans. These rice weevils
can be a serious pest of stored rice. These data show that 4- and 5-week-old individuals
would produce a larger infestation in a fixed amount of time than any other age groups.

As a final example, Newton and Rothery (1997) studied the life history of sparrowhawks in south Scotland. The birth rate and reproductive value schedule are shown
in Figure 8.4. Some hawks reach sexual maturity at age 1 but most do not reach sexual
maturity until age 3. Since the reproductive value at age x sums the reproductive contributions for age x and beyond, we see that the V_x curve lies above the b_x curve. Note
also that the peak of V_x occurs earlier (at ages 3 and 4) then the peak of b_x (at age 5).

The reproductive value at any age x may be further decomposed into the sum of two simpler parts by taking the first term in the summation, $t = x$, outside the summation side and then summing the rest from $x + 1$ to the maximum age:

$$V_x = \frac{\lambda^x}{l_x}\left(\lambda^{-x}l_xb_x + \sum_{t=x+1}^{max}\lambda^{-t}l_tb_t\right)$$

$$= b_x + \frac{\lambda^x}{l_x}\sum_{t=x+1}^{max}\lambda^{-t}l_tb_t \qquad (8.3)$$

$$= b_x + \frac{s_x}{\lambda}V_{x+1}$$

This last expression has two parts:

> 1. **present** reproductive value, b_x;
> 2. **future** (or **residual**) reproductive value, $V_x^* = (s_x/\lambda)V_{x+1}$.

(8.4)

Thus we can write a recursion formula for reproductive value of age x based on the reproductive value of the next oldest age, $x + 1$:

$$V_x = b_x + \frac{s_x}{\lambda}V_{x+1} \qquad (RV\ 3)$$

The simplest way to calculate reproductive value therefore is first to determine V_{max}, which is simply b_{max}, and then work backward successively in age to the very first age, applying Eq. (RV 3) at each step. An example will illustrate this technique. Consider the following vital rates.

Age group, x	s_x	b_x	l_x
0	1.0	0.0	1
1	0.5	0.0	1
2	0.5	0.0	0.5
3	0.4	3.2	0.25
4	0.0	1.111	0.1

For this life history, $\lambda = e^r = 0.970633$ and $\lambda^{-1} = 1.0302$. Using Eq. (RV 3), we start with the maximum age, $x = 4$, and work our way backward:

$$V_4 = b_4 + 0 = 1.111 + 0 \qquad\qquad\qquad = 1.111$$

$$V_3 = b_3 + s_3\lambda^{-1}V_4 = 3.2 + (0.4)(1.0302)(1.111) \quad = 3.6578$$

$$V_2 = b_2 + s_2\lambda^{-1}V_3 = 0 + (0.5)(1.0302)(3.6578) \quad = 1.8843$$

$$V_1 = b_1 + s_1\lambda^{-1}V_2 = 0 + (0.5)(1.0302)(1.8843) \quad = 0.97063$$

$$V_0 = b_0 + s_0\lambda^{-1}V_1 = 0 + (1)(1.0302)(0.97063) \quad = 1.00$$

In summary, we have the following.

Age group, x	Reproductive value, V_x (using Eq. RV 4)
0	1.0
1	0.9706
2	1.8842
3	3.3657
4	1.111

In Figure 8.1 the reproductive values were based on the Leslie matrix approach where the age class 0 is not explicit and the F_x terms are used instead of the b_x terms. If we were given the F_x terms, we could apply Eq. (RV 2), or, equivalently, we can take the reproductive values just calculated from the b_x terms, simply cast out V_0, and renormalize the others ($x = 1–4$) by dividing each by V_1. By this method, then, we arrive at

$$V_1 = 1$$

$$V_2 = 1.8843/0.9706 = 1.941$$

$$V_3 = 3.6578/0.9706 = 3.768$$

$$V_4 = 1.11/0.9706 = 1.144$$

Exercise: Verify that these same values can be reached by the application of formula RV 2.

For this life history, three-year-olds contribute the most to future population growth. If these data were from an endangered species, we might want to concentrate our conservation efforts on factors that might increase the survival and fecundity of this age group.

REPRODUCTIVE VALUE AND OPTIMAL REPRODUCTIVE EFFORT

Reproductive value is an important concept in population biology for another reason: it helps us to predict optimal reproductive effort when different ages have different potentials for reproduction and survival. The life history strategy of an organism can be characterized by the series of reproductive efforts at each age: $\boldsymbol{\theta} = (\theta_0, \theta_1, \theta_2, . . . , \theta_{max})$. The optimal strategy is the set of θ_x that maximizes $\lambda(\boldsymbol{\theta})$ for $x = 1$ to max.

The fecundity F_x of age class x depends on the reproductive effort at age class x, or θ_x. Additionally, it could depend on the reproductive efforts at previous age classes. Higher reproductive efforts at ages less than x may lead to lower amounts of energy available for body growth and energy storage and thus lower potential fecundity at age class x. Hence we may write

$$F_x = F_x(\theta_0, \theta_1, \theta_2, . . . , \theta_x).$$

In words, "Fecundity at age x is a function of present and previous reproductive effort but not future reproductive effort." In cases where there is extended parental care, this assumption may be violated since a mother's reproductive effort now affects the reproductive effort of the offspring under her care.

Survival to age class x, or l_x, depends on all the reproductive efforts that have gone before to age $x – 1$. Hence we may write

$$l_x = l_x(\theta_0, \theta_1, \theta_2, . . . , \theta_{x-1}).$$

In words, "Survivorship to age x is a function of all the reproductive efforts that have gone before, but not present and future reproductive effort." Again, in cases with extended parental care, this assumption could be violated.

Now we focus on age x and its reproductive effort, or θ_x. We can write the Lotka–Euler equation—split into two parts—as

$$1 = \underbrace{\sum_{t=1}^{x-1} F_t l_t \lambda^{-t}}_{\text{Ages before } x} + \underbrace{\sum_{t=x}^{max} F_t l_t \lambda^{-t}}_{\text{Ages } x \text{ and after}} \tag{8.5}$$

The first sum in Eq. (8.5) depends on θ_x only through its effect on $\lambda(\theta_x)$, since previous ages' birth and survival are not influenced by present reproductive effort, θ_x. Present reproductive effort (i.e., at age x) does affect the second sum, both through $\lambda(\theta_x)$ and through effects on $F_t(\theta_x)$ and $l_t(\theta_x)$ for all ages $t = x$ to max.

The second sum bears a relationship to the formula for reproductive value, Eq. (RV 2):

$$(\text{Second sum})\left(\frac{\lambda^{x+1}}{l_x}\right) = V_x,$$

and thus the second sum = $V_x l_x \lambda^{-(x+1)}$. Substituting this last expression for the second sum into Eq. (8.5) gives

$$1 = \sum_{t=1}^{x-1} F_t l_t \lambda^{-t} + V_x \lambda^{-(x+1)} l_x. \tag{8.6}$$

To find an optimal θ_x we need to differentiate Eq. (8.6) with respect to θ_x:

$$0 = \underbrace{\sum_{t=1}^{x-1} -t F_t l_t \lambda^{-t-1} \frac{\partial \lambda}{\partial \theta_x}}_{\text{From first sum}} + \underbrace{l_x \left[-(x+1)\lambda^{-x-2} \frac{\partial \lambda}{\partial \theta_x} V_x + \lambda^{-(x+1)} \frac{\partial V_x}{\partial \theta_x} \right]}_{\text{From second sum}}. \tag{8.7}$$

We can rearrange Eq. (8.7) to give an expression for the partial derivative of V_x with respect to θ_x in terms of the partial derivative of λ with respect to θ_x:

$$\frac{\partial V_x}{\partial \theta_x} = \frac{\displaystyle\sum_{t=1}^{x-1} t F_t l_t \lambda^{-t-1} \frac{\partial \lambda}{\partial \theta_x} + l_x \left[(x+1)\lambda^{-x-2} \frac{\partial \lambda}{\partial \theta_x} V_x \right]}{l_x \lambda^{-(x+1)}}.$$

After we collect terms, this last expression becomes

$$\frac{\partial V_x}{\partial \theta_x} = \left[\frac{\lambda^x}{l_x} \sum_{t=1}^{x-1} t F_t l_t \lambda^{-t} + (x+1)\lambda^{-1} V_x \right] \frac{\partial \lambda}{\partial \theta_x}. \tag{8.8}$$

The term in brackets in Eq. (8.8) is always positive. **Therefore, $\partial V_x/\partial \theta_x$ has the same sign as $\partial \lambda/\partial \theta_x$ and when $\partial \lambda/\partial \theta_x$ is at 0 representing a maximum for λ, $\partial V_x/\partial \theta_x$ will also be at zero.**

Equation (RV 3) shows the trade-off involved in maximizing V_x. Note how similar this equation is to the equation for λ for a perennial life history with instant adult vital rates that we used in Chapter 7 (Figure 7.14). We write them side by side to emphasize this similarity:

$$\lambda = F + s_a = s_0 b + s_a \quad \text{and} \quad V_x = b_x + \frac{s_x}{\lambda} V_{x+1}.$$

A similar expression for reproductive value shows this similarity even better:

$$V_x = \lambda^{-1}(F_x + s_x V_{x+1}). \tag{RV 4}$$

In all three equations, the first term is a measure of birth rate at the present age, and the second term is a measure of the value of living to later ages to reproduce again.

A greater investment in births at any one age will generally be at the expense of maintenance and growth, leading to a potential decrease in survival to the next age, s_x, and possibly reduced fecundities at future ages. How detrimental this decrease in survival to the next age is on total fitness, λ, depends on the reproductive value of individuals at the next age (the V_{x+1} term). This is why the product of s_x and V_{x+1} is evident in Eq. (RV 4). **All else being equal, the lower the future reproductive value, V_{x+1}, the more important (in terms of V_x and thus in terms of λ) is present reproduction, F_x, thus favoring a higher present reproductive effort, θ_x.**

To find the optimal θ_x for a particular age x we search for a maximum in V_x as a function of θ_x:

$$V_x(\theta_x) = \lambda^{-1}[F_x(\theta_x) + s_x(\theta_x)V_{x+1}(\theta_x)],$$

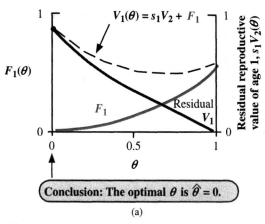
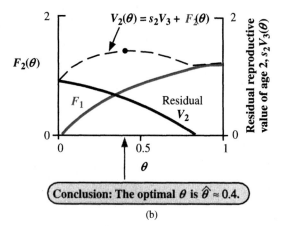

Figure 8.5

Reproductive value is the sum of present and future (or residual) reproductive value: (a) age 1 and (b) age 2.

or, equivalently, we could use Eq. (RV 1) if we had b_x data instead:

$$V_x(\theta_x) = b_x(\theta_x) + \frac{s_x(\theta_x)}{\lambda} V_{x+1}(\theta_x).$$

Note that λ^{-1} in Eq. (RV 4) is multiplying both terms in parentheses and is always positive. Thus the θ_x that maximizes reproductive value at age x will also be the θ_x that maximizes the sum of present and future reproductive value, or $F_x(\theta_x) + s_x(\theta_x)V_{x+1}(\theta_x)$. This will also be the θ_x that maximizes V_x and that locally maximizes λ (i.e., for age x). Figure 8.5 shows two examples of hypothetical trade-off curves between present reproductive value (i.e., fecundity, F_x), and future reproductive value, V_{x+1}.

SOLVED PROBLEM

1. (Advanced) Find the reproductive value schedule for a life history where reproductive maturity occurs at age 1, $s_0 = 0.6$/year and each adult age has survival rate $s = 0.8$/year and fecundity $b = 2$/year.

Answer:

Recall from Chapter 7 that this simple life history has $\lambda = s + s_0 b = 0.8 + (0.6)(2) = 2$/year. The survivorship from birth to age x is $l_x = s_0 s^{(x-1)}$ for $x > 0$ and 1 for age $x = 0$. Therefore, using Eq. (RV 1),

$$V_x = \frac{\lambda^x}{l_x} \sum_{t=x}^{max} \lambda^{-t} l_t b_t,$$

we may write for all ages $x > 0$,

$$V_x = \frac{\lambda^x}{s_0 s^{(x-1)}} \sum_{t=x}^{\infty} \lambda^{-t} s_0 s^{(t-1)} b.$$

After we remove constant terms from the summation,

$$V_x = \frac{s_0 \lambda^x b}{s_0 s^{(x-1)}} \sum_{t=x}^{\infty} \lambda^{-t} s^{(t-1)}$$

and cancel out the s_0, we have

$$V_x = \frac{\lambda^x b}{s^{(x-1)}} \sum_{t=x}^{\infty} \lambda^{-t} s^{(t-1)} \qquad (8.9)$$

It is not immediately obvious how to evaluate the sum of this infinite series given by Eq. (8.9). We note that s is less than or equal to 1, so as t (or x) gets larger, the terms get successively smaller, approaching 0. In Chapter 4 we presented three useful identities involving infinite sums for a fractional element S when $S < 1$. One of these, Infinite sum 1, is

$$\sum_{x=0}^{\infty} S^x = \frac{1}{1-S}$$

The next step is to try to push Eq. (8.9) into a form resembling Infinite sum 1. To do so, we bring everything but the constant b into the summation:

$$V_x = b \sum_{t=x}^{\infty} \lambda^{x-t} s^{(t-1-x+1)}$$

$$= b \sum_{t=x}^{\infty} \frac{s^{t-x}}{\lambda^{t-x}}.$$

To give the last expression a visual appearance more like that of Infinite sum 1, we make the substitution, $y = t - x$ for all ages, which will yield a 0 below the summation sign as in Infinite sum 1. Now we sum beginning at age $y = t - x = 0$:

$$V_x = b \sum_{y=t-x=0}^{\infty} \left(\frac{s}{\lambda} \right)^y \quad \text{for ages } x > 0. \tag{8.10}$$

We note also that $s/\lambda < s < 1$. (Recall that $\lambda = s + s_0 b$; therefore λ must be greater than s since b and s_0 are both positive.) Therefore we can apply Infinite sum 1 to Eq. (8.10), noting that in this case $S = s/\lambda$:

$$V_x = \frac{b}{1 - \frac{s}{\lambda}} \quad \text{for ages } x > 0 \tag{8.11}$$

This is a remarkable result. Since everything to the right of the equals sign is a constant, we see that the reproductive value does not depend at all on age x for this life history; all the age subscripts (x, t, and y) disappeared with cancellation. How can we explain this? Examining Eq. (RV 1) we see that the summation term must diminish for older ages since the sum begins at a later age, but the term outside the summation, λ^x/l_x increases with age in a way that exactly offsets this decline. The net result is that the reproductive value of all ages (beyond zero) are exactly equal. Moreover, since s_x, b_x and V_{x+1} are equal for all adult ages, the optimal reproductive effort is the same for all adult ages for this life history. Hence the reproductive value of ages at and beyond reproductive maturity (age 1) in this life history do not decline with age as they did in Figures 8.2, 8.3, and 8.4 for humans, weevils, and sparrowhawks, respectively. What is the numerical value for this constant reproductive value? Since we know λ, we can easily calculate it. Thus

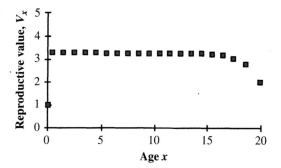

Figure 8.6
Reproductive value when all ages but the first and the last have the same vital rates. Here, all ages but newborns have the same vital rates, $b_a = 2$ and $s = 0.8$ (except that everyone dies at age 20; $s_{20} = 0$). Newborns have survival $s_0 = 0.6$.

$$V_x = \frac{b}{1 - \frac{s}{(s + s_0 b)}} \quad \text{for } x > 0,$$

and $V_0 = 1$. Using the values for s, s_0, and b, we get

$$V_x = \frac{2}{0.6} = 3.333\dots \quad \text{for } x > 0.$$

Any senescence that differentially diminishes b_x or s_x in later ages would, however, create a reproductive value schedule that declines with older ages. Suppose, for example, that senescence happens abruptly in the 20th year. Figure 8.6 shows the resulting reproductive value schedule. Note that reproductive value begins to decline even before age 20.

THE EVOLUTIONARY THEORY OF AGING (ADVANCED)

Have you ever wondered why life is not immortal? Why do people die? Clearly there are catastrophic events that can wipe people out, like fires or gunshots, but why does the probability of dying from natural causes increase as people get older? As people age, they accumulate knowledge and experience; for this reason alone we might, if anything, expect higher survival rates at older ages. Yet susceptibility to certain diseases increases at older ages. In short, people seem to deteriorate as they get older.

The formal definition of **aging** is a persistent decline in the age-specific fitness components of an organism due to internal physiological deterioration (Rose 1991). The **evolutionary theory of aging** posits that aging evolves through the action of natural selection because the force of natural selection generally diminishes with age and therefore any deleterious genes that increase mortality at late ages (e.g., like heart disease or Huntington's disease) are not strongly countered by natural selection. Thus these genes can increase in frequency either through genetic drift or because they might have side effects (pleiotropic effects), which may be slightly advantageous in the same individual when it is younger. That is, the bad effects in later ages are tolerated because of good effects at earlier ages. What does the Lotka–Euler equation have to say about this? Here we use Eq. (LE 3) from Chapter 4.

$$1 = \sum_{x=0}^{max} b_x l_x \lambda^{-x}$$

Hamilton (1966) argued that the proper assessment of selection against aging is by evaluating the effects of age specific perturbations of the vital rates, s_x and b_x, on r ($\lambda = e^r$), since r is often the appropriate measure of fitness in age-structured populations. Basically he sought to evaluate

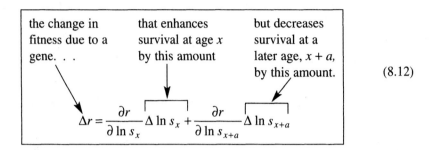

$$\Delta r = \frac{\partial r}{\partial \ln s_x} \Delta \ln s_x + \frac{\partial r}{\partial \ln s_{x+a}} \Delta \ln s_{x+a}$$

(8.12)

and

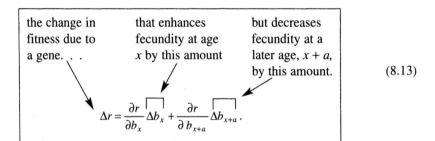

$$\Delta r = \frac{\partial r}{\partial b_x} \Delta b_x + \frac{\partial r}{\partial b_{x+a}} \Delta b_{x+a}.$$

(8.13)

To determine the selection on such genes, we need to know the magnitude of their effects at different ages (i.e., the Δb_x, Δb_{x+a}, Δs_x, and Δs_{x+a} in Eqs. 8.12 and 8.13 and the sensitivity of the fitness of the life history, which we take to be r, to changes in b_x and s_x, for different ages x, or, in other words, $\partial r / \partial \ln s_x$ and $\partial r / \partial b_x$. (The change to a ln transform for s_x is simply for mathematical convenience.) To determine these sensitivities we begin by taking the partial derivative of both sides of Eq. (LE 3) with respect to the change of the vital rate of interest (e.g., b_x). The partial derivative of the constant 1 with respect to b_x is, of course, 0 leaving us with just the chore of evaluating the partial derivative of the right-hand side

$$\frac{\partial 1.0}{\partial b_x} = 0 = \frac{\partial \left[\sum_{t=0}^{max} b_t l_t \lambda^{-t} \right]}{\partial b_x}.$$

Move the partial operation inside the summation

$$0 = \sum_{t=0}^{max} \frac{\partial (b_t l_t \lambda^{-t})}{\partial b_x}.$$

Next evaluate the partial derivative on the right

$$0 = l_x \lambda^{-x} - \sum_{t=0}^{max} t\, b_t l_t \lambda^{-t-1} \left(\frac{\partial \lambda}{\partial b_x} \right).$$

Rearrange this equation to solve for $\partial \lambda / \partial b_x$

$$\frac{\partial \lambda}{\partial b_x} = \frac{l_x \lambda^{-x}}{\sum_{t=0}^{max} t\, b_t l_t \lambda^{-t-1}}$$

Figure 8.7
The force of selection on the vital rates based on the life history in the Solved Problem. Reproductive maturity occurs at age 1, $s_0 = 0.6$/year, and each adult age has survival rate $s_a = 0.8$/year and fecundity $b = 2$/year.

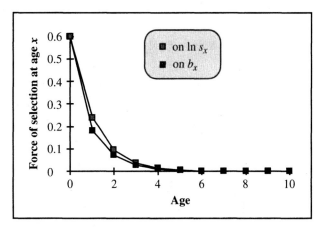

Finally, express this last result in terms of the intrinsic growth rate r rather than the discrete growth rate λ

$$\frac{\partial r}{\partial b_x} = \frac{\partial \ln \lambda}{\partial b_x} = \frac{\partial \lambda}{\lambda \partial b_x} = \frac{l_x \lambda^{-x}}{\sum_{t=0}^{\max} t\, b_t l_t \lambda^{-t}}$$

$$= \frac{l_x \lambda^{-x}}{A},$$

(8.14)

where the term A in the denominator is just another measure of generation time—one that is slightly different from the previous definition of T introduced in Chapter 4. It is the mean age of mothers of a set of newborns in a population with a stable age distribution.

Similarly, the parallel result for survival rate at age x, s_x, is made easier by taking natural logarithms:

$$\frac{\partial r}{\partial \ln s_x} = \frac{\partial \ln \lambda}{\partial \ln s_x} = \frac{\sum_{t=x+1}^{\max} b_t l_t \lambda^{-t}}{A}.$$

(8.15)

We learned in Chapter 4 that it is generally impossible to find a closed-form solution for r in terms of the vital rates, nevertheless, we have now found closed-form solutions given by Eqs. (8.14) and (8.15) for the precise way that r varies with changes in each of the vital rates b_x and s_x.

Figure 8.7 shows how $\partial r/\partial b_x$ and $\partial r/\partial \ln s_x$ change with age in the perennial population considered earlier in the Solved Problem ($s_0 = 0.6$, $s_a = 0.8$, and $b = 2$). Recall that for this life history the vital rates do not change with age; all adult ages have the same s_x and b_x. Nevertheless, the force of selection on s_x and b_x declines sharply in older ages. Once reproduction begins, the magnitude of the sum in the numerator of Eq. (8.15) must decrease with age because fewer positive terms are included.

As you can see in Figure 8.7, the force of selection is relatively weak at older ages. Any mutations that might have deleterious effects at these older and relatively "worthless" ages, could increase in the population if their negative effects late in life were compensated for by even slight positive effects earlier in life. Referring back to Eqs. (8.12) and (8.13), we can see that, as long as Δb_x and Δb_{x+a} are approximately equal in absolute value or when $\Delta \ln s_x$ and $\Delta \ln s_{x+a}$ are approximately equal in absolute magnitude, natural selection will favor alleles that enhance early survival or reproduction at the expense of later survival and reproduction. For birth rates, only if Δb_{x+a} (which is negative) is many times greater in absolute value than Δb_x (which is positive), would

Figure 8.8
The force of selection on ln s_x when all adult ages have the same s_x; $s_0 = 0.6$.

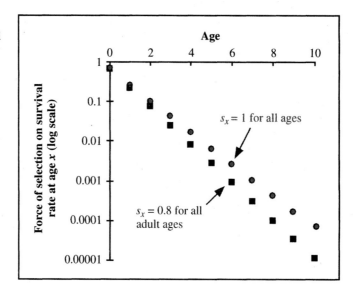

Figure 8.9
Reproductive value and the force of selection based on a more complex life history. The birth rate, b_x, is 0 until reproductive maturity at age 4, then increases gradually to a maximum of 2.0 at age 7, and later declines gradually to 0 at age 10. The individuals are assumed to be immortal ($s_x = 1$) for all ages.

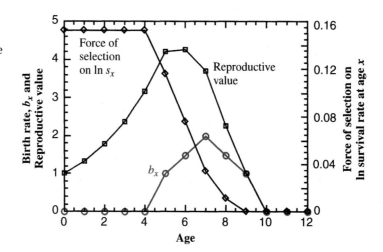

this tendency be counteracted. Similarly for survival rates: as aging evolves, the l_x curve swings downward some more, in turn weakening selection still more in later ages and thus creating a positive evolutionary feedback loop. Figure 8.8 shows the force of selection on ln s_x when $s_x = 1$, as well as when $s_x = 0.8$ for all ages x except newborns (as in Figure 8.7) to illustrate this comparison.

Any deleterious mutations that affect just later ages will be opposed by natural selection, albeit with weak force, unless they occur only in adults that no longer reproduce. However, if these same mutations have positive pleiotropic effects at earlier ages by enhancing reproduction or early survival, they could still increase in frequency because the net effect on r is positive. Mildly deleterious genes can also increase in frequency simply by random genetic drift, although this would happen only in relatively small populations. Yet, as selection becomes weaker, as it does at later ages, deleterious mutations could accumulate even in relatively large, but still finite, populations. So under both conditions, senescence might evolve.

Figure 8.9 shows a more complex life history. Superimposed is the force of selection on survival at each age, which stays at a maximum until reproductive maturity and then declines. The force of selection on s_x is most potent at weeding out deleterious mutations in the earliest ages and weakens progressively through life past reproductive maturity.

You may have noticed by now that the evolutionary theory of aging includes the assumption that genes exist that affect specific age classes. If genetic differences pro-

Figure 8.10
The number of eggs laid per year as a function of hen age for different strains of domestic chickens with different laying rates (from Pianka 1994 after Romanoff and Romanoff 1949).

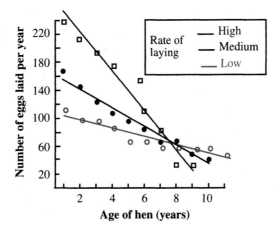

duce better or worse functioning for an organism at every stage of its life and for all life history components, then natural selection would not mold the rate of aging. All the vital rates would then be positively correlated, evolving together. The validity of this assumption is an open question and is subject to empirical tests. The best evidence that age-specific genetic differences exist is provided by selection experiments. The tiny nematode worm *C. elegans* has a particularly simple body plan and a rapid generation time and fecundity. Thus it is an ideal organism to use for isolating single mutations and studying their effects. An "age" mutation in the gene age-1 increases mean life span by about 65% and maximum life span by 110% by conferring resistance to extrinsic stress due to oxidizing agents (Lithgow and Kirkwood 1996). The mutants appear very similar to wild-type strains in their developmental rates, but they do show a loss in reproductive output (Johnson 1990). Thus even a single gene may exhibit pleiotropic effects between survival and reproductive output. Artificial selection on hens also reveals this trade-off. Selection for higher and earlier egg-laying in chickens results in more severe deterioration of egg production later in life, as illustrated in Figure 8.10.

One prediction from the curve for the force of selection on s_x in Figure 8.9 is that, if the age of reproductive maturity is postponed experimentally, then natural selection should remain at full intensity later into life. Over a number of generations, this might lead to the postponement of aging, as evidenced by higher survival rates for older ages. Michael Rose and colleagues conducted several experiments of this type with *Drosophila*. In one long-term experiment, five groups of flies were allowed to reproduce early in life and five groups of flies were kept from reproducing until late in life (Hutchinson and Rose 1990). The age at which females contributed eggs in the late reproducing lines was progressively increased, from generation to generation, until it was 70 days, compared to 14 days in the early reproducing lines. The mean lifetimes of these lines after 25 generations are shown in Figure 8.11. Note the increase in longevity of the late reproducing lines. This difference in longevity at the end of the experiment still exists if the flies are kept as virgins throughout their lives; therefore it seems that the physiology of the flies has really been altered by selection.

Moreover, after comparing these selected *Drosophila* lines that they had produced to be different in longevity, other characteristics associated with fitness were also apparent between them. The lines that were experimentally prevented from early reproduction and had evolved greater longevity showed declines in early fecundity, early metabolic rate, and early locomotor activity. On the other hand, these same lines had evolved greater fecundity at older ages and greater starvation and desiccation resistance. These changes in fecundity follow predictions based on the optimal reproductive effort model. **Increases in older adult survival rates selects for decreases in earlier reproductive effort.** The observed increase in later reproductive effort would also be expected if the s_x of these older ages was greater than somewhat younger ages—producing an l_x curve that declined abruptly in the terminal ages. This indeed was the case.

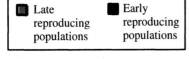

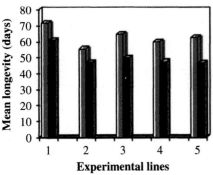

Figure 8.11
Mean longevities (days from pupal emergence) of early reproducing and late reproducing *Drosophila melanogaster* populations after 25 generations of selection (from Hutchinson and Rose 1990).

PROBLEMS

1. If you have access to mathematical software like Matlab©, the calculation of reproductive values is very easy. We have already shown that, for a Leslie matrix **L** (or any transition matrix for that matter), the dominant eigenvalue of **L** is equal to λ and the associated eigenvector (once scaled so that each element is divided by the sum of all the elements) gives the stable age structure. It turns out that the reproductive values are given by the dominant eigenvector (once divided by V_1) of the matrix which is simply the transpose of **L**.

What are the reproductive values associated with this Leslie matrix for the hypothetical deer population in Chapter 3 (Table 3.1)? Recall that F_1 is the first element in the first row.

$$\mathbf{L} \text{ (before)} = \begin{bmatrix} 0 & 0.42 & 0.468 & 0.6 \\ 0.9 & 0 & 0 & 0 \\ 0 & 0.9 & 0 & 0 \\ 0 & 0 & 0.5 & 0 \end{bmatrix}$$

Answer:

Recall that this Leslie matrix has $\lambda = 1.0$. The transpose of **L**(before) also has a dominant eigenvalue of 1.0. The associated eigenvector is given by Matlab as

0.5604   0.6227   0.4304   0.3362

We divide each element by the first $V(1) = 0.5604$, to calculate the reproductive values:

$$\mathbf{V}(x) = \begin{bmatrix} 1.0000 \\ 1.1112 \\ 0.7680 \\ 0.5999 \end{bmatrix}$$

Do you reach these same values for ages 1 to 4 using formula RV 2?

2. Consider a hypothetical organism with the relationship between residual reproductive value and present fecundity, F_x, shown in Figure 8.12.

    a. Determine the approximate sequence of optimal fecundities for each age.

    b. What is the optimal age of reproductive maturity for this organism?

    c. What is the maximal life span of this organism assuming optimal reproductive effort?

    d. Is this organism optimally semelparous or iteroparous?

3. Two similarly sized and closely related beetles live in the same habitat. The first beetle (species A) completes its entire life cycle on the leaves of deciduous trees. Adult beetles overwinter in the soil and then oviposit on leaves when they leaf out next spring; then the adults die. The other beetle (species B) completes its entire life cycle beneath the safety of bark of these same trees. These adults potentially live several years. Which beetle probably has the earliest age at reproductive maturity? Which beetle probably has the highest reproductive effort at age 1?

4. You want to become a green chinchilla rancher. Green chinchillas reach sexual maturity at age 2. Their yearly survivorship is about 0.8 and their yearly fecundity is $b_x = 2$ during the ages 2 to 5. After year 5, $b_x = 0$. Green chinchillas cost \$1000/pair, regardless of their age. To get your ranch profitable as soon as possible, what age chinchillas should you buy?

5. In many fish species, larger individuals have higher market value and thus fishing is concentrated on the larger individuals. These will be the fastest growing and/or older individuals in the population. This bias is sometimes even mandated by fishing regulations which restrict the catch based on a body size limit. What will the indirect evolutionary consequences be to the fish life history based on such a harvesting scheme?

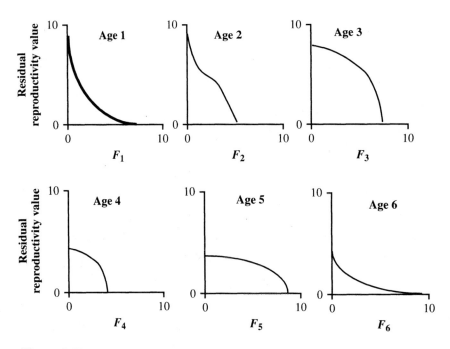

Figure 8.12

9

Density Dependent Selection on Life History Traits

We have shown how trade-offs between birth and death rates can influence the optimal life history strategy. In this chapter, we add two new features to our exploration of life histories. First, instead of simply relying on optimization arguments, we develop a simple explicit population genetic framework to explore the change in the frequency of genes controlling life history characters. Second, we examine the evolution of life history parameters when population growth is density dependent. What, if any, is the relationship between fecundity and population density? Darwin wrote in the *Origin of Species:*

> *The condor lays a couple eggs and the ostrich a score, and yet in the same country the condor may be the more numerous of the two: the fulmar petrel lays but one egg, yet is believed to be the most numerous bird in the world. One fly deposits hundreds of eggs and another like, Hippobosca, a single one; but this difference does not determine how many individuals of the two species can be supported in a district.*

Darwin's view is much like that expressed in the contiunous time logistic equation: the parameter r influences the rate of approach to carrying capacity but not the final population size, which is controlled solely by K. However, even if we accept this model as empirically correct, we may still wonder if population density affects the evolution of fecundity and whether natural selection will inevitably lead to increases in population size, even if, as we showed in Chapter 8, it does not inevitably lead to increases in birth rate.

EVOLUTIONARY ECOLOGY

A simple-minded difference between ecology and population genetics is that the former is concerned with changes in population size while the latter is concerned with changes in gene frequency. In this vein, ecology deals with equations like $N_{t+1} = f(N_t)$, where N_t is population size at time t, and a goal is to uncover the function f to be able to predict future population dynamics. Population genetics deals with equations like $p_{t+1} = g(p_t)$, where p_t is gene frequency at time t. Genetic evolution is the change in gene frequencies over time. The function g must contain the dynamics of natural selection, and it must also describe the features of genetic reassortment produced by meiosis and the fusion of new gametes each generation.

The subject of **evolutionary/ecology** links ecology and population genetics by recognizing that a population's growth rate may be influenced by the frequency of different genotypes in the population and that the fitness of different genotypes may, in turn, be functions of population size and gene frequency. Evolutionary ecology thus deals with coupled equations:

$$N_{t+1} = f(p_t, N_t) \quad \text{and} \quad p_{t+1} = g(p_t, N_t). \tag{9.1}$$

Figure 9.1
λ_g is the absolute fitness of a genotype g. It may or may not be influenced by population density or gene frequency. (a) The absolute fitness of genetype g, λ_g, is a constant, independent of density, N, and gene frequency, p. (b) The absolute fitness of a genotype , λ_g, is density dependent but not frequency dependent. Even if all geneotypes do absolutely worse at high densities than low, some genotypes may be less sensitive than others, as is 2 compared to 1. Thus the relative fitness of 2 compared to 1 increases with total population density. (c) Fitness, λ_g, is density independent but is frequency dependent. One example of frequency dependence is caused by predators that form a search image for the more common prey types. This can create a relative advantage for low-frequency genotypes in the prey population. (d) Fitness, λ_g, is both density and frequency dependent. In this example, genotype AA's absolute fitness is greatest at low total density and when it occurs in those populations at low frequency (p = frequency at allele A).

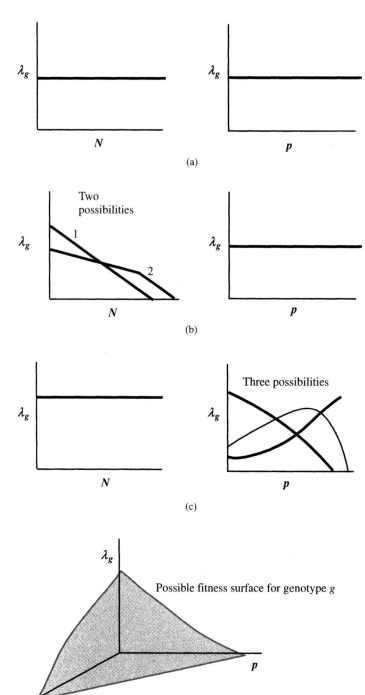

The mathematical model for several interacting species will be pairs of coupled equations for the numbers of each species and for the gene frequencies in each. Finally, several different genes may influence fitness within each species. If we know what the f and g functions are, then we should be able to project not only the changes in population size but also how changes in gene frequencies, in turn, influence population growth and density. To proceed we first must understand a few fundamentals of evolution by natural selection.

Darwinian fitness is the currency of natural selection. In lay terms *fitness* refers to properties of individuals that confer an advantage, but strictly applied, Darwinian fitness is an attribute of genes or combinations of genes (genotypes), not of individuals. Two individuals may be identical genetically—one, by chance, gets hit by lightning and

Figure 9.2
The life cycle has discrete, nonoverlapping generations.

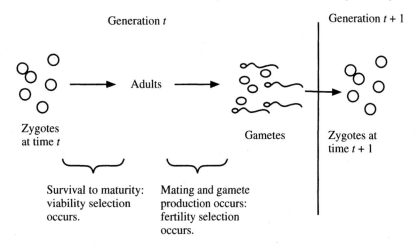

dies young. This has nothing to do with fitness—rather, it's bad luck. Fitness measures the average success of a particular genotype; it is averaged across all the individuals with that genotype. The cause of the fitness advantage may be lower mortality, higher birth rates, or more access to mates. Typically, fitness is expressed in relative terms— the success of a genotype compared to the average individual. *Relative* is a key word because it is not how well a gene or genotype does in an absolute sense that determines its increase in frequency over time, but only how well it does relative to other genotypes in the same population.

Potentially, natural selection may act in one of four ways with respect to population size and gene frequency. To keep the argument simple, simply consider a single locus with two alleles (a and A) in a diploid outbreeding sexual species. Let p be the frequency of allele A in the population censused at one specified time in the life cycle, say, at the zygote state. The absolute fitness of diploid genotype g is λ_g (where $g =$ AA, Aa, aa), as illustrated in Figure 9.1.

THE CASE OF NO DENSITY OR FREQUENCY DEPENDENCE

No density or frequency-dependent is the simplest case. To develop a model, we must first divide the life cycle of the organism into distinct phases and consider changes in gene frequency at the same point in the cycle from one time period to the next. Since selection can potentially operate in different phases of the cycle, this construction will prevent confusion and give us a common currency for comparison of fitness. While we have shown that fitness is a relative measure, it is best to start by defining terms describing natural selection in absolute terms (the absolute increase in a genotype's numbers from one generation to the next). Then, later, we can convert absolute fitness to relative fitness by dividing each genotype's absolute fitness by either mean fitness (averaged over all the genotypes) or by the fitness of the genotype with the maximum fitness. Either scaling converts fitness into relative terms.

Natural Selection and the Life Cycle

Imagine a semelparous life history with discrete nonoverlapping generations. The organism is a sexual diploid whose life cycle is illustrated in Figure 9.2. Adults die after they reproduce and the zygotes at $t + 1$ form the next generation.

Fertility selection could itself be broken into different components: mating success, birth rates, gamete survival, and so on. Similarly, males and females may differ. We assume random mating and ignore potential differences between the sexes. Also we assume that differences in gamete survival, if they occur, depend only on parental genotype, not on gamete genotype.

As we demonstrated in Chapter 1, without density-dependence, populations grow geometrically at some rate λ in discrete time. For a sequential life history like that shown in Figure 9.2, the only individuals that give birth are those that have survived to maturity from zygotes. Thus, if the fraction of the *homozygote* genotype AA that survives to maturity is $s_{0,AA}$ and the number of gametes produced per surviving AA individuals is b_{AA}, then an appropriate measure of absolute fitness of genotype AA is

$$\lambda_{AA} = s_{0,AA}b_{AA}. \tag{9.2}$$

Of course, for males, which produce millions of sperm, this measure of fitness would be huge. Most of these sperm are irrelevant since they quickly die. A more reasonable measure of fitness is therefore to equate b_g to the production rate of surviving gametes—those that survive to end up in zygotes; we assume that gamete survival is some constant incorporated in b_g and that it does not differ between genotypes. The composition of the zygote pool at time t (i.e., those zygotes on the far left in Figure 9.2) that has just been formed from random mating is

Genotype	Frequency
AA	p^2
aA	qp
Aa	pq
aa	q^2

Here, p = frequency of the A allele in the zygote pool and $q = 1 - p$ is the frequency of the a allele. Since it doesn't usually make any difference in fitness as to which chromosome (i.e., the one derived from the mother or from the father) contains the A or the a gene, the *heterozygotes* aA and Aa may be lumped into a single category with frequency $2pq$. These frequencies follow from the random mating and fusion of gametes as summarized by the Hardy–Weinberg principle. They depend on the population being very large in size (in fact infinite).

Given a total of N_t zygotes at time t, the absolute numbers of each genotype in the zygote pool can be deduced from the frequency of the p gene at time t.

Time	AA	Aa	aa
t	$p_t^2 N_t$	$2p_t q_t N_t$	$q_t^2 N_t$

At this point it is tempting to conclude that the numbers of each genotype in the zygote pool at time $t + 1$ will simply be the numbers at time t multiplied by the fitness of each genotype

Time	AA	Aa	aa	
$t + 1$	$p_t^2 N_t \lambda_{AA}$	$2p_t q_t N_t \lambda_{Aa}$	$q_t^2 N_t \lambda_{aa}$	(Guess 1)

However, this logic is wrong unless this organism reproduces asexually. An adult of genotype AA produces not just AA young, which is the assumption implied by Guess 1. The gametes from AA are all A, but they may fuse with other gametes to produce genotypes of both AA and Aa. We must build this genetic reassortment into the model.

The entries in Guess 1 can lead us to the correct expression, however, since Guess 1 tells us the gamete production rate by each genotype near the end of generation t and before new zygotes are formed:

$$\begin{bmatrix} p_{t,s} \text{ in the gamete pool (i.e., the frequency} \\ \text{of allele A after one bout of viability and} \\ \text{fertility selection)} \end{bmatrix} = \begin{bmatrix} \dfrac{\text{number of A alleles after selection}}{\text{number of all alleles after selection}} \end{bmatrix}.$$

Since each adult has two alleles, this expression reduces to

$$p_{t,s} = \frac{2p_t^2 N_t \lambda_{AA} + 2p_t q_t N_t \lambda_{Aa}}{2p_t^2 N_t \lambda_{AA} + 4p_t q_t N_t \lambda_{Aa} + 2q_t^2 N_t \lambda_{aa}} \tag{9.3}$$

The 2's and the *N*'s will cancel from the numerator and denominator of Eq. (9.3), giving

$$p_{t,s} = \frac{p_t^2 \lambda_{AA} + p_t q_t \lambda_{Aa}}{p_t^2 \lambda_{AA} + 2 p_t q_t \lambda_{Aa} + q_t^2 \lambda_{aa}} \qquad (9.4)$$

Note that the denominator of Eq. (9.4) is the average absolute fitness across all individuals (and genotypes). We denote this average (or mean) absolute fitness in the denominator as $\bar{\lambda}_t$. The subscript *t* indicates that the mean fitness of the population is changing over time—not because the genotypic fitnesses are changing (since these are constants) but because gene frequency and hence the genotypic composition of the population is changing. Defined in this way, $\bar{\lambda}_t$ also measures the total rate of population growth between *t* and *t* + 1. Thus

$$N_{t+1} = N_t(p_t^2 \lambda_{AA} + 2 p_t q_t \lambda_{Aa} + q_t^2 \lambda_{aa}) = \bar{\lambda}_t N_t. \qquad (9.5)$$

Finally, since we assume that the gametes fuse randomly to make new zygotes and that all gametes have the same survival rate, zygote production results only in a redistribution of the same relative numbers of alleles present in the gamete pool. Thus the gene frequency of allele A in zygotes at time *t* + 1 is identical to that in Eq. (9.4) for gametes, or

$$p_{t+1} = p_{t,s} = \frac{p_t^2 \lambda_{AA} + p_t q_t \lambda_{Aa}}{p_t^2 \lambda_{AA} + 2 p_t q_t \lambda_{Aa} + q_t^2 \lambda_{aa}} = \frac{p_t^2 \lambda_{AA} + p_t q_t \lambda_{Aa}}{\bar{\lambda}_t}. \qquad (9.6)$$

Consequently the absolute number of AA genotypes, for example, in zygotes at time *t* + 1 is

$$N_{AA,t+1} = p_{t,s}^2 N_{t+1} = P_{t+1}^2 \bar{\lambda} N_t$$

which is *not equal* to $p_t^2 N_t \lambda_{AA}$, the quantity in Guess 1, because $p_{t+1}\bar{\lambda}$ **(given by Eq. 9.6) is not equal to** $p_t^2 \lambda_{AA}$ **(as given by Guess 1).** This distinction is important and often a source of confusion. Review the development beginning with Figure 9.2 and Eq. (9.2) to reinforce this distinction.

In summary, Eq. (9.5) and (9.6) provide the functions *f* and *g* presented at the beginning of this section in Eq. (9.1) for the simple case where genotypic fitness depends neither on population density nor gene frequency. More complicated life histories, and the incorporation of frequency- and density-dependence in the fitness functions, can be handled by using the same techniques that we have just developed; we do some of this later in this chapter. Note that, even though the fitness of each genotype is independent of *N* and *p* (e.g., λ_{AA} is a constant, not a function of *N* and *p*), the change in *N* and the change in *p* each generation are very much functions of the frequency of genes in the population, as indicated by Eq. (9.4). **Thus *frequency independent* refers to genotypic fitness, λ_g, being independent of *p*; it does not mean that the total per generation change in *p* is independent of *p*. The two concepts are not identical.**

For **overlapping generations,** but still pulsed reproduction, the analog to iteration Eq. (9.5) for *N* and Eq. (9.6) for *p* are unchanged, except that time *t* now represents unit time periods (e.g., consecutive years) and λ is expressed in units of per year, rather than per generation. However, instead of Eq. (9.2), (that is, $\lambda_g = s_{0,g} b_g$, giving the fitness for each genotype *g* in terms of the life history parameters, we would substitute the λ_g derived from the Lotka–Euler equation for the life history of each genotype *g* (see Chapters 3 and 4). For example, in Chapter 3 we found that, when all adult ages have the same yearly survival rate, s_a, and birth rate, *b*, and when young mature at age 1 and have survival rate s_0, then the intrinsic rate of growth is quite simple: $\lambda = s_a + bs_0$. The absolute fitness of each genotype *g* (*g* = AA, Aa, aa) is thus $\lambda_g = s_{a,g} + bs_{0,g}$. If the life history schedule is more complicated, then we could form the Leslie matrix for each genotype and determine the dominant eigenvalue of that Leslie matrix and use that value for the gentoypic fitness. Iteration Eqs. (9.5) and (9.6) are still used to track changes in N and p with this substitution for the calculation of fitness. The assumption is that mating is random and that frequency dependence is absent. A further assumption is that selection is relatively weak. If selection is not weak, then two assumptions may be violated: the population may be far from Hardy–Weinberg equilibrium, and the population may be knocked out of the stable age structure. In the latter case, the λ_g's (which are constant only at stable age structure) will not approximate the change in numbers of each genotype by selection (see Chapters 3 and 4).

How Does Natural Selection Affect the Genetic Composition of the Population?

The coupled equations for the dynamics of population size and gene frequency can be used to explore the effect of different fertility and survivorship curves on the population dynamics and evolution of a population.

At this point it is useful to simplify our iteration equations, Eqs. (9.5) and (9.6). Recall that fitness has been defined so far in absolute terms, the λ_g's. However, in general, it is more revealing to describe fitness in relative terms. Note that in Eq. (9.6) the λ's appear in both the numerator and the denominator. This means that each λ can be multiplied by a constant (say, k) and that the constant will divide out. The exact values of the λ's are not important; the relative magnitude of one to the other is what counts. One way to make the relative relationship more apparent is to set one of the λ's to 1 and scale the others accordingly. For example, let's set $\lambda_{AA} = 1$. Then we have

$$w_{AA} = \frac{\lambda_{AA}}{\lambda_{AA}} = 1, \quad w_{Aa} = \frac{\lambda_{Aa}}{\lambda_{AA}}, \quad \text{and} \quad w_{aa} = \frac{\lambda_{aa}}{\lambda_{AA}}.$$

These w's may be called relative fitnesses and may be any number greater than 0. After substituting these $w_g\lambda_{AA}$ terms into Eq. (9.6) for each of the λ_g terms and then canceling all the λ_{AA} terms, we reach the equivalent expression,

$$p_{t+1} = \frac{p_t(p_t + q_t w_{Aa})}{p_t^2 + 2p_t q_t w_{Aa} + q_t^2 w_{aa}}. \tag{9.7}$$

The only advantage of Eq. (9.7) over Eq. (9.6) is that it has one less parameter since there is no w_{AA} term. There are several other ways to convert from absolute to relative fitness; one is to divide genotypic fitness by the mean fitness in the population at the time, which represents the denominator of Eq. (9.7). However, if gene frequencies change over time, as they certainly will under natural selection, then so must the mean fitness of the population. Consequently, the relative fitnesses of the genotypes defined by this scaling, can be changing over time even though the absolute fitnesses are constants. We'll stick for now with normalizing the fitnesses by the absolute fitness of genotype AA.

By definition an equilibrium gene frequency for p will occur when p no longer changes over time—in other words, when $p_{t+1} = p_t$. There are several ways this can happen: if $p_t = 0$, then p_{t+1} will be 0 by Eq (9.7). Similarly, if $p_t = 1$, then $p_{t+1} = 1$ and we have an equilibrium. However, a more interesting *interior* equilibrium (i.e., one with p_t at some value between 0 and 1, can also emerge for some values of the relative fitnesses, w_g. When we divide both sides of Eq. (9.7) by p_t and assume that an equilibrium exists (i.e., we assume that $p_{t+1}/p_t = 1$), we get

$$\frac{(\hat{p} + \hat{q}w_{Aa})}{\hat{p}^2 + 2\hat{p}\hat{q}w_{Aa} + \hat{q}^2 w_{aa}} = 1.$$

The hats over the p's and q's indicate that these aren't just any values of p and q, but rather are the equilibrium values. Our goal now is to solve for these equilibrium values. Rearranging gives

$$\hat{p} + \hat{q}w_{Aa} = \hat{p}^2 + 2\hat{p}\hat{q}w_{Aa} + \hat{q}^2 w_{aa}. \tag{9.8}$$

Rearranging Eq. (9.8) and recognizing that $q = 1 - p$, we arrive at an expression for the interior equilibrium gene frequency:

$$\hat{p} = \frac{w_{Aa} - w_{aa}}{(w_{Aa} - w_{aa}) + (w_{Aa} - 1)}. \tag{9.9a}$$

An equivalent formula based on the absolute fitnesses is:

$$\hat{p} = \frac{\lambda_{Aa} - \lambda_{aa}}{2\lambda_{Aa} - \lambda_{aa} - \lambda_{AA}}. \tag{9.9b}$$

When $\hat{p}$ is maintained at some intermediate level in a population, we say that a **polymorphism** exists at this locus.[1]

As we showed in Chapter 5, another important issue in dynamics is the evaluation of the stability of an equilibrium point once it has been identified. Equation (9.9) allows for two ways to get a feasible interior equilibrium gene frequency, $\hat{p}$: when the heterozygote Aa has a higher fitness than both homozygotes or when the heterozygote Aa has a lower gene frequency than both homozygotes. Do both situations lead to a stable situation—one that evolution would reach? The answer is no!

A full analytical stability analysis is beyond the level of this primer, but the results can be made understandable by simulating some sample dynamics, using the iteration Eq. (9.5) for population size, N, and Eq. (9.6) for gene frequency, p. Figure 9.3 shows the dynamics of N and p for some different combinations of λ_g's. In each case, the population is initiated with several different values of the gene frequency of allele A, which is p_0. In each case, the population goes to an equilibrium but both alleles persist at a stable equilibrium only when there is a heterozygote advantage. In the case of a heterozygote disadvantage, there are two alternative stable outcomes: $\hat{p} = 1$ when p_0 is initiated above a certain threshold determined by the relative magnitude of the λ_g's, and $\hat{p} = 0$ when p_0 is initiated below this threshold. Only if p_0 is at exactly the value of the unstable equilibrium frequency, $\hat{p}$, will p remain at this unstable equilibrium ($\hat{p} = 0.25$ for the example in Figure 9.3d).

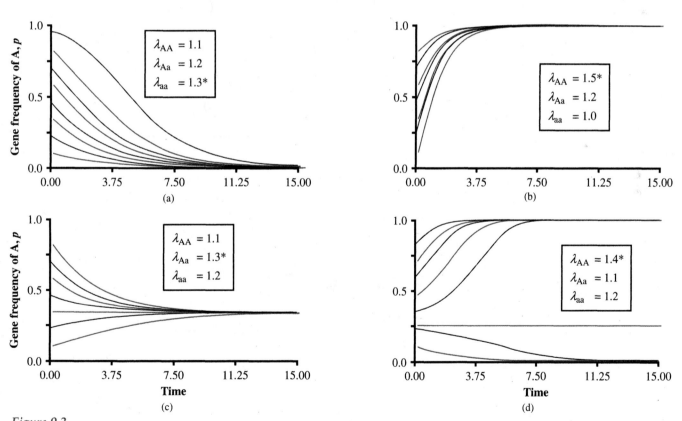

Figure 9.3
The dynamics of the frequency of allele *A, p,* for some different combinations of genotypic fitnesses and initial gene frequency. An asterisk indicates the genotype with the highest fitness.

---

1. If you're having difficulty getting from Eq. (9.8) to Eq. (9.9), don't feel too bad. It's a bit of a strain. You can find derivations in population genetics textbooks. See, for example, Roughgarden (1979) or Crow (1986).

Figure 9.4 shows the coupled changes in population size, N and frequency, p, over time for four selected cases. Figure 9.4(b) is based on Figure 9.3(b) (the homozygote AA is superior and the heterozygote Aa is intermediate). Figure 9.4(c) is based on Figure 9.3(c) (the heterozygote Aa is superior) and Figures 9.4(a) and (d) are based on Figure 9.3(d) (the heterozygote is inferior). In each case illustrated in Figure 9.4, as natural selection proceeds, the population size increases; the rate of population growth also increases until about the time that the gene frequency reaches its equilibrium value. However, comparing parts (a) and (d), both with the heterozygote inferior, note that in part (a), when the a allele becomes fixed at equilibrium, the final rate of population growth is still not as high as that achieved by natural selection in part (d), when the alternative A allele is fixed. Also, in the case shown in Figure 9.4(c), where the heterozygote has the highest fitness, the population consists of only a proportion of individuals that are heterozygote, because homozygotic offspring AA and aa are produced every generation through sexual reproduction. In this example this proportion at equilibrium is $0.1/0.3 = 1/3$ by Eq. (9.9b). Thus, natural selection tends to maximize population growth under some circumstances, but only within the limits of the genetic variability that it has to work with and given the constraints of sexual reproduction.

For age-structured populations, the λ_g of a particular genotype is a measure of its Darwinian fitness as long as selection is not too strong, mating is random, and fitness is not frequency dependent (Charlesworth 1994). This means that under these conditions it will be possible to evaluate natural selection as it affects the evolution of life histories in terms of the geometric growth rates, λ_g. Those genotypes whose life histories

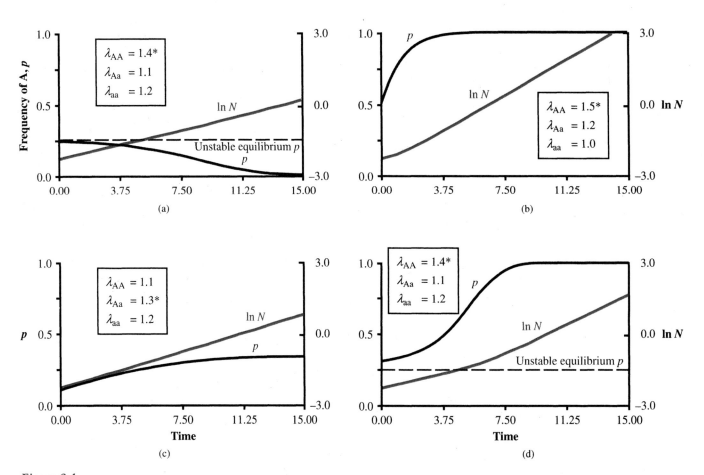

Figure 9.4
Change in the ln of population size, ln N, and the frequency, p, of the A allele when genotypic fitnesses are constant with the values shown. An asterisk indicates that genotype with the highest fitness.

have the highest λ_g, will increase in frequency relative to those with lower λ_g. In fact, it can be shown that for two alleles at a single locus (and with the assumptions previously stated), the equilibrium gene frequency that natural selection reaches is one that produces the highest average λ for the population, as long as there is one and only one stable equilibrium, $\hat{p}$. This latter caveat excludes the case where the heterozygote has the lowest fitness since, as we've shown, this produces two alternative stable equilibria, at $\hat{p} = 1$ and $\hat{p} = 0$; which one is reached depends on the initial conditions, not just on fitness. If the heterozygote has the highest fitness, p is adjusted by natural selection until mean fitness $p^2\lambda_{AA} + 2pq\lambda_{Aa} + q^2\lambda_{AA}$ is maximized. This result even extends to multiple alleles at a single locus; the vector of allele frequencies maximizes average population fitness. We now turn to the case of density dependent fitness.

DENSITY DEPENDENT NATURAL SELECTION

With this groundwork, we are now in a position to explore case 2 of Figure 9.1, namely, the interesting and ecologically relevant case where a genotype's fitness depends on population density (Roughgarden 1971, Charlesworth 1971). Because you are already familiar with the discrete logistic equation, we use it to describe density-dependence; each genotype is specified according to the R and K values that it would have in a homogeneous population of just its own genotype. Everything else in the development is the same, so equations

$$N_{t+1} = N_t(p_t^2\lambda_{AA} + 2p_tq_t\lambda_{Aa} + q_t^2\lambda_{aa}) = \bar{\lambda}_tN_t$$

and (9.6) still form the dynamical core, except that now instead of the λ_g (g = AA, Aa, aa) being constant over time, these fitnesses will be functions of total population size, N_t. In other words, we write $\lambda_g(N_t)$. The fitness of genotype g, λ_g, is a function of population size, N, which is itself changing over time t. Using the discrete logistic equation (see Chapter 5) to specify this particular density dependence gives

$$\lambda_g(N_t) = 1 + R_g\left(1 - \frac{N_t}{K_g}\right). \tag{9.10}$$

Since $\lambda_g(N_t)$ is a function of total population size, N_t, and since N_t is changing over time, the values of λ_g for each genotype change each time step. Nevertheless, we may simply iterate the coupled difference equations, Eqs. (9.5) and (9.6), with the substitution of Eq. (9.10) for the fitness of each genotype at time t to obtain

$$N_{t+1} = N_t\left(p_t^2\lambda_{AA}(N_t) + 2p_tq_t\lambda_{Aa}(N_t) + q_t^2\lambda_{aa}(N_t)\right) = \bar{\lambda}_t(N_t)N_t$$

and

$$
\begin{aligned}
p_{t+1} &= \frac{p_t^2\lambda_{AA}(N_t) + p_tq_t\lambda_{Aa}(N_t)}{p_t^2\lambda_{AA}(N_t) + 2p_tq_t\lambda_{Aa}(N_t) + q_t^2\lambda_{aa}(N_t)} \\
&= \frac{p_t^2\lambda_{AA}(N_t) + p_tq_t\lambda_{Aa}(N_t)}{\bar{\lambda}_t(N_t)}.
\end{aligned}
\tag{9.11}
$$

As we did for density independent selection, the two Eqs. (9.11) can be used to solve for the potential equilibrium of N and p. The next step for a fuller understanding of the dynamics is to evaluate the stability of that equilibrium. However, this exercise is quite tedious, and instead we simply simulate Eqs. (9.11) for different combinations of K_g and R_g. Figure 9.5 shows three examples when K varies across genotypes but R is constant.

Not too surprisingly, the genotype with the highest K ultimately goes to fixation unless it is the heterozygote. In the latter case a polymorphism is maintained at a stable gene frequency. The R_g values influence the rate of approach to the equilibrium but not its

Figure 9.5
Changes in population size, N, and gene frequency, p, when genotypic fitnesses are given by the density dependent discrete logistic equation.
(a) Homozygote AA has the highest K.
(b) Homozygote aa has the highest K.
(c) heterozygote Aa has the highest K, and genotype aa has the next highest K.

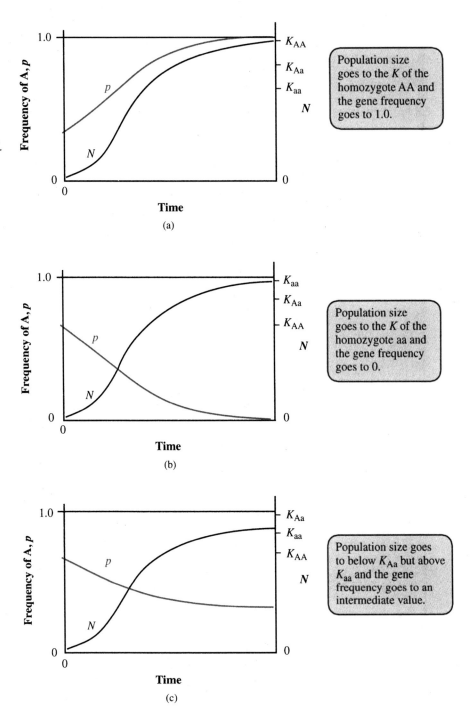

Population size goes to the K of the homozygote AA and the gene frequency goes to 1.0.

Population size goes to the K of the homozygote aa and the gene frequency goes to 0.

Population size goes to below K_{Aa} but above K_{aa} and the gene frequency goes to an intermediate value.

stability unless they are very high in a discrete time model. In that case the population dynamics can emit limit cycles and even chaos, as we demonstrated in Chapter 5. The dynamics of gene frequency can parody the complicated dynamics of population size.

Now let's take a closer look at the situation in which an inverse relationship exists between a genotype's R and its K, such that the genotype with the highest R has the lowest K, as depicted in Figure 9.6. Where will natural selection and population growth take population size and gene frequency?

At low densities (see the region of N spanned by the bar labeled low N) the AA genotype has the highest fitness and will grow faster than the other genotypes. Yet as the population size increases, genotype aa has the fitness advantage and the higher the population size, the greater is its advantage. Consequently, in this example the population will ultimately consist of just genotype aa—the genotype with the highest carrying

Figure 9.6
A possible specification for three genotypes with R and K inversely related.

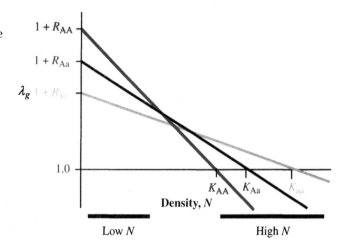

Figure 9.7
Simulation for a case similar to that described in Figure 9.6.

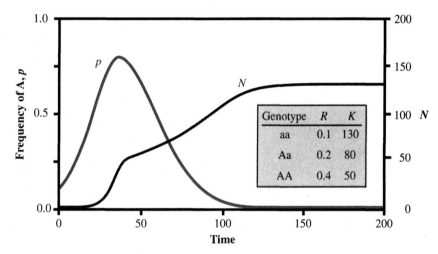

Genotype	R	K
aa	0.1	130
Aa	0.2	80
AA	0.4	50

capacity-and population size will reach K_{aa}. In summary, genotype aa, with the K advantage, will ultimately prevail and p will go to zero. This result is illustrated in the simulation presented in Figure 9.7, which is based on Eq (9.11). The population started at low numbers, and hence genotype AA, with a high R but low K, had the initial advantage and quickly increased. The overall size of the population gradually increased as p increased. Yet as the population size grew larger, density-dependence became stronger. So genotype AA, with the lowest K, began to decline and the frequency of the A allele, p, declined.

If the heterozygote has the K advantage, then, as for density independent selection, p will reach an intermediate level and population size will grow to a level intermediate between K_{Aa} and that of the homozygote with the next highest K (see Figure 9.5). The value of R_g has little effect on the long-term behavior of a population in a constant environment, but it does influence the transient behavior on the way to this equilibrium.

As we have discussed, real populations are often buffeted by density independent factors, too, which can knock the population down to levels far below the carrying capacity that it would otherwise achieve. To illustrate, imagine that the population described in Figure 9.7 experiences a random catastrophe (perhaps a cyclone or a fire) with a probability of 0.2 every year. Suppose further that, if the catastrophe hits, the population size of each genotype is equally reduced by 60%. If the catastrophe doesn't hit in a year, then the population and gene frequency grow according to Eq. (9.11). The results of such a simulation are shown in Figure 9.8. Under the threat of this severe but relatively infrequent catastrophe, the AA genotype, with the highest R (but lowest K), becomes nearly fixed in the population.

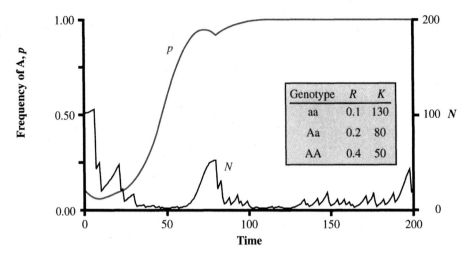

Figure 9.8
The situation illustrated in Figure 9.7, but now the population also experiences a random catastrophe.

How great does the density independent mortality need to be to shift the competitive dominance toward the high *R* species? Armstrong and Gilpin (1977) explored this question with a discrete logistic equation that includes one extra term, *f(t)*, for the fraction of individuals killed in a density independent manner. The fraction may vary from one time period to the next; hence the notation *f(t)*. For example, if 10% of the individuals die in 1990, then *f(t = 1990)* = 0.1. First, let's consider an asexual model. The change in numbers of a clone *g* is

$$\Delta N_g(t) = \left(\lambda_g(N_t) - 1\right)\Delta t = \left(R_g\left(1 - \frac{N_t}{K_g}\right) - f_g(t)\right)\Delta t \quad \text{for all genotypes } g. \quad (9.12)$$

Consider two clones, 1 and 2, growing according to Eq. (9.12). As densities increase, the growth rates of the clones decrease, but one clone might be able to continue to increase at densities for which the other clone is already declining. To determine which clone has the advantage, we need to determine their relative sensitivities to density, N_t, when density levels are at equilibrium. Both clones respond to N_t in the same way; this is why total population density does not have a *g* subscript. Thus the growth rates for each clone given by Eq. (9.12) can be used to produce two separate equations for N_t at equilibrium—one for each clone. For clone 1, we get

$$R_1\left(1 - \frac{N_t}{K_1}\right) - f_1(t) = 0.$$

After rearranging to solve for N_t we get

$$\frac{R_1 - f_1(t)}{\dfrac{R_1}{K_1}} = K_1\left(1 - \frac{f_1(t)}{R_1}\right) = N_t. \quad (9.13a)$$

For clone 2, we get

$$\frac{R_2 - f_2(t)}{\dfrac{R_2}{K_2}} = K_2\left(1 - \frac{f_2(t)}{R_2}\right) = N_t. \quad (9.13b)$$

Over time, *f(t)* may be approximated by $E(f_g)$, the time average of f_g. Consequently, the clone with the highest *average* per capita population increase is that with the highest

$$K_g\left(1 - \frac{E(f_g)}{R_g}\right). \quad (9.14)$$

Figure 9.9
Simulation results and predicted results based on the Armstrong–Gilpin criterion applied to sexual populations. The genotype with the highest R has the lowest K. The density independent mortality rate each time period, $f(t)$, is drawn from a random distribution, which is different for the three cases. The genotype with the maximum value of $K(1 - E(f)/R)$ is indicated by an arrow.

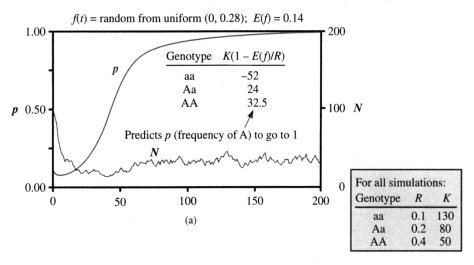

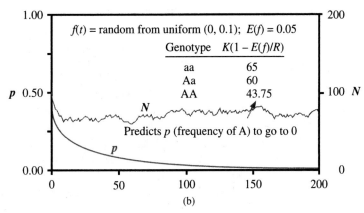

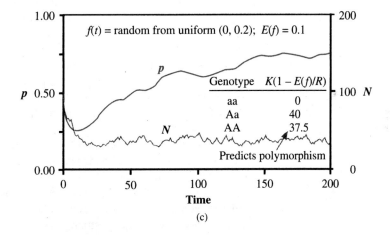

When the degree of density independent mortality is assumed to be equal for all clones, we can drop the subscript g for f and the criterion of fitness given by Eq. (9.14) reduces to

$$K_g\left(1 - \frac{E(f)}{R_g}\right). \tag{9.15}$$

When density independent mortality is zero (i.e., $E(f) = 0$), K alone determines the ultimate outcome. As average density independent mortality, $E(f)$, increases, the relative advantage of clones with higher R is increased precisely as specified by Eq (9.15).

To extend this to a sexual diploid population, we need to add random mating each generation. Figure 9.9 shows three simulations for density dependent selection

Figure 9.10
Grime's (1979) conceptualization of
life history strategies in plants. Typical
population densities are given by
shades of red; darker red indicates
higher density. Plant strategies
associated with the different
environmental conditions are ovals in
this two dimensional environmental
space.

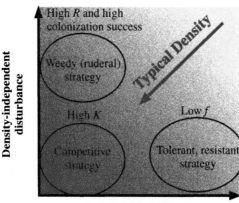

according to Eq. (9.12) applied to a sexual outbreeding species. Each time period a
random fraction, $f(t)$, of individuals die; $f(t)$ is drawn from a uniform random distribution and the resulting $f(t)$ is applied to all three genotypes. The fitness criterion of
Eq. (9.15) is calculated for each genotype.

Note that, at the intermediate level of the average density independent mortality,
$E(f)$, (Figure 9.9), the heterozygote is predicted to have the highest fitness. The simulation results thus show the emergence of a polymorphism, suggesting the Armstrong–
Gilpin criterion also extends to density dependent selection with density–independent
"noise" in sexual diploid species.

To the extent that a genotypic trade-off exists between R and K, the high-K genotypes are expected to prevail in environments that are relatively constant and crowded,
while the genotypes with the higher R's will prevail in environments that fluctuate due
to strong density independent forces knocking N's down to levels far below carrying
capacity. Density dependent forces that cycle—perhaps from the actions of predators
and parasites—could act like the catastrophes in the preceding model. This suggests an
additional life history criterion that influences the competitive outcome: the susceptibility of a genotype to forces imposing density independent mortality, f_g. In Eq. (9.14),
fitness is determined by the interaction of all three parameters.

Grime (1979) suggested that, to fully understand the life history strategies among
plants, we must characterize their environment according to two independent factors:
(1) the degree of disturbance (e.g., floods, fires, etc.), which we may translate into density independent mortality—for plants disturbance kills individuals but it also frees up
open space for new recruits; and (2) the severity of living conditions (the degree that
plants experience chronic shortages of water, nutrients, and light). The impact of such
factors will likely depend on the species experiencing them. In Grime's scheme, individual plant species fit into these two dimensions according to three central strategies,
as illustrated in Figure 9.10: a weedy or ruderal strategy when disturbance is high and
severity is low; a competitive strategy when population density is high, as it might be
when both disturbance and severity are low; and a tolerant or resistant strategy when
severity is high but disturbance is low. The weedy strategy necessitates a large R along
with good colonization abilities to get to new open space before others. The competitive strategy demands a high K, and the tolerant strategy implies a low f and an ability
to survive predictable but harsh conditions.

Self-Generated Population Cycles

In the preceding examples, external environmental factors led to variability in population size, which in turn altered the competitive balance among genotypes. We have also
shown that population cycles can be generated internally, through the discrete logistic
equation with $R > 2.0$, which creates the opportunity for an evolutionary positive feed-

back in R. Imagine an environment that initially favors high R genotypes. If these R's are large enough to produce internally driven population cycles or even chaotic population dynamics, then they could produce larger population fluctuations. This, in turn, might favor the evolution of still higher R's in a positive feedback manner (Southwood et al. 1974).

Is an R–K Trade-Off Biologically Reasonable?

Before moving on, we should ask whether it is even reasonable to expect a trade-off between R and K among a range of genotypes or competing species. Let's break R into its demographic components: birth and death rates. Figure 9.11 shows three possible cases for the evolution of density dependent birth and death rates with different assumptions about possible trade-offs between birth and death rates.

In Figure 9.11(a), the birth rate, $b(N)$, doubles through evolution at each density, leading to an increase of $R = b(0) - d(0)$ and a sizable increase in carrying capacity (from K_1 to K_2). If this diagram accurately depicts the evolution of R, then R and K are expected to be positively correlated. We would not expect a trade-off between R and K across genotypes under this model; in fact if this were generally the case, then contrary to Darwin's view expressed at the beginning of this chapter, species with high fecundities would also exhibit high abundance. This pattern has been seen in some selection experiments. Luckinbill (1978) subjected bacteria populations of *E. coli* to selection regimes involving low density and high density. He found that the K-adapted genotypes were clearly superior in competition under both high- and low-density regimes. Adaptation to high density and thus K selection resulted in the selection of genotypes with a higher growth rate *and* a higher K. Moreover, Luckinbill found that these K-adapted genotypes secreted an inhibitory substance that restricted the growth of r-selected strains. Luckinbill (1979) performed a similar experiment with the protozoan *Paramecium,* except in that case four strains were selected for high R. This selection resulted in an increase in both R and K. Thus in both experiments, the fundamental trade-off between R and K assumed by theory was not observed.

In Figure 9.11(b), a twofold evolutionary increase in $b(N)$ is also accompanied by a twofold increase in death rates $d(N)$. This scenario is more in keeping with the energy allocation models of Chapter 7, which imply that any increase in energy allocated to reproduction will result in a decrease in the energy that otherwise could be devoted to survival. Here, the resulting K is approximately unchanged, given this trade-off. This situation implies the kind of uncoupling between fecundity and typical population size that Darwin felt was the norm. It arises because high birth rates are associated with high death rates. In fact, Darwin (1859) anticipated exactly this condition. He stated that

> *A large number of eggs is of some importance to those species, which depend on a rapidly fluctuating amount of food, for it allows them rapidly to increase in numbers. But the real importance of a large number of eggs or seeds is to make up for much destruction at some period of life; and this period in the great majority of cases is an early one.*

It is quite possible, of course, that K could decrease with the evolution of higher birth rates if death rates increased disproportionately with increases in birth rates, as shown in Figure 9.11(c). In species where intraspecific interference plays a strong role in competitive ability, this might be expected. For example, in *Tribolium castaneum,* artificial selection that produces highly productive (high R) strains also leads to low population sizes and weak competitive abilities when paired with the similar species *T. confusum.* Recall that cannibalism dominates intra- and interspecific competition in these flour beetles (Wade 1980).

In conclusion, the existence of an R and K trade-off can be translated to imply a positive correlation between density dependent birth rates and death rates (or equivalently an antagonistic trade-off between birth rates and survival rates). Given what we presented about life history trade-offs in Chapter 8, a trade-off between birth rates and

Figure 9.11
Three possible cases for the coupled
evolution of per capita death rates, *d*,
birth rates, *b*, and carrying capacity, *K*,
under density-dependence in birth
rates and death rates.

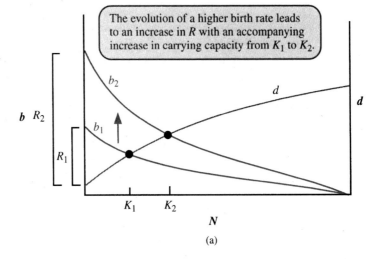

The evolution of a higher birth rate leads
to an increase in *R* with an accompanying
increase in carrying capacity from K_1 to K_2.

(a)

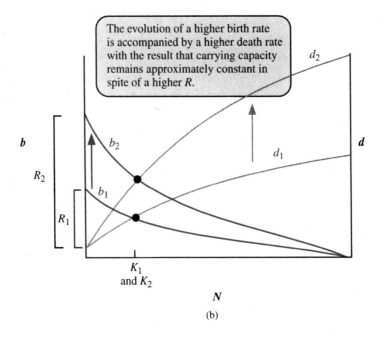

The evolution of a higher birth rate
is accompanied by a higher death rate
with the result that carrying capacity
remains approximately constant in
spite of a higher *R*.

(b)

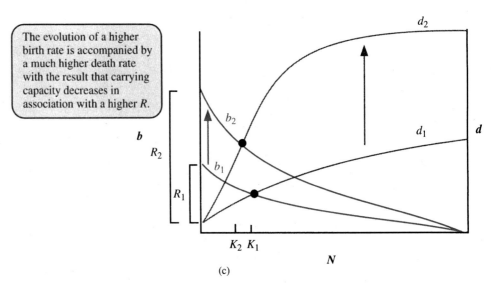

The evolution of a higher
birth rate is accompanied by
a much higher death rate
with the result that carrying
capacity decreases in
association with a higher *R*.

(c)

survival rates seems reasonable, but as shown in Figure 9.11, the trade-off has to be severe to keep R and K from being independent or even positively correlated in a density dependent context.

DENSITY AND FREQUENCY DEPENDENT COMPETITION

We move on now to situations combining both density and frequency dependent selection. The only assumptions that need to be altered are the equations governing fitness. As an example, we modify Eq. (9.10) so that each genotype's fitness depends more critically on its own population density than on that of the other genotypes. This might be the case if similar phenotypes—and hence similar genotypes—compete more with one another for limiting resources than with dissimilar phenotypes and genotypes. One possibility for genotype AA is the following:

Now the density dependent effect (multiplying total population size, N_t) is adjusted according to the frequencies of the different genotypes in the population at time t

$$\lambda_{AA}(N_t, p_t) = 1 + R_{AA}[1 - (1.0\,p_t^2 + (0.5)\,2p_tq_t + (0.5)\,q_t^2)N_t/K_{AA}]. \qquad (9.16)$$

The weighting factor for AA's effect on its own growth rate, λ_{AA}, is 1.0. . .

. . . but the weighting factor for Aa's and aa's effect on λ_{AA} is only 0.5.

Here the competitive effect of genotypes aa and Aa on AA's growth is only one-half the competitive effect of AA on itself. The fitness equations for the two other genotypes, λ_{Aa} and λ_{aa}, can be written in a similar form; self-genotype competition is twice that of nonself competition. The simulations illustrated in Figure 9.12, using Eq. (9.16), show the results when all genotypes have the same R but genotype AA has the highest K and genotype aa has the lowest K. Each population begins at a low population size ($N = 1$), and five different initial gene frequencies are simulated.

Note that allele A reaches a stable equilibrium where it is more frequent than allele a. However, allele A does not go to fixation even though $K_{AA} > K_{Aa} > K_{aa}$. The reason is that, as AA individuals become more common, they depress their own growth rate more severely than the growth rate of the other genotypes. The same effect holds for the other two genotypes. Common genotypes, regardless of which genotype it is, experience a disadvantage in this model such that neither allele can become fixed in

Figure 9.12

A simulation of the situation where self-competition is greater than nonself competition based on Eq (9.16). A stable polymorphism is maintained at $p = 0.8$ (the analytical calculation of this equilibrium value for p is difficult). The frequency of genotypes in the zygotes at equilibrium is given by the Hardy–Weinberg equilibrium as AA $= p^2$; Aa $= 2p(1 - p)$, and aa $= (1 - p)^2$.

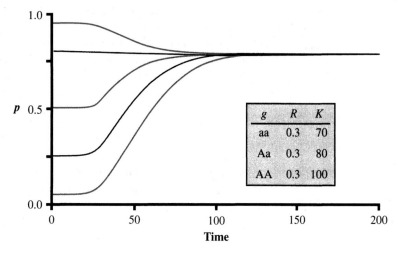

the population. The frequency-dependence in this model favors low-frequency genotypes. The result is a stable polymorphism, and all genotypes are maintained in the population.

Problem: For the example in Figure 9.12, the total population size also reaches a stable equilibrium, which is about 122.64. How can the total population reach a size greater than any of the genotypic K's?

In this example, a stable equilibrium point was reached. However, for other possible situations with frequency dependent fitness, the dynamics could be even more complicated. From one generation to the next, genotypes may change frequencies according to their present fitnesses $\lambda_{g,t}$, but as their numbers and frequencies change, so will their fitnesses and the fitnesses of all the other genotypes with which they interact. This has the potential effect of switching competitive advantages over time (as Bob Dylan said, paraphrasing Matthew 19:30, "The winner now will later be last.").

Recall the child's game: scissors, paper, rock: Scissors cuts paper, paper covers rock, but rock breaks scissors. These interactions form an intransitive loop; that is, the order of success forms a loop.

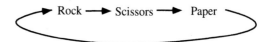

In such cases, the fitness of each of these three strategies depends entirely on the strategy your opponents are playing. Under conditions like these all hell can break loose and cyclic solutions or multiple alternative stable equilibria can emerge. There is some evidence that such intransitive loops exist among the competitive relationships of colonial organisms encrusting coral heads (Buss and Jackson 1979) and in wood decaying fungi (Fryar 1998). These organisms are capable of growing over one another and killing each other in the process. In pairwise species bouts, usually one of the pair is dominant, as in the scissors, paper, rock game, yet several species can apparently coexist because of intransitive relationships

A variety of ways can be used to alter the genotypic fitness functions to incorporate possible frequency dependence. Not all of them will necessarily promote the coexistence of both alleles, either by leading to stable equilibrium polymorphisms or by leading to oscillations that favor whichever genotype happens to be rare at the moment. If common genotypes have a disproportionate advantage over rare genotypes, all else being equal, then if one allele happens to become more common than the others, even if for entirely accidental reasons, then that allele will ultimately go to fixation.

One conceptual framework for analyzing situations where fitnesses are frequency dependent is called a *boundary stability analysis* or an *invasion analysis*. Figure 9.13 depicts the logic. The idea here is to ask which genotypes (or species) might be able to invade combinations of other genotypes if they occurred at an equilibrium. By stringing together different possible invasion scenarios, we could determine whether some combinations of genotypes are resistant to invasion by all others. If a genotype can invade, however, it might not lead to a new equilibrium point for the analysis of future invasions. Rather, the population and gene frequencies might reach a limit cycle around an unstable equilibrium point. Consequently, an invasion analysis is not a complete substitute for a full dynamical analysis, but it at least gives a rough indication of the direction of trajectories starting with low numbers.

Figure 9.13
An illustration of boundary (or invasion) analysis.

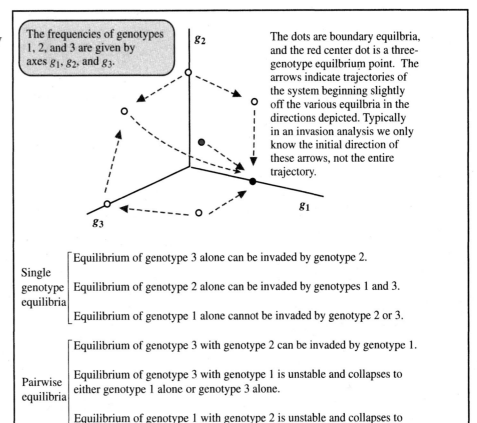

The frequencies of genotypes 1, 2, and 3 are given by axes g_1, g_2, and g_3.

The dots are boundary equilbria, and the red center dot is a three-genotype equilbrium point. The arrows indicate trajectories of the system beginning slightly off the various equilbria in the directions depicted. Typically in an invasion analysis we only know the initial direction of these arrows, not the entire trajectory.

Single genotype equilibria

Equilibrium of genotype 3 alone can be invaded by genotype 2.

Equilibrium of genotype 2 alone can be invaded by genotypes 1 and 3.

Equilibrium of genotype 1 alone cannot be invaded by genotype 2 or 3.

Pairwise equilibria

Equilibrium of genotype 3 with genotype 2 can be invaded by genotype 1.

Equilibrium of genotype 3 with genotype 1 is unstable and collapses to either genotype 1 alone or genotype 3 alone.

Equilibrium of genotype 1 with genotype 2 is unstable and collapses to genotype 1 alone.

Three-genotype equilibria

Equilibrium of all three genotypes together (the red center point) is unstable and collapses to genotype 1 alone.

Conclusion: Genotype 1 alone is stable, and all the other equilibrium points are unstable or open to invasion by the missing genotypes. Therefore we might expect all possible initial conditions to evolve to genotype 1 by itself.

Exercise: Suppose that the boundary analysis in Figure 9.13 revealed that none of the gene frequency equilibria, including zero for all three genotypes, was locally stable. What could you conclude about the dynamics of the system and the long-term coexistence of the genotypes?

Natural Selection, Adaptedness, and Population Density

Adaptedness refers to the ability of the carriers of a gene or genotype to survive and reproduce in a given environment (Dobzhansky 1967). The process of natural selection usually leads to an increase in a species adaptedness to the environment. As an organism becomes more adapted to its environment, a higher population density can usually be supported. For example, we showed with density dependent selection that carrying capacity, K, evolves to higher levels, unless there is a trade-off between R and K in combination with external forces keeping populations below K. However, natural selection

Figure 9.14
The population size of laboratory populations of *Tribolium castaneum* at day 37 over nine generations of natural selection (□). At the end of each generation, 16 adults are randomly chosen to begin the next generation. The decline in population size over time is not due to inbreeding, but most likely to natural selection acting on cannibalism rates. Standard errors are based on 48 replicates. (Based on the "C" line of Wade 1977). The pupal mortality estimates (shown in red) from adults come from assays of the stock of beetles used to initiate the experiment (the "S" stock) and at the end of the experiment (the final C line). The bars represent ± standard error based on sample sizes of $n = 20$ for the initial stock and $n = 10$ at the end. (After Wade 1979.)

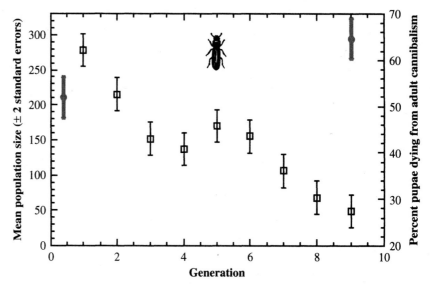

and adaptedness may not always go hand in hand. Under frequency dependence it is possible for fitness and adaptedness (and thus population size) to evolve in opposite directions. To see this is to appreciate the importance of the word *relative* in the definition of fitness, as the following example illustrates.

Imagine a gene that caused its bearers to eat young other than their own offspring, regardless of the genotype of those young. Imagine that this food source conferred more food for survival and reproduction. Bearers of the cannibal gene would produce more offspring than nonbearers. Both types would suffer equally the detrimental effects of the cannibalism. The relative fitness of the cannibal genotypes would be greater than that of the noncannibals and the cannibal gene would increase in frequency in the population even though the population size would probably decline as a consequence. Thus in this hypothetical example adaptedness and population size have declined through the action of natural selection. An example of this phenomenon seems to be provided by the decline of flour beetle numbers in replicated laboratory populations allowed to undergo natural selection (Wade 1977, 1979, McCauley 1978). As shown in Figure 9.14, the numbers (and population density) decline over time and a measure of cannibalism increases. Selection is probably acting on other demographic parameters as well, so cannibalism is only part of the story.

Similarly, in predator–prey systems, which we take up in more detail in Chapter 12, if some genotypes of a predator species have higher prey capture rates than other genotypes, then these high-capture-rate genotypes will theoretically be favored by Darwinian selection. This occurs even though it may lead to declines in the population size of both prey and predator and even the eventual extinction of both (see, e.g., Gilpin 1975a).

DENSITY PLUS FREQUENCY DEPENDENT SELECTION IN A LOTTERY MODEL (SUPPLEMENTAL)

One dubious feature of the *R–K* selection model just explored is the assumption that *R* and *K* are constants and that the discrete logistic equation is an accurate description not only of population growth but also of density dependent fitness. An obvious question is whether the conclusions about density dependent fitness are robust to different model formulations. If they aren't, then the *R–K* trade-off might not be very useful in accounting for patterns of life history evolution.

Let's now take a more mechanistic tact in exploring *R* and *K* selection. This new model allows us to develop some predictions for categorizing life history strategies. We begin with Figure 9.15.

Figure 9.15
Space is depicted in two dimensions. Each organism occupies one grid cell. One species is shown in red and the other in black.

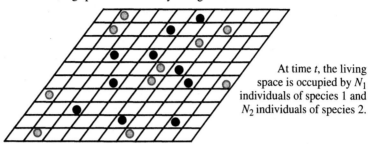

Lottery competition for space-based resources

The universe is a checkerboard divided into K equally sized living spaces. Each living space can hold only a single individual.

At time t, the living space is occupied by N_1 individuals of species 1 and N_2 individuals of species 2.

Each year, some living space, of the K possible slots, is made available by the death of some of the former occupants. Thus the total living space available for new recruits of either species at $t + 1$ is

$$\begin{bmatrix} \text{Available} \\ \text{space} \\ \text{at } (t+1) \end{bmatrix} = [K - N_1(t) - N_2(t) + dN_1(t) + dN_2(t)] = [K - sN_1(t) - sN_2(t)],$$

where d is the mortality rate and s is the survival rate of adults, assumed to be the same for both species.

If each adult produces B offspring in a year, the total offspring produced is simply $B(N_1(t) + N_2(t))$. If the total offspring produced is less than the available space in a year, then all young can recruit into the population. However, as time goes on and as space becomes filled to capacity, the number of new recruits into the living space by species 1 at $t + 1$ will be given by

$$R_1(t+1) = \begin{bmatrix} \text{Available} \\ \text{space } (t+1) \end{bmatrix} \begin{bmatrix} \text{Relative success} \\ \text{of species 1 at} \\ \text{acquiring the} \\ \text{available space} \end{bmatrix}$$

If the two species have identical competitive abilities, then the relative success of species 1 will simply equal its relative number of offspring, or

$$\begin{bmatrix} \text{Relative success} \\ \text{of species 1 at} \\ \text{acquiring the} \\ \text{available space} \end{bmatrix} = \begin{bmatrix} \dfrac{\text{Births of species 1}}{\text{Total births}} \end{bmatrix} = \dfrac{BN_1(t)}{BN_1(t) + BN_2(t)}.$$

Therefore, recruitment by species 1 is

$$R_1(t+1) = [K - sN_1(t) - sN_2(t)] \left[\dfrac{BN_1(t)}{BN_1(t) + BN_2(t)} \right].$$

And thus the total number of individuals of species 1 at $t + 1$ is

$$N_1(t+1) = \begin{bmatrix} \text{Surviving adults of} \\ \text{species 1 from time } t \end{bmatrix} + \begin{bmatrix} R_1(t+1) = \text{New} \\ \text{recruits of} \\ \text{species 1 at } t+1 \end{bmatrix}$$

$$= sN_1(t) + [K - sN_1(t) - sN_2(t)] \left[\dfrac{BN_1(t)}{BN_1(t) + BN_2(t)} \right]$$

A similar expression can be written for species 2 simply by changing subscripts.

This model is called the lottery model because offspring are like lottery tickets: the more offspring a species produces, the more lottery tickets it gains, and the greater its chance of winning available living space (Chesson 1986). As you can imagine, this lottery model does not allow the stable coexistence of two equivalent species.

Box 9.1

Suppose that the two species have different birth and survival rates, B and s; then once space is saturated, we have for species 1,

$$N_1(t+1) = s_1 N_1(t) + \left[K - s_1 N_1(t) - s_2 N_2(t)\right]\left[\frac{B_1 N_1(t)}{B_1 N_1(t) + B_2 N_2(t)}\right]$$

$$= N_1(t)\left[\frac{s_1\left(B_1 N_1(t) + B_2 N_2(t)\right) + B_1\left(K - s_1 N_1(t) - s_2 N_2(t)\right)}{B_1 N_1(t) + B_2 N_2(t)}\right]. \qquad \text{(a)}$$

The final expression in brackets gives the per capita growth rate of species 1 at time t and is therefore $\lambda_1(t)$. Note that $\lambda_1(t)$ is both density and frequency dependent. Multiplying through the parentheses and canceling terms in the numerator, we get

$$\lambda_1(t) = \frac{B_1 K + s_1 B_2 N_2(t) - s_2 B_1 N_2(t)}{B_1 N_1(t) + B_2 N_2(t)}. \qquad \text{(b)}$$

The expression for species 2 is obtained by changing all subscripts 1 to subscripts 2. To see which species has the competitive edge based on the birth and surival rates, B and s, we need to eliminate any advantage that might be due to an initial headstart in numbers because of frequency-dependence. Therefore we set the two species to the same density, $K/2$. Furthermore, since the value of K is arbitrary, for simplicity we set $K = 1$. If species 1 is competitively equal or superior to species 2, then $\lambda_1(t) \geq \lambda_2(t)$ and therefore, if both species are at $N_1 = N_2 = K/2 = 1/2$, after substituting Eq. (b) into $\lambda_1(t) \geq \lambda_2(t)$, we find the condition

$$2B_1 + s_1 B_2 - s_2 B_1 \geq 2B_2 + s_2 B_1 - s_1 B_2.$$

This in turn implies that

$$B_1 - s_2 B_1 \geq B_2 - s_1 B_2. \qquad \text{(c)}$$

And since the discrete death rate of species i is $1 - s_i$, Eq. (c) reduces to

$$\boxed{\frac{B_1}{d_1} \geq \frac{B_2}{d_2},} \qquad \text{(d)}$$

which represents the condition for the competitive equality or superiority of species 1 in the lottery model.

However, what if the two species are not identical—how can we predict who wins? If one species has a higher birth rate, B, or a lower death rate, d, than the other, then all else being equal, that species will be the winner (from any initial densities), and the other species will be the loser. If both survival and birth rates are different between the species, the winning species is that with the highest ratio of B/d, as shown in Box 9.1.

An example of the dynamics for a situation in which species 1 has the largest ratio of B/d is shown in Figure 9.16.

Box 9.1 shows that the lottery model contains both frequency- and density-dependence. But what is the nature of the frequency-dependence? Does it favor the rare species and thus allow the perpetual coexistence of both species? To explore the direction of the frequency-dependence, we assume that the population sizes of the two species have grown to saturate all the available space so that the sum of the numbers of species 1 plus species 2 equals K; that is, $N_1(t) + N_2(t) = K$. Again we will set $K = 1$ to look at proportional changes in density. Once space is saturated, the total number of individuals remains constant but the proportion of species 1 and 2 may change over time. We can rewrite Eq. (a) in matrix form:

$$\begin{bmatrix} N_1 \\ N_2 \end{bmatrix}(t+1) = \begin{bmatrix} 1 - d_1 + d_1 p_1(t) & d_2 p_1(t) \\ d_1 p_2(t) & 1 - d_2 + d_2 p_2(t) \end{bmatrix}\begin{bmatrix} N_1 \\ N_2 \end{bmatrix}(t), \qquad (9.17)$$

where $p_1(t)$ is the proportion of offspring of species 1 produced, $p_1(t) = B_1 N_1/(B_1 N_1 + B_2 N_2)$ and, $p_2(t) = B_2 N_2/(B_1 N_1 + B_2 N_2)$. The resulting dynamics can be seen by plotting Ricker

Figure 9.16
A simulation of the lottery model for a situation in which species 1 has the largest ratio of B/d; $K = 200$.

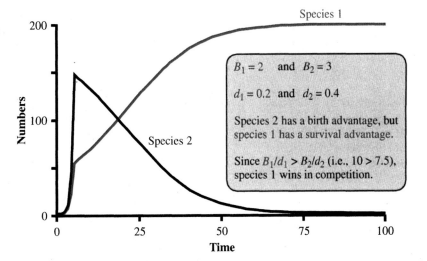

$B_1 = 2$ and $B_2 = 3$

$d_1 = 0.2$ and $d_2 = 0.4$

Species 2 has a birth advantage, but species 1 has a survival advantage.

Since $B_1/d_1 > B_2/d_2$ (i.e., $10 > 7.5$), species 1 wins in competition.

Figure 9.17
An illustration of population dynamics under the lottery model, using a Ricker diagram approach. (a) Species 1 wins. (b) Species 2 wins.

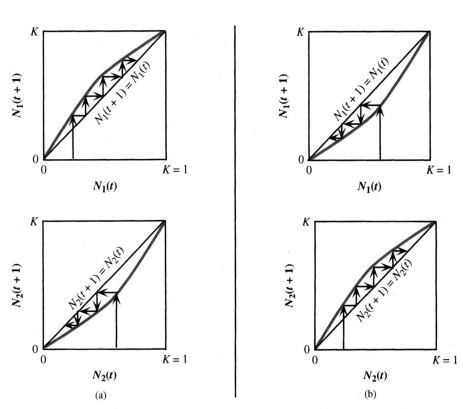

diagrams, as shown in Figure 9.17, for two different cases—one where species 1 wins and one where species 2 wins.

It is evident that the frequency-dependence in this model never reverses the relative advantage of the two species; that is, the $N(t + 1)$ curves never cross the diagonals. If species 1 has the highest ratio of B/d, then its fitness will be greater than that of species 2, regardless of the abundances of the two species.

Problem: Derive Eq. (9.17) from Eq. (a), Box 9.1.

Of course, nature rarely allows species continually to saturate the environment. Density independent catastrophes, like fires, hurricanes, and floods, can devastate populations while creating new living space. We have already explored the effect of density independent mortality in the discrete logistic equation. We now add this to the lottery

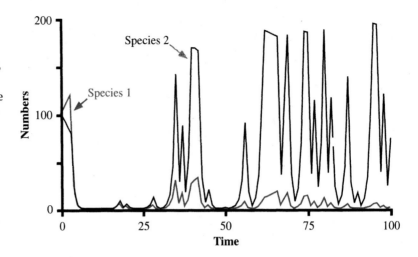

Figure 9.18
This simulation is based on the same parameters for *s* and *B* as in Figure 9.16, but now catastrophes have a 30% chance of occurring in any year. If a catastrophe occurs, it knocks down the numbers of both species by 80%, thus freeing much open space. Species 2 is now winning over species 1 because species 2, with the highest birth rate, has the highest fitness when there is much open space. As time goes on, species 1 eventually declines to extinction.

model. As before, if a catastrophe occurs, it knocks down the numbers of both species proportionally; hence the catastrophe alone does not directly favor either species. However, by freeing new open space, the species grow initially without much density-dependence. Hence the initial growth rate of the two competing species can be approximated by exponential growth, or $\lambda = B + s$. Thus catastrophes will indirectly favor the species with the highest $B + s$. Returning to the example shown in Figure 9.16:

$$\lambda_1 \text{ (at low total densities)} = 2 + 0.8 = 2.8$$

and

$$\lambda_2 \text{ (at low total densities)} = 3 + 0.6 = 3.6.$$

Under low population densities, species 2, with its much higher birth rate, is superior to species 1. Survival rates are confined to a maximum of 1, but birth rates have no upper bounds. Consequently, at low densities, the most fecund species might often have the highest fitness given by $B + s$. The results of such a simulation are shown in Figure 9.18 for the same birth and death rates as in Figure 9.16.

Because growth rates are both density and frequency dependent in the lottery model, it is not possible to base an interpretation of the outcome of competition simply on plots of per capita growth of the two species as a function of total density, as we did in Figure 9.6 for the discrete logistic equation of density-dependence. However, we can explore the potential trade-off between R and K in a different way. The first step is to introduce different K's into the lottery model for the two species. If individuals in the two species have different space requirements, perhaps because they have different body or territory sizes—or root coverages if they are plants—then the carrying capacity will be inversely proportional to body size. Bigger individuals will have larger territories and thus the larger species is expected to have a smaller carrying capacity, K. Then it can be shown that if just K and B vary between the two species, the condition for species 1 to be competitively equal or superior to species 2 becomes $B_1K_1 \geq B_2K_2$. The winner of the competition is the species with the largest product, BK. (Of course, this assumes that the species are equivalent in respects other than B and K). An example of such competition for space between species with different K's is shown in Figure 9.19.

Another example is shown in Figure 9.20. Here, species 2 has the K advantage and again wins in competition because it has the largest product, BK. However, if density independent factors operate to keep the population far below carrying capacity, then the highest BK rule will not indicate the highest growth rate. Hence it is again possible for the species with the highest $B + s$ to monopolize space even if it has the lower product, BK. In Figure 9.20, we had

$$\lambda_1 \text{ (at low total densities)} = 4 + 0.6 = 4.6$$

and

$$\lambda_2 \text{ (at low total densities)} = 3 + 0.6 = 3.6.$$

Figure 9.19
Another simulation of the lottery model. (a) The dynamics of competition. (b) Per capita growth rates as a function of population size. Initially at low densities, the per capita growth is just $B + s$. As space gets filled, however, the recruitment rate declines. Species 1 has a higher carrying capacity, but nevertheless loses in competition to species 2. Thus species 2 wins because it has the largest product of BK; k_i is the ratio K_i/K; $K = 200$.

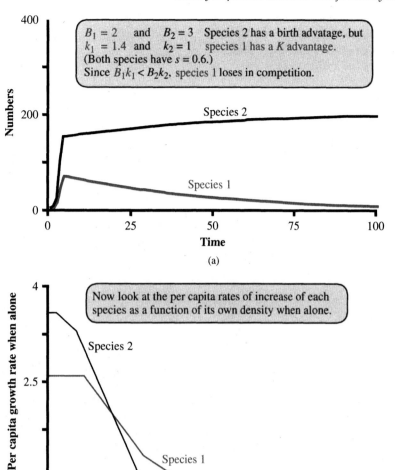

(a)

(b)

Figure 9.20
Species 2 wins because it again has the largest product of BK; k_i is the ratio K_i/K; $K = 200$.

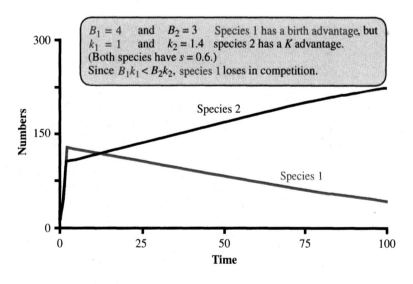

At these low densities, species 1, with its much higher birth rate, is now superior to species 2.

When catastrophes occur (as modeled in Figure 9.18), species 1 (with the lower K) now wins over species 2 with the higher K, as illustrated in Figure 9.21.

When all three life history parameters (birth rate, carrying capacity, and death rate) differ, the winner in competition in a constant environment is that species with the

Figure 9.21
This is the same example as in Figure 9.20, but now environmental catastrophes occur as in Figure 9.18 with a 30% chance. The winner in competition is reversed.

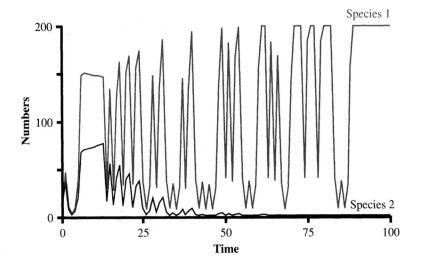

largest term, *BK/d*. However, as before, if densities are kept low, the competitive advantage can shift to those species with the higher birth rate.

In the Lottery Model, Independent Temporal Fluctuations in Birth Rates Can Enhance Coexistence

Are there any situations in the lottery model which might allow for species coexistence? So far we have imagined environmental variation as a source of density independent mortality and, for this case, only one species will prevail. However, birth rates, *B*, might fluctuate over time (Chesson 1986). Suppose that species 1 and 2 are identical in birth and survival rates, except that the birth rates, *B*'s, of the two species fluctuate independently from year to year. That is, what constitutes "a good year" for species 1's reproductive success does not necessarily constitute a good year for species 2's.

As time goes on, and breeding conditions fluctuate, a good year might occur for both species at the same time. In this case, their recruitment rates will be relatively similar and thus their relative abundances will not change (they both will increase by about the same amount). Similarly, if it's a bad year simultaneously for both species, both species will decrease by about the same amount. So far, temporal variability is not enhancing coexistence between these two species.

Now suppose that one species (say, species 1) happens by chance to have become relatively rare compared to species 2 over a period of years; that is, by chance there was a sequence of years that were relatively unfavorable for species 1's reproduction. Suppose also that the next year is good for species 1, the relatively rare species, but bad for species 2. Both species share the same population ceiling. Since species 1 is rare, its possible degree of increase to this ceiling is large. If the reverse occurs—that is, if it turns out to be a bad year for species 1 but a good year for species 2—then species 2 will increase and species 1 will decrease. However, since species 2 is already common and close to the ceiling set by space, the relative amount that it can increase is severely limited. If it's a good year for one species but a bad year for another, the proportionate increase for a rare species that happens to get the good year will be greater than the proportionate increase for a common species that happens to get the good year. In this way, a rare species is favored in the lottery model when birth rates fluctuate in each species *independently* over time or if they have negative covariation in their response to environmental fluctuations. The latter means that a good year on average for one species tends to be a bad year for the other (Chesson and Huntley 1997). An example of the effect of independent fluctuations in B is shown in Figures 9.22 and Figure 9.23. Of course, if environmental fluctuations are extreme and populations sizes are small, then extinction may result, so this process may not enhance coexistence in small, closed populations (Chesson and Huntley 1997).

Figure 9.22
The lottery model without temporal variation in birth rates. In this simulation the two species have the same life history parameters except for their birth rates, which slightly favor species 2. ($s = 0.9$, $k_1 = k_2 = 1$).

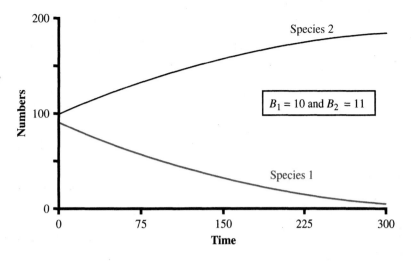

Figure 9.23
This simulation uses the same life history parameters as in Figure 9.22, but lets the birth rates fluctuate around their means (mean $B_1 = 10$ and mean $B_2 = 11$) from a uniform distribution. Conclusion: These two species can now coexist.

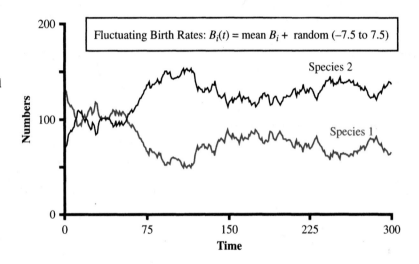

PROBLEMS

1. Assume that density dependent fitness follows the discrete logistic equation. Draw three trajectories for the case that three genotypes, AA, Aa, and aa, all have the same R but that their carrying capacities are in the order $K_{Aa} < K_{AA} = K_{aa}$. Begin at the three points p_1, p_2, and p_3 in the following diagram. Extend the trajectories so that an equilibrium gene frequency for the a allele is reached by generation 100.

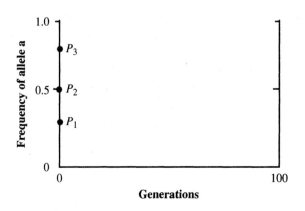

2. Two species compete for living space according to the lottery model. They have identical death rates ($d = 0.2$/year), but species A has a birth rate that is 3/year while that of species B is 2.5/year. The two species have different demands for space such that the carrying capacities are $K_A = 100$ and $K_B = 150$. In the following diagrams plot the qualitative change in numbers of species A and B that would result if you started with 10 individuals in each case.

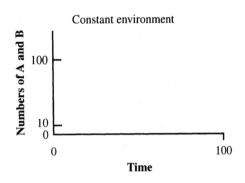

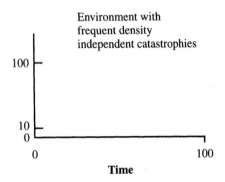

3. Three genotypes, AA, Aa, and aa, have the growth parameters for R and K shown in Figure 9.9. If the average density independent mortality, $E(f)$, is the same for all three genotypes and equals 0.02, what will be the ultimate gene frequency, p? What if $E(f) = 0.02$ for genotypes AA and Aa, but $E(f) = 0.05$ for genotype aa?

4. Assuming geometric growth, what will be the equilibrium gene frequency if $\lambda(AA) = 1.1$, $\lambda(Aa) = 1.5$, and $\lambda(aa) = 1.2$?

5. If the frequency of allele A is 0.5 at time t, what will it be, under random mating, in the next generation if $\lambda(AA) = 1.1$, $\lambda(Aa) = 1.5$, and $\lambda(aa) = 1.2$?

10

Exploited Resources

THE EXAMPLE OF GRAZING SHEEP

We introduce this chapter with a simple example drawn from practical ecology. Sheep eat grass and ultimately convert some of that grass to wool. Let's say that we can manage the number of sheep on a pasture and that we would like to graze sheep in a way that maximizes the long-term yield of wool. To manage the pasture sensibly, we need to consider sheep numbers, grass abundance, and grass productivity. We begin by exploring what happens to grass abundance as we change the number of sheep on the pasture. We need to know two things:

1. The amount of grass that an individual sheep eats per unit time and how that amount changes as the abundance of the grass increases. This is the **functional response** of the sheep (Solomon 1949).
2. The rate at which the grass grows under different levels of foraging. We will assume a discrete logistic curve for grass growth—thus grass will eventually reach a carrying capacity, *K,* set by its own resource levels (e.g., water, fertilizer, and sunlight) in the absence of any sheep. We subtract from this logistic growth rate the amount of grass mortality (per unit time) due to grazing. This mortality is the product of the number of sheep times the functional response of each sheep.

The Consumer's Functional Response

First, we need some labels. Let's call grass density R (for resource) and call the sheep's grass consumption rate simply "consumption rate" (with units of mass of grass eaten per sheep per unit time).

At relatively low grass densities, sheep must spend much time searching for patches of grass. As grass levels increase, search times for sheep decrease with the result that overall consumption rates increase.

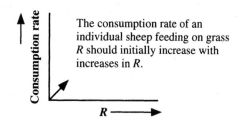

The consumption rate of an individual sheep feeding on grass R should initially increase with increases in R.

223

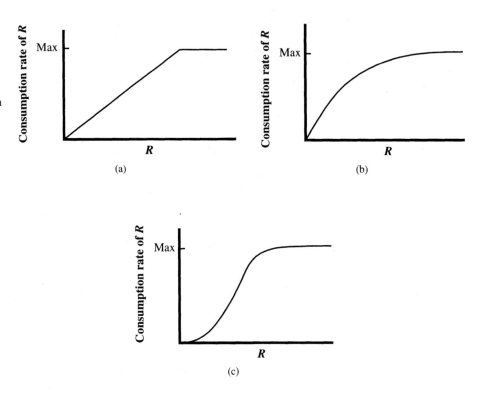

When grass is very abundant, locating it is no longer the bottleneck for sheep consumption. Instead sheep will reach a maximum consumption rate set by how quickly they can chew, swallow, and digest grass. We can imagine three different approaches to this maximum, as depicted in Figure 10.1.

In Chapter 11 we derive these different functional responses based on features of the consumer's behavior. For now, we simply note that these different shapes are possible for different kinds of animals. For example, a small mammal or a warbler feeding on cocoons and larvae of moths shows an S-shaped, or type 3, functional response, but a wasp that lays its eggs in sawfly cocoons shows a type 2 functional response, as do wolves responding to caribou density. These varying responses are shown in Figure 10.2.

Innate Growth of the Resource Population

Recall from Chapter 5, where we introduced single-species density dependent growth, that the continuous logistic growth rate for the total population as a function of its own density is a hump-shaped curve. Thus we would get something like the curve shown in Figure 10.3, which plots the grass growth rate, dR/dt, as a function of grass density, R.

Dynamics of the Resource Population under Exploitation

The next step is to combine the plot of resource growth from Figure 10.3 with the plot of resource consumption by sheep—the functional response in Figure 10.1. Since the plots have the same axes, we can superimpose them to examine the changes in grass numbers from both sources. Let's assume a type 3 functional response for sheep as a starting point. The total consumption of grass depends also on how many sheep are grazing. This introduces an additional consideration: Assume that sheep have **additive effects** on grass consumption such that 10 sheep consume grass 10 times faster than a

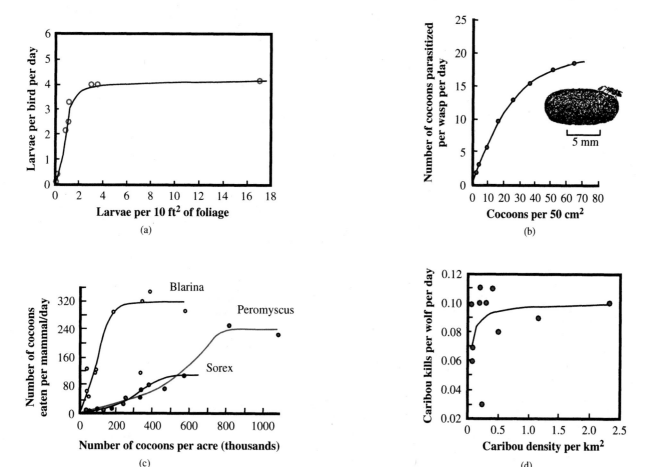

(a)

(b)

(c)

(d)

Figure 10.2

Various functional responses. (a) Functional response of baybreasted warblers to changes in the density of spruce budworm (a moth) larvae in New Brunswick (Mook 1963). (b) The chalcid wasp *Dahlbominus fuscipennis* parasitizing cocoons of the sawfly *Neodiprion setifer* in laboratory cages (Burnett 1956). (c) Small mammals feeding on sawfly cocoons; *Peromyscus* is a mouse, and the other two species are shrews (Holling 1959). (d) Wolves feeding on caribou (Dale et al. 1994).

Figure 10.3

This dashed curve shows how grass would grow in the absence of any grazing if its population growth rate were roughly a continuous logistic.

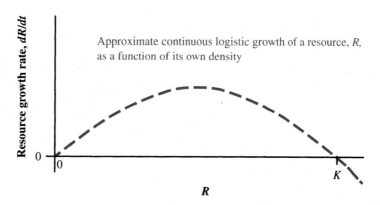

Approximate continuous logistic growth of a resource, *R*, as a function of its own density

single sheep for each grass level, and 100 sheep consume 100 times more grass per unit time than a single sheep, and so on. While this seems like a reasonable assumption, it would be violated if sheep interact with each other in ways that influence their ability to forage. For example, if the average sheep spends more time fighting with other sheep at high sheep numbers, then this would certainly cut into the time it has available for feeding on grass. In this case, 100 sheep might eat far less than 100 times the amount that a single sheep would eat if it were alone on the same pasture. Conversely, a single sheep scanning for predators might forage less efficiently than a small tight flock with many vigilant eyes. Thus there are several sound biological reasons why the additive assumption might not be realistic. For now we put these complications aside and deal with the additive case. (We return to possible sheep interference in Chapter 12.) The total consumption of multiple sheep is illustrated in Figure 10.4 for three quantitative levels of sheep: one sheep, a few sheep, and more sheep. The latter two curves are simply increasing multiples of the first.

Let's take a closer look at the one-sheep case. First, we extract a curve for the net growth rate of grass—the difference between the growth of grass and its consumption by sheep—as shown in Figure 10.5. Then we plot the net growth of grass—the stippled region in Figure 10.5—which is the difference between recruitment and consumption, as shown in Figure 10.6. In other words, with a single grazing sheep and the dynamical rates for gain and loss plotted in Figure 10.6, the grass eventually reaches a stable equilibrium at R^*, not far below K, the carrying capacity of the grass.

Figure 10.4
Total resource consumption by different numbers of sheep. Total resource consumption is the functional response (which has units of per consumer) multiplied by the number of consumers. Thus it represents the rate of loss of the resource (grass) and can be compared to the rate of gain given by the dashed continuous logistic curve.

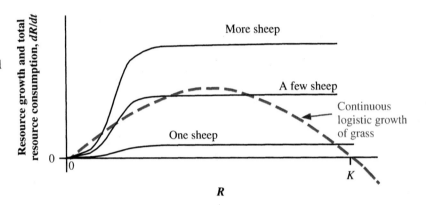

Figure 10.5
The stippled region shows the *net* growth of grass—the difference between the grass recruitment curve (logistic) and the consumption of grass by a single sheep.

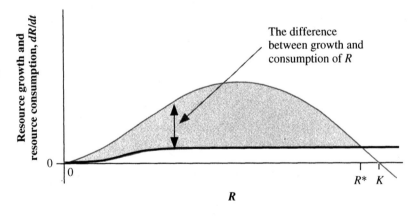

Figure 10.6
The net resource growth rate (gain minus loss) from Figure 10.5. By definition, an equilibrium occurs when the net growth rate is 0. Two resource levels yield an equilibrium here: zero grass, which is unstable, and R^*, which is a stable equilibrium point.

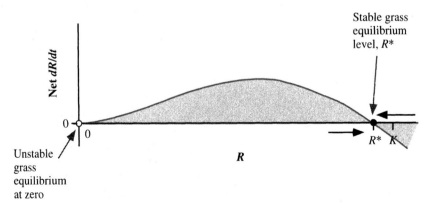

In Figure 10.7 we next explore the situation for a few sheep. To plot *net* grass growth we modify Figure 10.7 to plot only the difference between grass recruitment and its consumption by sheep, as shown in Figure 10.8. In Figure 10.8 there are three non-trivial equilibrium points labeled, 1, 2, and 3, and the trivial equilibrium point at $R = 0$. The equilibrium points labeled 1 and 3 are stable (dark circles) but number 2 is unstable (open circle) since resource levels will tend to grow away from it, either to increase to the level of stable grass equilibrium 3 or to decrease to stable grass equilibrium 1. The arrows show the direction of resource growth in the vicinity of the different equilibria. An arrow pointing to the right means that resources increase for that level of R since recruitment exceeds consumption (plotted on the y axis). Conversely, an arrow pointing to the left means that resources decrease at that level of R since consumption exceeds resource recruitment.

A similar curve can be constructed for the more sheep case. It shows only a single feasible equilibrium point, which is stable. In summary, we get the following three cases for the three qualitative levels of sheep.

Problem: Be sure that you understand how Figure 10.9 was constructed for the case of more sheep. Then answer the following questions.

1. Why are the stable equilibrium grass levels less than the value of K for grass?
2. How is it possible to determine the stability of each equilibrium point simply by looking at the plots of net dR/dt?
3. Why does the plot for a few sheep have two humps for the net recruitment rate but the plots for more sheep and one sheep have only one hump?

Now let's consider the effect of a continuous change in the number of sheep rather than simply three discrete values. We return to Figure 10.4 and imagine not just three consumption curves but many, each representing a different number of sheep; that is, we mentally fill in the intermediate levels of sheep between the three curves. In this way

Figure 10.7
As in Figure 10.5 but for the case of a few sheep.

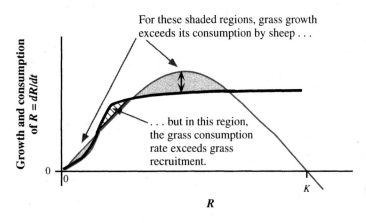

Figure 10.8
As in Figure 10.6 but for a few sheep. Four levels of grass now yield an equilibrium. Equilibrium points 1 and 3 are stable, but point 2 is unstable. The horizontal arrows indicate either increasing or decreasing R, based on whether dR/dt is positive or negative.

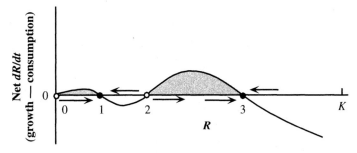

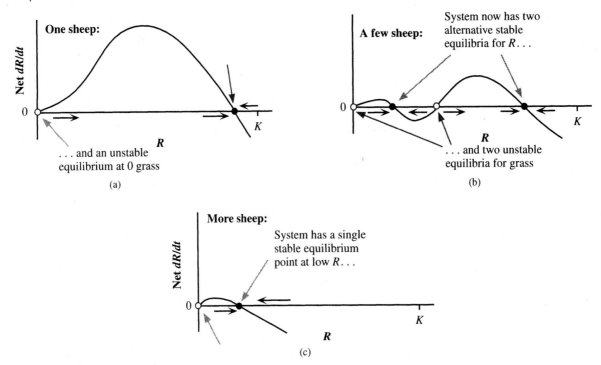

Figure 10.9
A summary of the changing dynamics for resources with different numbers of sheep on the pasture: (a) one sheep, (b) a few sheep, and (c) more sheep. Stable equilibrium points for R are indicated by filled dots and unstable points by open dots.

Figure 10.10
Similar to the case of Figure 10.9, but now the equilibrium levels of the resource (grass) are plotted for continuous changes in the number of sheep. Any point in the stippled region represents a particular combination of sheep and grass that leads to positive net growth for the grass. The curve itself yields zero net growth, by definition. Above the curve (in the unshaded region), grass growth is negative. For example, if grass and sheep were at point P and the number of sheep were fixed, grass levels would increase until they reached point R_P.

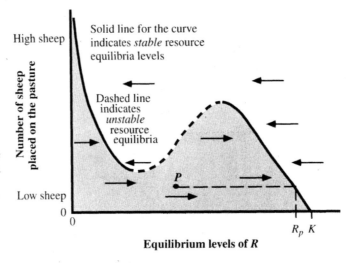

we can explore the continuous decrease in *equilibrium* grass abundance as sheep numbers increase. Starting with zero sheep, the grass is at K, and, as sheep are added, the combined consumption curve intersects the recruitment curve to the left of K. The equilibrium grass levels drop as more sheep are added. This relationship is plotted as a smooth curve in Figure 10.10.

While it might seem more natural to plot the number of sheep on the x axis in Figure 10.10, since it is the independent variable in this thought experiment, the latter development (in Chapter 12) will be simplified by switching axes. Then the x axis represents the equilibrium resource level that is reached for each number of sheep on the y axis. Between the T_1 and T_2 levels of sheep, the multiple equilibrium points correspond to the multiple equilibrium points illustrated in Figure 10.9 for the case of a few

Figure 10.11
As in Figure 10.10, but now the initial grass and sheep numbers, *P*, are outside the positive growth region for grass. Grass declines until it comes to rest at R_P^*.

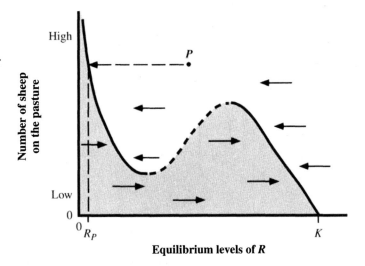

Figure 10.12
This is simply Figure 10.10 plotted again to highlight the two transition levels of sheep. The number of sheep at T_1 marks the threshold sheep level that begins the transition between a single nontrivial equilibrium point for grass to the three equilibrium points illustrated in Figure 10.9. As the number of sheep increase still more, a new threshold number of sheep is reached at T_2. Above T_2 sheep, the grass population returns to a single stable equilibrium at a low grass level. The sheep levels S_1, S_2, and S_3 loosely correspond to those of one sheep, a few sheep, and more sheep, respectively, in Figures 10.4 and 10.9.

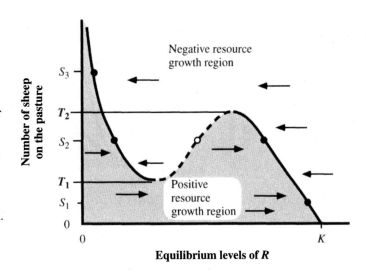

sheep. For all points below the curve (i.e., in the stippled region), the number of sheep is low relative to resource recruitment rates, so resources can increase. For example, when there are no sheep, grass will increase until it reaches a stable equilibrium at *K*. Imagine that the system starts at point *P* in Figure 10.10. At this point, the number of sheep is relatively low and grass, *R*, is at about *K*/2. Since *P* falls in the stippled region, resources will increase, as shown by the arrows pointing to the right, until they hit the equilibrium curve for this low number of sheep. But, if the number of sheep is high and the initial resource level is again at *K*/2, then the starting point *P* would be outside the stippled region. Grass levels would then decrease, as shown by the horizontal arrows pointing to the left, until resources reach the equilibrium curve for that particular number of sheep, as shown in Figure 10.11.

Threshold Consumer Levels for a Type 3 Functional Response

Figure 10.12 is a plot of Figure 10.10 again but special emphasis is added to show the levels of sheep marking a transition in the stability properties of the equilibria; these sheep levels are labeled T_1 and T_2. The number of sheep labeled T_1 marks the threshold for the transition from a situation wherein a single stable equilibrium resource level exists (low sheep numbers, e.g., sheep level S_1) to those intermediate sheep numbers that yield three nontrivial equilibrium resource levels (e.g., sheep level S_2) indicated by

Figure 10.13
The total sheep consumption curves for the two threshold sheep numbers, T_1 and T_2, overlaid on the grass recruitment curve, as in Figure 10.4.

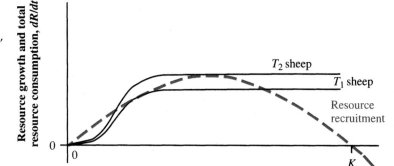

Figure 10.14
The solid dots correspond to stable equilibria, and the open dot corresponds to the unstable equilibrium when there are five sheep. These equilibria are projected to the y axis to show the values of dR/dt for these R values.

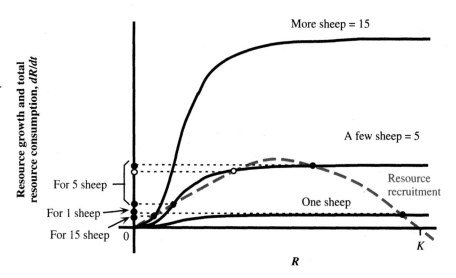

the solid dots. The number of sheep T_1 marks the threshold sheep level that begins the transition between these two behaviors. The total consumption associated with T_1 sheep is depicted in Figure 10.13. Finally, as sheep levels increase still more (i.e., climbing up the y axis), a new threshold number of sheep is reached at T_2, which marks the next and final transition. Above T_2 sheep, there is a return to a single stable, but low level, of resource. The total consumption associated with T_2 sheep is also depicted.

Maximizing Yield of a Managed Resource

Let's now suppose that we want to optimize the number of sheep on the pasture for maximized wool production. If too few sheep are grazed, then there won't be many sheep to produce wool. If too many sheep are grazed, then the grass will be overexploited; the sheep will have little to eat and therefore each sheep's ability to convert grass into wool will be diminished. Presumably, some intermediate number of sheep will be best, but how can we solve for this number? We specify wool production per sheep as proportional to food consumption per sheep times the number of sheep, which equals the total food consumption of the population of sheep grazing on the pasture. That is,

$$\text{Wool production rate} \propto \text{total grass consumption}, \frac{dR}{dt}.$$

The right-hand side of this expression is the y axis on the plots in Figures 10.4 and 10.9, but now we wish to plot dR/dt versus number of sheep and look for a maximum. Let's look at Figure 10.4 again, but let's now supply some actual numbers for the different levels of sheep to make things quantitative, as shown in Figure 10.14.

Figure 10.15
Equilibrium wool production as a function of the number of sheep grazing on the pasture. Two different curves are possible: (a) the pasture begins in a highly overgrazed state, and sheep are successively removed, or (b) the pasture begins in an ungrazed state, and sheep are successively added. (c) The two curves superimposed.

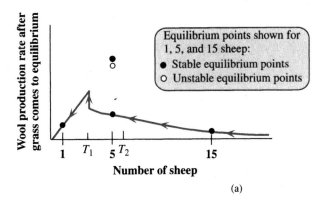

(a)

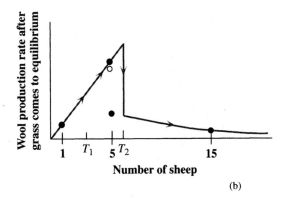

(b)

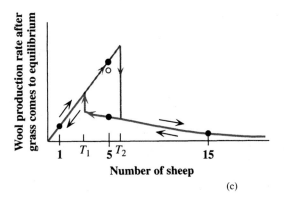

(c)

In Figure 10.15 we go one step further by plotting these equilibria dR/dt levels on the y axis of Figure 10.14 as a continuous function of the number of sheep on the x axis. We get two different curves, depending on whether we begin with no sheep and successively add sheep to the pasture or begin with an overstocked pasture and successively remove sheep. These alternative curves are a direct result of the emergence of the unstable equilibria points at intermediate numbers of sheep, indicated by the dashed portion of the curve in Figure 10.10. To the left of T_1, the two curves superimpose, and to the right of T_2, the two curves superimpose. Since an unstable equilibrium cannot be reached, there is no corresponding wool production rate for the five-sheep example (the open dot in Figure 10.14).

The optimum number of sheep to maximize wool production is exactly T_2. However, the important message here is that the dynamical behavior of resource populations under exploitation can be complex. Thresholds or *breakpoints* which, if exceeded, can abruptly and discontinuously move resources to new and undesirable levels. Also, *hysteresis* may occur with the curves for sheep addition and sheep subtraction not being identical. Each contains a breakpoint, but these break points can be at different levels. Consequently, it is not always easy to find the optimum level of sheep on the pasture. If the optimum number of sheep, T_2, in Figure 10.15(b) is exceeded by even a single sheep, then wool production falls drastically, to a much lower stable level. If that single sheep is then removed, the system does not return to its previous level, but instead rides the thin red curve in Figure 10.15(c). Only modest increases in wool yield are produced as more sheep are removed from the pasture until the number of sheep drops to T_1. When the number of sheep is below T_1, the wool production takes a giant leap upward. However, wool production still does not reach the level of production that was achieved at exactly T_2 sheep.

HARVESTING FISH

An Example of the Peruvian Anchovy

Peruvian anchovy live in the cool, upwelling, nutrient-rich waters along the coast of Peru and northern Chile. Reproductive maturity is reached at about 1 year of age and the typical anchovy lives about 3 years. Anchovy occur in large schools and are caught near the surface with nets. This was the largest fishery in the world until it collapsed in 1972 in association with an El Niño that disrupted these cold currents and diminished productivity. Fishing was suspended to allow the stocks to recover, but there was no immediate return of the anchovy. Populations of seabirds that feed on the anchovy also remained low. Since anchovy have a rapid generation time, we can assume that there has been adequate time for the population to return to its former level since the fishing suspension. But, as shown in Figure 10.16, it hasn't. Why hasn't it?

In the sheep-grazing example, we had complete control over the number of sheep that were grazed on the pasture. When we harvest trees from a forest or fish or whales from the sea, natural forces beyond the control of the forester or the fisher or the whaler influence the size of the resource available for harvesting. However, the theory that we have already developed will help us see possible outcomes of different resource harvesting strategies. Often the goal is to achieve **the maximum sustainable yield (MSY).** The word *sustainable* is an important qualifier since the maximum yield in any single year is to simply harvest all the resource that exists at that time. However, this is not sustainable since next year there may be no resource. A bit later, we show how to estimate the MSY, but for now we accept that the estimated MSY for Peruvian anchovy was about 9 to 10 million tons/year, based on average conditions for several years before the El Niño episode in 1972. Note that the harvest exceeded MSY in 1970 and in 1966–67.

Two common ways of regulating natural resources like fisheries are explained in Box 10.1.

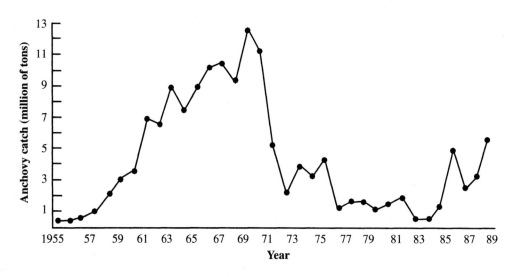

Figure 10.16
Total catch for the Peruvian anchovy fishery. After the collapse in 1972, fishing was greatly reduced but the population had only slightly recovered by 1989. After Krebs 1994.

Box 10.1 Methods of Regulating a Fishery or Other Dynamic Natural Resources

As with the sheep, we begin by setting up a growth and loss expression for fish:

$$\begin{matrix} \text{Net growth of a} \\ \text{fish population in} \\ \text{a fishing season} \end{matrix} = \underbrace{rN\left(\frac{K-N}{K}\right)}_{\substack{\text{logistic} \\ \text{recruitment} \\ \text{curve of the fish} \\ \text{(per season)}}} \underset{\text{minus}}{-} \underbrace{L}_{\substack{\text{total losses} \\ \text{from} \\ \text{fishing} \\ \text{(per season)}}} \quad \text{(a)}$$

Now we focus on the loss part of this equation, which is the part that governments can regulate. We may distinguish two methods for regulating the losses due to fishing.

1. **Fixed quota.** Total losses to fishing, L, are regulated by setting a limit, Q, on the number of fish harvested per season, regardless of the number of fish in the population, as illustrated in the following diagram.

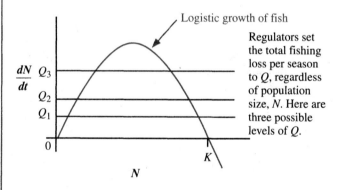

Regulators set the total fishing loss per season to Q, regardless of population size, N. Here are three possible levels of Q.

2. **Fixed effort quotas.** We can expand the loss part of Eq. (a) as follows:

$$L = (\text{fishing effort})(C)(N). \qquad \text{(b)}$$

Here the losses, L, are a product of three factors: the amount of fishing effort (which increases with the amount of time spent fishing, the number of fishing boats, the area the boats cover per day, etc.), a parameter of "fish catchability," C, and the size of the fish population, N. Fish catchability measures the efficiency of each unit of fishing effort. Fish catchability therefore has units of fish caught per effort expended per fish. We assume that C is a constant. Then, Eq. (b) says that all else being equal, the more fish in the population, the more effort expended trying to catch those fish, and the easier the fish are to catch, then the more fish will be caught in a given time period.

Equation (b) also says that the harvested losses represent a type 1 functional response by the consumers—the people doing the fishing. The numbers of fishers and the "foraging" effort of each fisher are combined into a single term: the "fishing effort." Thus the analog of Figure 10.14 in this case, where the number of fish caught increases linearly with the fish population size (as by Eq. b), looks something like the following.

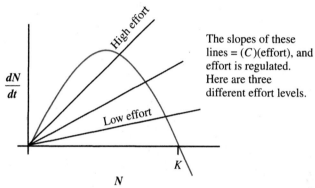

The slopes of these lines = $(C)(\text{effort})$, and effort is regulated. Here are three different effort levels.

Regulation is effected by limiting the "effort" term in Eq. (b). This may be done by restricting the number of fishing permits, the length of the fishing season, or restricting regions where fishing is allowed. Variations in fishing effort from year to year will lead to different catches, which will trace out the recruitment curve (assuming that it remains constant from year to year). For example, the following diagram shows five different efforts in five different years applied to a fish population and the resulting catches for each.

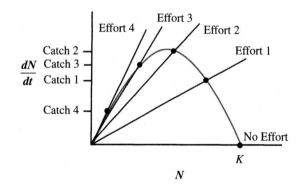

The following diagram shows those same fish catches plotted against the effort values instead of against fish population size, N.

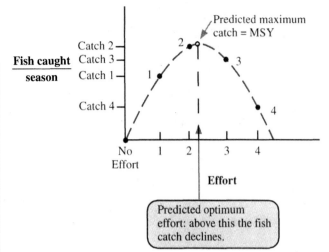

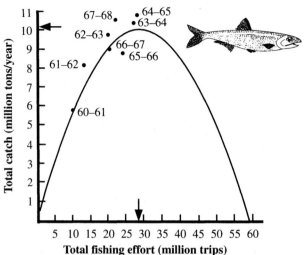

Figure 10.17
The relationship between total fishing effort and total catch for the Peruvian anchovy fishery. The effects of humans and sea birds are combined in these data. The parabola is based on a continuous logistic equation fitted to these data. Arrows indicate the estimate for MSY and appropriate fishing effort. After Krebs (1994) from Boerema and Gulland (1973).

This plot also yields a hump-shaped curve, like the logistic curve. This is a key result since this figure shows that, even without knowing the absolute fish population size, *N*, the optimum effort and the MSY can still be estimated simply by the relative position of the yield–effort curve. The optimum effort and the MSY are shown with arrows in the preceding figure. If, additionally, the catchability constant, *C*, is known, then fish population size, *N*, and the corresponding *r* and *K* may also be estimated from the yield–effort curve (we work through some examples later in this chapter).

The catch versus Peruvian anchovy data and a fitted curve are shown in Figure 10.17. Note that during this period the anchovy did not appear to be severely overexploited.

Obviously several complications may affect all this. Fishing effort is normally modified from year to year—but with climatic changes, the recruitment curve, and the *r* and *K* for the fish population are also changing yearly. Additionally, the fish population may not grow according to a logistic equation. A full model might require knowledge of the age structure of the fish population. Density-dependence

might be different for different ages, and the functional form of the density-dependence could contain several higher-order terms. Recall the more complicated single-species models that we explored in Chapter 5. Finally, the catchability, *C*, may not be a constant but instead a function of effort and/or fish population size, *N*. In this case, we would not have type 1 functional response curves but perhaps type 2 or type 3 curves. As we have already shown, the latter can yield break points and alternative stable states. Finally, the fish population is also being exploited by several natural predators, and we usually have little knowledge of their abundance and functional responses.

The task for government resource regulators is to prevent a **tragedy of the commons** by keeping the fishing industry from "killing the goose that laid the golden egg." It is in the best interest of each individual to prevent the overexploitation of the common fishery that they collectively depend on, yet at any particular time it is in the short-term economic best interest of each fisher to maximize his or her catch. Thus we have the classic problem of what is best in the short term for the individual, if followed by all individuals, leads to the collapse of the fishery and thus a bad situation for all who depend on it for their livelihood.

As we have indicated, one possible way to prevent the collapse of the fishery is for governments to impose a fixed-quota harvest (Box 10.1). This imposes a limit on short-term profits so as to maximize longer term profits and the continued viability of the natural resource. For example, regulatory agencies could put a limit on the number of fish that can be caught per season. The fishing season would close after the limit is reached.

Figure 10.18 illustrates the impact of fixed quota harvesting on a resource population with intrinsic logistic growth. We seek to maximize the fish yield in terms of *dR/dt*. There are two equilibria in Figure 10.18. The one to the right of the hump's peak is sta-

Figure 10.18
Growth of a logistically growing
resource under a fixed-
harvesting quota.

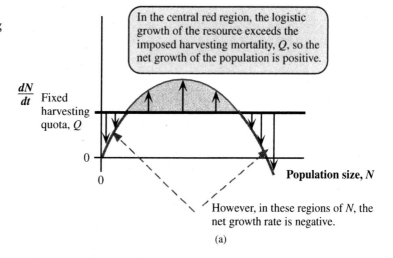

In the central red region, the logistic
growth of the resource exceeds the
imposed harvesting mortality, Q, so the
net growth of the population is positive.

$\frac{dN}{dt}$ Fixed
harvesting
quota, Q

0

0

Population size, N

However, in these regions of N, the
net growth rate is negative.

(a)

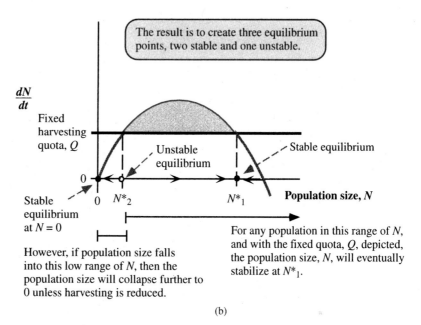

The result is to create three equilibrium
points, two stable and one unstable.

$\frac{dN}{dt}$

Fixed
harvesting
quota, Q

0

Unstable
equilibrium

Stable equilibrium

0   N^*_2

N^*_1 **Population size, N**

Stable
equilibrium
at $N = 0$

However, if population size falls
into this low range of N, then the
population size will collapse further to
0 unless harvesting is reduced.

For any population in this range of N,
and with the fixed quota, Q, depicted,
the population size, N, will eventually
stabilize at N^*_1.

(b)

ble, and therefore the yield at N_1^* is sustainable. However, the one to the left is unstable, and therefore the yield associated with N_2^* would be unsustainable at the same quota level. If the quota were increased, the horizontal line would march up the y axis, giving higher allowed yields. The MSY occurs when the quota is raised to the point that it just equals the peak of the logistic hump, as depicted in Figure 10.19.

For logistic recruitment, the population size will be at $K/2$ when the MSY is reached. Therefore the MSY will equal

$$\text{MSY} = \frac{dN}{dt} \quad \text{at} \quad \frac{K}{2} = \frac{rK/2}{K}\left(K - \frac{K}{2}\right)$$

$$= \frac{rK}{4}. \tag{10.1}$$

It shouldn't take long to see a major problem with regulation for a fixed-quota level Q_{max}. If Q_{max} is exceeded in a single year, perhaps because it was misidentified by government regulators or because environmental conditions change so that the recruitment parabola becomes lower, then resource levels will be depressed the following year to a level below $K/2$. Now, assuming the same quota level, the fish population will

Figure 10.19
A much higher fixed-quota harvest. This harvest will produce the MSY.

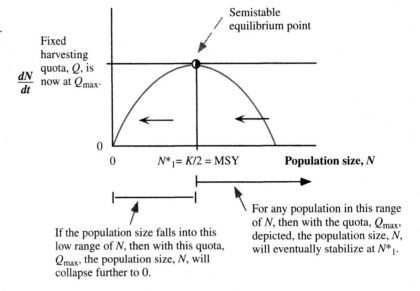

Semistable equilibrium point

Fixed harvesting quota, Q, is now at Q_{max}.

$\dfrac{dN}{dt}$

0

0          $N^*_1 = K/2 = $ MSY   **Population size, N**

If the population size falls into this low range of N, then with this quota, Q_{max}, the population size, N, will collapse further to 0.

For any population in this range of N, then with the quota, Q_{max}, depicted, the population size, N, will eventually stabilize at N^*_1.

Figure 10.20
Harvesting at the maximum sustainable level for four different initial population sizes: 90, 99, 120, and 200. For this example, $r = 1$, $K = 200$, and Q_{max} is 50. Since $K/2 = 100$, any population slipping below $N = 100$ eventually crashes unless Q is reduced.

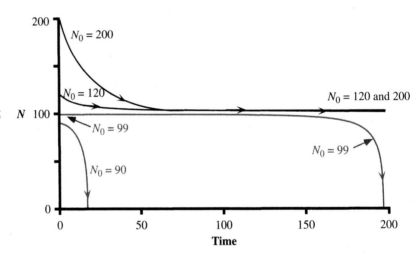

$N_0 = 200$

$N_0 = 120$ $N_0 = 120$ and 200

N 100

$N_0 = 99$ $N_0 = 99$

$N_0 = 90$

Time

approach the stable equilibrium of zero, sliding the fish population into extinction. Therefore harvesting will need to be suspended or greatly reduced until the resource level can return to a level greater than R^* so that the quota Q_{max} will again produce a stable equilibrium. An example of a logistic population being harvested at the MSY is shown in Figure 10.20 for $K = 200$. The initial fish population size is shown at four alternative levels, and for each the population is followed over time.

In this example, the two populations that dipped below $N = 100$ ($= K/2$) fell to extinction. The MSY can be approached only from population sizes above $K/2$. If regulators take quick action and reduce the harvest rate as soon as they notice that the fish population is being overexploited, it will rebound, as Figure 10.21 illustrates.

Why didn't the Peruvian anchovy population rebound like this after fishing was suspended? The anchovy had not gone extinct but were at very low levels. One of our assumptions may be wrong, or there may be more going on than is accounted for by this simple logistic model of fish growth. We will see how this question might be answered a little later in the chapter.

In real systems we typically have little knowledge of the actual level of the resource population, let alone the exact shape of the recruitment curve, and yearly variations in climate influence the shape of this curve anyway. Thus it is easy to overestimate Q_{max}, setting off a population decline and thus producing less than optimal long-term yields (May et al. 1979). A fixed-effort harvesting scheme avoids some of the stability pitfalls of the fixed-quota method, as demonstrated in Boxes 10.2 and 10.3). A fixed-effort

Figure 10.21
This is the same example as in Figure 10.20 for an initial population size of 90. The fish population is on its way to a crash. Now, however, at time 15, the harvest rate is reduced from 50 to 15. With this adjustment the fish population recovers.

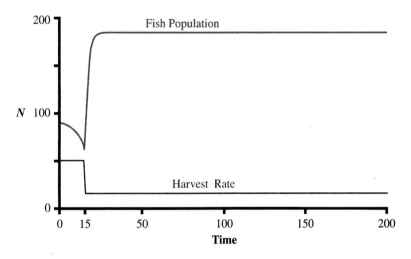

Box 10.2

In Figure 10.22, we develop a curve similar to that in Figure 10.15 for a type 1 functional response instead of a type 3 response. This situation corresponds to the fixed-effort regulation of a fishery discussed in Box 10.1. We label the resource population size R and the consumer population as predators P. For example, governments could regulate the number of fishermen, P, as a means of controlling fishing effort.

Solution:

Now we plot the relationship between the levels of R at equilibria (on the x axis) and the number of predators, P, in

Figure 10.23. The dots on the x axis in Figure 10.22 are carried over here for each level of prey, and projected (by dashed lines) to the appropriate number of predators. Then a line has been drawn to connect these dots. Figure 10.23 is the analog of Figure 10.12 for a type 1 functional response.

The final step is to produce the analog of Figure 10.15. The open squares on the y axis in Figure 10.22 show the equilibrium level of dR/dt associated with the indicated predator numbers on each total consumption curve. These values are plotted on the y axis in Figure 10.24 against the corresponding predator numbers on the x axis.

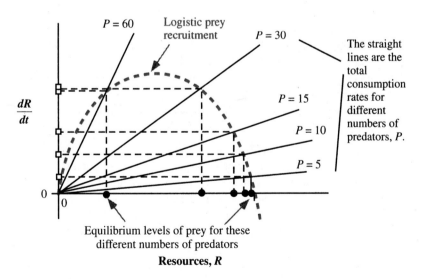

Figure 10.22

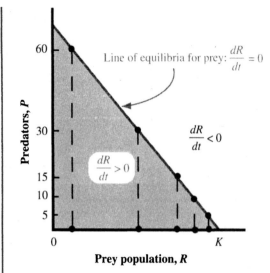

Figure 10.23

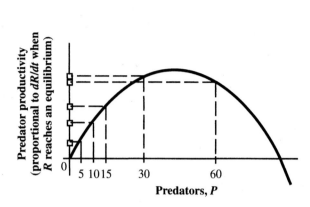

Figure 10.24

Box 10.3

Derive an algebraic expression for the graphical relationship between dR/dt and predator numbers that was developed in Box 10.2.

Solution:

Given a type 1 functional response of predators, with a population size P and a resource population R that grows logistically in the absence of harvesting, we may write

$$\frac{dR}{dt} = rR\left(1 - \frac{R}{K}\right) - aPR$$

The first term is the logistic equation for R.

The second term is the consumption rate of R by P predators, each with encounter rate a: a type 1 functional response.   (c)

We let R^* be the equilibrium level of R and solve for R^* by setting Eq. (c) to 0.

$$rR^*\left(1 - \frac{R^*}{K}\right) = aPR^*.$$

Rearranging, we get

$$R^* = K\left(1 - \frac{aP}{r}\right).$$

At R^*, consumer productivity $= aPR^*$ (from Eq. (c)), and thus the turnover of resources at this equilibrium is

$$\text{Consumption rate of all } R \text{ at } R^* = aPK\left(1 - \frac{aP}{r}\right). \quad (d)$$

Equation (d) describes the **parabola** plotted in Figure 10.24.

method is based on the assumption that the harvesters have a type 1 functional response. Unlike the situation for a type 3 functional response, a type 1 functional response produces no unstable equilibrium points and thus no hysteresis. However, overexploitation of the resources (i.e., diminishing returns with increasing effort) is still possible, as we have already shown in Box 10.1.

More about Fixed-Effort Harvesting

Because of the desirable stability features of the type 1 functional response (Box 10.3), the regulation of total resource harvesting can be accomplished most successfully by a

Figure 10.25
Fixed-effort harvesting of a resource, *R*. The medium harvesting effort produces the MSY and an equilibrium resource level of *K*/2.

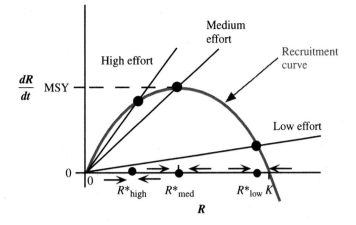

Figure 10.26
The only equilibrium *R* for the very high effort is zero resource.

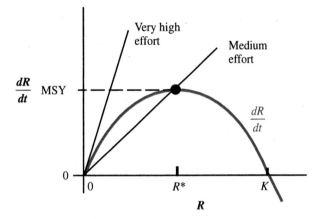

fixed-effort regulation of harvesting. Recall from Box 10.1 that the total yield (or catch) of resource to all harvesters is the product of three different terms:

$$\text{Total yield} = \begin{bmatrix} \text{capture effort expended} \\ \text{by all harvesters} \end{bmatrix} [\text{catchability}][\text{resource level}]. \qquad (10.2)$$

Regulating the first term sets the total yield, and the resource level adjusts accordingly. An example of the overall harvest in terms of *dR/dt* based on three different levels of capture effort per resource (assuming that catchability remains constant) is shown in Figure 10.25. Here, regulatory agencies do not fix the total quota per season, but instead try to regulate the amount of capture effort expended. To do this successfully, they must be able to obtain a census of the number of fish, *R,* and adjust effort accordingly so that the MSY can be reached.

In Figure 10.25, the medium effort corresponds to the MSY. Note, however, from the arguments developed in Box 10.1, that the equilibria fish numbers corresponding to all three efforts are stable. Moreover, unlike the situation with fixed-quota harvesting, if the medium effort is being maintained through regulation, but the resource level temporarily falls below R^*_{med}, the recruitment rate still exceeds the harvest rate and resource levels will increase back to R^*_{med} when conditions return to normal. Resources will not become extinct unless very high capture efforts are maintained, as illustrated in Figure 10.26.

With fixed-effort harvesting, the total yield varies with the resource population size according to Eq. (10.2) If the population size were to drop below *R**, then the yield would drop below MSY. The appropriate response by regulators would be to reduce the allowable effort to let resource levels increase back to *R** or beyond. However, competition among the harvesters would act in the opposite direction. In the face of diminishing catch, harvesters will apply political pressure to increase allowable effort so that yields can go up again, restoring their profit margins. At the same time, with a diminshed supply of resources, the marketplace will often respond with increased prices. This provides an additional incentive to overexploit declining resources. If allowed, such pressures could ultimately ratchet up the effort, forcing the resource to still lower levels. Thus economic pressures interacting with biological harvesting can produce a positive feedback situation that can result in the extinction of the common resource, even with fixed-effort regulation.

As before, the MSY yield occurs at $K/2$ and therefore from Eq. (10.1), MSY = $rK/4$. Of course, r and K are largely invisible to us without a lot of careful field studies. However, the catch at the MSY is measurable and from it we can at least calculate the product rK for this exploited fish population:

$$rK = 4(\text{MSY}).\tag{10.3}$$

Then if we know the catchability of fish, we can extract more information about the fish population. Using Eq. (10.2), we get

$$\frac{\text{Total catch}}{\text{season}} = (C)(\text{effort})(R).$$

At the MSY, the effort is effort* and the resulting catch is $rK/4$; therefore

$$\frac{rK}{4} = (C)(\text{effort}^*)\left(\frac{K}{2}\right),$$

and

$$r = 2(C)(\text{effort*}).\tag{10.4}$$

From Eq. (10.3), this result implies that

$$K = \frac{4(\text{MSY})}{2(C)(\text{effort}^*)}.\tag{10.5}$$

Exercise: An effort of 10 boats per fishing season results in a total season's catch of 20,000 fish. The catchability constant $C = 0.001$/boat. What is the fish population size, N?

Answer:

$$N = \frac{\text{Catch}}{(C)(\text{effort})}$$

$$= \frac{20,000}{(0.001)(10)}$$

$$= 2 \text{ million fish.}$$

Exercise: The MSY of a population has been determined by trial and error to be 30,000 fish produced by an optimum effort of 25 boats fishing per season. What are r and K for this fish population?

Answer:

$$r = 2(0.001)(25 \text{ boats}) = 0.05/\text{season}$$

and

$$K = \frac{4(30,000)}{0.05} = 2.4 \text{ million fish.}$$

Exercise: A woman lives in the only home by a small lake. The bass population in the lake has a logistic recruitment curve with $r = 0.1$/day and a carrying capacity of 100. She decides to continue fishing every day until she catches 4 bass. What will happen to the bass population if she continues this policy? Suppose that she changes her fishing so that she fishes every day for just 2 hours, regardless of how many fish she catches. Assume that her catch is linearly related to the bass population size (i.e., her "handling time" is negligible), and, when the bass are at carrying capacity, her 2-hour fishing effort translates into an average catch of 8 fish. What will happen to the bass population if she continues this policy?

Fixed-effort harvesting may result in instabilities if the resource recruitment curve is lopsided because the strength of density-dependence for fish population growth is nonlinear, perhaps because of an Allee effect, as shown in Figure 10.27. The equilibrium to the left of the recruitment curve hump is unstable. If the fish population is reduced below the level corresponding to the peak of the hump, it will crash, even if harvesting is suspended or reduced to a very low level as the crash occurs.

Exercise: For the situation depicted in Figure 10.27, if effort is maintained such that the MSY is reached, will the resulting equilibrium be stable to temporal changes in the level of R^*?

In Figure 10.28, the fish population begins at the equilibrium level associated with a harvest rate of 0.45/fish/time. The fish recruitment curve is asymmetric like the curve in Figure 10.27. At time 25, an El Niño occurs and the fish population crashes, either by 10% or 40%. In response to this crash, regulators drop the harvest rate at time 35 by 80%, in an effort to protect the fishery.

As Figure 10.28 shows, for the milder population crash of 10%, this reduction in harvest was sufficient to resurrect the fish population, which rebounded to reach a new plateau higher than the first one because of the now reduced harvest rate. Under identical conditions, the greater population crash of 40%, even though followed by an equivalent reduction in harvest rate, was not enough to prevent the fish population from continually crashing to zero.

If the harvest rate were reduced to zero, instead of just 80% of its former level, then in both cases (and in fact in every case for this model), the fish population would rebound and reach its carrying capacity. However, with the Peruvian anchovy, the population behaved differently: it seemed to reach a new stable equilibrium at a very low population level following the crash. Note that there is only one nontrivial stable equilibrium point in Figure 10.27; hence this model does not account for the existence of an alternative stable state at low fish numbers. What could account for this discrepancy?

Figure 10.27
Instabilities in fixed-effort harvesting resulting from a lopsided resource recruitment curve.

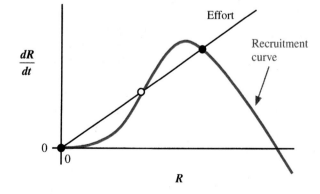

Figure 10.28
Time courses for a fish population, *N*, for two different scenarios of fixed-effort harvesting following a crash in the fish population. The harvest rate is the same for both scenarios—the harvest is reduced at time 35 by 80%.

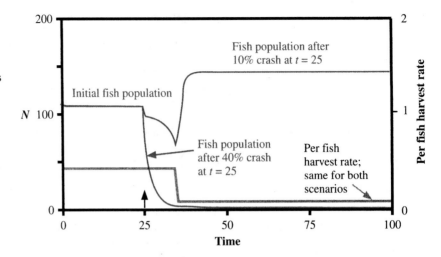

One possibility is that fishing boats have type 3 functional responses. Then a situation like that of intermediate sheep levels depicted in Figures 10.4 and 10.5 would result; two alternative stable equilibria would exist, one at high fish levels and one at low fish levels. However, most fishing boats spend much more time searching for fish than hauling in fish, and there is no evidence that they are usually at or close to some saturation point in harvesting rates that would yield a type 3 functional response.

A more likely alternative recognizes that fish and people are not the only players in this ecological interaction. The exploited fish have many nonhuman predators, including larger fish, sea birds, and marine mammals. If these predators have a type 3 functional response to their fish prey, then it is entirely possible that human overexploitation of fish might force the fish population to the lower of one or several alternative stable equilibria maintained by their natural enemies. To fully understand the dynamics of interacting prey and predators, we need to understand the numerical response of natural predator populations, as well as the within-predator functional response. In Chapter 12 we explore this theme in some detail.

PROBLEMS

1. You have a herd of milk cows and an overgrazed 10-acre pasture. Assume that the functional response of cows to grass on the pasture is a type 2. What is the relationship between cow numbers (*y* axis) and equilibrium grass levels (*x* axis)? Also describe graphically the total milk production as you graze more and more of your herd and state why it has the shape that you draw.

2. In the following diagram, the recruitment rate in the absence of predators and the death rate due to predator consumption is plotted for a particular prey species, *R*. The consumption curve is the functional response (showing total consumption of *R*) by a fixed number, *x*, of predators.

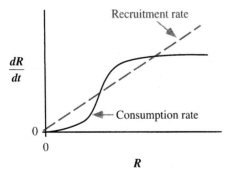

Draw the *net* growth curve of *R* for the prey species in the presence of this fixed number of consumers. Label all equilibrium points for the prey and state whether they are stable or unstable. Now answer the following questions:

    a. What type of functional response do the predators have?

    b. Do the resources (prey) show intraspecific density dependence?

    c. Sketch the curve of equilibrium prey levels as a function of predator numbers.

3. An effort of 20 boats per fishing season results in a total season's catch of 10,000 fish. The catchability constant $C = 0.001$/boat. What is the fish population size, *R*? The MSY of this population has been determined by trial and error to be 20,000 fish produced by an optimum effort of 30 boats fishing per season. What are *r* and *K* for this fish population?

4. Suppose sheep interfere with each other at higher sheep densities, such that the combined functional responses of multiple sheep increases less than linearly with the number of sheep on the pasture. How would you qualitatively modify Figure 10.14 to incorporate this behavior and how would this change the general shape of the wool production curve shown in Figure 10.15?

11 The Mechanics of Predation

As a prey population grows, predators respond locally in two ways. First, each predator potentially increases its rate of prey consumption (the predator's **functional response**). Second, with more to eat, predators have more energy and nutrients for reproduction, ultimately leading to higher predator numbers (the predator's **numerical response,** which we take up in Chapter 12). The functional response is more immediate while the numerical response requires some time to take effect; this time depends on the generation time of the predator. Here we develop a microscopic and economic look at predation by deriving the so-called **disk equation** and considering some choices that predators face in trying to maximize their prey intake.

THE DISK EQUATION

Our goal is to derive the shape of the functional response based on behavioral attributes of the predator. To begin, we describe an individual predator's foraging path as if the predator were a moving disk; the disk represents the two-dimensional sensory field of the predator as it searches for prey. The radius, r, of this sensory field will depend on the sensory acuity of the predator and also on the "visibility" of the prey. For example, large, brightly colored, smelly, and noisy prey species will be more obvious to the predator than small, cryptic, less smelly, and silent prey species. Consider the situations illustrated in Figures 11.1 and 11.2.

For a single foraging predator, what is the relationship between prey density, V (prey/area), and the consumption rate of those prey by the predator? We introduced three types of functional responses in Chapter 10 (illustrated in Figure 10.1). Now we derive the shape and formulas for these functional responses based on biological attributes of the predator.

We take a geometric approach. As Figures 11.1 and 11.2 illustrate, the area searched by the predator = shaded area in the figures = (predator speed)(sensory diameter)(time spent searching) = $s2rT_s$. (Units are (distance/time)(distance)(time) = (distance)^2 = area.) Finally, not all prey within this area may be detected by the predator. Some prey could be hidden at the time the predator moves through. We therefore need to introduce another constant, the detectability fraction, k (ranging from 0 to 1) to express this limitation.

For notational simplicity we lump the constants $ks2r$ into a single constant a, **the prey encounter rate,** which expresses the number of prey encountered by a single predator per time period that it spends searching. Then we have

$$\begin{bmatrix} \textbf{The number} \\ \textbf{of prey} \\ \textbf{encontered} \end{bmatrix} = \begin{bmatrix} \textbf{detectability} \end{bmatrix}\begin{bmatrix} \textbf{area searched} \end{bmatrix}\begin{bmatrix} \textbf{prey per} \\ \textbf{unit area} \end{bmatrix}.$$

243

Figure 11.1
A predatory beetle searching for randomly distributed prey. The area searched is determined by the predator's velocity and its sensory field.

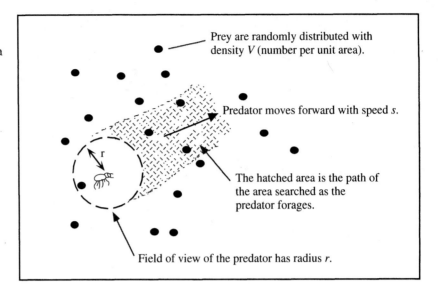

Prey are randomly distributed with density *V* (number per unit area).

Predator moves forward with speed *s*.

The hatched area is the path of the area searched as the predator forages.

Field of view of the predator has radius *r*.

Figure 11.2
(a) A predator with a narrow sensory radius. (b) A predator with a wide sensory radius, (c) a slower moving predator. Increasing the spatial sensory field leads to a larger swath of area searched per unit time. Similarly, all else being equal, faster search velocities lead to a larger area searched.

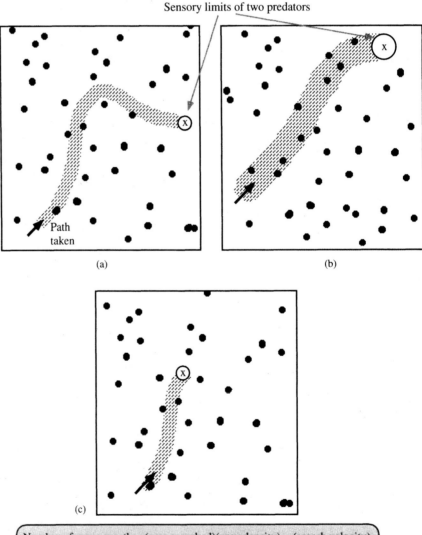

Sensory limits of two predators

Path taken

(a)

(b)

(c)

Number of prey on path = (area searched)(prey density) = (search velocity) (sensory diameter)(time spent searching)(prey density)

Or, symbolically, using V for prey density and V_a for the number of prey encountered during a search of duration T_s, we get

$$V_a = aT_sV. \tag{11.1}$$

We'll assume that the predator attempts to pursue all the prey that it detects, so V_a is also the number of prey chased or attacked. The predator, however, may be successful in subduing only some fraction f of the prey that it attacks, so the number of prey successfully captured, V_c, is

$$V_c = fV_a = faT_sV. \tag{11.2}$$

When all prey that are encountered are subdued, perhaps, because they put up no resistance, then $f = 1$.

The next step is to find an expression for the time spent searching, T_s, in terms of the total time available for feeding, T_t, in a day. The more prey that are encountered and attacked, the more time is devoted to chasing and then manipulating prey and the less time is available for the predator to search for new prey. We can express this trade-off in time allocation by allowing the time available for searching to be diminished by the time that is spent chasing and manipulating prey. If T_c is the average chase time per prey attacked (V_a) and T_m is the average manipulation time per prey captured (V_c), then search time is

$$
\begin{aligned}
T_s &= T_t - T_cV_a - T_mV_c \\
&= T_t - T_cV_a - T_mfV_a.
\end{aligned}
$$

$$
\underset{\text{searching}}{\text{Time for}} = \underset{\substack{\text{total} \\ \text{time to} \\ \text{feed}}}{} - \underset{\substack{\text{time spent} \\ \text{chasing prey} \\ \text{once they are} \\ \text{encountered}}}{} - \underset{\substack{\text{time spent} \\ \text{manipulating} \\ \text{prey if they} \\ \text{are subdued.}}}{} \tag{11.3}
$$

Note that by this definition the total time feeding is simply partitioned into time spent searching and time spent pursuing and manipulating prey. We may also write Eq. (11.3) as

$$T_s = T_t - V_a(T_c + fT_m). \tag{11.4}$$

With the assumption that f is a constant, we may combine the chase and manipulating times to produce an overall prey handling time of $T_h = T_c + fT_m$. Then Eq. (11.4) becomes

$$T_s = T_t - V_aT_h. \tag{11.5}$$

Substituting T_s from Eq. (11.5) into Eq. (11.1), we arrive at a new equation for the number of prey attacked, V_a, per total time available per day (rather than just per search time):[1]

$$V_a = aV(T_t - T_hV_a). \tag{11.6}$$

To complete the development Eq. (11.6) needs to be solved for V_a :

$$V_a = aVT_t - aVT_hV_a,$$

$$V_a + aVT_hV_a = aVT_t,$$

and

$$V_a = \frac{aVT_t}{1 + aT_hV}. \tag{11.7}$$

---

1. There is an additional complexity that we gloss over here. Every prey attacked and consumed leaves one less prey available; thus the density of prey changes over time. Moreover, looking at the paths of the predator in Figure 11.2, you can see that, over a stretch of time, the trail of a randomly foraging predator would circle back on itself. The area searched the second time around would contain less prey. In discrete time models, these factors require careful adjustment of the equations. However, here we assume infinitesimally small time units and differential equations, where the density, V, is effectively constant.

Figure 11.3
A type 2 functional response
approaches an asymptote, which is the
reciprocal of the handling time, T_h.
The shorter the handling time, the
higher is the maximum rate of food
intake.

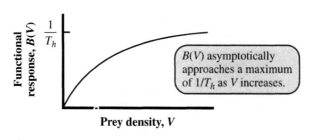

Figure 11.4
The effect of encounter rate, a, and
handling time, T_h, on the shape of a
type 2 functional response. (a) The
encounter rates are the same for all
these curves. All the functional
responses have the same initial slope,
but they become asymptotic at
different prey capture rates. (b) All the
handling times are the same ($T_h = 0.1$),
but the encounter rates differ. All the
curves will eventually reach the same
asymptote, but they do so at different
rates.

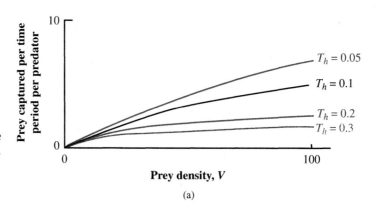

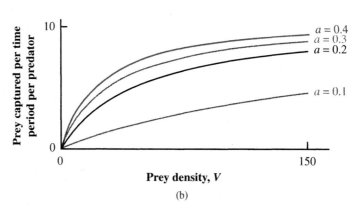

Equation (11.7) is called the **Holling disk equation** (Holling 1959) because it is based
on a "moving" disk representing the predator's moving sensory field (Figure 11.1). The
functional response of a predator is the number of prey captured per predator per unit
time, so it is simply Eq. (11.7) divided by T_t, or

$$\text{Functional response} = B(V) = \frac{V_a(V)}{T_t} = \frac{aV}{1 + aT_h V}. \qquad (11.8)$$

Equation (11.8) describes a type 2 functional response derived from "microscopic"
considerations of predator behavior and time accounting. The parentheses indicate that
the functional response is a function of prey density, as graphed in Figure 11.3.

To show that the maximum consumption rate is $1/T_h$, we divide both numerator and
denominator of Eq. (11.8) by aV (the numerator). This yields

$$B(V) = \frac{1}{1/aV + T_h}. \qquad (11.9)$$

As prey density, V, climbs to infinity, the first term in the denominator goes to zero, so
the entire expression approaches $1/T_h$. The effect of different encounter rates and handling times on the rate that prey are consumed is illustrated graphically in Figure 11.4.

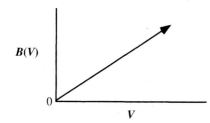

Figure 11.5
Linear (type 1) functional response. Ultimately, at high V, $B(V)$ may hit a maximum where it abruptly becomes flat.

Modifications to the Disk Equation

A type 1 functional response can be derived from the disk equation by assuming that the handling time is zero ($T_h = 0$). Then $B(V)$ will linearly increase with prey density, V. To see this, look at Eq. (11.9), but set T_h to zero to get

$$B(V) = \frac{1}{1/aV} = aV. \tag{11.10}$$

The functional response expressed in eq. (11.10) has a y intercept of 0 and increases linearly with slope a, as graphed in Figure 11.5. Even without any finite handling time, we might ultimately expect the consumption rate to hit some ceiling simply because the predator's gut will eventually become full.

For a type 3 functional response we assume that the encounter rate, a, is not constant across all prey densities but instead is an increasing function of V, so we write $a(V)$. This might be the case if predators gain experience with a particular prey when those prey are common, and this experience leads to increased prey detection rates. Predators might need the frequent reinforcement of high prey densities to recall their **search image** for the prey. With such a mental image, their sensory field effectively increases. The sigmoid curve of a type 3 functional response is achieved when $a(V)$ is a simple linear function, or $a(V) = aV$ in Eq. (11.8). Then

$$B(V) = \frac{V_a(V)}{T_t} = \frac{aVV}{1 + aVT_hV} = \frac{aV^2}{1 + aT_hV^2}. \tag{11.11}$$

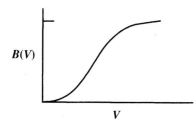

Figure 11.6
Type 3 functional response, as described by Eq. (11.11).

Note the similarity of Eq. (11.11) to Eq. (11.8) for the type 2 functional response. Yet the V^2 terms result in a functional response that is sigmoidal as depicted in Figure 11.6. Another reason that the encounter rate itself may increase with prey density is due to changes in the prey's detectability, k, with prey density, V. If prey hiding spots are limited, then as prey density increases a larger proportion of prey may be more exposed to the predator's sensory field. Again the result might be something like the curve shown in Figure 11.6.

> **Problem:** For a type 3 functional response, derive the asymptotic value of $B(V)$ from Eq. (11.11).

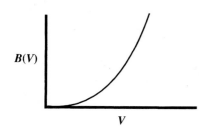

Figure 11.7
This functional response increases with the square of prey density.

The next functional response that we consider doesn't have a name in the ecological literature. It is probably biologically implausible since it never becomes asymptotic. It is worth considering, however, because it represents the case where $a(V) = aV$ (as for type 3) but the handling time $T_h = 0$ (as for the type 1). This functional response provides a very stabilizing situation for predator–prey dynamics because, as prey numbers climb, the predator's kill rate climbs even faster, as graphed in Figure 11.7. Thus prey outbreaks can be reduced by such a predator's functional response, even without any numerical response by the predator population.

So far, we've been looking at prey eaten from the predator's point of view. From the prey's point of view, we can define the probability that a *particular* prey is captured by a predator per unit time. This risk per prey can be calculated by dividing the functional response (which is in units of all prey eaten) by the prey density, V. An example of parasitism by a wasp on sawfly cocoons is shown in Figure 11.8.

Figure 11.9 summarizes the prey risk relationship for each functional response. For type 2 and 3 functional responses, the rate of prey capture increases with prey density but an *individual* prey has a lower probability of being captured at very high prey densities since the predator's functional response becomes maxed-out. Each predator's consumption cannot keep up with the increases in prey density and many prey go uneaten. The type 3 response shows the most complex behavior, since the S-shape approach to a maximum consumption rate leads to increases in risk at low prey density followed by declines in risk as prey density increases still more.

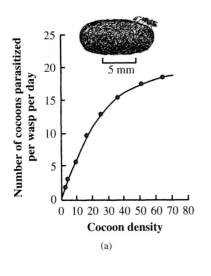

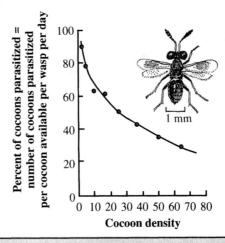

Conclusion: When cocoon density is relatively high, an individual cocoon is less likely to be parasitized because the wasps' time is tied up parasitizing other cocoons (i.e., the functional response approaches a maximum).

Figure 11.8
(a) The functional response of the chalcid wasp *Dahlbominus fuscipennis* parasitizing cocoons of the sawfly *Neodiprion setifer* by laying eggs on them in laboratory cages (Burnett 1956); sawflys are also a type of wasp. (b) The functional response from the vantage point of the cocoon: What is the risk of an individual cocoon being parasitized per day?

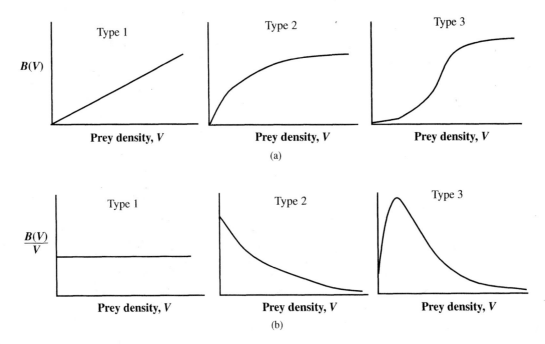

Figure 11.9
Summary of the relationship between prey consumption rates, $B(V)$, and prey consumption per prey available (individual prey risk): (a) Functional responses and (b) prey risk.

Problem: Show that a plot of prey risk for a type 1 functional response is equivalent to plotting the slope of the functional response at each prey density.

Figure 11.10
Another type of (a) functional
response and (b) corresponding prey
risk.

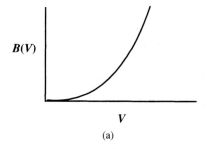

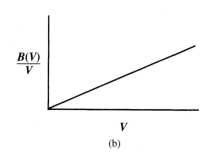

Figure 11.10
Another type of (a) functional
response and (b) corresponding prey
risk.

The comparable situation for the fourth type of functional response is shown in Figure 11.10. Unlike the other functional responses, as prey numbers increase, the prey's individual probability of being eaten continually increases.

MULTIPLE PREY ITEMS

What happens when a predator has multiple types of prey, each with a distinct density V_1, V_2, V_3, . . . ? Now each time the predator stops to eat any one prey item, the time lost in handling that prey detracts from its search time for *all* types of prey. Therefore the functional response for V_1, which we denote B_1, will be a function of the abundances of each prey item, not just prey 1. We define a vector **V** containing all the prey abundances as $\mathbf{V} = (V_1, V_2, V_3, . . . , V_n)$. Then

$$B_i(\mathbf{V}) = \frac{a_i(V_i)V_i}{1 + \sum a_j(V_j)V_j T_{hj}} \quad \text{for prey items } i = 1 \text{ to } n. \quad (11.12)$$

The notation of Eq. (11.12) allows a_i to be a constant or an unspecified increasing function of V_i, as would be the case for the type 3 functional response $a(V) = aV$.

Figure 11.11 shows an example of a functional response surface for prey 1 as a function of the densities of both prey 1 and prey 2 when a_i is a constant for both prey. Figure 11.11 is based on constant encounter rates; hence it yields a type 2 functional response surface. Note that the consumption rate of prey 1 increases with its own density but decreases with the abundance of prey 2. The more prey 2 available, the more prey 2 will be attacked, and therefore less time will be spent attacking prey 1.

Across the range of all possible densities of prey 1 and prey 2 we can also plot the *relative* consumption rates of the two prey species by the predator as a function of the preys' relative abundances rather than their absolute abundances. Figure 11.12 shows such a plot, using the parameters from the preceding example. Since prey species 2 is encountered more frequently than prey species 1 ($a_1 = 0.2$ and $a_2 = 0.4$), it is overrepresented in the diet of the predator.

Problem: Use Eq. (11.12) to show that the curves in Figure 11.12 do not depend on the absolute abundance of prey but only on their relative abundances. Also show that the curves are unaffected by the handling time values.

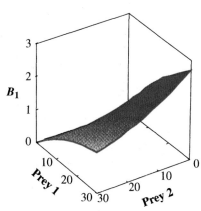

Figure 11.11
A functional response for prey 1 as a function of the abundance of prey 1 and prey 2 ($a_1 = 0.2$ and $a_2 = 0.4$; the handling times are $T_{h,1} = T_{h,2} = 0.2$).

Up to now we have assumed that encounter rates, a_i, are simply influenced by predator velocity, sensory field, and the conspicuousness of the prey. However, the preference for one mussel over another shown by *Thais* snails is not easily explained by such factors. There is also an element of choice shown by the predator, as illustrated in Figure 11.13. The predator may prefer some prey species over others because more nutritious, has fewer defenses, or requires less handling time. Over evolutionary time, encounter rates and handling times may coevolve.

Figure 11.12
A predator consuming two prey with a type 2 functional response for both. The predator has a higher encounter rate for Prey 2 compared to Prey 1 ($a_2 > a_1$) and thus is more common in the diet for all values of prey ratios present in the field.

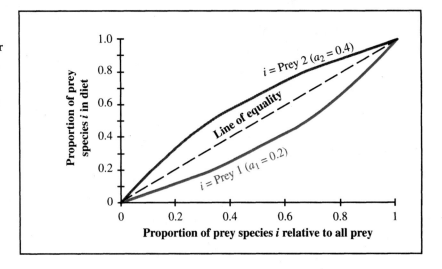

Figure 11.13
Predatory sea snails offered a diet of two mussels, *Mytilus edulis* and *M. californianus*. The average percent consumed (and standard errors) for each prey are shown. The snails prefer *M. edulis* regardless of its relative frequency in the environment (from Murdoch and Oaten 1975).

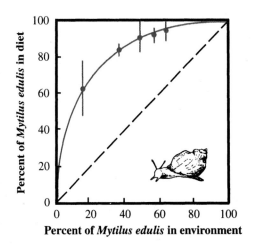

Figure 11.14
A predator with a type 3 functional response for two prey. Prey 2 is preferred over prey 1 ($a_2 > a_1$) and thus is more common in the diet for all values of prey ratios present in the field. For both prey, their consumption rate increases disproportionately as they become relatively more frequent up to the point that they are the most common prey in the diet.

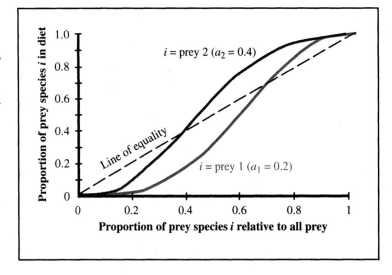

If the functional responses are of the type 3 variety, then with the same parameters as in Figure 11.13, the plot would look like the plot shown in Figure 11.14. If the two prey species had the same encounter rates ($a_1 = a_2$) but different handling times, this would not lead to any difference in the functional response for the two prey. The numerator of Eq. (11.12) would be identical for both species; since it contains no term for handling time and the denominator is the same for both species. The additional handling

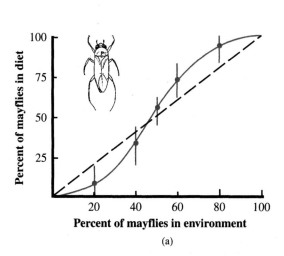

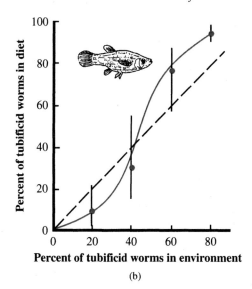

(a)

(b)

Figure 11.15
Switching predators. (a) Water bugs were presented with two types of prey—an isopod and larvae of the mayfly (Lawton et al. 1974). (b) Guppies feeding on a combination of fruitflies and tubificid worms (Murdoch and Oaten 1975).

time associated with one prey species subtracts equally from the search time available for hunting both prey species—hence identical plots.

An example of the S-shaped relative consumption curves from an experiment by Lawton et al. (1974) is shown in Figure 11.15. The investigators presented predacious water bugs with two types of prey, an isopod and larvae of the mayfly. Another example is provided by guppies feeding on a combination of fruitflies and tubificid worms (Murdoch and Oaten 1975).

An important ecological feature of type 3 functional responses for alternative prey items is that the predator takes a disproportionately larger proportion of whatever prey item tends to be the numerically most common in the environment (again refer to Figure 11.14 and compare the curves to the line of equality). This is called **predator switching.** Recall that we invoked the presence of search images to explain why encounter rates might increase with prey abundance. We now see that when this is the case, predator switching results. The community-level consequences are important. With switching, whenever one prey becomes rare relative to another—perhaps because it is an inferior competitor under present environmental conditions—the rare prey will experience relatively less predation than the more common prey. In this way, a relative competitive inferiority is compensated for by a relatively lower predation rate. This tendency for predator switching to give an advantage to the rarer of several competitors is very stabilizing in ecological systems since it may prevent one prey species from outcompeting the others (Roughgarden and Feldman 1976).

FOOD DEPLETION (ADVANCED)

How long will it take an animal with a type 2 functional response to completely deplete the food in a patch? The situation is graphed in Figure 11.16.

The equation of a type 2 functional response is

$$\text{Intake rate, } B(R) = \frac{aR}{1 + aT_h R}.$$

As food is consumed, its density, *R,* declines according to the rate of consumption:

$$\frac{dR}{dt} = -B(R).$$

(11.13)

Figure 11.16
Food intake rate is a type 2 functional
response to food density in initially
uncropped patches of food.

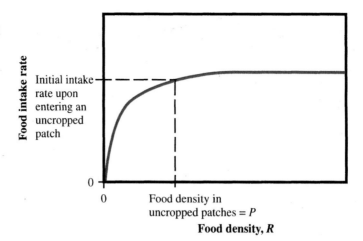

To find the time, T, it takes to reduce the patch from $R = P$ to $R = 0$, we need to solve
this differential equation. We begin by rearranging Eq. (11.13) to separate the terms:

$$\frac{dR}{-B(R)} = dt.$$

Now we integrate each side; the patch begins at $R = P$ and is consumed to 0 as time t
passes from 0 to T:

$$\int_P^0 \frac{dR}{-B(R)} = \int_0^T dt$$

or

$$\int_P^0 -\frac{1 + aT_h R}{aR} dR = T.$$

The left-hand side of this equation may be split into the sum of two integrals:

$$-\frac{1}{a}\int_P^0 \frac{1}{R} dR - \int_P^0 T_h dR = T.$$

After performing the integration we arrive at

$$T = \frac{\ln P - \ln 0}{a} + T_h (P - 0). \tag{11.14}$$

Since $\ln 0 = -\infty$, the time, T, needed to reduce the resources in the patch from P to 0 is
infinite.

This result initially seems paradoxical, since the patch contains a finite amount of
food and animals consume food at some finite rate. The problem is that the functional
response $B(R)$ lets the consumption rate go to zero as food levels approach zero. As
food density becomes smaller, the amount of food consumed per time unit becomes
smaller, approaching zero, and food levels can thus only go to zero asymptotically in
infinite time. This strange result is probably unrealistic. If an animal consumes the food
to some minimal level, R_{min}, greater than zero, then even if this level is very small—just
a few specks of food—the equation will now have a finite solution for T, the time to
reduce the food patch to this minimum level, as illustrated in Figure 11.17.

After all, we don't entirely "clean our plates" during a meal; we leave a tiny bit of
residual food or grease on the plate, even if it is of microscopic proportions. Why
should an animal spend increasingly longer times trying to reduce the food level in a
single patch to zero if it could increase its food intake rate by moving to another patch
where food had not yet been exploited?

Figure 11.17
Food intake rate as a function of food density. Beginning with uncropped patches of food having P food units, food can be harvested to a minimal level, R_{min}.

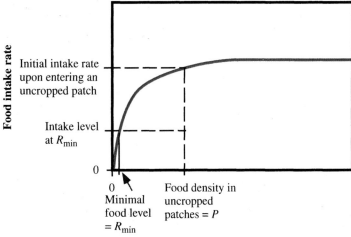

If an animal gives up searching for food in a patch when food levels in that patch fall to some minimum level, R_{min}, then the give-up time, T (the time it takes to reduce the food level in the patch from P to R_{min}), from Eq. (11.14), is

$$T = \frac{\ln P - \ln R_{min}}{a} + T_h \left(P - R_{min} \right). \tag{11.15}$$

Continuing with this line of logic we might imagine that animals behave in an optimal way with respect to the parameter R_{min}. An animal that is maximizing energy intake per unit time would want to leave a patch when its expectation of further food intake by leaving and finding a new patch with more food exceeds the expected food intake rate from staying in the same patch and depleting it still further. An optimal level of R_{min} could be determined if we knew the time it takes, on average, to find a new patch, the food density in each patch, and the shape of the functional response. In the next section we show how this optimal "give-up" time can be calculated.

OPTIMAL GIVE-UP TIMES (ADVANCED)

When should an animal stop foraging in a patch and move on to look for another? Consider the beetle foraging on discrete patches of food in Figure 11.18. Let's assume that the animal, through its previous experience in the habitat, knows the average rate of capture to expect in the entire habitat and that it mentally compares this to its current rate of food intake within the patch where it is now foraging. If it has these cognitive powers, then it should leave its present patch when its rate of food intake falls below the anticipated average rate of food intake over the entire habitat. This is called the **marginal value theorem** for optimal foraging (Charnov 1976). As an animal stays in a patch, its rate of food intake can only decline. But, if the animal moves, there will be a period when food intake is zero as it searches for a new patch. Once that patch is found, food intake rates can potentially increase.

To solve for the optimal give-up time, we first need to assess the rate of food intake in a patch as a function of the time spent foraging in that patch. Let's assume that the depletion rates from current foraging far outstrip the rate of new recruitment of food to the patch so that every bite of food consumed leaves that much less food available. Let us rewrite Eq. (11.15) in terms of how much food remains in a patch, $R(T)$, after a specified amount of time, T, foraging in the patch:

$$T = \frac{\ln P - \ln R(T)}{a} + T_h \left(P - R(T) \right). \tag{11.16}$$

Figure 11.18
A predator forages in a habitat
composed of distinct patches, each
with *P* food units.

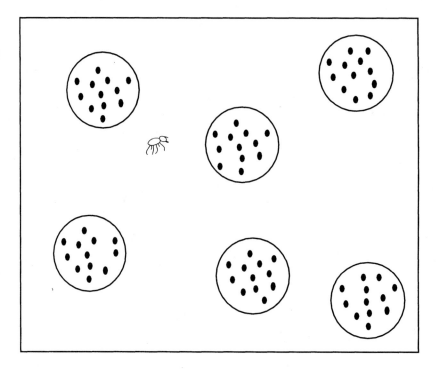

The cumulative amount of food consumed by the animal after T time units foraging in a patch is equal to the amount of food initially present, P, minus the amount remaining. We call this $G(T)$ (for gain) as a function of elapsed time T, or

$$\text{Cumulative food consumed after time } T = G(T) = P - R(T). \qquad (11.17)$$

We now try to solve for $G(T)$ as a function of time, T, in the patch. Unfortunately, we soon find that it is impossible to rearrange Eq. (11.16) to get $R(T)$ alone on the left-hand side of the equation. In other words, there is no analytical solution for the cumulative gain from this type 2 functional response. After rearranging Eq. (11.16) and collecting terms in $R(T)$, we can write

$$R(T)T_h + \frac{\ln R(T)}{a} = -T + \frac{\ln P}{a} + T_h P, \qquad (11.18)$$

which can be solved numerically for $R(T)$, given the parameters a, T_h, and P. We can also guess the general shape of $G(T)$. At $T = 0$, the cumulative gain will be 0 since the animal has not yet consumed any food. As time T progresses, the accumulated gain, $G(T)$, will increase but it must become asymptotic at P, since the animal cannot consume any more food than there is to begin with, even given an infinite amount of time as shown in Figure 11.19.

Generally, we would also get about the same shape for this gain curve for either a type 1 or type 3 functional response. The slope of this curve at any particular time gives the rate of gain. As time goes on, the rate of gain becomes zero, regardless of the exact shape of the functional response.

Exercise: Show that if a type 1 functional response is substituted for a type 2 in the development beginning with Eq. (11.13), that the closed-form solution for the cumulative gain $G(T)$ by the predator as a function of its time T spent foraging in the patch is:

$$G(T) = P - e^{\ln P - aT}$$

Now, if an animal stops foraging after a while and moves to another patch, its gain will be zero during the time it is in transit but then its gain will increase again when it enters a new patch. We seek the average rate of gain for this process, including time in

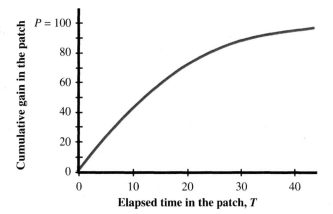

Figure 11.19
Cumulative gain curve, $G(T)$, for an animal foraging in a patch with an initial food density of $P = 100$. The gain curve is solved numerically by using a handling time of 0.1 and an encounter rate of $a = 0.1$.

transit and time foraging in a patch. While in a patch, the average rate of gain is $G(T)/T$, which of course varies with T. The average rate of gain, including the typical transit time, T_t, to a new patch, will depend on the spatial distribution of patches. The average rate of gain including both transit and foraging time combined is

$$A(T) = \frac{G(T)}{T + T_t}.$$

The optimal give-up time can be solved by taking the maximum of this expression. When elapsed time T is zero, $A(T)$ must also be zero since no food has been consumed. At the other extreme, when T is very large, the denominator becomes very large and $A(T)$ must therefore become small. Thus $A(T)$ has some intermediate maximum. From calculus, we can find this maximum, by taking the derivative of $A(T)$ with respect to T and setting it to 0:

$$\frac{dA(T)}{dT} = 0 \quad \text{or} \quad \frac{d\left(\dfrac{G(T)}{T + T_t}\right)}{dT} = 0.$$

We wish to find the optimal value of T—call it T^*. Applying the chain rule for derivatives gives

$$\frac{G'(T^*)}{T^* + T_t} - G(T^*)(T^* + T_t)^{-2} = 0,$$

and thus T^* satisfies the formula

$$G'(T^*) = \frac{G(T^*)}{T^* + T_t}, \tag{11.19}$$

where G' is the derivative of $G(T)$ with respect to T. Equation (11.19) has a particularly simple graphical solution, as shown in Figure 11.20, with the $G(T)$ curve displaced by T_t units.

Suppose, however, that patches contained not $P = 100$ units of food, but $P = 400$ units. Is the optimal give-up time affected? Figure 11.21 shows that, with a fourfold increase in patch food density, the optimal give-up time increases only by about twice.

The optimal give-up time increases if the travel time increases, as illustrated in Figure 11.22. It pays to keep foraging in the same patch if it's a long trip (without food) to the next patch.

Finally, the optimal give-up time decreases if the foraging rate within a patch increases, as shown in Figure 11.23.

Of course, all this assumes that, when an animal gets to a new patch, that patch contains P food items, and the animal's foraging rate returns to the level associated with P. If many predators are simultaneously foraging in the same area (and again assuming that the recruitment rate of new food to a patch is low), an individual predator that moves may only find patches that are as depleted as the ones that it just left. Any truly optimal give-up time must also take into account the activities of other predators foraging in the

Figure 11.20
A graphical method to find the optimal give-up time, T^*, in a patch. The typical travel time between patches is 10 time units so the $G(T)$ curve (in red) begins on the left at Time = 10. The dashed line from (0, 0) is tangent to a point on the $G(T)$ curve where the open dot identifies T^* on the x axis. The optimal give-up time here is about 18, corresponding to a cumulative gain rate of about 69.

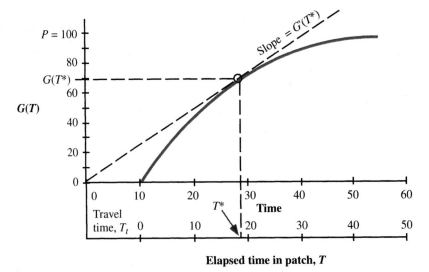

Note that T^* satisfies the relationship $G'(T^*) = \dfrac{G(T^*)}{T^* + T_t}$.

Figure 11.21
The effect of increasing the density in a food patch on the optimal give-up time. A fourfold increase in food per patch, compared to Figure 11.20, leads to about a doubling in the give-up time. The gain curve (in red) is solved numerically by using Eqs. (11.16) and (11.17). The optimal give-up time here is about 37.

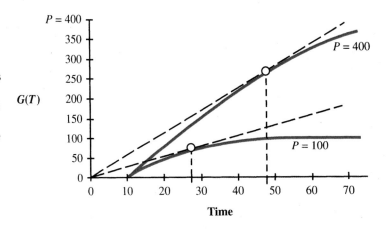

Figure 11.22
Increasing travel time between patches, which might be due to slower speed or to patches that are farther apart, leads to an increase in the optimal give-up time. The optimal give-up time for $T_t = 10$ is 28 − 10 = 18. The optimal give-up time for $T_t = 5$ is 17 − 5 = 12.

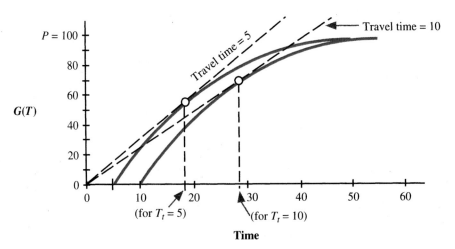

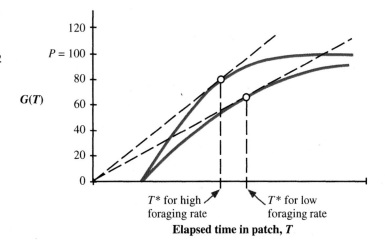

Figure 11.23
The optimal give-up times for two different foraging rates, $a = 0.1$ and $a = 0.3$. This figure is based on a type 2 functional response with $T_h = 0.1$.

same habitat and the recruitment rate of resources to patches. Such cognitive abilities may be beyond the abilities of most animals and may be one reason why they are instead territorial, defending their foraging space against intruders. In this way, they can control their return times to patches to synchronize them with the resource recovery times to level *P*, without concern that someone else will get to the patch before they do.

THE IDEAL FREE DISTRIBUTION

Different habitats may have different potential profitabilities for consumers. Consider two habitat types that differ in resource productivity and that are large enough to hold several foraging consumers simultaneously. The consumers are free to move back and forth between these two habitats. If only a single consumer were present, it would make sense for it to choose the high productivity habitat since its consumption rate would be higher than it would be in the low productivity habitat. As consumer numbers grow in this habitat, however, resources become depleted and consumers interfere with one another by aggressive behavior or by scaring away mobile prey items. An example of this effect is shown in Figure 11.24 for oystercatchers feeding on intertidal mussel banks exposed at low tide. The oystercatcher intake rate decreases as the number of other oystercatchers nearby increases. In this case, most of the decline is not due to resource depletion but rather to direct interference between individuals (Zwarts and Drent 1981). The two sites differ in their mussel density.

The average consumer will just be able to numerically replace itself when its consumption rate balances its maintenance and only the reproductive effort necessary to replace itself. Any additional consumption can be turned into a net excess of consumer births or could support the immigration of more consumers into the habitat, as shown in Figure 11.25.

As long as habitats have dissimilar consumption rates, consumers should move to the more profitable habitat. At the point that consumer density in the high-productivity habitat equals C_h. However, the next additional consumer will find that its consumption rate is equivalent (at Cons_h) in both habitat types. Therefore, additional individuals will settle equally in both habitats. Assuming that maintenance costs to the consumers are the same in both patches, the consumers should move and settle in the two habitats until they come to numerical equilibrium such that consumption rates equal maintenance costs. This equilibrium consumption rate will be the same for both habitats. The equilibrium density of consumers in the high-productivity habitat (C^*_{high}) will be greater than that in the low-productivity habitat (C^*_{low}). Extending this argument to multiple habitat types at equilibrium, all habitats, regardless of their innate productivities will have identical profitabilities because of the adjustments of consumer densities. They will also have identical densities of resources at equilibrium if the functional response of the predators is identical in the different habitats.

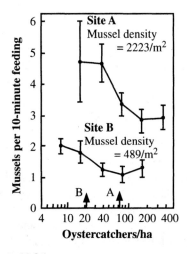

Figure 11.24
Oystercatchers feeding on mussels at two sites in the Netherlands. For each site, the feeding rate declines with increasing oystercatcher density. The higher density of mussels at site A, translates into higher feeding rates overall per oystercatcher densities (from Zwarts and Drent 1981). The vertical lines indicate one standard error. The mean density of oystercatchers at the two sites is indicated by the arrows at the bottom.

Figure 11.25
The per capita consumption rate for consumers as a function of consumer density, *C*, in two habitats. In high-productivity habitats the consumption rate is higher than in low-productivity habitats for all levels of *C*. At a consumer density of *C** consumption rates equal maintenance costs; high-productivity habitats can support more consumers than low-productivity habitats ($C^*_{high} > C^*_{low}$). Imagine that the habitats start out void of consumers. The first consumer to enter will maximize its energy gain by settling in the high-productivity habitat. At the point that consumer density in the high-productivity habitat equals C_h, the next additional consumer will find that its consumption rate is equivalent (at $Cons_h$) in both habitat types, and thus may settle in the low-productivity habitat. Additional consumers will settle such that consumption rates are equal in both habitats. At the limit, the high-productivity habitat will support a density of C^*_{high} consumers, and the low-productivity habitat will support a density of C^*_{low}.

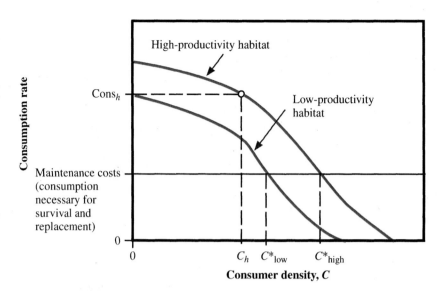

This process is called **ideal free distribution** (Fretwell and Lucas 1970). It is *ideal* because predators are assumed to have ideal knowledge of resource distributions and profitabilities across habitats. It is *free* because predators are assumed to be free to move in such a way that they can take advantage of any discrepancies.

Imagine a long, narrow aquarium containing six small fish. If food is added to one end at five times the rate that food is added to the other end, we expect five fish to be at the high-food end and one fish to be at the low-food end, as long as the total amount of food added does not exceed the maximum consumption rates of the fish. The results from just such an experiment are shown in Figure 11.26.

An important assumption of ideal free distribution is that the potential success of a new individual settling or being born in a given habitat is equal to the average of all the individuals already residing there. Many animals will not reach an ideal free distribution across habitats because they do not have ideal knowledge of the profitabilities of various habitats, because they do not have equal physical access to all habitats, or because some resident individuals have different functional responses than others or keep resources to themselves through aggressive dominance over newcomers (Fagen 1987). Finally, consumption rates are only one component of overall fitness. In some cases animals may find it desirable to aggregate even if consumption rates are reduced because groups have additional protection from predators or easier access to mates. The desirability of habitats will not usually be based solely on consumption rates. Protection from predators, nest site availability, and access to water and nesting materials will all be factors. A more general model of habitat selection would be based on overall fitness in each habitat type, as affected by density rather than just by consumption rates.

Ideal free distribution also contradicts the notion that in a landscape context some subpopulations may be net sources while others will be net sinks of individuals through dispersal (see Chapter 2). We have already demonstrated that, if movements are random (diffusive), some subpopulations (sinks) depend on others (sources) for their continued persistence. It is an empirical question of when and how often, with more directed movements of individuals, a system of subpopulations will maintain a source–sink relationship or will reach an ideal free distribution. A final complication arises if we allow dispersal rate to be an evolving attribute in a source–sink model. If individuals outside the source population experience lower fitness, on average, then dispersal rates will evolve to be lower. And if dispersal rates evolve to lower levels, sink populations will become extinct (Holt 1997), leaving only populations that are self-supporting. For these populations, the average individual would have the same fitness regardless of the patch it was in, assuming that all patches equilibrate at a mean fitness of $\lambda = 1$. Because dispersal usually involves some search costs, dispersal should evolve to zero. However,

Figure 11.26
Six sticklebacks (a species of small fish) are kept in a long aquarium and fed at the two ends. (a) When no food is added, the fish congregate in about equal numbers on each side of the aquarium. (b) When food (in the form of water fleas) is added at twice the rate on one end as on the other end, by the ideal free distribution, fish should aggregate until they are twice as common at the high-food end; this is generally the result observed. (c) When food is added at five times the rate on one end as on the other end, fish are generally five times more frequent at the high-food end than at the other end. After Milinski 1979.

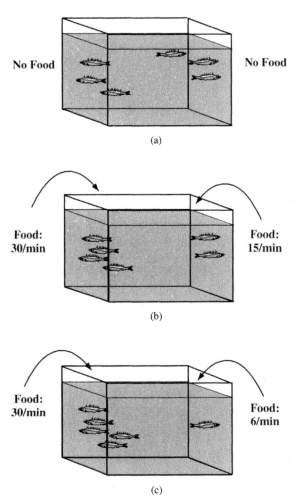

environmental fluctuations will create new opportunities and hazards that shift unpredictably across space. Thus it is likely that the assumption of fitness equilibration will not be met and that some level of dispersal is expected that would be proportional to the level of unpredictable environmental fluctuations.

PROBLEMS

1. Consider a predator that feeds on two prey, P_1 and P_2. The constants of prey consumption are $a_1 = 0.1$, $a_2 = 0.2$, $T_{h,1} = 1$, and $T_{h,2} = 1$. If the predator has a type 2 functional response, plot the functional response for the prey levels of 0, 5, 10, 15, and 20 for prey 1, assuming that it is the only prey item available. Repeat for prey 2. Now repeat for both prey, assuming the predator has a type 3 functional response.

2. Using the parameters in Problem 1, imagine that both prey are simultaneously available to the predator at the following densities for prey 1 and prey 2: (0, 20), (5, 15), (10, 10), (15, 5), and (20, 0). Thus there are a total of 20 prey in each case. Use Eq. (11.12) to determine the prey consumption rate for each prey at each of the five density combinations, first assuming a type 2 functional response and then a type 3 functional response. For each of the two functional responses, next make a plot of percent prey 1 available on the x axis versus percent prey 1 consumed on the y axis. How do the shapes of these plots differ for type 2 and 3 functional responses? What would these plots look like if you plotted percent prey 2 consumed on the y axis?

3. Again, consider a two-prey system. Now, however, the predator has a type 1 functional response to prey type 1 but a type 3 functional response to prey type 2. Can switching result?

4. In the development of the model to predict optimal give-up times, we assumed that the recruitment rate of resources into a patch was zero so that for each prey eaten there was one less prey available for future consumption. If we alter this assumption so that each prey consumed is replaced at a rate of 0.5 per hour and if the average transit time to go from one patch to another is about 2 hours, what will the optimal give-up time be? (Assume that there is only a single individual predator.)

5. In what way do the data in Figure 11.24, showing oystercatcher consumption rates at two sites, appear to violate the predictions of ideal free distribution? These two sites are on different islands that are separated by about 60 km.

12 *Predator–Prey Systems: Predator Dynamics and Effects on Prey*

Are natural consumers able to regulate the populations of their prey? In Chapter 10, the consumer numbers (e.g., sheep or fishing boats) were regulated externally by us or by the government. Prey numbers were decreased by predator consumption, but predator numbers did not necessarily reciprocate. We have shown that a predator's prey consumption rate (its functional response) often increases with prey abundance. This alone provides some measure of regulation of prey, but realistic functional responses ultimately become asymptotic at high prey densities. Thus prey populations, if they are large enough, are able to escape numerically from the controls of predators, based simply on the predators' functional responses. However, the functional response of an individual predator is only part of predator dynamics; the other part is the **numerical** response of the predator population to changes in prey abundance (Solomon 1949). That is, the predator's birth rate may increase as it captures more prey and its death rate may decrease with more food to eat. In open populations (i.e., with immigration and emigration of prey and predator), consumers may move into areas rich in resources and leave areas when resources are depleted. Consequently, the numerical response of the predator population will complement the predators' functional response but may be temporally delayed. In this chapter we explore the dynamics of coupled predator–prey systems and the implications for the stability and regulation of predator and prey numbers.

An example of a numerical response is shown for a species of jaeger in Figure 12.1. Jaegers are related to seagulls but primarily feed on land. Here the number of breeding pairs increases as the density of their prey, lemmings, increases.

COUPLED GROWTH EQUATIONS FOR PREDATOR AND PREY

The general form for the interaction between a predator (or consumer, *C,* to be more general) and its prey (or resource, *R*) may be written as

$$\frac{dR}{dt} = f_1(R,C) \quad \text{and} \quad \frac{dC}{dt} = f_2(R,C). \tag{12.1}$$

In Chapters 10 and 11 we explored the prey, or resource, growth, f_1, as it was affected by predator consumption. In this chapter we merge these considerations with those of the f_2 function. As predators consume prey, they give birth to new predators through the f_2 function, and these offspring, in turn, can consume more prey.

As usual we begin with a simple model. Imagine that the predators' numbers grow according to the following assumptions. Predator per capita birth and survival rates are functions only of resource abundance, not predator abundance. The more an individual consumer eats, the more offspring it can produce and the better its own chances of survival. You might think by this assumption that we are excluding predator–predator competition—yes and no. Predator exploitative competition for food is not excluded:

Figure 12.1
Numerical response of pomerine jaegers to lemming density. After Maher (1970).

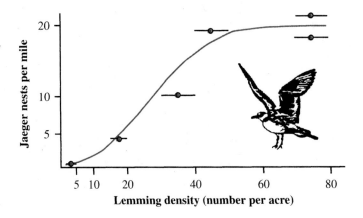

Figure 12.2
Regions of positive and negative growth for the predator population plotted in the phase space of predator and resource densities. The vertical line at $R = R^*$ shows all those values of C and R for which consumer population growth is exactly zero. At resource levels above R^* (the shaded region) the predator population increases in size. At resource levels below R^*, the predator population decreases.

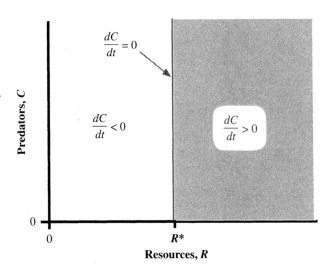

The more predators that are consuming resources, the lower the resource growth rate is and the fewer resources are available for consumers. Thus the predator–predator competition is an indirect one, mediated through the depletion of the shared resources, R. The only type of competition excluded by this assumption is direct interactions among predators, over and above those having an impact on resources. Predator fighting and other forms of within-species interference are excluded.

With this one assumption, we can now write the predator growth equation, Eq. (12.1), as

$$\frac{dC}{dt} = f_2(R,C) = C[b(R) - d(R)].\qquad(12.2)$$

The instantaneous per capita birth function, $b(R)$, and the instantaneous per capita death function, $d(R)$, depend only on R, not on C. The $b(R)$ function is positive and increases with R, since we expect predator birth rates to increase with more resources available. As food levels R increase, the predator death rate $d(R)$ is expected to decrease, and thus dC/dt will increase. The total predator population, C, multiplies the per capita terms within the brackets to give the total populational change of numbers of predators in continuous time as dC/dt.

At equilibrium, $dC/dt = 0$. Consequently, Eq. (12.2) implies that at equilibrium either $C = 0$ or $b(R) = d(R)$. Regardless of the functional form of $b(R)$ and $d(R)$, both the birth and death terms involve only resource numbers, R, and not predator numbers. Thus, given the assumptions of the model, the equilibrium level of C is determined only by R. This set of circumstances is depicted graphically in Figure 12.2.

In Figure 12.2 the abundances (or densities) of predators are plotted on the y axis, and the abundances (or densities) of prey plotted on the x axis. For each point in this

Figure 12.3
The functional response of a single consumer superimposed on its replacement requirements (threshold consumption rate).

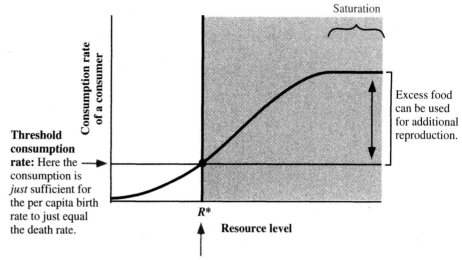

Threshold consumption rate: Here the consumption is *just* sufficient for the per capita birth rate to just equal the death rate.

R*

This is the food level where the average consumer's birth rate just balances its death rate, so per capita growth is zero. At higher resource levels, the consumer population will increase since births from additional consumption will exceed deaths.

C–R state space, we can determine the predators' growth rate from Eq. (12.2). The shaded region of the *C–R* **state space** (also called the **phase space**) is the set of points where the predator population initially increases according to Eq. (12.2). At high levels of resources (i.e., values greater than *R**), the predators' birth rate exceeds their death rate and their numbers will increase; at low resource levels (below *R**), the predators' numbers must decline since their death rate exceeds their birth rate. Finally, there exists some level of resources, *R**, at which the predators reach equilibrium, $dC/dt = 0$. Since we're plotting this in *C–R* phase space and since per capita consumer growth is not affected by the number of consumers, *C*, according to our assumptions, a straight vertical line, indicating no dependence on *C*, separates the two regions: increasing predator numbers on the right and declining predator numbers on the left.

It may seem strange that in this figure, when resources are at *R**, the predator population will be at equilibrium no matter how many predators there are. How can that be, since predators are consuming the resources, and the more predators, the more resources will be consumed? The explanation is that we are imagining that we are fixing resource levels and then letting predator numbers dynamically adjust to the amount of resource present. Recall that in Chapter 10, we fixed predator numbers (sheep) and let resource levels (grass) adjust. Figure 12.2 describes the reverse thought experiment. Surely 100 sheep will consume grass 100 times faster than a single sheep. But that means we must "experimentally" replace the grass 100 times faster to keep the amount of grass fixed at *R**. This is why the equilibrium level of resource (grass) does not depend on the number of predators (sheep). We ultimately want to explore the dynamics when neither predator nor prey is kept fixed; we get to that analysis shortly.

For now, let's explore the biological factors that influence the position of *R**, the threshold level of resource needed for positive consumer growth on the *R* axis. The dynamical Eq. (12.2) is posed in terms of effects of resource levels on predator per capita birth and death rates. As we discussed in the preceding paragraph, *R** is the same for any number of predators, so we can just consider a single predator's functional response—the amount of resource that an individual predator gathers per unit time. Some of the energy that it gathers from feeding must be devoted to maintenance costs simply so that it can survive and reproduce just enough to replace itself. Any additional harvested resources above those levels necessary for maintenance can be allocated to additional reproduction, as shown in Figure 12.3. If each individual predator produces more offspring than necessary just to replace themselves, then the population grows. But, if resource levels are below the maintenance threshold, *R**, then

Figure 12.4
A consumer (or predator) with a higher functional response curve can just break even on lower resource levels. The predator zero-isocline shifts to the left. Hatched or shaded regions show resource levels yielding positive per capita consumer growth.

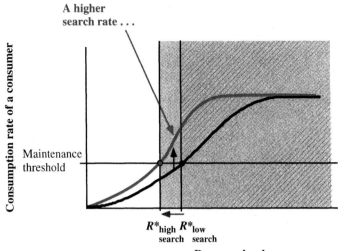

A higher search rate . . .

Consumption rate of a consumer

Maintenance threshold

R^*_{high} R^*_{low}
search   search

Resource level

. . . leads to a lower threshold resource level necessary for positive consumer growth rate.

(a)

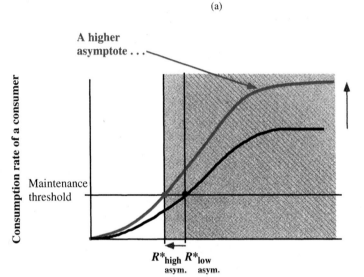

A higher asymptote . . .

Consumption rate of a consumer

Maintenance threshold

R^*_{high} R^*_{low}
asym.   asym.

Resource level

. . . leads to a lower threshold resource level necessary for positive consumer growth rate.

(b)

reproduction cannot keep up with maintenance needs, the average predator will die before it replaces itself, and a population of such predators decreases in size.

The result of a higher search rate for a predator is a lower threshold resource level necessary for positive predator growth, as shown in Figure 12.4.

If the predator requires additional energy just to maintain survival, the result will be a higher threshold resource level for positive consumer growth as shown in Figure 12.5.

With this understanding of the factors determining the position of the predator zero-growth line ($dC/dt = 0$) on the R axis, we can finally couple consumer dynamics and resource dynamics so that both are free to vary. While the $dC/dt = 0$ relationship is simply a vertical line, what does the $dR/dt = 0$ relationship look like in C–R space? This is exactly what we derived in Chapter 10 for the case of a type 3 functional response in the example of the grazing sheep and is plotted again in Figure 12.6 for convenience.

The joint equilibrium for both species occurs at those values of C and R that yield zero growth rates simultaneously for both species. The predator and prey dynamics are superimposed in Figure 12.7.

Figure 12.5
A consumer with greater survival costs will need higher resource levels just to break even. The predator zero-isocline shifts to the right.

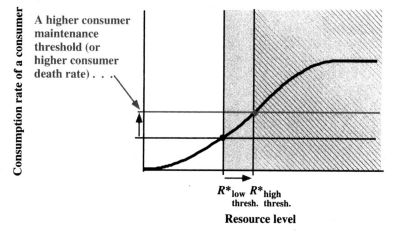

Figure 12.6
The prey (resource) zero growth curve from Chapter 10 (Figures 10.10 and 10.11). Points inside the curve (shaded area) support positive resource growth rates, and points outside the curve yield negative resource growth rates.

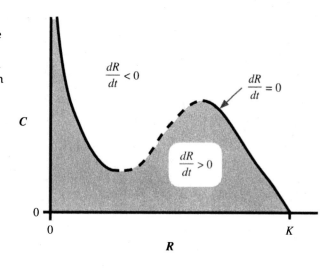

Figure 12.7
Superimposing the predator, P, and resource, R, dynamics. The predator region of positive growth is shown in red. The two zero-isoclines intersect at (C^*, R^*). This point is a feasible equilibrium since both C^* and R^* are greater than zero.

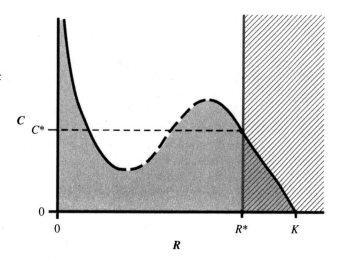

Figure 12.8

The regions of positive and negative growth for both prey, *R*, and predators, *C*, plotted in phase space. (a) The arrows show the qualitative trend of population growth for resources, *R*, and consumers, *C*. The black arrowheads are for resources, and the open arrowheads are for consumers. The overall qualitative direction of change in both *R* and *C* together is shown by the arrow in the middle of each arrow triplet. (b) State space may be divided into four regions based on whether the growth rate of each species is positive or negative.

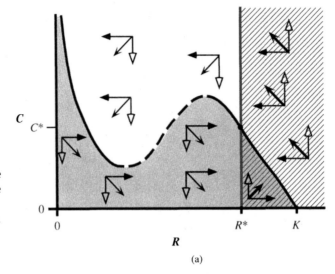

(a)

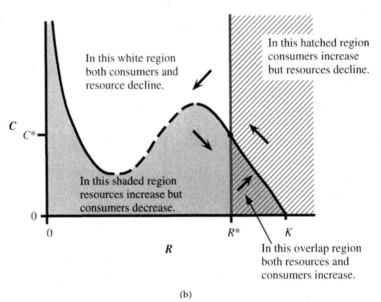

In this white region both consumers and resource decline.

In this hatched region consumers increase but resources decline.

In this shaded region resources increase but consumers decrease.

In this overlap region both resources and consumers increase.

(b)

We can categorize each point in phase space according to whether the predator and prey populations will initially increase or decline from that point as illustrated in Figure 12.8.

We may use these phase–space diagrams to evaluate qualitatively the position and stability of equilibrium points, as shown in Figures 12.8 and 12.9.

The zero growth curves for the two species populations are also called **zero-isoclines.** Recall from Chapter 10, that the dashed portion of the resource zero-isocline, the portion just to the left of the prey peak, yields unstable equilibrium points for the prey. The solid portions, on either side, correspond to stable equilibrium for prey. Figure 12.10 further explores the consequences of changing the point of intersection of the two zero-isoclines.

We expect trajectories to spiral inward to a stable predator–prey equilibrium point and spiral outward from an unstable equilibrium point. But in the latter case, if the equilibrium point is unstable, where will these trajectories ultimately end up? An example of the dynamics of a prey that grows logistically and its predator with a type 3 functional response is shown in Figure 12.11. In this case the parameters of the model are such that the zero-isoclines intersect to the left of the prey peak where the interior equilibrium point is unstable.

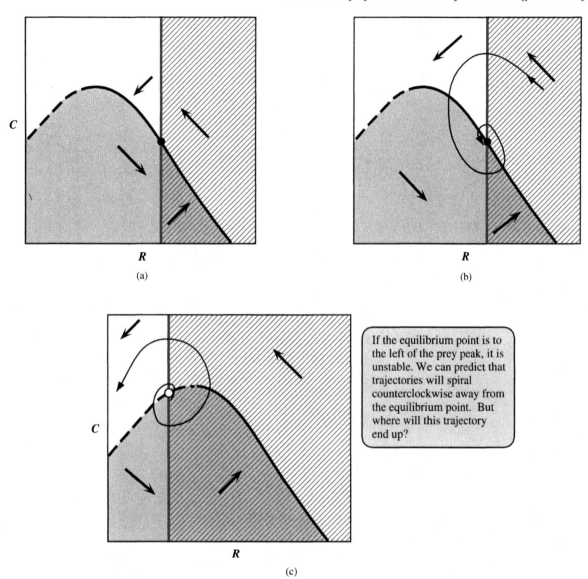

(a)

(b)

(c)

If the equilibrium point is to the left of the prey peak, it is unstable. We can predict that trajectories will spiral counterclockwise away from the equilibrium point. But where will this trajectory end up?

Figure 12.9
(a) and (b) A magnified part of Figure 12.8 in the vicinity of the feasible equilibrium point and a trajectory emanating from a point close to, but not exactly on, the equilibrium point. This equilibrium point falls to the right of the peak of the prey zero-isocline and thus yields a stable equilibrium point for R. Given the direction of the arrows, along with the knowledge that equilibria on the dashed line are unstable, we predict that trajectories will spiral counterclockwise, *inward* toward the equilibrium point. (c) The equilibrium point falls to the left of the prey peak, on the dashed portion of prey zero-isocline. The analysis in Chapter 10 showed that such an equilibrium point would be unstable for prey. Trajectories beginning near this equilibrium point will spiral counterclockwise *outward.*

The trajectory spirals out to an egg-shaped **limit cycle.** Predator and prey numbers continually oscillate, and the amplitude and period of the oscillation are fixed. This limit cycle is stable in the sense that it "attracts" trajectories from inside and outside. If the system is initiated at a different point, say, one closer to the unstable equilibrium point, then the trajectory will ultimately cycle out to the fixed cycle shown in Figure 12.11. For predator–prey systems, the cycle is **counter-clockwise** (with predators plotted on the y axis), as shown by the direction of the arrows in Figure 12.8. As indicated in Figure 12.12, this means that the prey population begins to decline while the predator population is still increasing. Then, at the bottom of the cycle, the prey population starts to increase while the predator population is still declining.

Figure 12.10

Different possible positions for the predator zero-isocline. Without knowing the exact relationship of the birth and death rates of the two species, we do not know exactly where the intersection of these two zero growth curves will lie. Parts (a)–(c) represent successively lower consumer death rates.

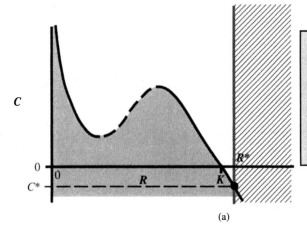

In this case the consumer cannot survive (it is has a negative equilibrium density). The interior equilibrium point is said to be **unfeasible**. The consumers will become extinct and the resources will grow to K, not to R^*.

(a)

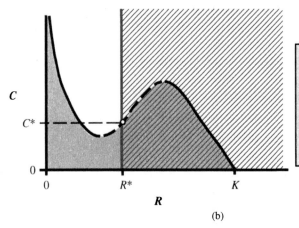

In this case the consumer/resource interior equilibrium is feasible, but the equilibrium point lies on the **unstable** portion of the resource zero-isocline. Hence this equilibrium point will be unstable.

(b)

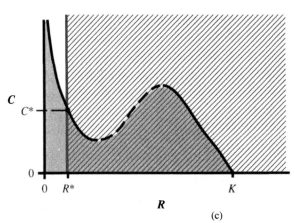

In this case the consumer/resource interior equilibrium is again feasible and stable.

(c)

Figure 12.11

Predator–prey limit cycle. These cycles arise when the interior equilibrium occurs to the left of the peak on the descending portion of the prey zero-isocline. The predator zero-isocline is red and the prey's is black. A sample trajectory (in black) beginning at the gray point is shown. You can look ahead to Box 12.1 to see the equations for this system, which are based on the development in Chapter 11. Other parameters are $a = 0.002$, $T_h = 4$, $k = 0.5$, $d = 0.1$, and $r = 0.5$.

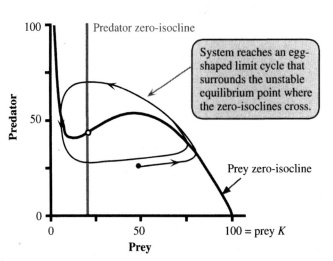

System reaches an egg-shaped limit cycle that surrounds the unstable equilibrium point where the zero-isoclines cross.

Figure 12.12
The limit cycle from Figure 12.11, indicating joint numerical trends around the cycle.

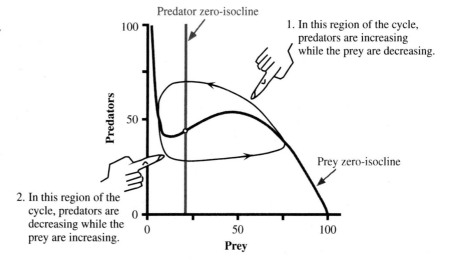

Figure 12.13
Predator and prey numbers over time for the trajectory depicted in Figure 12.12. The fingers point to the approximate positions of the cycle referred to in Figure 12.12.

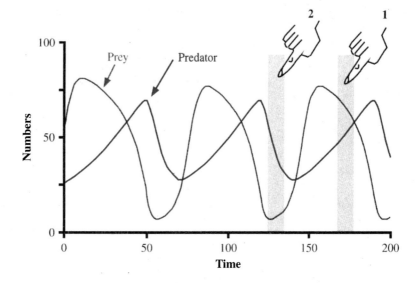

This situation is more easily seen in Figure 12.13, which shows the time series for the trajectory plotted in Figure 12.11.

In contrast, if the parameters create a predator zero-isocline that lies even farther to the left (e.g., like that in Figure 12.10c), then the interior equilibrium point lies on the ascending portion of the prey zero-isocline and the interior equilibrium point is stable as shown in Figure 12.14.

The equations describing this interaction—which were used to produce the simulations illustrated in the preceding figures—are shown in Box 12.1. The equation for a type 3 functional response was developed in Chapter 11.

In the next sections we add some complications that vary some of these assumptions and allow for additional biological realism. Before proceeding, be sure that you understand where the two zero-isoclines get their particular shapes, why the predatory–prey cycle is counterclockwise, and why equilibrium points to the left of the prey hump's peak produce unstable equilibrium points. You may want to review Chapter 10 to reinforce these points.

Exercise: There are two other (boundary) equilibrium points in Figure 12.11. Where are they? Are they stable?

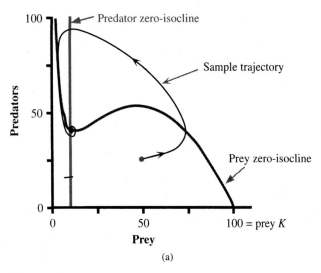

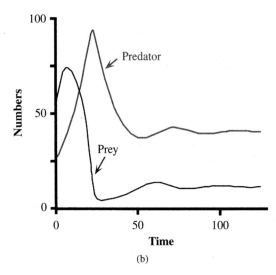

Figure 12.14
(a) The predator zero-isocline is shifted to the left (by decreasing the predator death rate to $d = 0.06$); now the interior equilibrium point is stable. (b) Time course for the trajectory shown in (a).

Box 12.1 *Predator–Prey Dynamics with a Type 3 Functional Response*

$$C = \text{predator numbers, and } R = \text{prey numbers.}$$

The predator equation:

$$\frac{dC}{dt} = \quad \text{births} \quad - \quad \text{deaths}$$

$$= kC\left(\frac{aR^2}{1 + aT_h R^2}\right) \quad - \quad dC. \qquad \text{(a)}$$

Consumption of resources is converted to reproduction of C according to the conversion rate, k.

Consumption follows a type 3 functional response with encounter rate a and handling time T_h. For total consumption by the entire population, multiply this by the number of predators, C.

Consumers have constant per capita death rate d independent of C or R.

The prey equation

$$\frac{dR}{dt} = \quad \text{recruitment} \quad - \quad \text{deaths to predators}$$

$$= rR\left(1 - \frac{R}{K}\right) \quad - \quad C\left(\frac{aR^2}{1 + aT_h R^2}\right). \qquad \text{(b)}$$

Logistic birth rate function

Death due to consumption by predators equals the functional response multiplied by the number of predators, C.

A TYPE 2 FUNCTIONAL RESPONSE

So far we have illustrated predator–prey dynamics with a type 3 functional response. This S-shaped functional response led to the creation of a prey zero-isocline with one valley and two peaks. What happens in predator–prey dynamics if the predator has a simpler type 2 functional response? The type 2 functional response has a positive slope that steadily decreases as prey numbers increase. In Chapter 11, we found that it could be expressed in terms of two parameters, an encounter rate, *a,* and the handling time, T_h. Box 12.2 presents the dynamical equations for a type 2 functional response.

By setting Eq. (c) and Eq. (d) to zero, you should be able to show that the zero-isoclines for predator and prey are given by the following equations.

$$\text{Prey zero-isocline:} \quad C = \frac{r}{a}\left(1 - \frac{R}{K}\right)(1 + aT_h R). \tag{12.3}$$

$$\text{Predator zero-isocline:} \quad R = \frac{1}{a\left(\dfrac{k}{d} - T_h\right)}. \tag{12.4}$$

Box 12.2 *Predator–Prey Dynamics with a Type 2 Functional Response*

$$C = \text{predator numbers, and } R = \text{prey numbers.}$$

The predator equation:

$$\frac{dC}{dt} = \quad \text{births} \quad - \quad \text{deaths}$$

$$= kC\left(\frac{aR}{1 + aT_h R}\right) \quad - \quad dC. \tag{c}$$

Consumption of resources is converted to reproduction of *C* according to the conversion rate, *k*

Consumption by predators follows a type 2 functional response with encounter rate *a* and handling time T_h. For total consumption by the entire population, multiply this by the number of predators, *C*.

Consumers have constant per capita death rate *d* independent of *C* or *R*.

The prey equation

$$\frac{dR}{dt} = \quad \text{recruitment} \quad - \quad \text{deaths to predators}$$

$$= rR\left(1 - \frac{R}{K}\right) \quad - \quad C\left(\frac{aR}{1 + aT_h R}\right). \tag{d}$$

Logistic birth rate function

Death due to consumption by predators equals the functional response multiplied by the number of predators, *C*.

Figure 12.15
Predator–prey zero-isoclines when the predator has a type 2 functional response and the prey grows logistically. Phase–space regions of positive growth for each species are shaded. The predator's region of positive growth is pink and the prey's is gray.

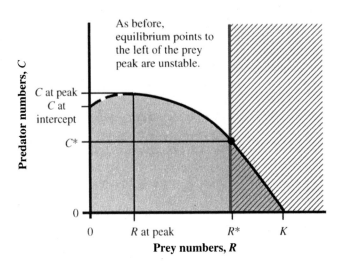

Note that the predator zero-isocline does not involve predator numbers and thus is a vertical line in the *C*–*R* phase plane. The prey zero-isocline is the equation of a truncated parabola. Figure 12.15 illustrates the zero-isoclines for both predator and prey.

Exercise: For this model, use Eqs. (c) and (d) of Box 12.2 to verify the following three algebraic expressions:

$$R^* = \frac{1}{a\left(\dfrac{k}{d} - T_h\right)}, \tag{12.5}$$

$$C \text{ at intercept } = \frac{r}{a},$$

and

$$R \text{ at peak } = \frac{K}{2} - \frac{1}{2aT_h}.$$

Sensitivity Analysis

From Eq. (12.5), for the equilibrium resource level, R^*, you can see that unless $k/d > T_h$, the equilibrium level of prey, R^*, is negative. In reality if this were the case, the predator would decline to extinction and the prey population would then climb to its carrying capacity, K. The larger the death rate of the predator or the larger the prey handling time, T_h, the greater will be the equilibrium prey level, R^*. Increasing the handling time causes the prey zero-isocline curve to stretch upward and shifts its peak to the right, as depicted in Figure 12.16.

The position of the predator zero-isocline also shifts to the right with increases in T_h as you can see by examining Eq. (12.4) or (12.5). Figure 12.17 makes the visual comparison.

Increasing prey carrying capacity, K, affects only the prey zero-isocline since K is not a parameter in Eq. (12.4) for the predator zero-isocline. Figures 12.18 and 12.19 illustrate this condition.

Increasing prey carrying capacity, K, increases predator numbers at equilibrium and can shift the interior equilibrium point to the left side of the peak of the prey hump. This, in turn, results in an unstable interior equilibria. As for a type 3 functional response (see Figure 12.10), limit cycles result. Moreover, the farther the unstable equilibrium is to the

Figure 12.16
The effect of different handling times on the prey zero-isocline. Other parameters from equations in Box 12.2 are $a = 0.02$, $k = 1$, $r = 0.2$, and $K = 100$.

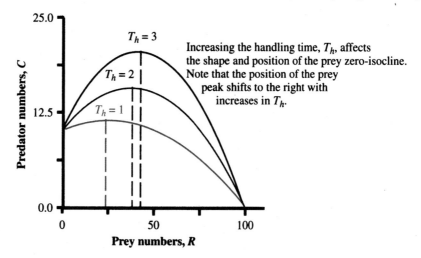

Increasing the handling time, T_h, affects the shape and position of the prey zero-isocline. Note that the position of the prey peak shifts to the right with increases in T_h.

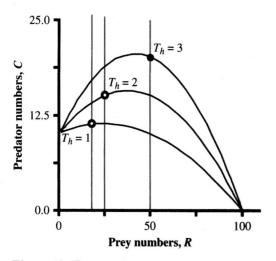

Figure 12.17
The handling time parameter influences both the predator and prey zero-isoclines. Other parameters as in Figure 12.16 and $d = 0.25$.

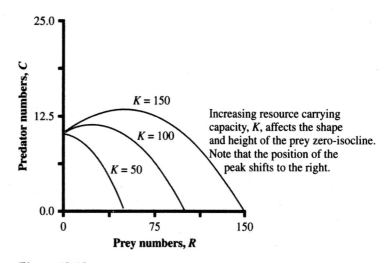

Increasing resource carrying capacity, K, affects the shape and height of the prey zero-isocline. Note that the position of the peak shifts to the right.

Figure 12.18
The effect of different prey carrying capacities, K, on the prey zero-isocline. Other parameters as in Figure 12.16.

Figure 12.19
Conclusion: Increasing prey carrying capacity can destabilize a stable predator/prey system. These three prey zero-isoclines differ only in K. A predator zero-isocline is superimposed on them. Higher K results in higher numbers of predators at the interior equilibrium point but does not change the equilibrium prey numbers. Higher K also shifts the relative position of the equilibrium point to the left of the prey zero-isocline's peak. Other parameters as in Figure 12.16 and $T_h = 1$, $d = 0.25$.

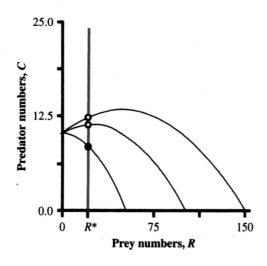

left of the prey hump, the larger the limit cycles become in amplitude, as shown in Figure 12.20. As these limit cycles become larger in amplitude, they come progressively closer to converging on one of the axes, with the subsequent extinction of a species.

Increasing the prey growth rate, r, as shown in Figure 12.21, results in higher predator numbers at equilibrium but does not destabilize the equilibrium point, at least in continuous time models.

Exercise: How would increasing the encounter rate, a, change equilibrium numbers of prey and predators? Is a predator–prey system of this type likely to be stabilized, destabilized, or unaffected by successive increases in the encounter rate?

Exercise: Three predator species eat the same prey. The predator species differ only in their death rates, d. Their zero-isoclines are as shown in the following diagram. Other parameters are $K = 200$, $T_h = 0.9$, conversion rate $k = 1$, and encounter rate $a = 0.02$. Only one predator will prevail. Which one and why?

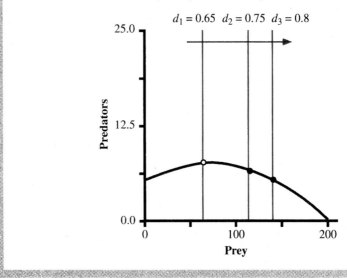

Figure 12.20

Two alternative prey with the same predator. One prey has $K = 150$, and the other has $K = 120$. All other parameters are the same for both predator and prey. Both equilibria fall to the left of the prey hump and so are unstable. However, the $K = 150$ equilibrium is relatively farther to the left Other parameters are $a = 0.02$, $k = 1$, $d = 0.4$, $r = 0.2$, and $T_h = 1$.

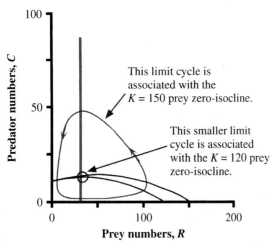

Figure 12.21

Increases in the prey intrinsic growth rate, r, affect only the prey zero-isocline; it increases the equilibrium abundance of predators but not of prey. Also, changes in r do not move the equilibrium point to the left of the prey zero-isocline peak.

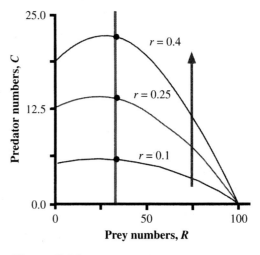

PREY REFUGES CAN MIMIC A TYPE 3 FUNCTIONAL RESPONSE

Another mechanism can lead to a prey zero-isocline somewhat like that for a type 3 functional response. Imagine a type 2 functional response but also refuges for the prey to hide where they are safe from predators (Rosenzweig and MacArthur 1963). Let's suppose that a maximum of H prey can hide in these refuges. On the one hand, if prey numbers are below H, then no matter how many predators are present, all the prey will be secure. On the other hand, when prey levels exceed the number of safe sites, some prey will be exposed to predation. Figure 12.22 illustrates this process for a refuge size of $H = 10$. Clearly, refuges can be stabilizing in the sense that they prevent the predator from completely overexploiting the prey.

> **Exercise:** Modify the equations for prey and predator growth rates, Eqs. (c) and (d) Box 12.2 to incorporate a refuge of size H. What is the new equation for the predator zero-isocline?

Laboratory Examples

Maly (1969) used a microcosm composed of the protozoan *Paramecium* as prey and the rotifer *Asplanchna* as a predator. He studied species interactions in very small 1 ml depressions in microscope slides. Maly placed different numbers of predators and prey in many different depressions and observed their initial growth rates over short periods

Figure 12.22
The effect of a prey refuge on the prey zero-isocline. (a) Type 2 functional response with a prey refuge of size $H = 10$. (b) Other parameters are $r = 0.25$, $K = 100$, $a = 0.1$, and $T_h = 3$.

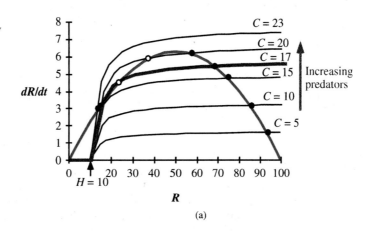

(a)

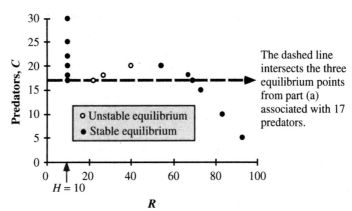

The dashed line intersects the three equilibrium points from part (a) associated with 17 predators.

(b)

Figure 12.23
The arrows show the initial direction of growth for *Paramecium* at different numbers and with varying numbers of a rotifer predator. Open circles indicate populations that changed less than 5% in 24 hours. After Ricklefs (1979) from Maly (1969).

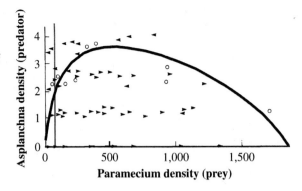

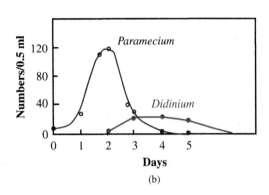

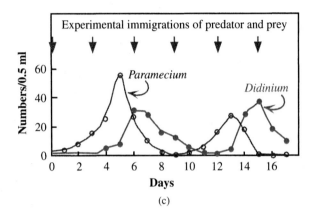

Figure 12.24
(a) *Didinium nasutum* devouring a *Paramecium caudatum*. (b) The prey *Paramecium* begins growing in a test tube at day 0. The predator *Didinium* is introduced at day 2. The predators rapidly multiply and eliminate all the *Paramecium* and then gradually die out themselves. (c) Same as (b), but Gause added one *Didinium* and one *Paramecium* to the test tube every third day. With this modification, the two species coexisted for 15 days, followed by the extinction of the *Paramecium*. After Gause 1934.

of time. In this way, he was able to construct the arrows in diagrams like those in Figure 12.8. From them he could infer the shape of the zero-isoclines, as depicted in Figure 12.23.

This predator–prey system seems to be unstable. Moreover, since the apparent equilibrium point lies far to the left of the prey hump and close to prey = 0, the limit cycle around this equilibrium is expected to be large and closely converging on the axes. Predator and/or prey are predicted to become extinct quickly—the result that Maly found in longer term experiments.

Another microcosm that has received substantial attention is that of *Paramecium* as prey and another protozoan, *Didinium*, as the predator. One of the pioneering studies by Gause (1934) explored the dynamics of this system. He found that *Didinium* was such an efficient predator that it overexploited the prey and that both species collapsed to extinction unless he took elaborate steps to keep this from happening. *Didinium* is smaller in size than the *Paramecium* that they consume, as shown in Figure 12.24(a), and they can keep dividing, becoming smaller and smaller for several days, even in the absence of any *Paramecium*. If *Paramecium* are now added, the *Didinium* are so numerous that they quickly overexploit their prey, as indicated in Figure 12.24(b).

Gause speculated that these two species coexist in nature in spite of local extinctions driven by overexploitation because these small *Paramecium* can quickly colonize new ephemeral water ponds where *Didinium* might be absent. (We explore this idea of regional coexistence in spite of local noncoexistence in Chapter 16.) Gause attempted a crude test of this idea by reintroducing one individual of each species every third day, as shown in Figure 12.24(c) to his test tubes. This periodic immigration allowed prolonged coexistence, but still only for two cycles.

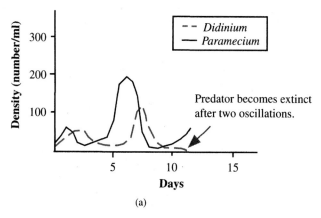

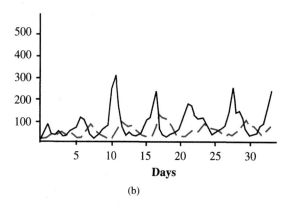

(a)

(b)

Figure 12.25
Protozoan predator–prey dynamics: (a) with methyl cellulose (b) With methyl cellulose and reduced food for *Paramecium*. After Luckinbill 1973.

Luckinbill (1973) took up the challenge. He reasoned (according to the theory that we have just reviewed) that it should be possible to get more prolonged coexistence in this system by reducing the encounter rate of the predator with their prey. To accomplish this he made the water medium more viscous by adding methyl cellulose. Methyl cellulose slowed the swimming speeds of both predator and prey to about 3% of their previous levels. One of his typical results is shown in Figure 12.25(a). He was able to get coexistence for two cycles without periodic immigration of the two species.

Luckinbill next lowered prey *K* by reducing the food (i.e., bacteria) levels for the *Paramecium*. This lower *K*, in combination with the methyl cellulose, gave prolonged cycles, as indicated in Figure 12.25(b). In terms of the zero-isocline depiction, both measures theoretically result in shifting the intersection of the zero-isoclines more to the right (see Figure 12.23). Empirically, both populations still oscillate, indicating that the interior equilibrium point for this system is still to the left of the peak of the hump of the prey zero-isocline, but not so far to the left that oscillations are extreme and trajectories "hit" an axis.

PROBLEMS

1. Modify the predator growth equation (Eq. c) of Box 12.2 to include migrants that come into the system from outside at a constant rate *m*. This yields

$$\frac{dC}{dt} = m + k\left(\frac{aRC}{1 + aT_h R}\right) - dC.$$

Sketch the predator zero-isocline for this modified model. Does this confer or detract from stability?

2. Sometimes the migration rate of predators into an area is in response to available food levels in that patch. Modify the predator growth equation (Eq. c) of Box 12.2 to include migrants that come into the system from outside at a rate proportional to the abundance of resources. This yields

$$\frac{dC}{dt} = m'R + k\left(\frac{aRC}{1 + aT_h R}\right) - dC.$$

Sketch the predator zero-isocline for this modified model. Does this confer or detract from stability?

3. Consider the model in Box 12.1 with a type 3 functional response. Show that the equilibrium level of resource, *R**, is modified from that in Eq. (12.3), namely,

The equilibrium resource level from Box 12.1 is

$$R^* = \frac{1}{\sqrt{a\left(\dfrac{k}{d} - T_h\right)}}.$$

4. Suppose that an unstable equilibrium point exists for this case of a type 3 functional response. Could increasing the prey carrying capacity, *K*, ever stabilize the system?

MULTIPLE PREDATORS ON A SINGLE PREY

In another classic laboratory experiment that provides a useful illustration of predator–prey theory, Utida (1957) examined a system comprising a bean weevil (*Callosobruchus chinensis*) as prey and a wasp (*Heterospilus prosopidis*), which lays eggs in the weevil larvae, as the predator. The eggs hatch into larvae that kill the weevil host. Parasites with this behavior are called **parasitoids.** After the wasp egg hatches, the wasp larvae devour the beetle larvae from the inside. The abundance of weevil hosts therefore influences the number of wasp larvae that develop and thus the number of adult wasps that emerge in the following generation. The prey weevils feed on and oviposit in beans, which Utida placed in petri dishes. Population densities of the weevil and wasp oscillated, with the parasitoid cycles lagging behind the cycles of the prey as expected by theory and as illustrated in Figure 12.26. Even though the physical experimental system is very simple and relatively homogeneous, the two species coexisted for nearly 30 generations (or four complete cycles)—at which time Utida discontinued the experiment.

Next Utida added a second wasp parasitoid (*Neocatolaccus mamezophagus*). Both wasps were forced to compete for the single host species. Populations fluctuated dramatically, but all three species coexisted in these simple petri dishes for more than 70 generations, as shown in Figure 12.27.

This result is exceptional because it is usually very difficult to achieve coexistence in the confined and homogeneous arrangements of most laboratory experiments. The result is interpretable, however, with a simple extension of predator–prey theory. The two wasps apparently had different functional responses so that one wasp did relatively better, as judged by the number of emerging wasps per generation, at prey densities below a density of about 200 weevils, while the other wasp did relatively better at high weevil densities, above 200, as plotted in Figure 12.28.

Figure 12.26
Predator–prey oscillations of the bean weevil (*Callosobruchus chinensis*) and a parasitoid wasp (*Heterospilus prosopidis*) in petri dishes. After Utida (1957).

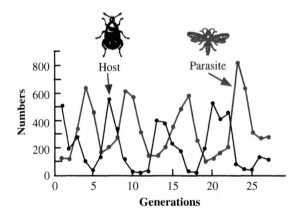

Figure 12.27
Two parasitoid wasps coexisting on a single host species of bean weevil. The total duration of this experiment was about 4 years. After Utida (1957).

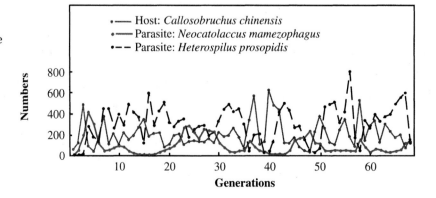

If the wasps are otherwise identical, this result implies that, if host densities can be kept experimentally at relatively high levels, then the second wasp should outcompete the first. Yet if host densities are kept at low prey levels, the first wasp should prevail. Finally, if weevil numbers fluctuate, sometimes being greater than the crossover density of 200 and sometimes being less, then it might be possible for the two wasps to coexist. But what might drive such a fluctuation? One reason that beetle numbers fluctuate is because the wasp/beetle interaction itself produces a limit cycle. We saw such oscillations with the one wasp system (see Figure 12.26). We next explore this situation within the context of the predator–prey theory that we have just developed, as illustrated in Figure 12.29.

The two predators compete for the same prey species, which grows logistically ($K = 200$, $r = 0.1$). Figure 12.30 shows the zero-isoclines for each predator when alone with the same prey under two different predator death rates. With the decreased death rates for the two predators (Figure 12.30b), the predators' zero-isoclines move to the left (review Figure 12.5) and become more similar in position. The different shape of the predators' functional responses (as determined by parameters a and T_h) causes the reversal in the relative position of their zero-isoclines. From equation 12.4, the position of the predator zero-isocline is at

$$R^* = \frac{1}{a\left(\dfrac{k}{d} - T_h\right)}.$$

Thus the position of the predator zero-isocline depends on its death rate, d, its handling time, T_h, and the encounter rate, a. Applying this formula to the situation shown in Figure 12.30(a), for predator 1, $R^* = 1/[0.04((1/0.75) - 1.2)] = 187.5$ and for predator 2, $R^* = 1/[0.02((1/0.75) - 0.9)] = 115.3$. In Figure 12.30(b) for predator 1, $R^* = 1/[0.04((1/0.65) - 1.2)] = 73.90$ and for predator 2, $R^* = 1/[0.02((1/0.65) - 0.9)] = 78.30$.

The time series for all three hypothetical species shows that for the case of lower death rates, limit cycles emerge and both predators can coexist on this single prey, as shown in Figure 12.31. Is the coexistence of Utida's beetles completely explained by

Figure 12.28
Total number of parasitic wasps of two species emerging from larvae of the host weevil plotted as a function of host weevil density. After Utida (1957) and Pianka (1994).

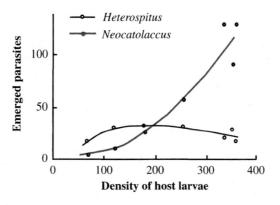

Figure 12.29
Two predators' functional responses differ so that predator 1 has a higher encounter rate and also a higher handling time (thus a lower asymptote) than predator 2.

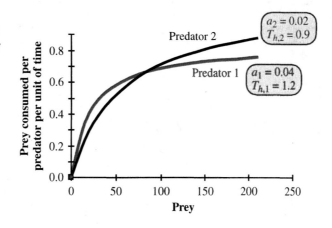

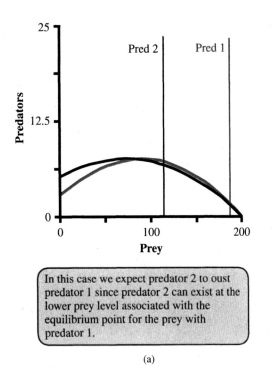

In this case we expect predator 2 to oust predator 1 since predator 2 can exist at the lower prey level associated with the equilibrium point for the prey with predator 1.

(a)

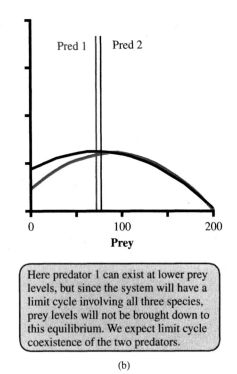

Here predator 1 can exist at lower prey levels, but since the system will have a limit cycle involving all three species, prey levels will not be brought down to this equilibrium. We expect limit cycle coexistence of the two predators.

(b)

Figure 12.30
Both parts of the figure show a zero-isocline for two alternative predators feeding on the same prey. The prey species is the same for both predators, but because the zero-isocline of the prey depends on the predator's encounter rate and handling time parameters, there are two prey zero-isoclines in each case. (a) The predators' death rates are relatively high at 0.75, and each predator, when alone with the prey, has a stable interior equilibrium point. (b) The two predators and prey are the same as in (a), but the predators' death rates have been lowered to 0.65. Consequently, the predators can now increase at lower prey levels (the predator zero-isoclines shift to the left). The prey and predator zero-isoclines intersect at an unstable equilibrium point, producing limit cycle dynamics. Other parameters are $k = 1$, $r = 0.1$, and $K = 200$.

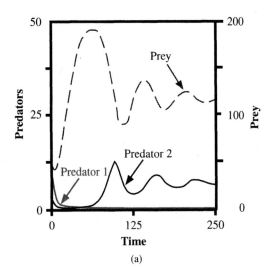

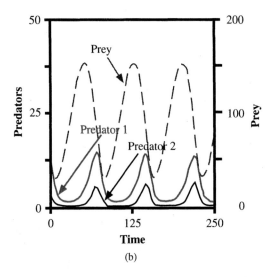

(a)   (b)

Figure 12.31
One prey with two predators, based on the situation depicted in Figure 12.30. (a) The predator death rates are high ($d_1 = d_2 = 0.75$), so prey levels are kept high; only one predator survives. (b) The predator death rates are lower ($d_1 = d_2 = 0.65$), and the equilibrium prey density (although unstable) is at a low level. Now both predators coexist with limit cycle oscillations.

this model? The results diverge from theoretical expectations in two ways. The erratic behavior of the three insects does not seem to show the regular limit cycles suggested by the model, although this may simply be due to some additional environmental noise superimposed on the system. More important, in Utida's experiment the fluctuations of the two wasps seemed to be largely out of phase with one another (see Figure 12.27), yet the theory predicts that the cycles of the two predators should be in phase (see Figure 12.31b). Although one wasp is relatively more successful than the other at low weevil densities while the reverse is true at high beetle densities, both wasps do *absolutely* better at high beetle densities (see Figure 12.28); hence theory predicts that the two wasps will positively covary. Perhaps the discrepancy between theory and these empirical results is because the two wasps respond differently to uncontrolled environmental features of this experiment, or, alternatively, they may directly interfere with each other. It is also important to keep in mind that this host/parasitoid system may not be perfectly analogous to predator–prey equations with a type 2 functional response.

PREDATOR INTERFERENCE, ALLEE EFFECTS, AND OTHER MODIFICATIONS TO PREDATOR–PREY INTERACTIONS

In summary, to help you understand the dynamics of interacting species, we introduced a method of plotting zero-isoclines in state space along with the regions of positive and negative growth for each species. Without actually solving the coupled differential equations of population growth, we have been able to make reasonable inferences about trajectory direction and the local stability of equilibria. This is valuable because the coupled differential equations that we have been exploring do not even have closed form analytical solutions. Yet it is important to remember that our instability determination for equilibrium points to the left of the prey zero-isocline peak has been based on an assumption of a **linear numerical response** for the predators. We have assumed, for example, that the consumption of five sheep is five times as great as the consumption of a single sheep and the per predator growth equation is multiplied by 5 to get the total growth rate for this five-predator population. If this linearity assumption is violated, then the stability of equilibria along the prey curve may change. In Box 12.3 we explore a possible departure from this assumption, involving intraspecific predator interference.

If we write the per capita form of the consumer growth equation, Eq. (e), we note that it involves a term in *fC*, the interference term, or

$$\frac{dC}{Cdt} = kB(R) - d - fC. \tag{12.6}$$

Box 12.3 *Predators with Intraspecific Interference*

$$\frac{dC}{dt} \quad = \quad k[B(R)C] \quad - \quad dC \quad - \quad \boxed{fC^2}. \tag{e}$$

| The growth rate of the consumer population | = | consumer birth rate. It is proportional to prey intake per consumer, B(R), (Type 2 functional response) times the number of consumers | minus | density independent consumer death rate | minus a new term for consumer interference. The death rate is density dependent. |

Figure 12.32
With predator self-interference the predator zero-isocline is no longer vertical.

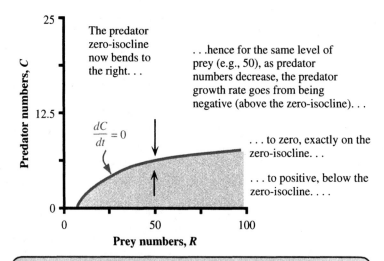

The predator zero-isocline now bends to the right. . .

. . .hence for the same level of prey (e.g., 50), as predator numbers decrease, the predator growth rate goes from being negative (above the zero-isocline). . .

. . . to zero, exactly on the zero-isocline. . .

. . . to positive, below the zero-isocline. . . .

$$\frac{dC}{dt} = 0$$

Conclusion: The fewer predators, the less time they spend squabbling with each other, and thus the more time they spend foraging. Greater food intake, in turn, leads to higher predator birth rates.

Figure 12.33
Predator (red) and prey (black) zero-isoclines.

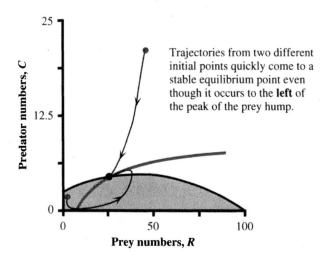

Trajectories from two different initial points quickly come to a stable equilibrium point even though it occurs to the **left** of the peak of the prey hump.

By setting Eq. (12.6) equal to zero, we may solve for the consumer zero-isocline,

$$C = \frac{kB(R) - d}{f}.$$

Now we plug in a type 2 functional response for $B(R)$ to get

$$C = \frac{1}{f}\left(\frac{kaR}{1 + aT_h R} - d\right). \tag{12.7}$$

The consumer zero-isocline bends to the right, as shown in Figure 12.32.

The effect of the incorporation of predator self-interference on the overall dynamics is stabilizing since now predators have a self-limitation in addition to the limitation set by resources (Figure 12.33).

This is not to say that predator interference necessarily leads to stable points and eliminates limit cycles but only that, if it is sufficiently strong, it has this potential to do so, as Figure 12.34 illustrates.

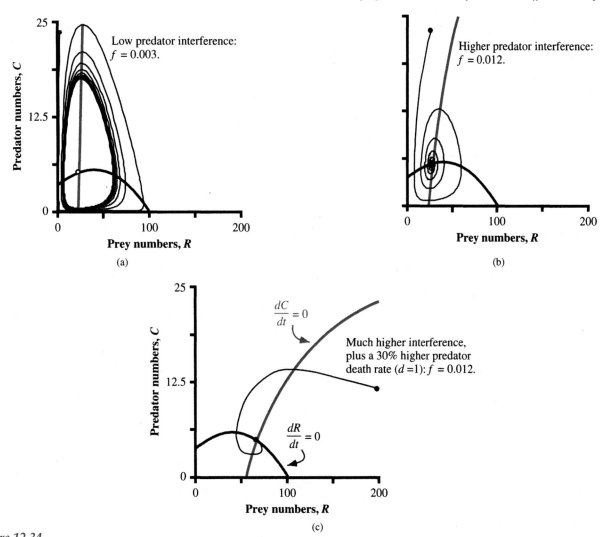

Figure 12.34
(a) Even with a low predator interference rate, f, the system has a stable limit cycle. (b) Predator interference is increased, which leads to a stable feasible equilibrium point even though this point falls to the left of the peak of the prey zero-isocline. (c) The predator death rate is increased by 30% so that the equilibrium point now falls to the right of the prey peak. The trajectory quickly reaches the feasible equilibrium point. All trajectories begin at the grey dots. Other parameters are $a = 0.06$, $k = 1$, $d = 0.7$, $r = 0.2$, $K = 100$, and $T_h = 0.7$. The predator zero-isocline is given by Eq. (12.7).

Exercise: We have been using a graphical argument based on the superimposition of a logistic prey recruitment curve and combined predator consumption curves (e.g. Figure 10.14), producing a picture of resource dynamics, which we then used to account for the instability of the two-species equilibrium point when it lies to the left of the peak of the prey zero-isocline. While this is valid when the predator zero-isocline is vertical, we have now seen that it falls apart when predator self-interactions are present, for example, if the sheep of Chapter 10 interfered with one another in grazing disputes. How does predator interference qualitatively allow for the stability of equilibrium points even if they lie to the left of the peak of the prey zero-isocline?

In Chapter 5, we explored the consequences of an Allee effect on the population dynamics of a single species. We now add an Allee effect to prey growth in the context of predator–prey interactions. To do this we adjust the prey's logistic growth so that per capita prey growth at low prey densities is negative, as illustrated in Figure 12.35.

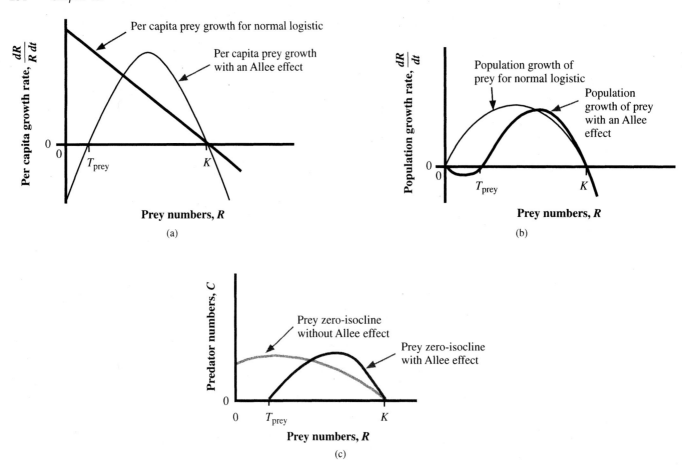

Figure 12.35
An Allee effect for the prey changes its zero-isocline. (a) The prey's per capita growth rate may be reduced at low prey numbers because it is harder to find mates, to forage solitarily, or to defend against predators. This is called an **Allee effect.** Prey numbers must exceed the threshold level, T_{prey}, for positive growth. At high densities, crowding leads to declines in per capita growth as before. (b) Population growth curves for the logistic compared to an Allee effect. The curve in (a) is multiplied by R. (c) When the Allee effect in (b) is combined with consumption by a predator with a type 2 functional response, the result is to shift the prey zero-isocline to the right. Prey growth is negative when prey numbers are below. T_{prey}

Figure 12.36 combines the prey zero-isocline in Figure 12.35 with a predator zero-isocline and shows some sample trajectories (again assuming a type 2 functional response). The result is a typical vertical predator zero-isocline with an unstable interior equilibrium point if it intersects on the left of the prey peak. Now, however, unlike the situation in Figure 12.20 without an Allee effect, here there is no stable limit cycle. All the trajectories lead to the extinction of both predator and prey. Point (0, 0) is globally stable.

But, as before, if the predator zero-isocline lies to the right of the prey peak, the interior equilibrium point is locally stable, as shown in Figure 12.37. But point (0, 0) is also locally stable. **In short, the incorporation of an Allee effect eliminates the possibility of limit cycles in this model.** Moreover, if the intersection of the predator and prey zero-isoclines occurs to the right of the prey hump, stable coexistence is possible, but if the intersection occurs to the left of the prey hump, the prey and predator crash to extinction. Once the predators bring the prey down to very low numbers, even if the predators subsequently become extinct, the prey still decline from the Allee effect; their numbers are now below the level necessary for positive growth.

Note that point (0, 0) is also stable in Figure 12.37. The prey's social facilitation (Allee effect) means that prey are more prone to extinction if their numbers become low. Even if this prey were reintroduced to an area where predators are absent, it would not be successful unless the number of introduced individuals was above the left-most

Figure 12.36
Prey have an Allee effect and the predators have a type 2 functional response. (a) In the phase–space diagram the interior equilibrium point occurs to the left of the prey peak. (b) Time course of the lowermost trajectory on the phase–space diagram in (a). Other parameters are $k = 0.8$, $a = 0.02$, $r = 0.2$, $K = 130$, $T_h = 1$, $d = 0.4$, and $T = 30$.

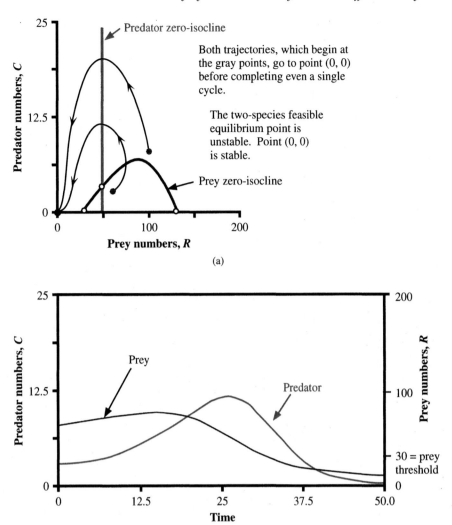

Both trajectories, which begin at the gray points, go to point (0, 0) before completing even a single cycle.

The two-species feasible equilibrium point is unstable. Point (0, 0) is stable.

(a)

(b)

Figure 12.37
The prey has an Allee effect, and the predator has a type 2 functional response. The predator has a lower encounter rate or a higher death rate than in Figure 12.36; hence its zero-isocline has shifted to the right. In this case, there is a stable feasible interior equilibrium. The point (0, 0) is still locally stable. Starting points are shown as dots. Trajectories are black or grey curves. Starting points that lead to point (0, 0) are shown in gray. Now, however, the interior equilibrium point represents an alternative domain of attraction. Starting points that lead to the stable interior equilibrium point, where the zero-isoclines cross, are shown in black. Other parameters are $k = 0.9$, $a = 0.02$, $r = 0.12$, $K = 160$, $T_h = 1$, $d = 0.66$, and $T = 60$.

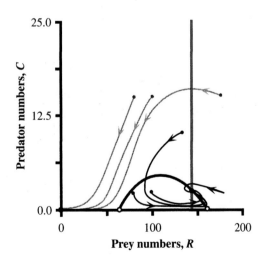

threshold of their zero-isocline (see Figure 12.37). Since point (0, 0) is locally stable, levels of prey below this threshold will only decline further to 0.

Figure 12.37 shows two locally stable equilibrium points. Trajectories that begin from the black dots spiral into the stable interior equilibrium point, while starting points beginning from grey dots move toward (0, 0). We have already discussed the possibility of multiple stable points, each with some domain of attraction (see Chapters 5 and 10). By simulating trajectories from many points, the state space can be divided into two regions: the basin of attraction for point (0, 0) and the basin of attraction for the interior equilibrium point. The curve that divides these two basins of attraction is called the **separatrix.** You can think of it as the watershed line of a mountain range. Runoff goes a different direction on each side of the watershed line, just as trajectories go in different directions on each side of the separatrix. For most ecologically plausible models, the separatrix will be difficult to calculate analytically, but it can be determined readily by simulation, as shown in Figure 12.38.

A model that combines an Allee effect in the prey with very strong predator self-interference (as in Box 12.3) is shown in Figure 12.39. This model has one stable and

Figure 12.38
The separatrix for the system in Figure 12.37, based on tracing many trajectories. The domain of attraction for the feasible interior equilibrium point is shaded grey. Trajectories outside this domain go to (0, 0).

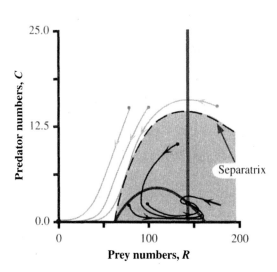

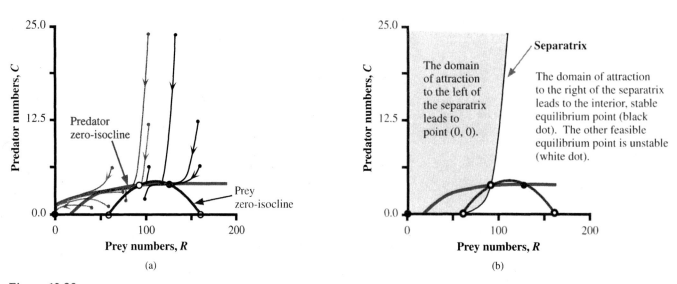

(a)

(b)

Figure 12.39
Predators have a type 2 functional response, and they also interfere strongly with one another; prey have an Allee effect. (a) Of the several trajectories shown, some (gray) end up at (0, 0) and others (black) end up at the stable interior equilibrium point (black circle). There is also an unstable interior equilibrium point (open circle). (b) As is generally the case, the separatrix between the two domains of attraction intersects the unstable equilibrium points. Other parameters are $k = 1$, $a = 0.08$, $r = 0.2$, $K = 160$, $T_h = 0.7$, $d = 0.5$, $T = 60$, and $f = 0.2$.

one unstable interior equilibrium point. Point (0, 0) is also locally stable. The separatrix runs through the unstable interior equilibrium point. Its exact shape is influenced by the values specified for several parameters in the model.

MUTUALISTIC RELATIONSHIPS

Many plants depend on insects for pollination. The insects get resources from nectar and sometimes pollen, and the plants are fertilized by the transfer of pollen from other plants. Each species in a plant–pollinator pair probably requires a minimum density of the other for existence. Each species can probably be satiated by the other; plants need only so many insects to enable successful sexual reproduction—and they may be limited by other factors (e.g., light, nutrients, and space). Insects may also be limited by resources other than nectar (or pollen supplies), such as larval food, warm days for growth, or suitable habitat space for breeding. The zero-isocline depiction in Figure 12.40 integrates these features.

Despite the positive interspecific feedback inherent in any mutualistic relationship, a stable feasible equilibrium point may exist due to negative (density dependent) intraspecific interactions. Note too the direction of the arrows in Figure 12.40, suggesting an absence of oscillatory trajectories in these mutualistic interactions.

Exercise: Consider Figure 12.40. In the absence of any plants, what is the equilibrium abundance of insects? In the absence of any insects, what is the equilibrium abundance of plants? If the maximum density of plants were doubled, what would happen to the size of the insect population? If the maximum density of plants were halved, what would happen to the size of the insect population?

Exercise: Compare the relative selective forces for specialization of an insect species to one flower species compared to the forces for specialization of a plant species to one insect pollinator species. Are these two forces about equal or are they lopsided in one direction?

An Example of a Predator-Prey Cycle in Nature

Snowshoe hares and their lynx predators are found primarily in the boreal zone of the northern United States. Charles Elton (1942) accumulated data on the trapping results of the Hudson Bay Company. We have already shown the 10-year lynx cycles in

Figure 12.40
A zero-isocline depiction for a plant–pollinator interaction. As usual the hatched regions yield positive growth rates. Each species has internal self-damping as well as a positive interspecific effect.

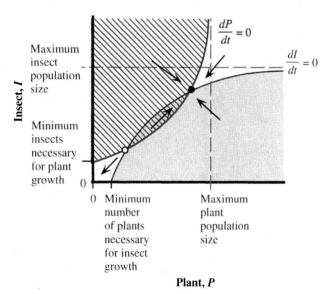

Chapter 6. In Figure 12.41, these data for lynx are superimposed on the population fluctuations of the snowshoe hare. Note the two different scales for the *y* axis. We cannot scale these numbers to absolute population densities because we do not know what area these pelt returns encompass and exactly how trap success for each species is related to population density. However, if we assume that hares and lynx have equal trapping rates, then relatively speaking, hare numbers are about twenty-fold higher than lynx numbers.

Are these hare–lynx cycles consistent with the limit cycles that we explored earlier in this chapter? For such limit cycles, it is the interaction of predator and prey that drives the cycle, rather than the predator simply riding out a cycle that the prey would make in their absence.

Some have argued that these cycles are not predator–prey limit cycles since the hare seem to cycle on a 10-year period in the absence of lynx on Anticosti Island, Quebec. However, observations indicate that the amplitude of the cycles there is not as great as on the mainland. Moreover, it could be that alternative predators on the island (e.g., fox) compensate for the absence of lynx and drive the hare cycle anyway. In fact, 10-year cycles are a common feature of vertebrate populations in the boreal zone and are evident in muskrats, foxes, coyotes, skunks, martens, mink, fishers, and wolverines (Finnerty 1980).

Gilpin (1972) plotted the lynx–hare data from 1875 to 1906 in predator–prey phase space. The results are shown in Figure 12.42; the different colors separate different cycles.

Our theory holds that predator–prey cycles should run counterclockwise. Surprisingly, the cycles here seem to be in the shape of a figure 8, and the biggest lobe of the 8 is running clockwise. This seems to imply the nonsensible result that hares eat lynx! Perhaps, then, this is not a predator–prey driven cycle. Yet, even if lynx numbers were

Figure 12.41
Hare–lynx cycles from fur returns to the Hudson Bay company. After Elton (1942).

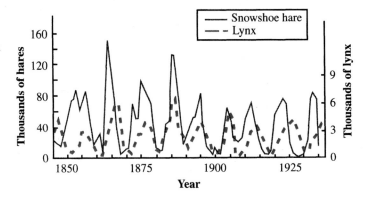

Figure 12.42
Snowshoe hare and lynx numbers from 1875 to 1906 plotted in phase space. After Gilpin (1972) based on data from Charles Elton (1942) on the fur returns to the Hudson Bay Company.

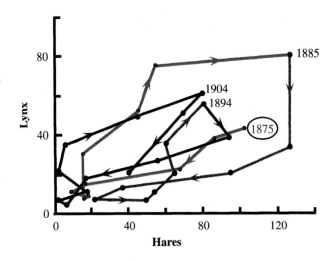

simply tracking the number of hares, the phase of the lynx cycles would still lag behind the phase of the hare cycles (not be ahead of them). Gilpin suggested that hare might be serving as vectors or reservoirs for diseases that were transmitted to lynx. Another possibility is that trappers wait for hare populations to increase before going out to trap either hares or lynx. This would bias the data, making it appear as if hares lag behind lynx. Some support for this explanation comes from more recent studies where hare and lynx counts through one cycle are determined by scientists not trappers (Boutin et al. 1995). The lynx clearly follow the hare cycle, and in phase space the cycle is counter-clockwise, as theory predicts and as shown in Figure 12.43.

Perhaps the most difficult feature of these cycles to reconcile with predator–prey theory is that so many different predators and prey have the same cycle period, 10 years, although they may be out of phase with each other by a few years. Even some forest moths show a similar 10-year cycle, as depicted in Figure 12.44. Myers (1988) suggests that these cycles may be caused by epidemics of disease. These different species with 10 year cycles have different generation times, death rates, birth rates and life histories, so it is not difficult to understand how the cycles may be out of phase.

However, in the predator–prey theory that we just developed, these same parameters also influence cycle period. Thus the paradox: Why are these very different predators all expressing approximately the same period? Perhaps some still unknown external force is synchronizing or entraining the cycles to bring these several different species into the same period. Sinclair et al. (1993) used evidence from tree rings and ice cores to show that 10-year solar-spot cycles may be a factor influencing weather and thus playing this role.

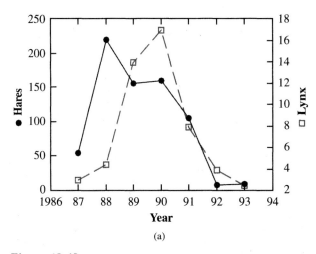

(a)

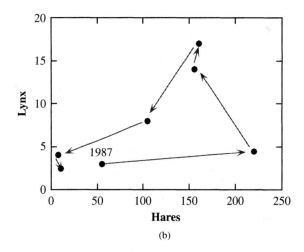

(b)

Figure 12.43
A hare–lynx cycle from 1986 to 1993 in the Yukon (from Boutin et al. 1995): (a) Time course and (b) trajectory in phase space. Densities are numbers per 100 km². Hare data are the average of three control grids, each 32.5 km² in area.

Figure 12.44
Population oscillations of the larch budmoth (*Zeiraphera diniana*) at two sites in the Engadine valley of Switzerland. After Myers (1988).

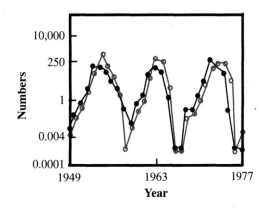

PROBLEMS

1. Suppose that the prey equation, Eq. (d) in Box12.2, included an additional term for a density independent death rate. How would this affect the interior equilibrium prey density, predator density, and their stability?

2. Suppose that the prey equation, Eq. (d) in Box12.2, had a θ term to the recruitment of prey, $1 - (R/K)^{\theta}$. Draw a qualitative sketch of the prey zero-isocline for θ greater than 1 and for θ less than 1.

3. Draw the predator and prey zero-isoclines for a situation where the predators have a type 2 functional response and a numerical response dominated by an Allee effect at low predator densities and a self-interference effect at high densities. The prey recruitment is logistic. Evaluate the stability of the various equilibrium points.

4. Consider these four alternative shapes for a predator's functional response (plotted as $B(V)$).

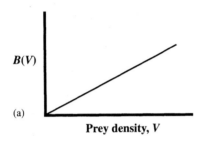

(a)

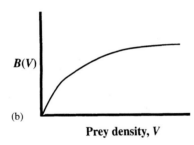

(b)

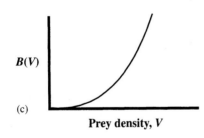

(c)

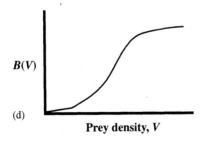

(d)

Answer the following questions.
   a. What are the units of $B(V)$?

b. Which of the four functional responses is most stabilizing for a predator–prey interaction? Circle one: A, B, C, D.

c. Which is a type 2 functional response? Circle one: A, B, C, D.

d. For which functional responses does the risk of predation for an *individual* prey drop to zero as prey numbers climb toward infinity? Circle the correct letter(s): A, B, C, D.

e. For which functional responses does the risk of predation for an individual prey increase as prey numbers climb toward infinity? Circle the correct letter(s): A, B, C, D.

f. If there were alternative prey for the predator, which functional response curve would most likely result in prey switching? Circle one: A, B, C, D.

g. For each of the four cases, draw the prey and predator zero-isoclines in state space. Assume that the prey population grows logistically in the absence of the predators and that the predator's per capita growth rate depends only on resource abundance (not predator abundance). Shade the regions of positive population growth and remember to label the zero-isoclines.

5. The following diagram depicts the zero-isoclines of a predator–prey system with three alternative predators, P_i feeding on the same prey.

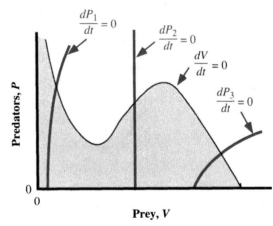

Rank the three alternative predators from highest to lowest in terms of:
   a. Predator density at the interior two-species equilibrium point.
   b. Predator self-interference rates.
   c. Stability of the two-species equilibrium point.
   d. Predator per capita death rate.
   e. Draw the time course for the prey population when predator 2 is introduced at very low numbers to the prey population at its carrying capacity. (All the other predators are completely absent.)
   f. Repeat the plot in (e), but now draw the time course for the prey population when predator 1 is introduced at very low numbers to the prey population at its carrying capacity. (All the other predators are completely absent.)

6. Suppose that you have been doing some experiments in petri dishes containing kidney beans, a bean weevil that lays its eggs in the beans, and a wasp species, A, that parasitizes the weevils. You do all your experiments in a carefully controlled temperature and humidity chamber. You observe that regardless of how many weevils and wasps that you start with, you always end up with an equilibrium number of about 100 weevils and 25 wasps. Now you find a new wasp species, B,

which can find and parasitize bean weevils at a faster rate. You repeat the experiments with this new wasp, B, (and not wasp species A), and observe that regardless of how many weevils and wasps you begin with, wasp species B's numbers and weevil numbers fluctuate with a fairly regular period and amplitude. Draw two zero-isocline diagrams, one for the weevil/wasp species A system and the other for the weevil/wasp species B system to explain these results.

7. Consider the following predator–prey zero-isoclines. Which of the six labeled points are equilibrium points? Which are locally stable? Which are globally stable? On the diagram, carefully draw the trajectory that begins at point 6.

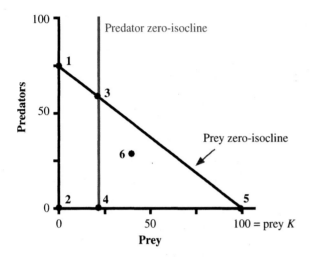

8. Consider a deer population, D, and a tick population, T, that feeds exclusively on these deer. Deer numbers are food limited and deer grow logistically with a carrying capacity K. The ticks feed on the blood of deer but cause no deer mortality or changes in deer birth rates, regardless of tick numbers. Tick reproduction is proportional to the food supply (i.e., deer numbers). The tick population requires at least S deer before it can increase. In addition ticks have an upper limit to their population size, U, set by their own predators and parasites. Tick numbers cannot exceed U, even if deer numbers are extremely high. Draw zero-isoclines for deer and ticks consistent with this description; put deer numbers on the x axis and ticks on the y axis. Shade the regions of positive growth and place the points U, S, and K on the figure. Are deer tick-limited? Are ticks deer-limited? Are deer numbers stable in the absence of ticks? Is there a feasible interior equilibrium point? Is it stable?

9. Only one of the following four limit cycles (a, b, c, or d) represents a mathematically feasible cycle for a predator–prey system. Which is it and why?

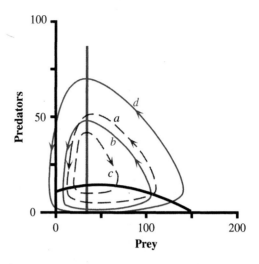

10. Three predator species eat the same prey. The predators differ only in their death rates, d. Their zero-isoclines and that of the prey are as shown. Suppose that this system is initiated at equal numbers of all three predators ($P_i = 2$ for all i) and prey = 200. As time goes on, which predator species will prevail? What will happen to the prey population? Suppose that the predators represented three different genotypes at the same locus $P_1 = AA$, $P_2 = Aa$, and $P_3 = aa$. What happens to the gene frequency of A as time goes on? What happens to the total abundance of predators and prey?

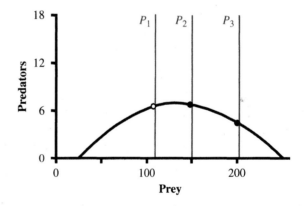

13 Stability of Predator–Prey Systems: Analytical Methods (Advanced)

In this chapter we develop general tools for the mathematical evaluation of the stability of equilibria in coupled population growth equations. While we develop these methods for two-species models, these techniques can be generalized to several interacting species including additional trophic levels (Chapters 14 and 15) or even additional age or stage groups. You might think that the graphical analysis of Chapter 12 is adequate for this task and that more sophisticated tools aren't needed. However, the graphical analysis works only when you can visualize the system in two or at most three dimensions, so models with more interacting species still need additional attention. Also, even in two dimensions, it is sometimes difficult to predict when an equilibrium point will be stable or unstable, based on graphical analysis alone. For example, take another look at Box 12.2. In Figure 12.34, with predator interference, the feasible equilibrium point can sometimes be stable and sometimes unstable even when it lies to the left of the peak of the prey zero-isocline. Without simulation we would not be able to predict which points are stable and which are not. The analytical techniques that we present in this Chapter can provide this ability.

THE CLASSICAL LOTKA–VOLTERRA PREDATOR-PREY EQUATIONS

We start with the simplest possible predator–prey system, known as the **classical Lotka–Volterra predator-prey equations**:

$$\frac{dV}{dt} = V[b - aP] = f_1(V, P) \tag{13.1}$$

and

$$\frac{dP}{dt} = P[-d + kaV] = f_2(V, P). \tag{13.2}$$

In the absence of a predator, the prey population, V, grows exponentially with an intrinsic growth rate, b. The prey death rate increases linearly with the number of predators. Predators have a type 1 functional response, aV, with a capture rate parameter per prey per predator of a. Prey captured are converted into new predator offspring at a rate determined by the proportionality constant, k. For example, if a predator must consume 10 prey before it has enough energy to produce one predator offspring, then $k = 1/10$ or 0.1. In addition, predators have a constant per capita death rate, d. Thus predator numbers decline exponentially (with exponent $-d$) in the absence of prey for them to eat. The per capita growth rate of prey and predators is plotted as a function of both prey and predator numbers in Figures 13.1 and 13.2.

293

Figure 13.1
The relationship between per capita growth of prey, *dV/V dt*, prey numbers, *V*, and predator numbers, *P*, for the classical Lotka–Volterra predator–prey model.

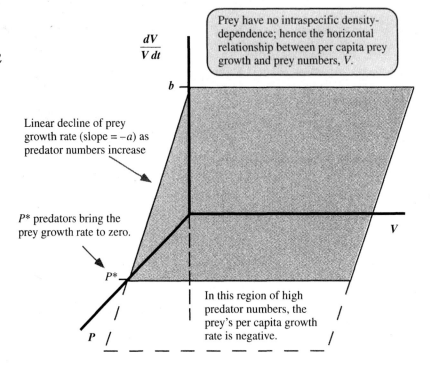

Prey have no intraspecific density-dependence; hence the horizontal relationship between per capita prey growth and prey numbers, *V*.

$\dfrac{dV}{V\,dt}$

b

Linear decline of prey growth rate (slope = −*a*) as predator numbers increase

*P** predators bring the prey growth rate to zero.

*P**

V

P

In this region of high predator numbers, the prey's per capita growth rate is negative.

Figure 13.2
Predator per capita growth, *dP/P dt* as a function of prey numbers, *V*, and predator numbers, *P*.

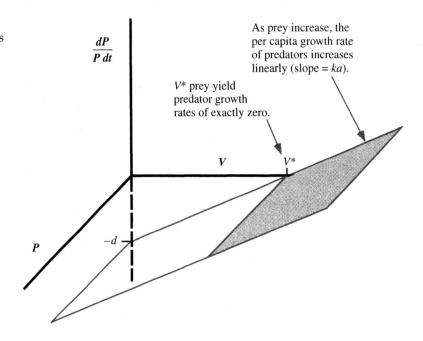

$\dfrac{dP}{P\,dt}$

As prey increase, the per capita growth rate of predators increases linearly (slope = *ka*).

*V** prey yield predator growth rates of exactly zero.

V

*V**

P

−*d*

Superimposing these two planes for the per capita growth of each species gives the situation in Figure 13.3.

The zero-isoclines for prey and predators can be found algebraically by setting the growth rates of the two populations, as given by eq. (13.1) and (13.2), to zero:

$$V[b - aP] = 0 \tag{13.3}$$

and

$$P[-d + kaV] = 0. \tag{13.4}$$

There are only two ways that each of these equations may become zero. Either the term outside the brackets equals zero (i.e., *V* = 0 or *P* = 0) or the terms inside the brackets,

Figure 13.3
The combination of Figures 13.1 and 13.2. The interior equilibrium point where the two growth planes cross on the plane of zero–zero growth is shown as a white dot.

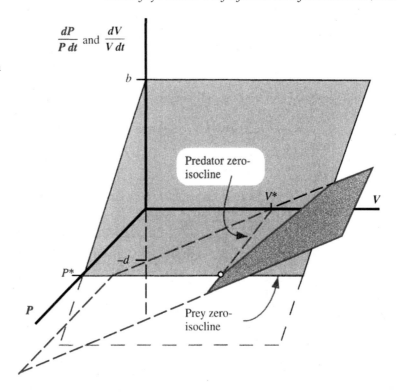

which are the per capita growth rates, equal zero. The latter condition for both species defines the interior equilibrium point. By setting the prey per capita growth equation, Eq. (13.3), to zero, we get

$$b - aP^* = 0,$$

which yields a solution for the predator's equilibrium density as

$$P^* = \frac{b}{a}.$$

This is also, by definition, the equation for the zero-isocline of the prey from Eq. (13.3). It is a horizontal line placed on the P axis (y axis) at $P = b/a$. If predators exceed this level, the prey population declines; if predators are below this level, the prey population increases.

By setting the predator growth equation to zero Eq. 13.4, we find that the prey's equilibrium density is

$$V^* = \frac{d}{ka}.$$

This also defines the equation for the predator zero-isocline from Eq. (13.4). It is a vertical line placed on the V axis (x axis) at $V = d/ka$. If the prey population exceeds this level, then the predator population increases; if prey are below this level, then the predator population must decline.

Phase space is the bottom P–V plane in Figure 13.3 flipped on its side. Figure 13.4 shows the zero-isoclines for the equations derived.

Exercise: Look at the point on the y axis in Figure 13.4, where $P^* = 15$ and $V = 0$. Why isn't this a boundary equilibrium point?

Figure 13.4
The predator and prey zero-isoclines for the classical Lotka–Volterra equations. Regions of positive growth rate, based on Eqs. (13.1) and (13.2), are shaded. The equilibrium point is (V^*, P^*).

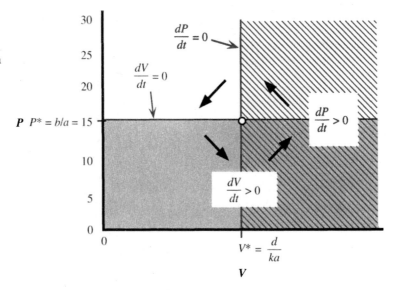

To evaluate the stability of an equilibrium point we need to nudge the system away from the equilibrium by changing the abundances of predator and/or prey by some small deviation. Then we determine how these deviations change over time (see Figures 5.42 and 5.43). For a two-species system with coupled growth equations, as for Eqs. (13.1) and (13.2), we may write the deviations of the two state variables, V and P, from their equilibrium values, V^* and P^*, as

$$v = V - V^* \qquad (13.5a)$$

and

$$p = P - P^*. \qquad (13.5b)$$

Substituting Eqs. (13.5a) and (13.5b) into Eqs. (13.1) and (13.2) and noting that $dv/dt = dV/dt$ and $dp/dt = dP/dt$ (because V^* and P^* are constants), we obtain

$$\frac{dv}{dt} = f_1(v + V^*, p + P^*) \quad \text{and} \quad \frac{dp}{dt} = f_2(v + V^*, p + P^*).$$

Here the f notation is shorthand for the actual equations, Eqs. (13.1) and (13.2); it also reminds us that while we are specifically analyzing Eqs. (13.1) and (13.2), the method being applied is general for any functions f_1 and f_2. To reach the linearized dynamics of the system around this equilibrium point, we perform a Taylor's expansion on f_1 and f_2 in the neighborhood of an equilibrium point. Recall from Chapter 5 that a continuous function $F(x)$ evaluated at an equilibrium point may be decomposed into a polynomial using a Taylor's expansion:

$$F(x) \approx F(x^*) + \frac{F'(x^*)(x - x^*)}{1!} + \frac{F''(x^*)(x - x^*)^2}{2!} + \frac{F'''(x^*)(x - x^*)^3}{3!} + \cdots.$$

For a function F of two variables, x and y, a similar expansion may be performed. Now, however, we need to evaluate the partial derivatives of $F(x, y)$ with respect to x and then again with respect to y. A graphical depiction of these partial derivatives is the slope of $F(x, y)$ in the two directions x and y as depicted in Figure 13.5.

The Taylor's expansion of $F(x, y)$ at point (x^*, y^*) is

$$F(x, y) \approx F(x^*, y^*) + \frac{1}{1!}\left[(x - x^*)\frac{\partial F(x^*, y^*)}{\partial x} + (y - y^*)\frac{\partial F(x^*, y^*)}{\partial y}\right]$$

$$+ \frac{1}{2!}\left[\left((x - x^*)\frac{\partial}{\partial x} + (y - y^*)\frac{\partial}{\partial y}\right)^2 F(x^*, y^*)\right] + \cdots. \qquad (13.6)$$

Figure 13.5
A graphical representation of the partial derivatives of a function, f, of two variables, x and y. The partial derivative of f with respect to x is positive, but the partial derivative of f with respect to y is negative.

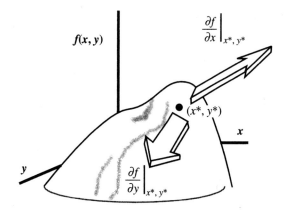

In this case the F functions are f_1 and f_2, which themselves are derivatives over time. Since at the equilibrium point (V^*, P^*) both f_1 and $f_2 = 0$, by definition the constant term in the Taylor's expansion, $F(x^*, y^*)$ is zero. Thus the change in the deviations from equilibrium over time as a first-order approximation are

$$\frac{dv}{dt} = \frac{\partial f_1(V^*, P^*)}{\partial V}v + \frac{\partial f_1(V^*, P^*)}{\partial P}p \quad \text{and} \quad \frac{dp}{dt} = \frac{\partial f_2(V^*, P^*)}{\partial V}v + \frac{\partial f_2(V^*, P^*)}{\partial P}p,$$

which can be expressed more compactly in matrix form as

$$\begin{bmatrix} \dfrac{dv}{dt} \\ \dfrac{dp}{dt} \end{bmatrix} = \begin{bmatrix} \dfrac{\partial f_1}{\partial V} & \dfrac{\partial f_1}{\partial P} \\ \dfrac{\partial f_2}{\partial V} & \dfrac{\partial f_2}{\partial P} \end{bmatrix}_{V^*, P^*} \begin{bmatrix} v \\ p \end{bmatrix}. \tag{13.7}$$

As usual, the partial derivative $\partial f_1 / \partial V$ means to take the derivative of f_1 with respect to the variable V, holding fixed the value of the other variable, P. These partial derivatives are then evaluated at the equilibrium point V^*, P^*. To simplify notation still more, we write Eq. (13.7), using boldface type to signify matrices and vectors:

$$\mathbf{n} = \begin{bmatrix} v \\ p \end{bmatrix} \tag{13.8}$$

and

$$\frac{d\boldsymbol{n}}{dt} = \boldsymbol{Jn}. \tag{13.9}$$

We now have two coupled first-order (i.e., linear) differential equations. The matrix of partial derivatives, **J**, in Eq. (13.9) is called the **Jacobian matrix.** It describes the dynamic forces acting around the equilibrium point. The local stability of the equilibrium point can be determined very simply from the Jacobian matrix. Recall that, for single species exponential growth, we also had a first-order differential equation,

$$\frac{dN}{dt} = rN, \tag{13.10}$$

whose solution is

$$N(t) = N_0 e^{rt}.$$

The resemblance between the scalar differential equation, Eq. (13.10), and the matrix differential equation, Eq. (13.9), is so striking that without the boldface type we could not distinguish them. This example suggests that the solution to the matrix differential equation, Eq. (13.9), might have a form similar to the solution for exponential growth, or

$$\mathbf{n}(t) = \mathbf{n}_0 e^{\mathbf{J}t}. \tag{13.11}$$

But what does it mean to raise e to the power of a matrix? Perhaps there is some number that can be extracted from the matrix that makes a sensible expression out of Eq. (13.11). Let's explore this line of thought. If there is such a number, let's call it λ, it will need to have the property that for some vector $\tilde{\mathbf{n}}$,

$$\mathbf{n}(t) = e^{\lambda t}\tilde{\mathbf{n}}. \tag{13.12}$$

Moreover, if Eq. (13.12) is indeed a solution to Eq. (13.9) for some particular values of λ and $\tilde{\mathbf{n}}$, then we could write

$$\mathbf{J}\,\tilde{\mathbf{n}} = \lambda\tilde{\mathbf{n}}. \tag{13.13}$$

The left-hand side of Eq. (13.13) is a matrix, $\mathbf{J}$, times a (2×1) vector; for the dimension of the product this yields $(2 \times 2)\,(2 \times 1) = (2 \times 1)$. The right-hand side is a scalar times a (2 by 1) vector and, of course, this also yields a (2×1) vector, so this equation is dimensionally consistent. If you return to Chapter 3, you will see that Eq. (13.13) is identical, except for notation, to Eq. (3.21), which we used to define the eigenvalues, λ, and the eigenvectors, $\tilde{\mathbf{n}}$, of a matrix. Equation (13.13) makes the proposition that the matrix multiplication, $\mathbf{J}\,\tilde{\mathbf{n}}$, can be simplified to a multiplication of a scalar times a vector, $\lambda\tilde{\mathbf{n}}$. So if a number λ and vector $\tilde{\mathbf{n}}$ exist that have the property specified by Eq. (13.13), we could use λ (a scalar), in place of $\mathbf{J}$ (a matrix) to follow the dynamics of the system via Eq. (13.12). As we demonstrated in Chapter 3, an $n \times n$ matrix will have n eigenvalues and n eigenvectors (although some may be duplicates of others).

In Chapter 3 we used eigenvalues to examine geometric population growth, whereas here we are looking at growth only in a small region around the equilibrium point. With the Jacobian matrix, we have rescaled the equilibrium point to be zero.

The simplest case is that the eigenvalues are real numbers. If the sign of each eigenvalue is negative, then $\mathbf{n}(t) = e^{\lambda t}\tilde{\mathbf{n}}$ will grow smaller and smaller, approaching zero (the equilibrium point) as time goes on. In other words, the deviations $\mathbf{p}$ and $\mathbf{v}$ grow smaller over time. Since perturbations are reduced over time, we can infer that the equilibrium point is locally stable. **Thus there is a connection between the sign of the eigenvalues of the Jacobian matrix and the local stability of the equilibrium point.** Since eigenvalues can be complex numbers, things can become a bit more complicated, but not much. A complex number $a + bi$ has real (a) and imaginary (bi), parts where i equals the square root of -1. Consider a general solution in the vicinity of the equilibrium point, like Eq. (13.12):

$$\mathbf{n}(t) = g_1\hat{\mathbf{n}}_1 e^{\lambda_1 t} + g_2\hat{\mathbf{n}}_2 e^{\lambda_2 t} + g_3\hat{\mathbf{n}}_3 e^{\lambda_3 t} \cdots g_n\hat{\mathbf{n}}_n e^{\lambda_n t},$$

when the g's are scalars determined by the initial conditions and the eigenvectors are $\hat{\mathbf{n}}_i$. If the eigenvalues are complex, $\lambda = a + bi$, then each e term can also be expressed as

$$e^{(a+bi)t} = e^{at}e^{bit}.$$

This last expression is the product of two parts—one real and the other imaginary. The number e raised to an imaginary number can also be expressed as

$$e^{bit} = \cos(bt) + i\,\sin(bt).$$

The graph of this last function for $t = 0$ to 21 is shown in Figure 13.6.

Note in Figure 13.6 that the imaginary exponential term e^{bit} neither winds in nor out, but simply oscillates on a circle; thus this term represents an **undamped oscillation** of unit amplitude ($r = 1$), regardless of the particular value of b, which only determines the frequency of the oscillation. In contrast the real exponential part, e^{at}, which multiplies the imaginary part, can exhibit a variety of behaviors, depending on whether a is positive, negative, or zero. **Local stability requires that the real part of each eigenvalue of J be strictly negative** so that perturbations decline exponentially to zero with time. In Box 13.1 we explore the consequences of the sign of the real part and the presence of an imaginary part in solutions.

Figure 13.6
The graph of $e^{bit} = \cos(bt) + i \sin(bt)$ in the imaginary–real plane with a choice of $b = 0.3$ and time t varying from 0 to 21. Line segments connect successive points in the time series. Note that all the points fall on a circle of radius 1. The value of b represents the frequency of oscillation around the circle.

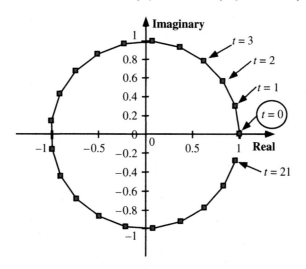

Box 13.1

Figure 13.7 presents examples of eigenvalues (red dots on the right) for a coupled pair of differential equations in the real–imaginary plane. Both eigenvalues of the Jacobian matrix must fall in the shaded region (real part < 0) for local stability. Samples trajectories for each case are shown on the left.

Figure 13.8 a topographical depiction, using a gravity metaphor for stability in two dimensions. The direction and steepness of the slope of the landscape are determined by the sign and magnitude of the real part of the two eigenvalues. In the case of neutral stability, the landscape is completely flat and the real part of both eigenvalues is zero.

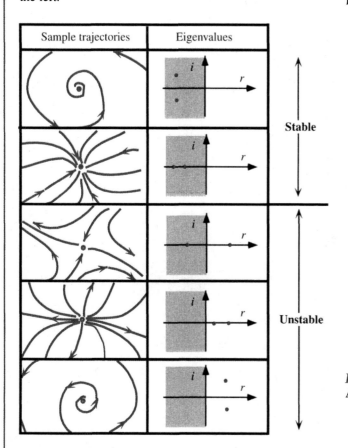

Figure 13.7
Eigenvalues and stability.

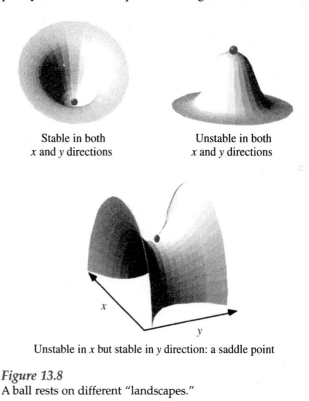

Unstable in x but stable in y direction: a saddle point

Figure 13.8
A ball rests on different "landscapes."

Let's now return to the Jacobian matrix, **J,** which contains the partial derivatives of the growth equations. We want to evaluate the forces of population growth at a particular equilibrium. The partial of the prey population growth, f_1, with respect to V, holding P fixed, is

$$\frac{\partial f_1}{\partial V} = b - aP.$$

In words: the rate at which the rate of increase of the prey population, f_1, changes with changes in their own numbers, V, equals the prey per capita birth rate, b, minus their per capita death rate from predation, $-aP$.

The other partial derivatives in the Jacobian matrix are

$$\frac{\partial f_1}{\partial P} = -aV \quad \text{(the effect of predators on prey growth rate)};$$

$$\frac{\partial f_2}{\partial V} = kaP \quad \text{(the effect of prey on predator's growth rate)};$$

$$\frac{\partial f_2}{\partial P} = -d + kaV \quad \text{(the effect of predators on their own growth rate)}.$$

In general, these partial derivatives are functions of V and P, so the changes in their magnitudes depend on the particular values of V and P at which the partial derivatives are evaluated. To test the local stability of an equilibrium point, we evaluate these partial derivatives (i.e., determine their magnitudes) at that equilibrium point. We begin by evaluating the stability of the interior equilibrium point.

The Jacobian matrix containing the terms that we just determined is

$$\mathbf{J} = \begin{bmatrix} b-aP & -aV \\ kaP & -d+akV \end{bmatrix}. \tag{13.14}$$

Now we substitute into Eq. (13.14) the particular values of P and V at the interior equilibrium point, $V^* = d/ka$ and $P^* = b/a$,

$$\mathbf{J} = \begin{bmatrix} b - \dfrac{ab}{a} & -\dfrac{ad}{ka} \\ \dfrac{kab}{a} & -d+\dfrac{akd}{ka} \end{bmatrix}.$$

After canceling terms, we have

$$\mathbf{J} = \begin{bmatrix} 0 & -\dfrac{d}{k} \\ kb & 0 \end{bmatrix}.$$

Now we solve for the eigenvalues of **J:**

$$\det \begin{bmatrix} 0-\lambda & -\dfrac{d}{k} \\ kb & 0-\lambda \end{bmatrix} = 0.$$

This leads to the characteristic equation

$$\lambda^2 + db = 0 \quad \text{or} \quad \lambda^2 = -bd,$$

and thus the solutions for the two eigenvalues describing the forces around the interior equilibrium point are

$$\lambda = 0 \pm i\sqrt{bd}.$$

The eigenvalues of **J** when evaluated at the interior equilibrium point are purely imaginary numbers—the real parts are zero. This explains why the forces on (V^*, P^*) are neutral and why undamped (or unmagnified) oscillations occur. (Recall the connection between imaginary numbers as exponents of e and sin and cos functions developed in Chapter 3.)

This neutral stability is somewhat artifactual for two reasons. First, it rests on differential equations. Converting this predator–prey model to a discrete time analog inevitably adds a time lag as we showed in Chapter 5. Even for the smallest conceivable finite birth and death rate terms, b and d, the equilibrium point now becomes unstable, not simply neutrally stable. Second, as we indicated in Chapter 12, the incorporation of more realistic assumptions about the growth and interaction of predator and prey creates models wherein the equilibrium point becomes either stable or unstable. We pursue some more complicated predator–prey systems a bit later.

Next, let's evaluate the stability of one of the boundary equilibrium points, namely, (0, 0). We return to the general form of the Jacobian matrix of Eq. (13.14), but this time we substitute the values $V^* = 0$ and $P^* = 0$ to get

$$\mathbf{J} = \begin{bmatrix} b & 0 \\ 0 & -d \end{bmatrix}.$$

The eigenvalues of this diagonal matrix are simply the diagonal elements themselves. Since

$$\det \begin{bmatrix} b - \lambda & 0 \\ 0 & -d - \lambda \end{bmatrix} = 0.$$

yields the characteristic equation

$$(b - \lambda)(-d - \lambda) = 0,$$

therefore $\lambda_1 = b$ and $\lambda_2 = -d$. These eigenvalues are pure real numbers with no imaginary parts. Since the eigenvalue associated with prey growth is positive, the (0, 0) equilibrium is unstable (thus verifying our graphical analysis). The other eigenvalue is negative, so the stability picture here is that of a saddle point. The absence of imaginary parts means that in the vicinity of (0, 0) the growth of the prey is exponential (not oscillatory), trending upward, while that of the predator is exponential, trending downward.

STABILITY EVALUATION APPLIED TO MORE REALISTIC MODELS

We can now apply the technique that we just developed to a slightly more elaborate predator–prey system:

$$\frac{dV}{dt} = f_1(V, P) = \frac{rV}{K}[K - V] - aVP \tag{13.15}$$

and

$$\frac{dP}{dt} = f_2(V, P) = P[-d + kaV]. \tag{13.16}$$

This predator–prey system is modified from the one that we just explored in that the prey now grow logistically, rather than exponentially, in the absence of any predators. The predator zero-isocline is found by setting the term in brackets in Eq. (13.16) to zero, which yields

$$V^* = \frac{d}{ka}, \tag{13.17}$$

and thus the interior equilibrium point for the prey is $V^* = d/ka$.

Figure 13.9
Predator and prey zero-isoclines based on Eqs. (13.17) and (13.18).

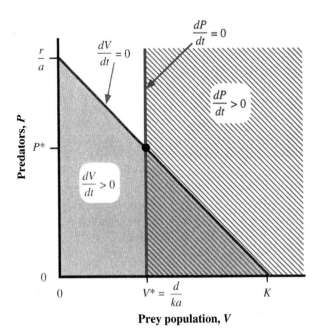

To determine the equilibrium number of predators, P^*, we first need to solve for the prey zero-isocline. Setting Eq. (13.15) to zero and solving gives

$$\frac{r(K-V^*)}{K} = aP^*.$$

Rearranging gives

$$P^* = \frac{r}{a}\left(1 - \frac{V^*}{K}\right). \tag{13.18}$$

Thus the prey zero-isocline is a line with a negative slope of $-r/(aK)$ and a y intercept of r/a, intersecting the horizontal prey axis at the prey K. The prey will reach K in the absence of any predators. Figure 13.9 displays the two zero-isoclines based on Eqs. (13.17) and (13.18).

The next step is to find an expression for the equilibrium abundance of predators, P^*. We substitute $V^* = d/ka$ from Eq. (13.17) into Eq. (13.18) and solve for P^*:

$$P^* = \frac{r}{a}\left(1 - \frac{d}{kaK}\right). \tag{13.19}$$

Before moving on to the rest of the stability analysis, it is useful to examine the graphical relationship between P^*, V^*, and some of the parameters of this model. Note, for example, that both zero-isoclines (Eqs. 13.17 and 13.18) contain the term for the capture rate, a. The equilibrium prey density, V^*, declines continuously as the capture rate, a, increases (Eq. 13.17), but the relationship between a and P^* is more complex in Eq. (13.19). Very high values of a will allow P^* to be negative, but clearly this is impossible, so we must discard any algebraic analysis based on these unfeasible values of P^*, as illustrated in Figure 13.10.

Equation (13.19) contains a term in a^{-2}, so this equation describes a humped relationship for P^* as a function of a. As the capture rate, a, increases from zero, P^* first increases from zero and later decreases as the capture rate increases still more. This effect is illustrated in Figure 13.11, which shows five different values of the capture rate, a, as it affects the position of the zero-isocline for both prey, V and predator, P.

Figure 13.10
An unfeasible equilibrium point in this case means that the predator will not be able to exist and that the prey goes to its carrying capacity, *K*.

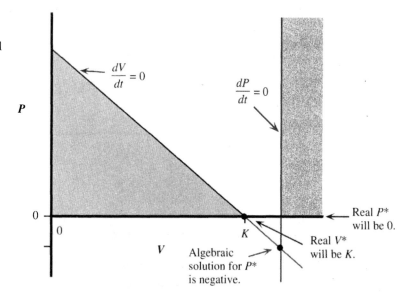

Figure 13.11
The effect of increasing the capture rate, *a*, on the predator and prey zero-isoclines and thus the position of the predator–prey interior equilibrium point.

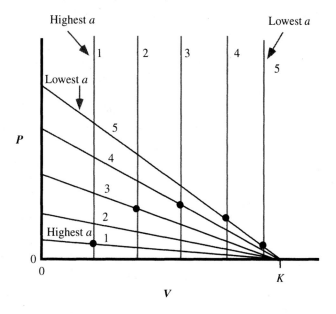

Note the equilibrium points (black dots) for this family of different prey and predator zero-isoclines. Here V^* continuously decreases as the capture rate, *a*, increases, but P^* rises and then falls. As predators become increasingly successful at capturing capture rate increases still more. This effect is illustrated in Figure 13.11, which shows five different values of the capture rate, *a*, as it affects the position of the zero-isocline for both prey, *V* and predator, *P*. Note the equilibrium points (black dots) for this family of different prey and predator zero-isoclines. Here V^* continuously decreases as the capture rate, *a*, increases, but P^* rises and then falls. As predators become increasingly successful at capturing prey, their own densities (at equilibrium) ultimately decrease as they overexploit their prey population. This trend is shown in Figure 13.12, which is based on Figure 13.11 but shows the effect of continuously varying *a*.

This up-then-down feature of P^* is common in predator–prey interactions, but it is not ubiquitous (see the Problems at the end of this chapter).

Figure 13.12
A summary of the results from Figure 13.11, showing an example of the relationship between P^* and the capture rate, a, for equations (13.15) and (13.16).

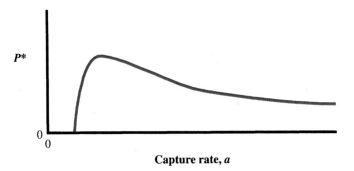

Exercise: If the prey population contained genetic variation for the capture rate, a, such that the different zero-isoclines drawn in Figure 13.12 were associated with different genotypes, which genotype would be most favored by natural selection? How might this affect the ultimate population sizes of the prey and predator populations and their long term persistence?

We return to the stability analysis for the predator–prey system of Eqs. (13.15) and (13.16). Using our expressions for the interior equilibrium point, we may form the Jacobian matrix **J**, to evaluate the stability of this equilibrium point. The partials are evaluated at V^* and P^*. We now fill in the elements of **J**, beginning with the effect of prey numbers on their own growth rate:

$$\frac{\partial f_1}{\partial V} = rK - \frac{2rV}{K} - aP.$$

At equilibrium, from Eq. (13.15)

$$rK - \frac{rV^*}{K} - aP^* = 0,$$

so $\partial f_1/\partial v$ reduces to

$$\frac{\partial f_1}{\partial V} = -\frac{rV^*}{K}.$$

Similarly, after solving for the other partial derivatives and evaluating each at (V^*, P^*), we reach

$$\mathbf{J} = \begin{bmatrix} -\dfrac{rV^*}{K} & -aV^* \\ kaP^* & 0 \end{bmatrix}.$$

At this point we could substitute the algebraic expressions for V^* and P^* from Eqs. (13.17) and (13.19), but we save this step for later. Instead, we go ahead and evaluate the eigenvalues of **J**, using

$$\det \begin{bmatrix} -\dfrac{rV^*}{K} - \lambda & -aV^* \\ kaP^* & 0 - \lambda \end{bmatrix} = 0.$$

The characteristic equation is

$$\lambda^2 + \left(\frac{rV^*}{K}\right)\lambda + a^2 kP^* V^* = 0. \tag{13.20}$$

Note that the sign of every term in Eq. (13.20) is positive as long as the equilibrium point is feasible. Thus, if we factor this quadratic into the product of two terms, it will look like this:

$$(\lambda_1 + \text{"something"})(\lambda_2 + \text{"something else"}) = 0,$$

where we must have plus signs in each of the factors in parentheses. This immediately tells us that the *real parts* of both λ's are negative and equal to "something" and "something else." Moreover,

If $(\lambda_1 + \text{"something"}) = 0$, then $\lambda_2 = -\text{"something else,"}$

and if $(\lambda_2 + \text{"something else"}) = 0$, then $\lambda_1 = -\text{"something."}$

Therefore without doing any more math we know that, as long as the interior equilibrium is feasible (i.e., $V^* > 0$ and $P^* > 0$), this equilibrium point is locally stable, since both eigenvalues for the Jacobian matrix will have negative real parts. To solve for "something" and "something else" we apply the quadratic root formula:

$$\lambda_1 = \frac{\dfrac{-rV^*}{K} + \sqrt{\left(\dfrac{rV^*}{K}\right)^2 - 4a^2 kP^* V^*}}{2}$$

and

$$\lambda_2 = \frac{\dfrac{-rV^*}{K} - \sqrt{\left(\dfrac{rV^*}{K}\right)^2 - 4a^2 kP^* V^*}}{2}.$$

From this last pair of equations, you can also see that, even for positive V^* and P^*, it's quite possible that the radical contains a negative number, implying that the eigenvalues may have imaginary parts. Referring to Box 13.1, you can see that this means that trajectories would spiral into the equilibrium point (since the real parts are both negative). Also, if there are complex numbers for roots to the characteristic equation, they come in pairs that are called **complex conjugates.** That is, they both have the same real part, and only the sign of their imaginary part is different. We may plot these complex numbers in the plane where the x axis represents the real part and the y axis the imaginary part. We then have, for example, one of the situations depicted in Box 13.1.

A complementary way to see that the interior equilibrium point of the system described by Eqs. (13.15) and (13.16) is necessarily stable, if it is feasible, is to apply a graphical analysis that parallels the development in Chapter 10, Figure 10.22.

Box 13.2 shows how the condition for stability of an equilibrium point needs to be modified for coupled difference equations instead of differential equations.

FISHING INTENSITY AND FISH POPULATION SIZES

A young Umberto D'Ancona worked in the fishery industry in Italy after World War I. He noticed that predatory fish seemed to have increased in the Adriatic Sea following the war, while smaller prey fish had decreased. Naturally the war had interrupted most commercial fishing. He interested his mathematician father-in-law, Vito Volterra, into trying to understand and explain this phenomenon. Thus began Volterra's interest in the

Box 13.2. *Discrete Time Predator-Prey Models*

A general form for a single predator, P and prey, V, species interacting according to a difference equation is

$$P_{t+1} = f_1(P_t, V_t) \quad \text{and} \quad V_{t+1} = f_2(P_t, V_t).$$

An equilibrium point occurs when

$$P_{t+1} = P_t \quad \text{and} \quad V_{t+1} = V_t.$$

An alternative notation that matches more the form of a differential equation is based on just the change in population size on the left-hand side of the equation, or

$$P_{t+1} - P_t = F_1(P_t, V_t) \quad \text{and} \quad V_{t+1} - V_t = F_2(P_t, V_t).$$

By this formulation, an equilibrium point occurs when

$$F_1(V_t, P_t) \quad \text{and} \quad F_2(V_t, P_t) = 0.$$

Contrasting these two formulations, we see that

$$f_1(P_t, V_t) - P_t = F_1(P_t, V_t).$$

The function minus P_t = the function for $\dfrac{\Delta P}{\Delta t = 1}$. for P_{t+1}

And, similarly,

$$f_2(P_t, V_t) - V_t = F_2(P_t, V_t).$$

Now let's go ahead and apply the Jacobian approach to the equations for $\Delta P/\Delta t$ and $\Delta V/\Delta t$, just as we did with differential equations:

$$\frac{\partial F_1}{\partial P} = \frac{\partial f_1}{\partial P} - 1$$

$$\frac{\partial F_2}{\partial V} = \frac{\partial f_2}{\partial V} - 1$$

$$\frac{\partial F_1}{\partial V} = \frac{\partial f_1}{\partial V}$$

$$\frac{\partial F_2}{\partial P} = \frac{\partial f_2}{\partial P}.$$

The Jacobian matrix therefore has the structure

$$\mathbf{J} = \begin{bmatrix} \dfrac{\partial f_1}{\partial P} - 1 & \dfrac{\partial f_1}{\partial V} \\[2mm] \dfrac{\partial f_2}{\partial P} & \dfrac{\partial f_2}{\partial V} - 1 \end{bmatrix}$$

Note that the diagonal terms have a value of 1 subtracted from them in contrast to the structure of the Jacobian matrix for differential equations. To find the eigenvalues of $\mathbf{J}$, we get

$$\det \begin{bmatrix} \dfrac{\partial f_1}{\partial P} - 1 - \lambda & \dfrac{\partial f_1}{\partial V} \\[2mm] \dfrac{\partial f_2}{\partial P} & \dfrac{\partial f_2}{\partial V} - 1 - \lambda \end{bmatrix} = 0. \quad (13.21)$$

Naturally, we want to evaluate these partials at the equilibrium point of interest. But regardless of what point that is, we can also write this last equation as

$$\det \begin{bmatrix} \dfrac{\partial f_1}{\partial P} - \lambda' & \dfrac{\partial f_1}{\partial V} \\[2mm] \dfrac{\partial f_2}{\partial P} & \dfrac{\partial f_2}{\partial V} - \lambda' \end{bmatrix} = 0, \quad (13.22)$$

where $\lambda' = 1 + \lambda$. In Chapter 3, we found that, for age-structured growth modeled as a difference equation, general solutions of difference equations have the form

$$\mathbf{n}(t) = g_1 \mathbf{x}_1 \lambda_1^t + g_2 \mathbf{x}_2 \lambda_2^t + g_3 \mathbf{x}_3 \lambda_3^t + \cdots + g_n \mathbf{x}_n \lambda_n^t.$$

For local stability the absolute value of each $\lambda_i < 1$, which implies that, based solely on the difference equation formulation (i.e., the f_1 and f_2 functions in the Jacobian of matrix Eq. 13.22) we have the requirement that

$$|\lambda'| < 1$$

dynamical modeling of ecological systems, beginning with the development of the predation equations that we have already discussed. Volterra reasoned that, without much commercial fishing during the war, predatory fish, P, experienced lower death rates while prey fish, V, probably had higher recruitment rates, r. The per capita capture rates, he reasoned, probably stayed about the same. While Volterra modeled these changes, using the artificial model of the classical Lotka–Volterra predator equations, the same conclusions emerge from more realistic models as well, as illustrated in Figure 13.14.[1]

---

1. Apparently, unbeknownst to Volterra, these equations had been analyzed earlier by Alfred Lotka (1920). However, his work contained some serious errors. For example, Lotka initially did not realize that the cycle's amplitude and frequency depended on the initial conditions. These errors were later corrected in Lotka's 1924 book *Elements of Physical Biology*, which came out 2 years before Volterra's apparently parallel discovery of the same predator–prey dynamics (Volterra 1926).

for each eigenvalue. The absolute value condition arises from the fact that a complex number, $\lambda = a + bi$, can be raised to the power t to reach

$$\lambda^t = r^t(\cos t\theta + i \sin t\theta), \qquad (13.23)$$

where r and θ are defined in the following diagram.

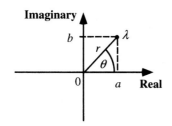

The **absolute value r** (also called the magnitude or modulus) of a complex number is

$$r = |\lambda'| = \sqrt{a^2 + b^2}.$$

Now the second part of Eq. (13.23), $\cos t\theta + i \sin t\theta$, is just an undamped oscillation with an amplitude of 1 and a frequency of b (see Figure 13.6). This term alone causes deviations from equilibrium to get neither closer to nor farther from equilibrium with time. However, this part is multiplied by r^t, where $r = |\lambda'|$ is the absolute value. **Thus, if and only if $|\lambda_i'| < 1$, for each eigenvalue, deviations from equilibrium become smaller over time.** The condition that $|\lambda'| < 1$ is plotted in Figure 13.13 and compared to the analogous condition for differential equations.

Difference equations with $\Delta t = 1$

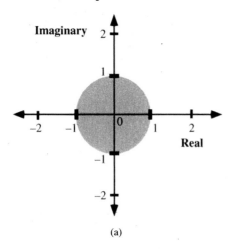

(a)

Differential equations with dt

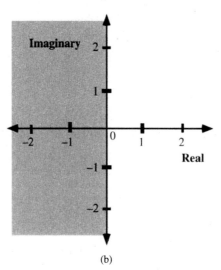

(b)

Figure 13.13
For local stability, all the eigenvalues of the Jacobian matrix must fall in the shaded regions. This figure should be compared to the analogous development in Box 5.3 for a single-species dynamical model described by either difference equations or differential equations and to Box 13.1.

A SHORTCUT FOR STABILITY EVALUATION OF 2-SPECIES SYSTEMS

We have seen that an equilibrium point will be stable if and only if the real parts of the eigenvalues of the Jacobian matrix are both negative. Consider the following general representation for a two-by-two Jacobian matrix:

$$\mathbf{J} = \begin{bmatrix} a & b \\ c & d \end{bmatrix}.$$

Its eigenvalues are found by solving the equation

$$\det \begin{bmatrix} a - \lambda & b \\ c & d - \lambda \end{bmatrix} = 0,$$

Figure 13.14
A way to explain the increase in large predatory fish and the decrease in smaller fish after World War I, as originally suggested by Volterra (1926) but illustrated here with different predator–prey equations.

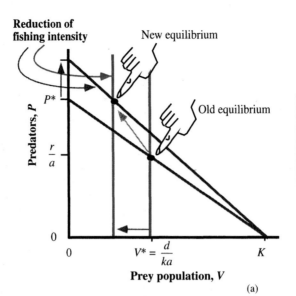

Conclusion: Reducing fishing intensity (lower r for prey fish and lower d for predator fish) leads to lower populations of prey fish and higher populations of predator fish.

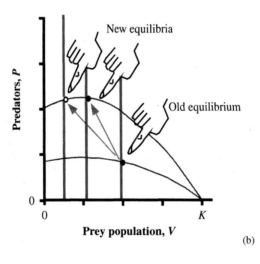

Conclusion: Reducing fishing intensity again leads to lower populations of prey fish and higher populations of predator fish but now can also destabilize the equilibrium point producing cycles.

which yields

$$\lambda^2 - a\lambda - d\lambda + ad - bc = 0.$$

Note that this last equation can also be written as

$$\lambda^2 - (\text{trace of } \mathbf{J})\ \lambda + (\det \text{ of } \mathbf{J}) = 0$$

where the **trace** of $\mathbf{J}$ is simply the sum of the diagonal elements $(a + d)$. For both roots of this last equation to be negative, the middle term, the trace of $\mathbf{J}$, must be negative and the determinant of $\mathbf{J}$ must be positive. Then we will have an equation of the form,

$$(\lambda + \text{something})\ (\lambda + \text{something else}) = 0.$$

Hence a two-species system will be stable if and only if the trace of $\mathbf{J}$ is negative *and* the determinant of $\mathbf{J}$ is positive.

While it is difficult to translate the condition on the determinant of $\mathbf{J}$ into graphical terms, the condition on the trace of $\mathbf{J}$ is readily evaluated from the depiction of the two zero-isoclines. Let's return to the more complicated predator–prey equations with a type 2 functional response explored in Chapter 12. In the absence of predator interference the Jacobian matrix has the sign structure

$$\begin{bmatrix} - \text{ or } + & + \\ - & 0 \end{bmatrix}.$$

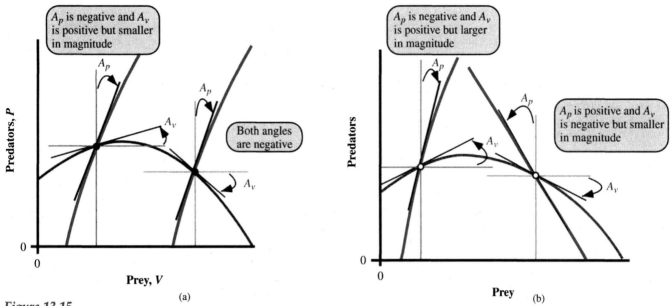

Figure 13.15

A graphical means of evaluating a necessary condition for stability. The trace of the Jacobian matrix will be negative when the sum of angles A_P and A_V is negative. The light gray lines are simply vertical and horizontal construction lines though the equilibrium point to facilitate the measurement of the critical angles A_P and A_V. The black lines are the tangent lines for the predator and prey zero-isoclines at the feasible equilibrium point. Angles are measured from the construction line to the tangent line. By this convention a clockwise angle is negative and represents a negative partial derivative on the diagonal of the Jacobian matrix. The two examples in (a) pass this necessary stability test while the two examples in (b) fail it.

The **J**(1,1) element, corresponding to the prey–prey interaction is ambiguous in sign, an expression of the rising and then falling shape of the prey zero-isocline. However, the determinant of this matrix is necessarily positive regardless of the ambiguous sign of **J**(1,1). On the other hand, the trace will only be negative if **J**(1,1) is negative. Graphically **J**(1,1) represents the slope of the prey zero-isocline at the equilibrium point. When this equilibrium point lies to the right of the peak of the prey zero-isocline this slope is negative, but it is positive when it falls on the left side of the hump.

Now let's tackle the case of systems with non-zero predator self-interactions. These matrices will have a sign structure for the Jacobian:

$$\begin{bmatrix} - \text{ or } + & + \\ - & - \text{ or } + \end{bmatrix}$$

The **J**(2,2) term will be negative for predator–predator interference and positive for predator–predator mutualism. Figure 13.15 shows four examples where we graphically evaluate the "trace is negative" test since it boils down to a condition on the angles of the zero-isoclines at the equilibrium point. The two examples in (a) pass this test while the two examples in (b) fail it. We are not assured of stability even for the two examples in (a), however, because we still have the "determinant is positive" test to evaluate, but we can at least see how it is possible that equilibrium points that lie to the left of the prey peak can be stable if predator interference is strong enough. The other side of the coin is that equilibrium points to the right of the peak can be unstable if predator mutualism is strong enough.

PROBLEMS

1. What are the eigenvalues associated with the equilibrium point $(0, 0)$ for the predator–prey system of eqs. (13.15) and (13.16)?

2. Robert MacArthur (1972) introduced the following predator–prey equations to describe a situation where the rate that resources, R, enter an area is independent of the number already there:

$$\frac{dR}{dt} = F - aRP \qquad (13.24a)$$

and

$$\frac{dP}{dt} = P(kaR - d). \tag{13.24b}$$

He called this system the "falling fruit" model since the resources, R, are like fruit falling from trees; the number of fruit on the ground, R, and thus available for consumption by consumers, P, does not directly affect the rate that new fruit will fall onto the ground, which is a constant, F, in Eq. (13.24a).

For this model, what is the relationship between the equilibrium levels of R and P (i.e., R^*, and P^*) and the capture rate, a? What is the relationship between R^* and P^* and the rate of fruit fall, F? Form the Jacobian matrix and answer the following True/False questions.

T F **a.** If the equilibrium point is positive for both R and P, it will be locally stable.

T F **b.** If the equilibrium point is positive for both R and P, it will be globally stable.

T F **c.** Resource growth is self-inhibited (i.e., density dependent).

T F **d.** Per capita consumer growth has a positive feedback term (intraspecific mutualism).

T F **e.** Consumers have negative growth in the absence of resources.

T F **f.** Increasing k will decrease the equilibrium (or standing crop) of resources.

T F **g.** In the absence of consumers, resources will grow exponentially.

T F **h.** As the capture rate, a, increases, the equilibrium density of consumers will necessarily increase.

3 The Nicholson–Bailey model of host–parasite interactions differs from all other models in this chapter in that it is expressed as discrete time difference equations (Nicholson and Bailey 1935), as shown in Figure 13.16.

Similar to the Lotka–Volterra predation equation, the prey grow geometrically in the absence of predators. The predators, however, decline immediately to zero in the absence of prey, unlike in the exponential decay of the Lotka–Volterra equation.

Show that the interior equilibrium point is $P^* = \ln(\lambda)/a$ and $H^* = \ln(\lambda)/[ca(\lambda - 1)]$. (*Hint*: $e^{z\ln a} = a^z$). This interior equilibrium is unstable. Small perturbations from equilibrium lead to oscillations of rapidly increasing amplitude.

4. Construct one set of linear two-species zero-isoclines consistent with all the following assumptions.

   a. Species 1 and 2 each have intraspecific density dependent population growth.

   b. Species 1 and 2 are interspecifically mutualistic (i.e., higher numbers of one species enhances the growth rate of the other species).

   c. Species 1 and 2 can coexist at a stable equilibrium point.

On your diagram, cross-hatch the regions of positive growth rate for both species; label the axes, zero-isoclines, and points of intersection of the zero-isoclines with each axis. Draw circles around each equilibrium point. Indicate which of these equilibria are stable and unstable.

Partial Answer:

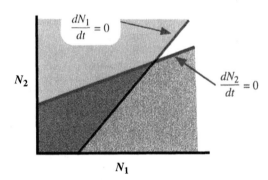

5. For the zero-isocline situation depicted in the following diagram, identify and label each equilibrium point. For each equilibrium point state whether it is or is not locally stable. (The shaded and hatched regions yield positive growth.)

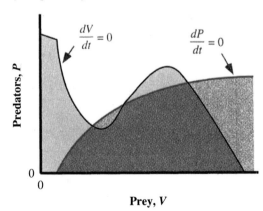

Answer the following True/False questions about the system.

T F **a.** Predators exhibit self-interference at high densities.

T F **b.** Predators have a type 1 functional response.

T F **c.** Prey experience intraspecific density dependence.

T F **d.** Prey have a carrying capacity.

T F **e.** This model has just one domain of attraction.

T F **f.** Point $(0, 0)$ is locally stable.

Nicholson–Bailey Model—Host numbers are H and parasite numbers are P:

$$H_{t+1} = \lambda H_t \ \exp(-aP_t).$$

The geometric growth rate of the hosts in the absence of parasitism | The fraction of hosts at time t that escape parasitism. Parameter a is the searching efficiency of the parasites. As P goes toward infinity, this term goes to zero. If $P = 0$, this term becomes 1.

Hosts grow geometrically in the absence of parasites ($P = 0$). Assume that only one parasite infects each infected host; then the parasite equation is

$$P_{t+1} = c\lambda H_t[1 - \exp(-aP_t)]$$

If all parasites emerge from infected hosts, then $c = 1$, but if some die then $c < 1$. | The fraction of hosts at time t that are parasitised

Figure 13.16
The Nicholson–Bailey host–parasite model.

14 *Competitors*

INTERSPECIFIC COMPETITION MAY BE OF TWO RELATED TYPES

Interspecific competition between two species occurs when individuals of one species suffer a reduction in growth rate from a second species due to their shared use of limiting resources **(exploitative competition)** or active interference **(interference competition).** In general, we do not expect interference competition to evolve unless this energetically costly squabbling potentially leads to some benefits to the aggressors (Case and Gilpin 1975). Such would be the case if an aggressor were able to protect some limited resource from being consumed in its territory by individuals of a competing species. For this reason, we might expect a correlation between the degree of interference competition between two species and the extent that they share critical resources. Moore (1978) studied interspecific aggression by mockingbirds inhabiting residential areas in South Carolina. In the fall and winter these birds defend territories containing trees and shrubs laden with small fruits. The level of aggression by mockingbirds to intruders in their territory is roughly proportional to the degree of frugivory in the diet of the intruding species, as indicated in Figure 14.1.

Typically, interference competition is asymmetric, with one species aggressively dominating the other. This dominance is often associated with larger body size (Lawton and Hassel 1981, Robinson and Terborgh 1995). In many vertebrates, interspecific aggression can lead to interspecific territoriality such that individuals defend their territories against intrusion from individuals of their own and other species. However, in other cases it can simply lead to the decline and local extinction of the inferior competitor across all habitats. Recall the example of the spreading Argentine ants in southern California from a study of Erickson (1971). In Chapter 2, we showed that the radius of their distribution across a field grew linearly with time. As they moved across this field and as their colony number increased, a native harvester ant, *Pogonomyrmex californicus,* declined, as illustrated in Figure 14.2. Since these two ant species eat different kinds of foods, the major cause of this competitive displacement can be ascribed to the dominance of the Argentine ant in interference competition. Here, however, this advantage isn't due to a larger body size—Argentine ants are substantially smaller than harvester ants. Rather the Argentine ants have larger colonies and a more coordinated fighting response in which several workers team up to defeat a single harvester ant. If these ants do not compete for food, why should they bother fighting? Perhaps because they compete for other resources such as available below-ground nest space.

LABORATORY EXPERIMENTS ON COMPETITION

We now turn to some observations taken from laboratory experiments involving simple two-species competitive systems. In the laboratory it is possible to manipulate species densities and control external environmental conditions to focus on the role

Figure 14.1
Mockingbirds are generally more aggressive to other bird species which, like them, eat a large proportion of small fruit (from Moore 1978).

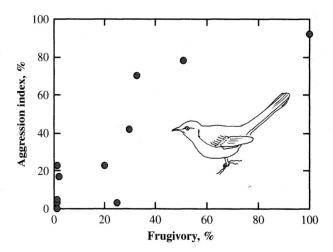

Figure 14.2
As the population of the introduced Argentine ants in an old field in Southern California increased, the seed-eating harvester ants, *Pogonomyrmex californicus*, declined (modified from Erickson 1971).

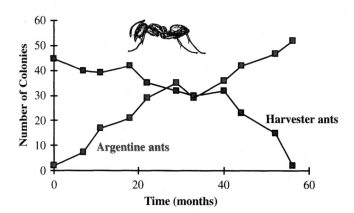

of particular interspecific interactions. These observations will guide us in developing a theory to predict the consequences of interspecific competition under different situations.

Paramecium

In Chapter 5 we showed—from the experiments of Gause (1934, 1936)—that population growth of the protozoan *Paramecium* provided a reasonable approximation to the continuous logistic equation. Gause and others also conducted experiments growing different species of *Paramecium* alone and then together. The time course of one set of experiments involving the species *P. aurelia* and *P. caudatum* when alone and grown together is shown in Figure 14.3.

The cell sizes of these two species are very different. Because *Paramecium caudatum* is 2.5 times larger than *P. aurelia*, its carrying capacity is less than one-half that of *P. aurelia*. In other experiments, Gause demonstrated that the carrying capacity of each species alone seemed to be a direct function of the amount of food (bacteria) added to the culture medium. Subsequently, Gause (1935) placed other species in competition, using somewhat different rearing conditions. In one set of experiments he found the results summarized in Figure 14.4.

As before, *P. caudatum* is competitively excluded in the presence of *P. aurelia*. However, it coexists with the equally small *P. bursaria*, which also has a larger carrying capacity, *K*. Clearly, *K* alone is not a predictor of competitive outcome. *Paramecium aurelia* can also coexist with *P. bursaria*. Upon closer inspection, Gause (1936) noticed that the coexisting species pairs subdivided space in the test tubes. *Paramecium bur-*

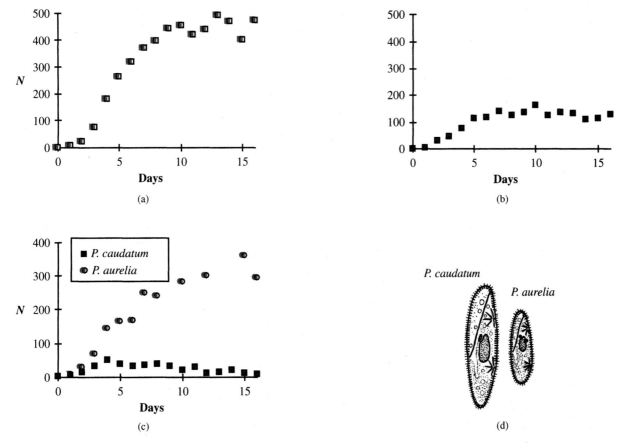

Figure 14.3
Paramecium numbers are expressed in numbers per 0.5 ml. Data and drawings are from Gause (1934, page 99 and Table 4).
(a) *P. aurelia* alone. (b) *P. caudatum* alone. (c) *Paramecium aurelia* out-competes *P. caudatum* when the two are together.
(d) *Paramecium caudatum* is about 2.5 times larger than *Paramecium aurelia* in cell volume.

Figure 14.4
Competitive outcomes with different species pairs of *Paramecium*. The diagonal gives estimates for *r* and *K* based on some single species experiments. The *K*'s are the number of individuals per 0.5 ml (recalculated from Gause 1935).

Relative cell volume = 1	0.39	0.41
P. caudatum	*P. aurelia*	*P. bursaria*

	P. caudatum	*P. aurelia*	*P. bursaria*
P. caudatum	*r* = 1.10 *K* = 80	P. aurelia wins	Coexist
P. aurelia		*r* = 1.00 *K* = 510	Coexist
P. bursaria			*r* = 0.94 *K* = 400

saria have tiny symbiotic green algae inside them. These algae, like all plants, produce oxygen that allows *P. bursaria* to live in the bottom of the tubes where oxygen levels are depleted by bacteria, which tend to sink. The other two *Paramecium* have higher oxygen requirements and are relatively more common in the upper parts of the tubes where oxygen levels are higher but food (bacteria) concentrations are lower.

For the species pairs that coexist, roughly the same equilibrium densities are reached regardless of the initial densities, as long as both species are simultaneously introduced into the medium. The four trajectories in Figure 14.5 all converge to the small region of phase space marked by the rectangle.

Vandermeer (1969) repeated some of these experiments with the same three species plus one other. His experimental conditions were somewhat different and, most important, he used a strain of *P. bursaria* that lacked symbiotic algae. He found conditions under which *P. caudatum* could coexist with *P. aurelia*, yet both would exclude *P. bursaria*, quite contrary to Gause's results. Altogether, these experiments indicate that food supply levels, oxygen levels, and the buildup of metabolic by-products in the culture media are important in limiting the growth of these species and determining competitive advantage or coexistence.

Tribolium Flour Beetles

Park (1954, 1962) conducted competition studies on flour beetles in the genus *Tribolium*. These small insects grow readily in vials containing common wheat flour and yeast. Their generation time is about 35 days, and population censuses were taken once a month. In one set of experiments Park explored the role of climatic conditions on the growth of the species when they were alone and then when they were together. Figure 14.6 presents the results. First, unlike the case with some *Paramecium*, Park

Figure 14.5
Four different initial conditions produce trajectories that converge on roughly the same equilibrium abundances of *Paramecium bursaria* and *P. caudatum*, the stippled box. The *K*'s show the carrying capacities of the two species when they are alone. After Gause (1936).

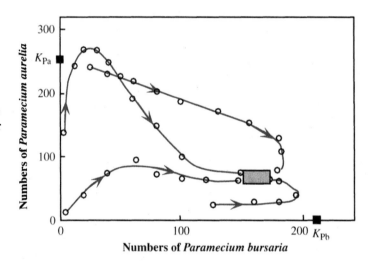

Figure 14.6
Flour beetle competition under different climatic conditions. The mean population size per gram of flour includes larvae, pupae, and adults, based on 15–40 replications of each experiment (standard errors are small and range from about 0.4 to 3.5 individuals). Each experiment was initiated with eight adults (four males and four females) of each species introduced into 8 grams of a flour/yeast mixture. After Park (1954).

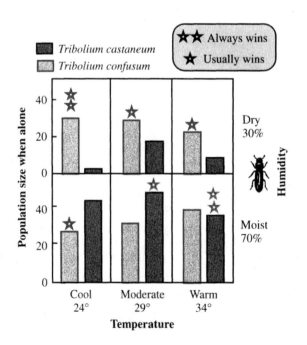

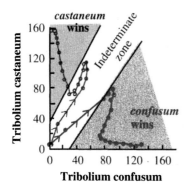

Figure 14.7
The head-start advantage of *Tribolium* competition. These results are for the cool/moist conditions shown in Figure 14.18. The outcome of competition is strongly influenced by the initial conditions. The regions display the qualitative outcome of competition starting from different initial abundances for the two species. In the *"confusum* wins" region, *confusum* outnumbers *castaneum* initially. When the two species are approximately equal in abundance, the outcome of competition is probabilistic. Two sample trajectories are shown, both from the same initial condition of eight beetles of each species. Each dot is a census 30 days apart. After Neyman et al. (1956).

found no conditions under which the two species could coexist, nor did both species become extinct. Instead, a clear winner and loser always emerged. Which species won, however, was not obviously related to its abundance or even relative abundance in the single species experiments. For example, under the cool/moist conditions, *T. castaneum* had a much higher mean density when alone but was the usual loser in competition. However, it occasionally would win, providing a degree of indeterminacy to the results.

This indeterminacy is typically seen when the initial numbers of each species are approximately equal as they were for this particular set of experiments (eight individuals of each species). For the same climatic conditions, the initial numbers make a big difference in determining the winner and loser. Either species, if given a strong numerical head start, becomes the favored species, as shown in Figure 14.7.

Several experiments by Park and others demonstrated that if the amount of flour is increased, the carrying capacity of each species is increased in a roughly linear fashion. Consequently, equilibrium densities (numbers per gram of flour) remain roughly constant for the same environmental conditions. Food, however, is not limiting in this system. These beetles are literally oversupplied with food continuously. Instead, intraspecific and interspecific cannibalism by adults and larvae eating eggs and pupae is responsible for the density-dependence (Park 1965, Wade 1979, 1980). At higher densities these deadly encounters happen more frequently. Also, Park (1965) demonstrated that overall interspecific predation rates were greater than intraspecific predation rates.

This suggests that if mutual predation could be eliminated in this system, more prolonged coexistence might be possible. Crombie (1946) found just this result. Two beetles, this time a *Tribolium* and a related genus, *Oryzaephilus,* show the same sort of competitive incompatibility found in Park's experiments with two species of *Tribolium:* no coexistence and *Tribolium* was always the winner. Crombie (1946) next added short lengths of capillary tubing to the flour mixture. These provided *Oryzaephilus* safe sites for pupation free from predation by *Tribolium.* Now the two species could coexist, as illustrated in Figure 14.8. *Tribolium* reached higher densities in spite of the now higher numbers of *Oryzaephilus,* suggesting that it was relieved of some cannibalism as well.

In summary, the ultimate outcome of competition depends on environmental conditions and sometimes on initial conditions. Some species can coexist, as with some species of *Paramecium,* but for other species or environmental conditions, they cannot, as with other *Paramecium* species pairs and the *Tribolium.* Those species that can coexist seem to utilize the environment in different ways or have a spatial refuge from the pressures of the other species (mutual cannibalism in the case of *Tribolium* and food and oxygen exploitation in *Paramecium*).

We now search for a conceptual framework for these observations. We'd like to uncover general rules about the assembly of competitors in communities—rules that might let us predict which species combinations will be compatible and which will not.

Figure 14.8
Competition between two flour beetles. After Crombie (1946). (a) Without glass tubing. (b) With glass tubing. Fine glass capillary tubing added to the flour provides safe sites for pupation free from predation, allowing the two species to coexist.

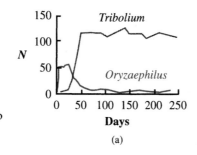

(a)

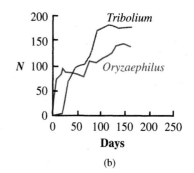

(b)

MODELING COMPETITION

As we have already shown with single-species growth models, logistic equations implicitly invoke intraspecific competition in the form of a carrying capacity term for a population. The experimental results discussed earlier show that competition can also occur between individuals of different species. The dynamical effects of interspecific competition may be modeled by using two different approaches, depending on whether resource dynamics are **explicitly** or **implicitly** considered, as depicted in Figure 14.9. The explicit approach considers the dynamics of the consumer–resource interactions. The arrows represent qualitative effects of one species on another's per capita growth rate with the sign indicated. Resources increase consumers' growth rates, so the sign on this arrow is positive. Consumers decrease resource growth rates, so the sign on this arrow is negative. A full explicit model would flesh this out by specifying differential or difference equations for the functions *f* and *g*.

Figure 14.9
Two approaches to modeling competition. Implicit versus explicit models for interspecific competition among two consumers, C_1 and C_2. The plus or minus sign by the arrowhead shows the effect on that species caused by the other species.

Intraspecific Competition

Implicit intraspecific competition within a consumer species, C.

$$\frac{dC}{dt} = f(C)$$

Explicit intraspecific competition of a consumer species, C, for three resources.

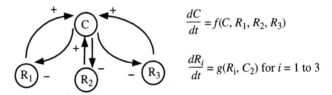

$$\frac{dC}{dt} = f(C, R_1, R_2, R_3)$$

$$\frac{dR_i}{dt} = g(R_i, C_2) \text{ for } i = 1 \text{ to } 3$$

Interspecific Competition

Implicit Models

Exploitation and interference cannot be distinguished, but the model is much simpler.

Explicit Models

Exploitation competition between two consumers for three resources. Resources have positive effects on consumer growth, and consumers have negative effects on resource growth.

Exploitation plus interference competition among consumers

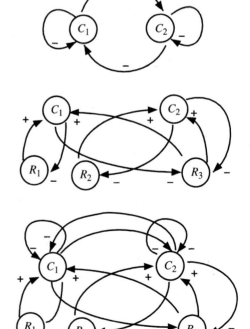

The implicit approach buries all these details into an implicit competition equation, just as the logistic equations ignored the mechanisms of intraspecific competition by introducing the notion of a carrying capacity for the population.

By modeling resource dynamics explicitly, we gain a better understanding of how resources influence competitive success and the mechanisms involved (Tilman 1986). In going from an explicit consumer–resource model to an implicit model, we necessarily lose important features of the dynamics of species interactions. The hope, though, is that we will not be seriously misled in our search for simplicity, at least regarding qualitative aspects of competition, such as whether two competitors can or cannot coexist. However, with implicit models we may lose the ability to predict accurately the actual trajectories and to understand the mechanism of competition.

The Lotka–Volterra competition equations, like the logistic equations, treat competition implicitly; they extend the logistic equations by incorporating inter- as well as intraspecific growth limitations. Moreover, in this formulation, the per capita effects of density on per capita growth rate are assumed to be linear for both intra- and interspecific effects, as shown in Figure 14.10. Also interference and exploitative competition are not distinguished. Figure 14.11 takes you through the qualitative dynamics to the development of one competitor's zero-isocline, just as we did with predator–prey systems in Chapter 12.

As we did with predator–prey systems, we have now developed the zero-isocline for a species in state space and have drawn arrows indicating the qualitative direction of growth for that species in different regions of state space. The zero-isocline for species i defines the set of points in the N_1–N_2 phase plane where species i's population neither increases nor decreases.

We can also overlay information about the magnitude of the growth rates at various points of the species 1–species 2 phase space, just as a topographical map overlays elevations on the two dimensions of latitude and longitude. Figure 14.12 shows a plot of the three-dimensional representation of species 1's growth, originally plotted in just two dimensions—the N_1–N_2 plane—with the different isoclines representing different "elevations."

Usually, however, as with predator–prey systems, useful information about the direction of initial trajectories and the stability of equilibrium points can be determined solely from qualitative information depicted in plots that show the zero-isoclines of each species and the regions of state space that yield positive or negative growth for each. These diagrams are also useful for another reason. Analytical mathematical

Figure 14.10

Two species Lotka–Volterra Competition. The red triangular plane shows the per capita growth rate of species 1 declining with increases in density of both species 1 and species 2. At high densities, species 1's growth rate even becomes negative.

How do the densities of species 1 (N_1) and species 2 (N_2) affect the growth rate of species 1? Plot species 1's per capita growth rate in the vertical direction as a function of N_1 and N_2.

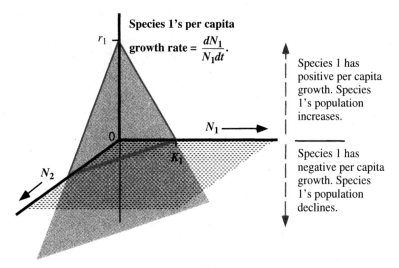

Species 1's per capita growth rate = $\dfrac{dN_1}{N_1 dt}$.

Species 1 has positive per capita growth. Species 1's population increases.

Species 1 has negative per capita growth. Species 1's population declines.

Figure 14.11
These three diagrams develop a zero-isocline for species 1.

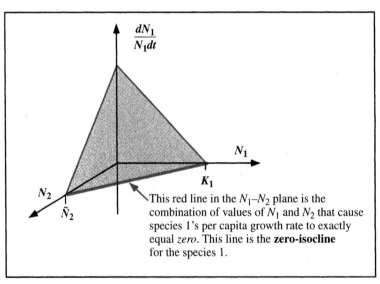

Here, for example, are two joint densities of N_1 and N_2 (the gray dots). Each is projected onto the red growth plane for species 1 to show the corresponding per capita growth rate for species 1 at these two points.

For these low values of N_1 and N_2, species 1 has positive per capita growth.

For these high values of N_1 and N_2, species 1 has negative per capita growth.

(a)

This red line in the N_1–N_2 plane is the combination of values of N_1 and N_2 that cause species 1's per capita growth rate to exactly equal *zero*. This line is the **zero-isocline** for the species 1.

(b)

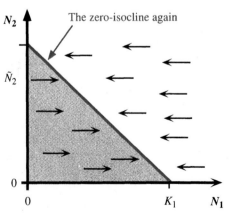

Now flip the bottom plane in (b) on its side and get rid of the per capita growth dimension so that you now look at just the N_1–N_2 **phase plane**.

Densities of the two species in the region below the zero-isocline—the shaded region—yield positive growth for species 1 and hence the increases in N_1, as indicated by the arrows pointing to the right.

The zero-isocline again

Densities of the two species above the zero-isocline yield negative growth rates for species 1 and hence the declines in N_1, as indicated by the arrows pointing to the left.

(c)

Figure 14.12
A "topographic map" of species 1's per capita growth rate as a function of density of both species 1 and 2.

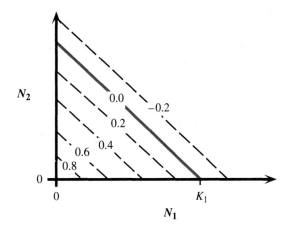

results for coupled differential equations are often impossible to achieve. For example, the predator–prey equations that you studied in Chapter 12 do not have a general closed-form analytical solution for $C(t)$ and $R(t)$ over time (as we found for exponential growth and logistic growth of single species). The best we can do is infer qualitative dynamics from approximate or numerical solutions or instead rely on qualitative information based on these phase–space pictures. Similarly, the Lotka–Volterra competition equations, which we develop next, as simple as they are, have no closed-form analytical solution.

LOTKA–VOLTERRA COMPETITION EQUATIONS

With this graphical exposition behind us, we move on to the equations that match the depictions in Figures 14.10 and 14.11. The assumptions behind these graphs may be encapsulated in a differential equation for species 1's growth:

$$\frac{dN_1}{dt} = \frac{r_1 N_1}{K_1}(K_1 - N_1 - \alpha_{12}N_2).$$

Except for this term, this equation is the continuous logistic equation.

(14.1)

The last term with the α_{12} converts species 2's numbers into an effect on species 1's per capita growth rate; α_{12} is the scaling factor that makes the conversion. Read the term α_{12} as the "effect of an individual of species 2 on the per capita growth rate of species 1." Its magnitude will be relative to species 1's effect on its own growth rate. Note that there is an implicit constant of 1 in front of the term N_1 (i.e., $\alpha_{11} = 1$). If, for every additional individual of species 2, the per capita growth rate of species 1 declines by half the amount that it would by the addition of an individual of species 1, then $\alpha_{12} = 0.5$. If species 1's per capita growth rate declines twice as fast with additions of species 2 compared to additions of species 1, then $\alpha_{12} = 2$. The relationship between per capita growth and numbers of both species is presented visually in Figure 14.13.

Now, armed with an actual equation for competition, we return to Figure 14.13 and label points of interest to arrive at the diagram shown in Figure 14.14.

Next, we need to add the second competing species to the picture. We do this by drawing its per capita growth rate as it is affected by species 1's density as well as its own density, as illustrated in Figure 14.15.

We abstract this three dimensional depiction to just the two dimensions of the N_1–N_2 phase plane by eliminating the quantitative information provided by the vertical growth rate axis. This leads us to the zero-isocline depiction shown in Figure 14.16, which is just the bottom plane in Figure 14.15 flipped on its side.

Figure 14.13
Lotka–Volterra competition equations.
Derivation of the equation for species
1's zero-isocline.

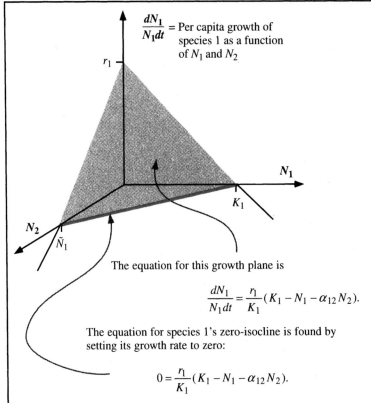

The equation for this growth plane is

$$\frac{dN_1}{N_1 dt} = \frac{r_1}{K_1}(K_1 - N_1 - \alpha_{12} N_2).$$

The equation for species 1's zero-isocline is found by
setting its growth rate to zero:

$$0 = \frac{r_1}{K_1}(K_1 - N_1 - \alpha_{12} N_2).$$

The right-hand side of the last equation could be zero if either r_1 is zero or if the term in
parentheses is zero. Since r_1 is not zero, the term in parentheses should be set to zero, or

$$K_1 - N_1 - \alpha_{12} N_2 = 0.$$

Since we have been plotting N_2 on the y axis, we rearrange the zero-isocline equation so that
N_2 appears alone on the left-hand side of the equation, so the equation for species 1's zero-
isocline is

$$N_2 = \frac{K_1 - N_1}{\alpha_{12}}.$$

Figure 14.14
Solving algebraically for the slope and
y intercept of species 1's zero-isocline.

Equation for species 1's zero - isocline: $N_2 = \dfrac{K_1 - N_1}{\alpha_{12}}.$

Rearrange it to get $N_2 = \dfrac{K_1}{\alpha_{12}} - \dfrac{1}{\alpha_{12}} N_1 .$

This is a straight line with $\underbrace{}_{\substack{y \\ \text{intercept}}}$ and $\underbrace{}_{\substack{ \\ \text{slope}}}$.

Hence we can draw the following diagram.

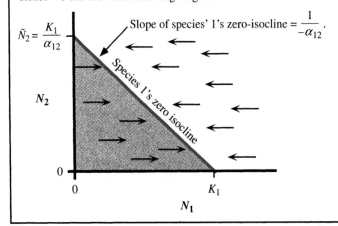

Figure 14.15
The decline in per capita growth rate for species 1 and 2, with increases in density of species 1 and 2. The axis lines meet at the origin (0, 0, 0).

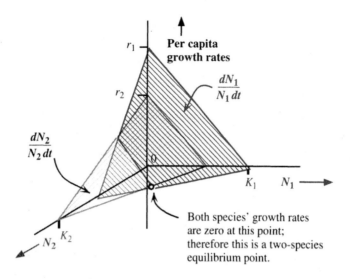

Both species' growth rates are zero at this point; therefore this is a two-species equilibrium point.

Figure 14.16
Zero-isoclines separate regions of positive and negative growth rates for each species. The black arrows in (b) indicate the qualitative direction of growth from a point in state space (i.e., initial values for N_1 and N_2) and are derived from the arrows for each species in (a). Refer to Appendix 1, Part 2 and Figure 5.8 for a review of the concept of stability and the meaning of a saddle point.

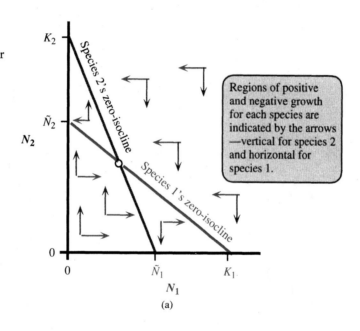

Regions of positive and negative growth for each species are indicated by the arrows —vertical for species 2 and horizontal for species 1.

(a)

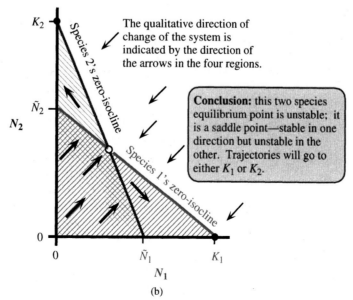

The qualitative direction of change of the system is indicated by the direction of the arrows in the four regions.

Conclusion: this two species equilibrium point is unstable; it is a saddle point—stable in one direction but unstable in the other. Trajectories will go to either K_1 or K_2.

(b)

Figure 14.17
Drawing the zero-isoclines with different slopes and *y* intercepts makes it possible to construct a phase–space diagram with zero-isoclines that produce a stable interior equilibrium point.

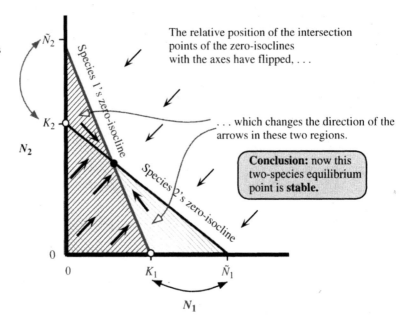

The relative position of the intersection points of the zero-isoclines with the axes have flipped, . . .

. . . which changes the direction of the arrows in these two regions.

Conclusion: now this two-species equilibrium point is **stable.**

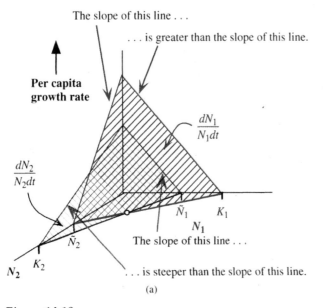

(a)

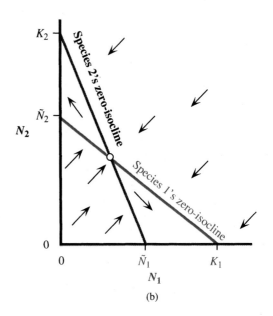

(b)

Figure 14.18
An explanation of the difference between stable and unstable competitive interactions in terms of per-capita effects—the unstable case. Intraspecific competition is *less* than interspecific competition. The growth rate declines shown in (a) produce the zero-isoclines shown in (b).

The particular way that we drew the growth rate response of the two species in Figure 14.15 was arbitrary. We could just as easily have drawn the two zero-isoclines somewhat differently as depicted in Figure 14.17.

In the **unstable** case (Figures 14.15 and 14.16), the effect of species 1's numbers on its own growth rate is *less* than the effect of species 2's numbers on species 1's growth rate. Similarly, the effect of species 2's numbers on its own growth rate is *less* than the effect of species 1's numbers on its growth rate. In other words, intraspecific competition is less than interspecific competition, as summarized in Figure 14.18.

In the **stable** case, the effect of species 1's numbers on species 1's own growth rate is *greater* than the effect of species 2's numbers on species 1's growth rate. Similarly, the effect of species 2's numbers on species 2's own growth rate is *greater* than the effect of species 1's numbers on species 2's growth rate. In other words, intraspecific competition is greater than interspecific competition, as summarized in Figure 14.19.

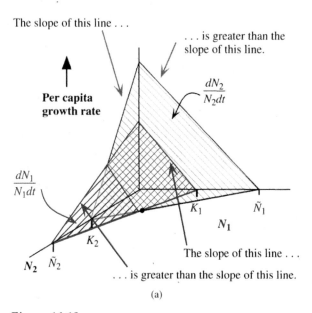

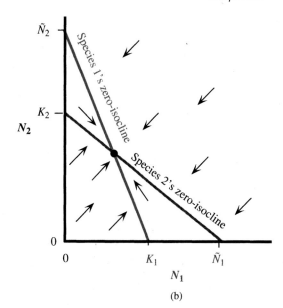

Figure 14.19
An explanation of the difference between stable and unstable competitive interactions in terms of per capita effects—the stable case. Intraspecific competition is *greater* than interspecific competition. The growth rate declines shown in (a) produce the zero-isoclines shown in (b).

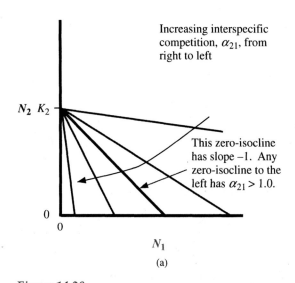

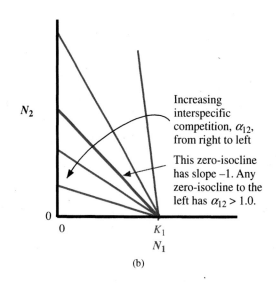

Figure 14.20
The intensity of interspecific competition affects the slope of the zero-isoclines. (a) Some possible zero-isoclines for species 2. (b) Some possible zero-isoclines for species 1.

In summary, the difference between the stable and unstable cases depends on the relative strengths of interspecific and intraspecific competition. Figure 14.20 illustrates how increasing interspecific competition affects the species' zero-isoclines.

The unstable case in Figure 14.18 corresponds to a situation where *both* α_{12} and α_{21} are greater than 1. We show later that this is not necessary for instability; instead it is necessary and sufficient that the product $\alpha_{12}\alpha_{21}$ be greater than 1. In addition to stability, we also need information on the feasibility of the equilibrium point: Are the species densities positive? Figure 14.21 explores the meaning of situations where the two zero-isoclines do not intersect in the positive orthant, leading to "negative" equilibrium species abundances.

Figure 14.21
If the zero-isoclines do not intersect in the positive orthant, then the mathematical "equilibrium point" will be negative for one of the species. In both (a) and (b), species 1 outcompetes species 2 because species 1's zero-isocline completely encloses species 2's. In (a), species 1 has a negative mathematical "equilibrium density," while in (b), species 2 has a negative "equilibrium density." Organism numbers, of course, can't be negative, so the real equilibrium density is the boundary solution where species 1 is at its carrying capacity K_1 and species 2 is at zero, for both situations. Thus algebraic negativity of N_i^* is not necessarily indicative of extinction.

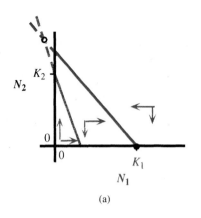

(a)

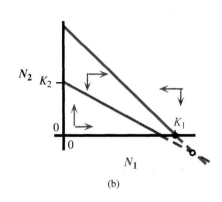

(b)

Figure 14.22
The algebraic conditions for stability and feasibility of an equilibrium point for two-species Lotka–Volterra competition.

Stability and Feasibility Conditions for an Equilbrium Point Are Separate Issues

Stability condition:

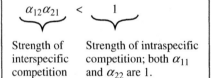

Strength of interspecific competition

Strength of intraspecific competition; both α_{11} and α_{22} are 1.

Feasibility condition:

Both N_1^* and $N_2^* > 0$.

The feasibility condition means that the two zero-isoclines must cross in the positive orthant of phase space.

Figure 14.22 provides a summary of the separate criteria of stability and feasibility. The combination of the possible stability and feasibility outcomes for two-species competition are cataloged in Figure 14.23.

Figure 14.24 shows sample trajectories for these four cases. Look at the trajectories with the open arrows in parts (b), (c), and (d). The loser in the competition can initially *increase* before it turns around and becomes extinct, as, for example, does species 2 in part (c) and species 1 in part (d).

BACK TO *TRIBOLIUM* AND *PARAMECIUM*

Figure 14.24(b) depicts the case of an unstable interior equilibrium point. Although the interior equilibrium point is unstable, two locally stable (but not globally stable) single-species equilibrium points exist at K_1 and K_2. This situation corresponds to the features of *Tribolium* competition. It is produced when interspecific competition is greater than intraspecific competition. Recall that this was the case with the main density regulating

Figure 14.23
Possible outcomes for two-species
Lotka–Volterra competition. In each
case, stable equilibrium points are
indicated by dark dots and unstable
equilibrium points by white dots.

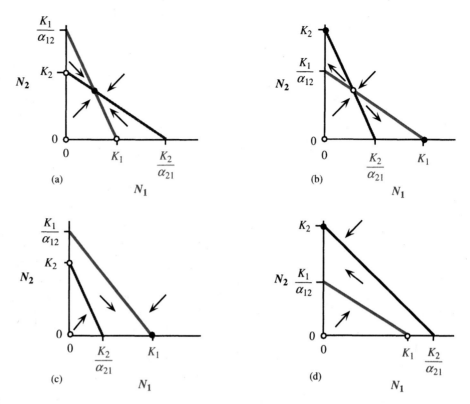

Figure 14.24
Sample trajectories for the four cases
of the two-species Lotka–Volterra
competition shown in Figure 14.23.
Arrowheads show the direction of
trajectories over time. Open
arrowheads mark trajectories in
which the losing species first
increases before later becoming
extinct ($r_1 = r_2 = 1$).

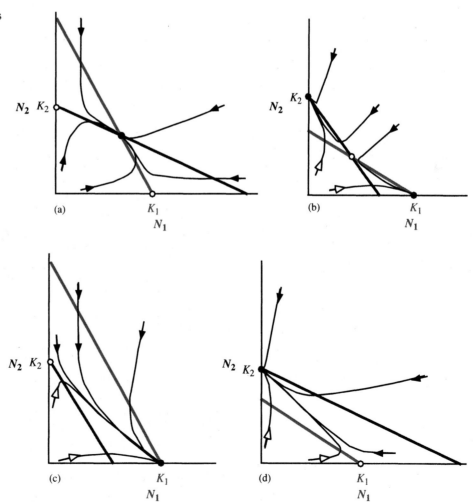

factor for these beetles—mutual cannibalism. This situation predicts a complete absence of coexistence; instead, it predicts alternative outcomes, depending on initial conditions—matching the empirical observations with *Tribolium.*

Figure 14.24(a) qualitatively matches the empirical pattern and mechanism of competition between either *P. caudatum* or *P. aurelia* and *P. bursaria* shown in Figures 14.3 and 14.5 or between *Tribolium* and *Oryzaephilus* once glass capillary tubing was added to the flour (Figure 14.8). Here, the separation of the species into different spatial regions means that intraspecific competition will be more intense than competition between these physically distant and functionally different competing species. Finally, without this separation, the situations depicted in Figures 14.24(c) or (d) parallel the competitive interactions observed between *P. caudatum* and *P. aurelia.* Indeed, the open arrow trajectories in Figure 14.24 show the competitively inferior competitor first increasing and then later decreasing to extinction; this pattern matches the abundance of *P. caudatum* when in competition with *P. aurelia* in Figure 14.3 and the trajectories for both *Tribolium* in Figure 14.7.

Changing environmental conditions (temperature, humidity, food levels, etc.) alter r and K values and thus can also influence the outcome of the competitive interaction between two species. However, a little exploring with the model should convince you that a feasible, unstable equilibrium point cannot be converted into a stable, feasible equilibrium with changes in r and K. Changing climatic conditions never allowed the two *Tribolium* to coexist. By changing α_{ij} terms by the addition of glass capillary tubes, however, the two species can be made to coexist, again matching the possibilities raised by the model. In conclusion, the model seems to provide a reasonable fit to the qualitative dynamic behaviors that are predicted for different sets of environmental conditions.

Exercise: In one set of experiments with *Paramecium aurelia* (Pa) and *P. caudatum* (Pb), Gause (1935) determined that

$$K_{Pa} = 265.2 \qquad K_{Pc} = 64$$

$$r_{Pa} = 1.244 \qquad r_{Pc} = 0.794$$

$$\alpha_{Pa,Pc} = 4.141 \qquad \alpha_{Pc,Pa} = 0.242$$

Paramecium aurelia always clearly won in competition in all replicates, but the calculation of the two zero-isoclines indicated that they were superimposed, within experimental error. (Verify this fact). Gause concluded that his method of counting the *Paramecium* every day may be contributing to the competitive advantage of *P. aurelia.* Gause had been removing one-tenth of the volume from each tube daily to record the progress of the populations. He therefore revised the competition equations to reflect the density independent mortality he had been imposing by counting the populations:

$$\frac{dN_i}{dt} = \frac{r_i N_i}{K_i}(K_i - N_i - \alpha_{ij}N_j) - dN_i.$$

For $d = 0.1$, draw the zero-isoclines for *P. caudatum* and *P. aurelia* based on this equation. Label the points of intersection with the axes. In terms of this equation, what is it about *P. aurelia* that gives it a competitive advantage over *P. caudatum* at higher levels of d?

The Separatrix

We introduced the separatrix in Chapter 12 (Figure 12.38). In situation (b) of Figure 14.23, the r's of the two species can influence the relative shape of the basin of attractions for the alternative equilibria K_1 and K_2, as shown in Figure 14.25. If the r's

Figure 14.25
The separatrix for an unstable interior equilibrium point. Species 1 or 2 alone at their respective carrying capacities, K_1 and K_2, are the two locally stable equilibrium points. The value of the growth rates, r, influences the shape and position of the separatrix, which divides the two domains of attraction. Trajectories above the separatrix go to K_2 and trajectories below it go to K_1. The separatrix goes through the unstable feasible equilibrium point. In this example, $\alpha_{12} = \alpha_{21} = 1.5$ and the K's are also equal. Only the r's of the two species differ; $r_1 > r_2$, and the separatrix bows toward the N_1 axis. If r_2 were greater than r_1, then the separatrix would bow toward the N_2 axis.

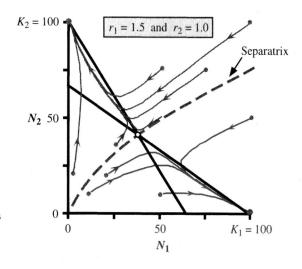

Figure 14.26
(a) Note that in (b), the stable interior equilibrium point falls above the dashed line connecting the carrying capacities, while the unstable interior equilibrium point in (a) falls below the dashed line. For Lotka–Volterra competition, species that stably coexist produce a combined density at equilibrium that is greater than the density of either species when alone. Species that cannot stably coexist have a combined density at equilibrium less than either species when alone.

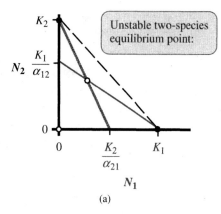

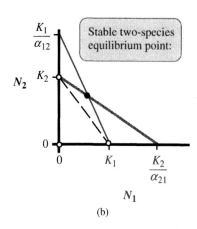

of the two species are equal, the separatrix will be a straight line through the two-species equilibrium point. If $r_2 > r_1$, the separatrix bows in the opposite direction from that shown.

NON-LOTKA–VOLTERRA COMPETITION

The condition for the stability of a feasible equilibrium point can also be assessed by a simple graphical rule: Draw a straight line connecting the two species' K's, and determine whether the equilibrium point lies above or below this line, as illustrated in Figure 14.26. This rule-of-thumb is fragile, however, to the assumptions of Lotka–Volterra competition. It breaks down if the competitive effects are nonlinear.

Gilpin and Justice (1972) found that *Drosophila* grown under laboratory conditions did not show the linear decline in growth rate with density, either of conspecifics or of heterospecifics expected from the logistic equation and the Lotka–Volterra equations. The intraspecific effects were not linear but concave for both *D. willistoni* and *D. pseudoobscura,* as shown in Figure 14.27(a). After fitting recruitment to the experimental data, the investigators found that the zero-isoclines for this two-species interaction bowed inward, as shown in Figure 14.27(b). With this modification, even though the two-species equilibrium point is stable (see the direction of the arrows), the interior equilibrium point lies below the line that would connect the two carrying capacities.

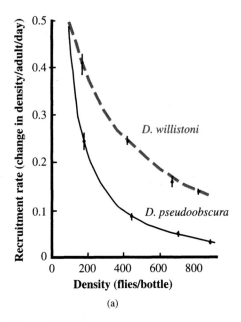

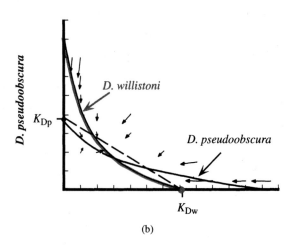

Figure 14.27
Two-species competition in laboratory *Drosophila*. After Gilpin and Justice (1972). (a) Both species display nonlinear per capita recruitment as a function of intraspecific density. (b) The zero-isoclines are concave.

Exercise: Use the θ**logistic** model of Problem 3 at the end of Chapter 5 to modify the Lotka–Volterra competition equations to produce nonlinear zero-isoclines like those observed for laboratory populations of *Drosophila*. For what values of θ will the zero-isoclines bow inward? For what values of θ will the zero-isoclines bow outward?

Exercise: Vandermeer's (1969) experiments with competition between species of *Paramecium* allowed the calculation of r and K for each species, as well as the α_{ij} terms for their interactions. The Lotka–Volterra equations proved to be a good fit to the data. The results for one species pair were

$$P.\ aurelia:\ r_a = 1.05,\ K_a = 671;$$

$$P.\ bursaria:\ r_b = 0.47,\ K_b = 230;$$

$$\alpha_{ab} = -2.0,\ \alpha_{ba} = 0.5.$$

(Yes, the competition coefficient for the effect of *P. bursaria* on *P. aurelia* is negative, meaning a positive interaction).

Sketch the zero-isoclines for this system (put *P. aurelia* numbers on the x axis). What do you predict regarding the coexistence of this species pair? What would happen if the carrying capacity of *P. bursaria* were doubled?

ANALYTICAL EXPOSITION OF LOTKA-VOLTERRA COMPETITION (ADVANCED)

We now derive the rule of Figure 14.21: Stable competition demands that $\alpha_{12}\alpha_{21} < 1.0$, which we developed earlier using a graphical argument. We also extend competition theory to more than two species. First we write the two-species Lotka–Volterra competition in algebraic form:

$$\frac{dN_1}{dt} = f_1(N_1, N_2) = \frac{r_1 N_1}{K_1}[K_1 - N_1 - \alpha_{12}N_2] \tag{14.2a}$$

and

$$\frac{dN_2}{dt} = f_2(N_1, N_2) = \frac{r_2 N_2}{K_2}[K_2 - N_2 - \alpha_{21} N_1]. \tag{14.2b}$$

At an equilibrium point, the population growth rates for both species must be zero, by definition. Equations (14.2) indicate that this can happen in two ways: either the abundance of the species is zero ($N_i = 0$) so that the term in front of the bracket is zero, or $N_i \neq 0$, but the term in the brackets—and hence the *per capita* growth rate—is zero. The latter condition for both species specifies the solution of the **interior equilibrium point:**

$$K_1 - N_1{}^* - \alpha_{12} N_2{}^* = 0 \tag{14.3a}$$

and

$$K_2 - N_2{}^* - \alpha_{21} N_1{}^* = 0 \tag{14.3b}$$

The asterisk on N_i indicates that this is a nonzero equilibrium value of N. Equations (14.3) are simply two linear equations in two unknowns, $N_1{}^*$ and $N_2{}^*$, and so are easy to solve. However, because we later want to expand this analysis to include more than two interacting species, it is useful now to write Eq. (14.3) in more general matrix form. (As usual, matrices and vectors are set in boldface type.)

$$\mathbf{K} = \boldsymbol{\alpha}\mathbf{N^*}, \tag{14.4}$$

where

$$\mathbf{K} = \begin{bmatrix} K_1 \\ K_2 \end{bmatrix}, \mathbf{N^*} = \begin{bmatrix} N_1{}^* \\ N_2{}^* \end{bmatrix}, \quad \text{and} \quad \boldsymbol{\alpha} = \begin{bmatrix} 1 & \alpha_{12} \\ \alpha_{21} & 1 \end{bmatrix}.$$

The alpha matrix gives the per capita interaction strengths. The intraspecific effects, which are all 1, are on the diagonal, while the interspecific effects are in the off-diagonal positions. Also, recall that the notation rule is **to subscript rows first and then to subscript columns.** Thus α_{12} means the element in the first row and second column of the $\boldsymbol{\alpha}$ matrix, and this represents the effect of species 2 *on* species 1.

To solve Eq. (14.4), we'd like to find the values of the unknowns, $N_i{}^*$, that fulfill the identity of Eq. (14.4), given the parameters in the $\boldsymbol{\alpha}$ matrix and the **K** vector. Two techniques for solving matrix equations like these are presented in Appendices 2 and 3; one method involves taking the inverse of the matrix such that

$$\mathbf{N^*} = \boldsymbol{\alpha}^{-1}\mathbf{K}. \tag{14.5}$$

The other method (and this really amounts to the same thing) is to apply **Cramer's rule** (see Appendix 3). To solve for the equilibrium level of $N_1{}^*$, we take the ratio of the determinants of two different matrices. In the numerator, we form the matrix $\boldsymbol{\alpha}(1)$, which is simply the matrix $\boldsymbol{\alpha}$ with the first column replaced by the **K** vector, and, in the denominator, we have the matrix of $\boldsymbol{\alpha}$:

$$N_1{}^* = \frac{\det \begin{vmatrix} K_1 & \alpha_{12} \\ K_2 & 1 \end{vmatrix}}{\det \begin{vmatrix} 1 & \alpha_{12} \\ \alpha_{21} & 1 \end{vmatrix}} = \frac{K_1 - K_2\alpha_{12}}{1 - \alpha_{21}\alpha_{12}}. \tag{14.6a}$$

The equilibrium density of species 2, $N_2{}^* = \det |\boldsymbol{\alpha}(2)| / \det |\boldsymbol{\alpha}|$, gives

$$N_2{}^* = \frac{\det \begin{vmatrix} 1 & K_1 \\ \alpha_{21} & K_2 \end{vmatrix}}{\det \begin{vmatrix} 1 & \alpha_{12} \\ \alpha_{21} & 1 \end{vmatrix}} = \frac{K_2 - K_1\alpha_{21}}{1 - \alpha_{21}\alpha_{12}}. \tag{14.6b}$$

The denominator of Eqs. (14.6a and b) is the determinant of $\boldsymbol{\alpha}$. Unless the determinant of $\boldsymbol{\alpha}$ is > 0 (i.e., $\alpha_{12}\alpha_{21} < 1$), the interior equilibrium will be unstable. To show

how this falls out analytically, we perform a local stability analysis like that developed in Chapter 13. You should review Chapter 13 to refresh your memory about Jacobian matrices and how they relate to the stability of an equilibrium point. For two-species Lotka–Volterra competition, the dynamics of small deviations (n_1 and n_2) away from equilibrium points N_1^* and N_2^* can be described by the linear matrix differential equation,

$$\begin{bmatrix} \dfrac{dn_1}{dt} \\ \dfrac{dn_2}{dt} \end{bmatrix} = \begin{bmatrix} \dfrac{\partial f_1}{\partial N_1} & \dfrac{\partial f_1}{\partial N_2} \\ \dfrac{\partial f_2}{\partial N_1} & \dfrac{\partial f_2}{\partial N_2} \end{bmatrix} \begin{bmatrix} n_1 \\ n_2 \end{bmatrix},$$

which can also be written as

$$\frac{d\mathbf{n}}{dt} = \mathbf{J}\,\mathbf{n},$$

where $\mathbf{J}$ is the Jacobian matrix and all the partial derivatives in $\mathbf{J}$ are evaluated at the equilibrium point. The local stability of the equilibrium point can be determined from the eigenvalues of the Jacobian matrix. We now calculate the Jacobian matrix for the interior equilibrium point for two-species Lotka–Volterra competition. Let's begin with the term in the upper left-hand corner:

$$f_1\left(N_1, N_2\right) = r_1 N_1 - \frac{r_1}{K_1} N_1^2 - \frac{r_1}{K_1} \alpha_{12} N_1 N_2.$$

Taking the partial derivative of f_1 with respect to N_1 (keeping N_2 fixed) gives

$$\frac{\partial f_1}{\partial N_1} = r_1 - \frac{2 r_1 N_1}{K_1} - \frac{r_1}{K_1} \alpha_{12} N_2. \tag{14.7}$$

We want to evaluate this partial derivative at the equilibrium point $\mathbf{N^*}$, so

$$\frac{\partial f_1}{\partial N_1} = \frac{r_1}{K_1}(K_1 - 2N_1^* - \alpha_{12} N_2^*). \tag{14.8}$$

Note that the term in parentheses is identical to Eq. (14.3a), except for the 2 in front of N_1^*. Thus we may also write Eq. (14.8) as

$$\frac{\partial f_1}{\partial N_1} = \frac{r_1}{K_1}(K_1 - N_1^* - \alpha_{12} N_2^* - N_1^*),$$

and, since $K_1 - N_1^* - \alpha_{12} N_2^* = 0$ at the interior equilibrium, we reach

$$\frac{\partial f_1}{\partial N_1} = \frac{-r_1 N_1^*}{K_1}. \tag{14.9}$$

The text term in the first row of $\mathbf{J}$ is

$$\frac{\partial f_1}{\partial N_2} = -\frac{r_1}{K_1} \alpha_{12} N_1^*. \tag{14.10}$$

The second row of $\mathbf{J}$ will have the same form as the first row so that, finally, we arrive at

$$\mathbf{J}(\text{interior}) = \begin{bmatrix} -\dfrac{r_1 N_1^*}{K_1} & -\dfrac{r_1}{K_1} \alpha_{12} N_1^* \\ -\dfrac{r_2}{K_2} \alpha_{21} N_2^* & -\dfrac{r_2 N_2^*}{K_2} \end{bmatrix}.$$

The next step is to realize that, for simplicity, $\mathbf{J}(\text{interior})$ may also be written as the product of two matrices:

$$\mathbf{J}(\text{interior}) = \begin{bmatrix} -\dfrac{r_1 N_1^*}{K_1} & 0 \\ 0 & -\dfrac{r_2 N_2^*}{K_2} \end{bmatrix} \begin{bmatrix} 1 & \alpha_{12} \\ \alpha_{21} & 1 \end{bmatrix}. \tag{14.11}$$

The first matrix in Eq. (14.11) is a diagonal matrix, and each element is a negative constant if the equilibrium point is feasible, or $N_i^* > 0$. The second matrix is simply the original α matrix. One useful aspect of Eq. (14.11) is that it simplifies the determination of the eigenvalues of $\mathbf{J}$. It can be shown that for any two $\times$ two matrix like $\mathbf{J}$(interior)—which is the product of an all-negative diagonal matrix times another matrix α—all the eigenvalues of $\mathbf{J}$ will have negative real parts if, and only if, the eigenvalues of α are all positive. Thus, if the eigenvalues of α are all positive, the eigenvalues of $\mathbf{J}$ are all negative, and the interior equilibrium point is locally stable. Let's now evaluate the eigenvalues of α. We set up the characteristic equation:

$$\det \begin{vmatrix} 1-\lambda & \alpha_{12} \\ \alpha_{21} & 1-\lambda \end{vmatrix} = 0$$

or

$$\lambda^2 - 2\lambda + (1 - \alpha_{21}\alpha_{12}) = 0. \tag{14.12}$$

At this point, we could go ahead and solve for the two λ's, but simply by the sign structure of Eq. (14.12) we can see that it can be factored to have the form

$$(\lambda - \text{something})(\lambda - \text{something else}) = 0.$$

and therefore the two eigenvalues are positive if, and only if, the determinant of α is positive.

Applying the quadratic formula (see Appendix 4),

$$x = \frac{-b \pm \sqrt{b^2 - 4ac}}{2a},$$

where $a = 1$, $b = -2$, and $c = 1 - \alpha_{12}\alpha_{21}$, we reach

$$\lambda_i = \frac{2 \pm \sqrt{4 - 4 + 4\alpha_{21}\alpha_{12}}}{2}$$

$$= 1 \pm \sqrt{\alpha_{21}\alpha_{12}},$$

which can also be written as

$$\lambda_i = \sqrt{1} \pm \sqrt{\alpha_{21}\alpha_{12}}.$$

From this expression we may again infer that both λ's of α will be positive if, and only if, $\sqrt{1} - \sqrt{\alpha_{21}\alpha_{12}} > 0$. This, of course, means that $1 - \alpha_{21}\alpha_{12}$ must be positive. Thus stability demands that **intraspecific competition ($\alpha_{11}\alpha_{22} = 1$) be greater than interspecific competition ($\alpha_{12}\alpha_{21}$).**

The Stability of a Boundary Equilibrium Point

We next evaluate the stability of one of the boundary equilibria. As an example, consider the boundary point where species 1 is at its carrying capacity, K_1, and species 2 is at zero. This is an equilibrium point, but is it stable? We have already evaluated the form of the partial derivatives for $\mathbf{J}$, but now we need to substitute the boundary equilibrium point $(K_1, 0)$ instead of the interior equilibrium point (N_1^*, N_2^*). We rewrite Eq. (14.7) with this substitution to get

$$\frac{\partial f_1}{\partial N_1} = \frac{r_1}{K_1}(K_1 - 2K_1 - 0) = -r_1,$$

and, from Eq. (14.10),

$$\frac{\partial f_1}{\partial N_2} = -\frac{r_1}{K_1}\alpha_{12}N_1.$$

This last partial derivative evaluated at $N_1 = K_1$ is simply $-r_1\alpha_{12}$.

Continuing with the second row of **J**, we have

$$\frac{\partial f_2}{\partial N_1} = \frac{-r_2\alpha_{21}N_2}{K_1} = 0 \quad (\text{since } N_2 = 0),$$

and, finally,

$$\frac{\partial f_2}{\partial N_2} = \frac{r_2}{K_2}(K_2 - 2N_2 - \alpha_{21}N_1)$$

$$= \frac{r_2}{K_2}(K_2 - \alpha_{21}K_1).$$

Putting this all together, we get for the Jacobian matrix **J**,

$$\mathbf{J}(N_1 = K_1, N_2 = 0) = \begin{bmatrix} -r_1 & -r_1\alpha_{12} \\ 0 & \dfrac{r_2(K_2 - \alpha_{12}K_1)}{K_2} \end{bmatrix}. \tag{14.13}$$

The eigenvalues of **J**(boundary) are easy to find since **J** is a *triangular* matrix; that is, all the elements below the diagonal are zero. The eigenvalues of a triangular matrix are simply given by the elements on the diagonal

$$\lambda_1 = -r_1$$

and

$$\lambda_2 = \frac{r_2(K_2 - \alpha_{12}K_1)}{K_2}. \tag{14.14}$$

Compare this second eigenvalue to the Lotka–Volterra per capita growth rate equation for the missing species 2:

$$\frac{dN_2}{N_2 dt} = \frac{r_2}{K_2}(K_2 - N_2 - \alpha_{21}N_1).$$

If N_2 is negligible, as it would be if species 2 were invading at low density, and if N_1 were at its carrying capacity, then the growth rate of species 2, when rare, is identical to this eigenvalue. Since r is always positive, the sign of 2 from Eq. (14.14) is given simply by the term in parentheses, which we call g_2 for convenience:

$$g_2 = K_2 - \alpha_{12}K_1. \tag{14.15}$$

Hence this boundary equilibrium is stable (the eigenvalue λ_2 has a negative real part) if species 2 cannot invade species 1 at its carrying capacity. This will be true as long as $g_2 < 0$, which is pretty intuitive. After all, how could the boundary equilibrium lacking species 2 be stable if species 2 can invade species 1 and increase.

Finally, just for notational completeness, the Jacobian matrix $\mathbf{J}(N_1 = K_1, N_2 = 0)$ can be written as the product of a diagonal matrix and the original α matrix, as we did for the Jacobian matrix of the interior equilibrium point, Eq. (14.11), but now we must add a third matrix:

$$\mathbf{J}(N_1 = K_1, N_2 = 0) = \begin{bmatrix} \dfrac{-r_1N_1}{K_1} & 0 \\ 0 & \dfrac{-r_2N_2}{K_2} \end{bmatrix} \begin{bmatrix} 1 & \alpha_{12} \\ \alpha_{21} & 1 \end{bmatrix} + \begin{bmatrix} 0 & 0 \\ 0 & \dfrac{r_2g_2}{K_2} \end{bmatrix}. \tag{14.16}$$

Comparing Eq. (14.16) for this boundary equilibrium to Eq. (14.11) for the interior equilibrium point, we see that the Jacobian matrices are formed identically from the same three matrices (with suitable interpretation for the N_i values) in the first matrix. However, the last matrix—when the interior equilibrium point is zero—drops out of consideration. As we show in the next section, this feature generalizes to the structure of the Jacobian matrix for any boundary equilibrium involving subsets of arbitrary size from n competing species.

> **Problem:** Verify that the matrix multiplication specified by Eq. (14.16) with $N_1 = K_1$ and $N_2 = 0$, does yield **J**, as given in Eq. (14.13).

MORE THAN TWO COMPETING SPECIES

We can generalize the concept of zero-isoclines to more or less than two interacting species. Figure 14.28 makes the comparison, and Figure 14.29 summarizes this development.

Figure 14.28
Isopoints and isosurfaces.

1. When it comes to defining isoclines, there is nothing special about two-species systems. We have already discussed the concept in the logistic equations for single species population growth.

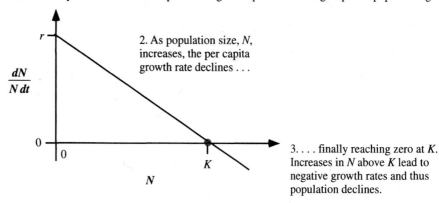

2. As population size, N, increases, the per capita growth rate declines . . .

3. . . . finally reaching zero at K. Increases in N above K lead to negative growth rates and thus population declines.

4. Removing the y dimension and just looking at numbers of N, . . .

5. . . . now the state space is one-dimensional, and we have a **zero-isopoint** at K.

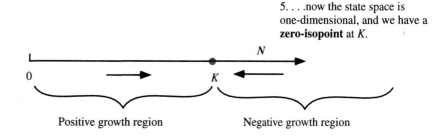

Positive growth region   Negative growth region

6. Similarly for three species, we could define a **zero-isosurface** as in . . .

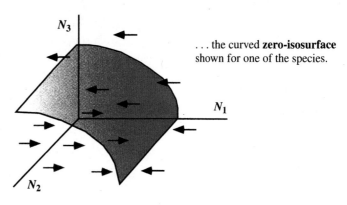

. . . the curved **zero-isosurface** shown for one of the species.

Figure 14.29
The isospace is one dimension less than the state space represented by the number of species, *n*, because it is an intersection of two *n*-space shapes.

Number of species, *n*	Isospace
One	Zero: a point
Two	One: a line or curve
Three	Two: a plane or surface
.	.
.	.
X	*X* – 1: a **hypersurface**

Figure 14.30
A zero-isoplane for species 3 in a three-species Lotka–Volterra competition system. Above the plane species 3 has a negative growth rate, and below the plane (i.e., towards the origin) it has a positive growth rate.

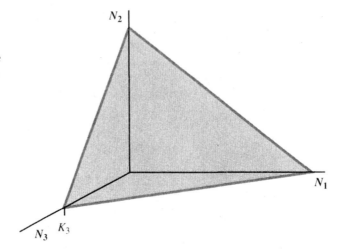

For three species of Lotka–Volterra competitors the state space will be a cube and the zero-isosurface of each species is a plane, as depicted in Figure 14.30.

By superimposing isoplanes, one for each species, we can examine graphically the nature of the various equilibria, as illustrated in Figure 14.31. However, it is difficult to draw and to interpret these cluttered diagrams, so an analytical approach is preferred and, of course, is absolutely necessary for more than three species. The needed mathematics is a straightforward extension of that already examined for two-species competition; we develop it a bit later. First, however, let's examine a more fundamental question.

Imagine a community of three fish species, A, B, and C, existing together in a large aquarium. By raising populations of each of these species alone, we could find their respective intrinsic growth rates, *r*, and carrying capacities, *K*. However, this would not tell us anything about their relative competitive abilities, the α_{ij} terms. To determine these, we could put the species together in each possible pairwise combination of species: AB, BC, and AC. From the resulting dynamics and previous knowledge of *r* and *K* from single-species populations, we would then be able to calculate α_{ij} for each pairwise combination. We could now put these α_{ij} terms into a 3 × 3 matrix. An open question is whether this **α** matrix, along with the **K** and **r** vectors, enables us to predict the behavior and outcome of the full three-species system. The answer depends on whether the interaction terms, α_{ij}, which we estimated in pairwise contests, fully express all the interactions between the species in the full three-species community. This would not be the case if the pairwise α_{ij} terms were altered by the presence or absence of additional species. If some species combinations formed coalitions against other species, the total competitive effect could be greater than or less than the sum of

Figure 14.31

Two examples of three-species Lotka–Volterra competition systems. The pairwise equilibrium points are indicated by dots—solid for stable and open for unstable. The intersections of the zero-isoplanes are shown as dashed lines. (a) All pairwise species sets are locally stable, and all three species coexist at a stable equilibrium point—where the dashed lines for the intersection of each pair of isoplanes cross. It is stable because it falls above the plane connecting the species K's (not shown). (b) The three-species equilibrium point falls outside the positive orthant, so it is unfeasible. There are two locally stable equilibrium points: species 3 alone at its carrying capacity and species 1 and 2 together.

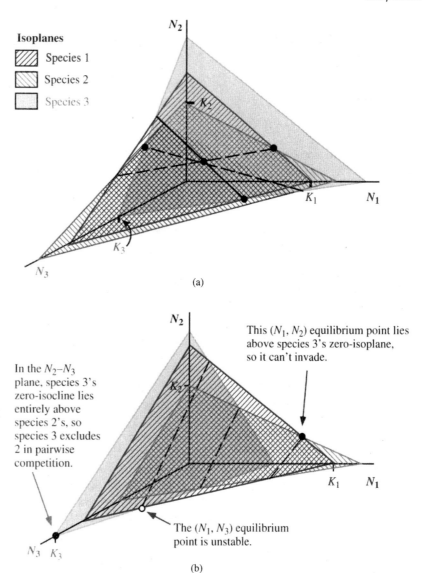

Isoplanes

Species 1
Species 2
Species 3

This (N_1, N_2) equilibrium point lies above species 3's zero-isoplane, so it can't invade.

In the N_2–N_3 plane, species 3's zero-isocline lies entirely above species 2's, so species 3 excludes 2 in pairwise competition.

The (N_1, N_3) equilibrium point is unstable.

(b)

each species' interaction when competition occurs only in pairs. In ecology, such "non-additive" behavior is called **higher-order interactions** or **interaction modifications** (Billick and Case 1994, Wootton 1994). The presence of species C might increase or moderate the competitive impact of A on B (also see Chapter 6). One particularly interesting type of interaction modification is where one species' mere presence "intimidates" others, such that they alter their behavior and thus the strength of their interaction (Beckerman et al. 1997). **Trait mediated indirect effects** are these more rapid changes in the interactions between species. They are those *not* mediated by numerical changes in population size of intervening species in the food chain but simply by changes in the behavioral reaction of a species due to another species (Abrams 1993). For example, Peacor and Werner (1997) found that the presence of predacious dragonfly larvae, even if caged and restricted from eating prey, caused small tadpoles to reduce their foraging efforts and this, in turn, led to a reduction in the predation rate of these tadpoles by another predator. Similarly, Beckerman et al. (1997) glued shut the jaws of spiders so that they could not kill or consume grasshopper prey. Nevertheless, the mere presence of such spiders reduced grasshopper foraging activity, leading to decreased rates of exploitation of grass by the grasshoppers.

Vandermeer (1969) searched for higher-order interactions in competition among four species of ciliates (three species of *Paramecium* and *Blephanaria*). He found that the extrapolation of the r, K, and α_{ij} terms from the one- and two-species experiments

Figure 14.32

A simulation of Lotka–Volterra competition for four interacting protozoan species, using the parameter values calculated from the one- and two-species competition experiments. The actual range of densities for each species at the end of the four-species competition experiments is also shown for day 32. After Vandermeer (1969).

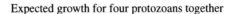

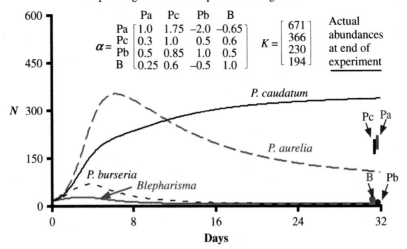

fit the dynamics of the four-species community qualitatively. The four-species community experimentally collapsed to just two species: *P. aurelia* coexisting with *P. caudatum,* as predicted. However, as shown in Figure 14.32, the abundances of the surviving species were off by about 40%.

Additional simulations from an assortment of different initial densities reveal that this boundary equilibrium is globally stable. In summary, based on an assumption of no higher-order interactions, this four-species system is predicted to collapse to a particular two-species boundary solution (Pc with Pa), which is globally stable. This behavior is, in fact, what is observed experimentally with these four species. In surely one of the most exhaustive searches yet for higher-order interactions, Fryar (1998) compared the growth rates of seven species of wood-rotting fungi, with the convenient property that they can be grown readily in the laboratory in petri dishes. Fryar conducted all possible single-species, pairwise, and three-species sets and compared their growth rates. In 18 out of a possible 35 three-species combinations, she found a nonadditive interaction, based on pairwise performance. In 13 of the 18 nonadditive interactions the direction of the higher-order interaction was to increase coexistence, that is, to allow the three species to coexist when extrapolations of two-species interactions predicted just the opposite.

Boundary Equilibria

Is there an analytical way to determine the stability of the various boundary equilibria? If we assume that higher-order interactions are absent, a first step in attempting to do so is to enumerate the locations of all the various equilibrium points and next determine which, if any, are stable. Let's consider a system with three species, like those in Figure 14.31, and the following equilibria.

Number of species	Number of equilibrium points
Zero species	1 (the (0, 0, 0) point)
Single species	3 (each *K*)
Two-species	3 (each pairwise combination of the three species)
Three-species	1 (the interior equilibrium point)

To obtain the local stability of each of these points by calculating eigenvalues for each of the relevant Jacobian matrices **J** is straightforward but tedious and not intuitive. We can, however, make use of some theorems dealing with the stability properties of matrices.

One appealing aspect of Eqs. (14.5) and (14.11) is that they generalize to *n* species. The interior equilibrium point is given by

$$\mathbf{N}^* = \boldsymbol{\alpha}^{-1}\,\mathbf{K},$$

and the stability matrix for this interior equilibrium point is

$$\mathbf{J}(\text{interior}) = \mathbf{D}\,\boldsymbol{\alpha}, \tag{14.17}$$

where $\mathbf{D}$ is an $n \times n$ matrix of diagonal elements $-r_i N_i^*/K_i$, and $\boldsymbol{\alpha}$ is the $n \times n$ interaction matrix with 1's along the diagonal (the α_{ii} terms) and α_{ij} elsewhere. By evaluating the sign of the real parts of the eigenvalues of $\mathbf{J}(\text{interior})$, we will know whether this equilibrium point is stable. Recall that for two species we could short-circuit this by simply evaluating the eigenvalues of $\boldsymbol{\alpha}$. When n is greater than 2, for reasons that are a little too complicated to explain here, we lose this ability. Instead, the eigenvalues of the Jacobian matrix $\mathbf{J}(\text{interior})$ must be determined. The sign of the eigenvalues of $\boldsymbol{\alpha}$ are not necessarily indicative of the sign of the eigenvalues of $\mathbf{J}$ as they were for two species; even changes in r can alter the stability of an equilibrium point in three space (Strobeck 1973). We present an example of this in the last section of this chapter. However, for one class of a matrix there is the same correspondence as for the two-species case; namely, symmetric $\boldsymbol{\alpha}$ matrices. Moreover, even if $\boldsymbol{\alpha}$ itself is not symmetric, but is formed as the product of a symmetric matrix and some diagonal matrix,

$$\boldsymbol{\alpha} = \mathbf{D}\,\mathbf{S}, \tag{14.18}$$

where $\mathbf{S}$ is a symmetric matrix and $\mathbf{D}$ is a diagonal matrix, then the sign of the eigenvalues of $\boldsymbol{\alpha}$ will again be fully indicative of the sign of the eigenvalues of $\mathbf{J}$. (The proof of this is far beyond the scope of this book but may be found in Case and Casten 1978). You may think that this is a pretty worthless shortcut: Why should nature contrive species interactions that happen to follow the rather strict formula in Eq. (14.18)? In Chapter 15, on niche theory, we show that, under some broad assumptions, this could be the case if competitive interactions are based on niche overlap.

The use of the Jacobian matrix to evaluate various boundary equilibria can be extended in a simple way for any number of competing species because the Jacobian of a boundary solution has a special simple form for Lotka–Volterra competition. To illustrate, let's imagine a four-species system, with species 1 and 2 at equilibrium with each other but in isolation from species 3 and 4. We denote the equilibrium abundance of species 1 and 2 in the absence of species 3 and 4 as $N_1^*(3, 4)$ and $N_2^*(3, 4)$. Importantly, these values are not the same as N_1^* and N_2^*, the equilibrium abundances of species 1 and 2 in the full four-species community. The equilibrium values for the two-species system (1, 2) are given by Eq. (14.6a). The structure of the Jacobian matrix for the boundary equilibrium with species 3 and 4 absent is

$$\mathbf{J}(3, 4) = \mathbf{R}\,\boldsymbol{\alpha} + \mathbf{G}.$$

where $\mathbf{R}$ is a 4×4 diagonal matrix with terms $-r_i N_i/K_i$ (as in Eq. 14.16), $\boldsymbol{\alpha}$ is the original $\boldsymbol{\alpha}$ matrix, and $\mathbf{G}$ contains zeros everywhere except for the growth-when-rare terms on the diagonal for the missing species ($i = 3$ and 4):

$$\mathbf{G} = \begin{bmatrix} 0 & 0 & 0 & 0 \\ 0 & 0 & 0 & 0 \\ 0 & 0 & \dfrac{r_3 g_3}{K_3} & 0 \\ 0 & 0 & 0 & \dfrac{r_4 g_4}{K_4} \end{bmatrix}. \tag{14.19}$$

By a simple extension of Eq. (14.15), we get the condition for positive growth of the invader, i, g_i, as

$$g_i = \left[K_i - \sum_{j=1}^{2} \alpha_{ij} N_j^*(3, 4) \right] \quad \text{for } i = 3 \text{ to } 4.$$

Since $N_i^* = 0$ at the boundary for each missing species $i = (3, 4)$, the matrix $\mathbf{R}\,\alpha$ becomes

$$\mathbf{R}\alpha = \begin{bmatrix} \dfrac{-N_1^*(3,4)r_1}{K_1} & -\alpha_{12}\dfrac{N_1^*(3,4)r_1}{K_1} & -\alpha_{13}\dfrac{N_1^*(3,4)r_1}{K_1} & -\alpha_{14}\dfrac{N_1^*(3,4)r_1}{K_1} \\[2mm] -\alpha_{21}\dfrac{N_2^*(3,4)r_2}{K_2} & \dfrac{-N_2^*(3,4)r_2}{K_2} & -\alpha_{23}\dfrac{N_2^*(3,4)r_2}{K_2} & -\alpha_{24}\dfrac{N_2^*(3,4)r_2}{K_2} \\[2mm] 0 & 0 & 0 & 0 \\[2mm] 0 & 0 & 0 & 0 \end{bmatrix},$$

and the full Jacobian matrix $\mathbf{J}(3, 4)$ is

$$\mathbf{J}(3,4) = \mathbf{R}\alpha + \mathbf{G} = \begin{bmatrix} \dfrac{-N_1^*(3,4)r_1}{K_1} & -\alpha_{12}\dfrac{N_1^*(3,4)r_1}{K_1} & -\alpha_{13}\dfrac{N_1^*(3,4)r_1}{K_1} & -\alpha_{14}\dfrac{N_1^*(3,4)r_1}{K_1} \\[2mm] -\alpha_{21}\dfrac{N_2^*(3,4)r_2}{K_2} & \dfrac{-N_2^*(3,4)r_2}{K_2} & -\alpha_{23}\dfrac{N_2^*(3,4)r_2}{K_2} & -\alpha_{24}\dfrac{N_1^*(3,4)r_2}{K_2} \\[2mm] 0 & 0 & \dfrac{r_3 g_3}{K_3} & 0 \\[2mm] 0 & 0 & 0 & \dfrac{r_4 g_4}{K_4} \end{bmatrix}.$$

The boundary equilibrium is locally stable if (a) the two-species equilibrium point for species 1 and 2 together is stable in the N_1–N_2 orthant and (b) if species 3 and 4 cannot invade this boundary equilibrium (g_3 and g_4 are then negative). This will ensure that all the eigenvalues of $\mathbf{J}(3, 4)$ have negative real parts. Condition (a) ensures that the first two eigenvalues of $\mathbf{J}$ have negative real parts, and condition (b) ensures that the second two eigenvalues have negative real parts, as demonstrated in Box 14.1.

In a similar vein, we can determine the structure of the Jacobian matrix for any boundary solution and for any number of species present or absent from some larger species pool. The ordering of the species in the interaction matrix is immaterial. You can always rearrange the order of the species by rearranging the rows and columns of the interaction matrix. If the full species pool is n and you wish to evaluate the boundary solution involving, say, m particular missing species, you can place all the missing species at the end of the rows and columns of the interaction matrix (the ordering of the rows and columns must be the same). Then the bottom m rows of the Jacobian matrix $\mathbf{J}$(boundary) will be zero, except for the diagonal elements, which are $r_i g_i / K_i$ (for $i = n - m$ to n species). Finally, note that the calculation of the Jacobian matrix as $\mathbf{R}\,\alpha + \mathbf{G}$, also applies to the interior equilibrium point. In this case $\mathbf{G}$ simply contains all zeros, and the $\mathbf{R}$ matrix contains the N_i^* terms for the interior equilibrium point in its diagonal elements.

Invasions

What happens when one species invades a community? Will the invasion be resisted or will the invader increase? If it increases, what happens to the resident species? Will they persist or become extinct? Answers to these questions can be easily visualized with two resident species and one invader. Let's call the invader species 3 and the resident species 1 and 2. In the absence of the invader, the two resident species are at their equilibrium densities, $N_1^*(3)$ and $N_2^*(3)$. The condition for the growth rate of the invader to be positive is

$$g_3 = [K_3 - \alpha_{31} N_1^*(3) - \alpha_{32} N_2^*(3)] \geq 0.$$

Box 14.1   *The Determinant and Eigenvalues of a Block Matrix*

Consider a matrix **A** as being composed of four blocks:

$$\mathbf{A} = \begin{bmatrix} \mathbf{A}_{11} & \mathbf{A}_{12} \\ \mathbf{A}_{21} & \mathbf{A}_{22} \end{bmatrix}.$$

The only restriction on the dimensions of the blocks is that blocks $\mathbf{A}_{11}$ and $\mathbf{A}_{22}$ must be square (since we're going to take their determinants). A theorem in linear algebra provides a convenient expression for the determinant of **A** in terms of the blocks. The determinant of **A**, denoted $|\mathbf{A}|$ is given by

$$|\mathbf{A}| = |\mathbf{A}_{22}| \, |\mathbf{A}_{11} - \mathbf{A}_{12} \, \mathbf{A}_{22}^{-1} \, \mathbf{A}_{21}|. \qquad (14.20)$$

In our case the Jacobian matrix, **J**, is 4×4 and each block is a 2×2 submatrix. Moreover, the $\mathbf{A}_{21}$ block contains only zeros, so we have

$$\mathbf{J} = \begin{bmatrix} \mathbf{A}_{11} & \mathbf{A}_{12} \\ 0 & \mathbf{A}_{22} \end{bmatrix}.$$

Hence the determinant of $\mathbf{J}(3, 4)$ is

$$|\mathbf{J}| = |\mathbf{A}_{22}| \, |\mathbf{A}_{11}|.$$

Now, to get the characteristic equation of **J**, we must take the determinant of the matrix **J**, with λ_i subtracted from each diagonal term, and then set this determinant to zero, or

$$|\mathbf{J} - \lambda \mathbf{I}| = |\mathbf{A}_{22} - \lambda \mathbf{I}| \, |\mathbf{A}_{11} - \lambda \mathbf{I}| = 0. \qquad (14.21)$$

Here **I** is the identity matrix, which is 4×4 on the left-hand side but only 2×2 in each term on the right-hand side. The solutions for λ are the eigenvalues of **J**. Because of the special form of Eq. (14.21), the four eigenvalues of **J** will be equal to the two eigenvalues $\mathbf{A}_{22}$ plus the two eigenvalues of $\mathbf{A}_{11}$. Since $\mathbf{A}_{22}$ is a diagonal matrix, its eigenvalues are simply the terms on the diagonal, which in our case are $r_i \, g_i / K_i$. The eigenvalues of $\mathbf{A}_{11}$ will have negative real parts as long as that community subset coexists at a stable and feasible equilibrium point.

Figure 14.33
Invasion criterion for a third competitor, species 3, invading species 1 (red) and species 2 (gray) at their two-species equilibrium point (the solid dot). The dashed lines are given by Eq. (14.23).

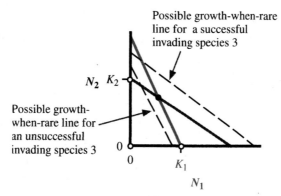

The growth rate of the invading species 3 for any combination of densities of species 1 and 2 is

$$K_3 - \alpha_{31} N_1 - \alpha_{32} N_2 = 0. \qquad (14.22)$$

Equation (14.22) can be rearranged and then plotted as a line in $N_1 - N_2$ space, or

$$N_2 = \frac{K_3}{\alpha_{32}} - \frac{\alpha_{31}}{\alpha_{32}} N_1, \qquad (14.23)$$

and compared to point $[N_1{}^*(3), N_2{}^*(3)]$ on the N_1–N_2 plane.

Figure 14.33 shows two cases: an unsuccessful invader and a successful invader. An unsuccessful invader is represented by a line given by Eq. (14.23) that falls below the $[N_1{}^*(3), N_2{}^*(3)]$ equilibrium point. The line for a successful invader falls above this equilibrium point. However, we cannot determine from Figure 14.33 alone whether the invader will supplant one or more of the resident species or coexist with them at a three-species equilibrium point. This would require a full three-species depiction like those shown in Figure 14.31. For more than two resident species, the resulting community composition after the invasion must usually be determined by simulation since general analytical techniques are lacking (Case 1990).

> **Problem:** Show that the growth-when-rare line for species 3 as given by Eq. (14.23) is just the projection of species 3's zero-isoplane onto the N_1–N_2 plane. (**Hint:** Write the equation for species 3's zero-isoplane and then set species 3 to zero to get the line in just the N_1–N_2 plane.) Examine species 3's zero-isoplane in Figure 14.31 as it hits the N_1–N_2 plane. In Figure 14.31(a), species 3 can invade, but in Figure 14.31(b), it cannot.

Competitive Limit Cycles

With three or more Lotka–Volterra competitors the interior equilibrium point may be unstable, but all species still indefinitely coexist because they follow a limit cycle. An example of such a limit cycle for three Lotka–Volterra competitors is shown in Figure 14.34. This limit cycle *is* stable and results from differential equations (not difference equations). The parameters that lead to the particular cycle shown in Figure 14.34 (from Strobeck 1973) are

$$\alpha = \begin{bmatrix} 1 & 2 & 4 \\ 1/3 & 1 & 2 \\ 1/3 & 1/3 & 1 \end{bmatrix}, \quad K = \begin{bmatrix} 19 \\ 9 \\ 4 \end{bmatrix}, \quad \text{and} \quad r = \begin{bmatrix} 19 \\ 3/2 \\ 12 \end{bmatrix}. \tag{14.24}$$

Note how asymmetric the competition is here (e.g., $\alpha_{13} = 4$, but $\alpha_{31} = 1/3$). Also note that r_1 and r_3 are relatively huge compared to r_2. In fact, it is quite difficult to find situations that yield a limit cycle for three Lotka–Volterra competitors (Gilpin 1975). With a little work, it is possible to show that species 2 and 3 can coexist stably in the absence of species 1, that species 1 and 3 cannot coexist with each other since species 3 ousts species 1, and that species 1 and 2 can coexist at a stable equilibrium point in the absence of species 3. Thus the limit cycle shown here is an emergent property of all three competitors' interactions. None of the three species pairs cycle alone; the cycle emerges only when all three competitors are present. Box 14.2 summarizes some observations about limit cycles in ecological models.

Figure 14.34
The time course for three competitors following a limit cycle specified by the parameters of Eq. (14.24).

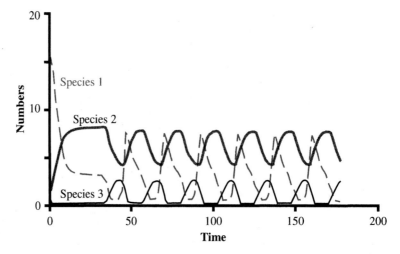

Box 14.2   *A Few More Words about Limit Cycles*

- For a single-species model, limit cycles can emerge with differential equations that contain time lags or with discrete time models.
- For two interacting species, limit cycles may also occur in continuous time models (i.e., without time lags) but only if the zero-isoclines are nonlinear.

  The existence of a two-species limit cycle requires an unstable interior equilibrium point inside the cycle. Also, neither species can increase without bounds in the presence of the other.

- For more than two interacting species, limit cycles may emerge in differential equations even when zero-isosurfaces are linear. For example, limit cycles can occur with three Lotka–Volterra competitors and with two prey and one predator with a type 1 functional response (see Chapter 15).

When the zero-isoclines are linear, the time average of the abundance of each species over the period of the limit cycle is equal to the abundance of that species at the unstable interior equilibrium point (*Note:* This average is not the same as the geometric center of the limit cycle since the trajectory moves at an uneven speed.) When the zero-isoclines are nonlinear, as in some two-species predatory–prey systems, the time average does not necessarily equal the equilibrium abundance.

Exercise: For the system described by Eq. (14.24), what is the interior equilibrium density for the three species? Next, write the Jacobian matrix to evaluate the stability of the interior equilibrium point. Use Eq. (14.11). Verify that the answer is

$$N^* = \begin{bmatrix} 3 \\ 6 \\ 1 \end{bmatrix} \quad \text{and} \quad \mathbf{J} = \mathbf{D}\alpha = \begin{bmatrix} -3 & 0 & 0 \\ 0 & -1 & 0 \\ 0 & 0 & -3 \end{bmatrix} \begin{bmatrix} 1 & 2 & 4 \\ 1/3 & 1 & 2 \\ 1/3 & 1/3 & 1 \end{bmatrix}.$$

How can you show that this interior equilibrium point is unstable?

With three or more competitors, it is possible to alter the local stability of the interior equilibrium point by simply altering the r's (Strobeck 1973). Show what happens to the stability of the interior equilibrium point if the $\mathbf{r}$ vector is changed to

$$\mathbf{r} = \begin{bmatrix} 19/3 \\ 3/2 \\ 4 \end{bmatrix}.$$

PROBLEMS

1. Assume that the population dynamics of rabbits (r) and hares (h) on a 100-acre plot follows the Lotka–Volterra competition equations. Suppose that you have determined that a 100-acre plot can support either 70 rabbits alone or 90 hares alone or 46 rabbits and 49 hares together.

  a. What are α_{hr} and α_{rh}?
  b. Is $(\alpha_{hr})(\alpha_{rh}) < 1.0$?

Plot the zero-isoclines. Label your axes and points of intersection with each axis. Indicate whether the joint densities form a stable equilibrium point.

2. The competition equations for two species are

$$\frac{dN_1}{dt} = \frac{0.5 N_1}{90}(90 - N_1 - 0.5 N_2)$$

and

$$\frac{dN_2}{dt} = \frac{N_2}{200}(200 - N_2 - N_1).$$

  a. What are the equations for the N_1 and N_2 zero-isoclines?
  b. With N_2 plotted on the y axis and N_1 on the x axis, what is the y intercept for species 1's zero-isocline? _____ for species 2's?_____

c. What is the slope of the zero-isocline for species 1? _____ For species 2? _____

d. Plot the zero-isocline for species 1 and shade the regions of positive growth.

e. Plot the zero-isocline for species 2 and shade its region of positive growth.

f. Draw the approximate trajectory from point (25, 25).

g. What will be the ultimate outcome of this competition?

Now suppose that K_1 is increased to 150.

h. Write the zero-isocline equations for both species and plot them again.

i. Hatch the areas of positive growth; draw the arrows for growth in each different area.

j. Draw the approximate trajectory from point (25, 25).

k. What will be the ultimate outcome of this competition?

3. In the Lotka–Volterra equations for two species, what is the biological meaning of

a. a situation in which $\alpha_{12} = \alpha_{21}$? In which $\alpha_{12} = \alpha_{21} = 1$?

b. What outcome is predicted when $\alpha_{12} = \alpha_{21} = 1$ and $K_1 = K_2$?

c. What ecological reasons might cause α_{12} to be greater than 1.0?

4. Consider the 2-species system described by the following isocline representation.

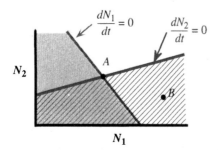

The hatched regions indicate positive growth.

a. Is the qualitative effect (i.e., the sign) of species 1 on its own per capita growth rate positive or negative? _____

b. Is the qualitative effect of species 1 on 2's per capita growth rate positive or negative? _____

c. Is the equilibrium point labeled *A* stable? _____

d. If initial densities were at point *B*, sketch the resulting trajectory describing the dynamical behavior of the two species.

5. Given a 2-species system described by the zero-isocline shown—in which the hatched regions indicate positive population growth—circle the single correct answer in each of (a)–(c).

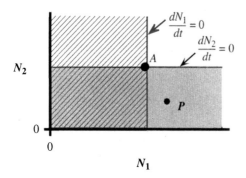

a. The qualitative effect (i.e., the sign) of species 1 on its own per capita growth rate is (i) positive, (ii) negative, (iii) neutral, (iv) variable in sign.

b. The qualitative effect of species 1 on species 2's per capita growth is (i) positive, (ii) negative, (iii) neutral, (iv) variable in sign.

c. The qualitative effect of species 2 on species 1's per capita growth is (i) positive, (ii) negative, (iii) neutral, (iv) variable in sign.

d. Is the equilibrium point labeled *A* stable? _____

e. Is the equilibrium point (0, 0) stable? _____

f. If the initial densities of the two species were at point *P*, draw the resulting trajectory in phase space describing the dynamic changes in N_1 and N_2.

6. Consider Lotka–Volterra two-species competition with $\alpha_{12} = 1.5$ and $\alpha_{21} = 1.5$; both species have equal carrying capacities, but species 1's r is larger than species 2's. How will the competitive outcome depend on initial abundances?

7. For the equation

$$\frac{dN_1}{dt} = \frac{0.5N_1}{100}(100 - N_1 - 0.5N_2 - 1.2N_3)$$

solve for the zero-isosurface and plot it in the state space of N_1, N_2, and N_3. (Let N_3 be the vertical axis).

8. Construct a three-species Lotka–Volterra competition system that has three locally stable boundary equilibria, one for each of the species at its carrying capacity. Give the α matrix and **K** vector.

9. Write the discrete-time analog of the two-species Lotka–Volterra competition equations. What will be the effect of R on the stability of the system? Do you expect limit cycles under some conditions?

10. Do you agree with the following statement? "The relative magnitude of the r's in Lotka–Volterra competition can influence which species will beat the other in competition in some situations." If you agree, give an example of the circumstances under which this statement is true.

11. Consider the following zero-isoclines for two-species competition.

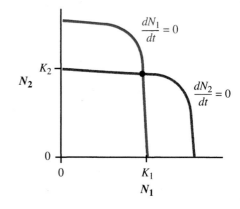

Suppose that these two species have the densities of the interior equilibrium point. You do a field experiment to see if these two species compete. You remove species 1 and wait until species 2 reaches a new equilibrium. You also do the reciprocal experiment in some other plots by removing species 2. What results would you find? What would you conclude about the extent of competition between these two species? How would these results compare to a second set of experiments wherein you doubled the size of species 2's population in some plots while in others you doubled the size of species 1's population?

12. Experimentally, you obtain the following result for the outcome of competition between two species based on many different initial abundances of the two.

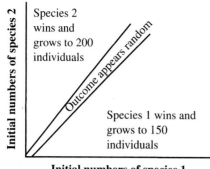

Initial numbers of species 1

In the region labeled "outcome appears random", only one species persists, but sometimes it's species 1 and sometimes it's species 2 in an unpredictable fashion. Draw a plausible scheme for the zero-isoclines for this system and label all points of intersection with each axis. Also, draw small circles around each equilibrium point and indicate which are stable and which are unstable.

13. The following zero-isocline diagram incorporates an Allee effect into two-species competition for the case where the interior equilibrium point is stable (see Figure 14.23a).

Note how many new equilibrium points this adds. Stable equilibria are indicated by a black dot and unstable equilibria by a white dot. Construct depictions of the other three cases previously shown in Figure 14.23 by incorporating this type of Allee effect.

14. Consider three Lotka–Volterra competitors. All have the same K, and the α matrix is

$$\alpha = \begin{bmatrix} 1 & 1.5 & 0 \\ 0 & 1 & 1.5 \\ 1.5 & 0 & 1 \end{bmatrix},$$

and the determinant of α is positive.

a. Assuming any initial numbers where all species have positive density, how many domains of attraction are there for this three-species system? What are they?

b. In the absence of species 3, can species 1 and 2 coexist? If not, who wins?

c. In the absence of species 2, can species 1 and 3 coexist? If not, who wins?

d. If species 1 is at its carrying capacity, can species 2 invade? Can species 3 invade?

e. Interpret the symbolism in the following diagram.

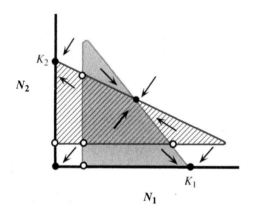

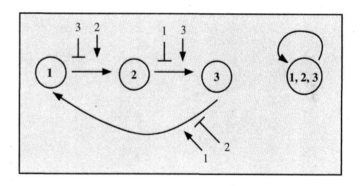

15 *Multispecies Communities*

GENERALIZED LOTKA–VOLTERRA EQUATIONS

In Chapter 14 we extended two-species competition models to three and more species. In the process, we showed that with matrix algebra the analysis was easily generalized. In this chapter we explore interactions among several species at multiple trophic levels. The method of solving for and evaluating the stability of equilibrium points for n interacting species, using matrix mathematics, is particularly straightforward for systems of equations called the generalized Lotka–Volterra equations. **A generalized Lotka–Volterra (GLV) system** is a community model in which the dynamics of the species can be described by equations with a linear *per capita* form, such as

$$\frac{dN_i}{N_i dt} = k_i + \sum_{j=1}^{n} a_{ij} N_j .$$ (15.1a)

Here k_i and a_{ij} are constants with different biological meanings for different trophic levels. Equation (15.1a) may be written in matrix form as

$$\frac{dN}{N\,dt} = k + A\,N,$$ (15.1b)

where A is a square matrix containing all the species interaction terms, a_{ij}. Matrix A is sometimes called the interaction matrix or the **community matrix.**[1] Note, for example, that the two-species Lotka–Volterra competition equation for species 1 is typically cast as

$$\frac{dN_1}{N_1 dt} = \frac{r_1}{K_1}\left(K_1 - \alpha_{11} N_1 - \alpha_{12} N_2 \right).$$

It may then be rearranged to match the form of Eq. (15.1a):

$$\frac{dN_1}{N_1 dt} = r_1 + \frac{-r_1 \alpha_{11} N_1}{K_1} + \frac{-r_1 \alpha_{12} N_2}{K_1}.$$

Similarly, for an arbitrary number of competitors, say, n, we could write

$$\frac{dN_i}{N_i dt} = r_i + \sum_{j=1}^{n} \frac{-r_i \alpha_{ij} N_j}{K_i} \quad \text{for } i = 1 \text{ to } n.$$ (15.2)

---

1. This terminology is not necessarily standard. Some ecologists call the Jacobian matrix, the "community matrix" (see e.g., Yodzis 1989). With this usage, the terms of the community matrix change, depending on the equilibrium point under consideration. In our usage, the "community matrix" is invariant, but is undefined except for GLV systems.

In this form the a_{ij} terms of Eq. (15.1a) are given by α_{ij} multiplied by $-r_i/K_i$, and the k_i of Eq. (15.1a) equals r_i. Thus an n-species Lotka–Volterra competition community can also be written in matrix form as

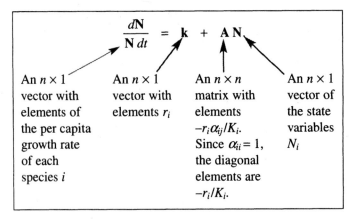

$$\frac{d\mathbf{N}}{\mathbf{N}\,dt} = \mathbf{k} + \mathbf{A}\,\mathbf{N}.$$

An $n \times 1$ vector with elements of the per capita growth rate of each species i

An $n \times 1$ vector with elements r_i

An $n \times n$ matrix with elements $-r_i\alpha_{ij}/K_i$. Since $\alpha_{ii} = 1$, the diagonal elements are $-r_i/K_i$.

An $n \times 1$ vector of the state variables N_i

At equilibrium, the per capita growth rate of each species is zero, so

$$\mathbf{A}\,\mathbf{N}^* = -\mathbf{k}.$$

The interior equilibrium point $\mathbf{N}^*$, an $n \times 1$ vector, can be solved as

$$\mathbf{N}^* = -\mathbf{A}^{-1}\mathbf{k}, \tag{15.3}$$

where $\mathbf{A}^{-1}$ is the inverse of the $n \times n$ community matrix $\mathbf{A}$.

> **Problem:** Show that Eq. (15.3), with $a_{ij} = -r_i\alpha_{ij}/K_i$ and $k_i = r_i$ is equivalent to the expression used in Chapter 14, $\mathbf{N}^* = \mathbf{A}^{-1}\mathbf{k}$, where now $a_{ij} = \alpha_{ij}$ and $k_i = K_i$.

Predator–prey systems with linear zero-isoclines also fit the form of the generalized Lotka–Volterra equations (see Chapters 12 and 13). Here are the coupled equations from Chapter 13 where the prey, V, grows logistically and the predator, P, has a type 1 functional response:

$$\frac{dV}{V\,dt} = \frac{r}{K}(K - V) - aP \quad \text{and} \quad \frac{dP}{P\,dt} = -d + kaV,$$

where a is the encounter rate with prey, d is the per capita death rate of the predator, and k is the predator's conversion efficiency. This model may be arranged into the generalized Lotka–Volterra matrix form as

$$\begin{bmatrix} \dfrac{dV}{V\,dt} \\ \dfrac{dP}{P\,dt} \end{bmatrix} = \begin{bmatrix} r \\ -d \end{bmatrix} + \begin{bmatrix} \dfrac{-r}{K} & -a \\ ka & 0 \end{bmatrix} \begin{bmatrix} V \\ P \end{bmatrix}. \tag{15.4}$$

The community matrix $\mathbf{A}$ is

$$\mathbf{A} = \begin{bmatrix} \dfrac{-r}{K} & -a \\ ka & 0 \end{bmatrix}$$

and that the vector of constants $\mathbf{k}$ is

$$\mathbf{k} = \begin{bmatrix} r \\ -d \end{bmatrix}.$$

> **Problem:** Algebraically solve for the equilibrium values P^* and V^* in Eq. (15.4), using Cramer's rule (see Appendix 3).

A number of extensions are possible with this basic framework. The predator–prey system may be modified to allow for additional interspecific interactions. For example, if two prey species compete directly with one another, the terms become

$$\mathbf{A} = \begin{bmatrix} \dfrac{-r_1}{K_1} & \dfrac{-r_1\alpha_{12}}{K_1} & -a_1 \\ \dfrac{-r_2\alpha_{21}}{K_2} & \dfrac{-r_2}{K_2} & -a_2 \\ kw_1a_1 & kw_2a_2 & 0 \end{bmatrix} \quad \text{and} \quad \mathbf{k} = \begin{bmatrix} r_1 \\ r_2 \\ -d \end{bmatrix}. \tag{15.5}$$

Direct interspecific competition between the two prey species is described by α_{12} and α_{21}. We also slipped a new parameter into Eq. (15.5): w_j, which weights the different prey according to their importance to predator growth. One prey may be bigger or more nutritious than another. An assumption here is that these weights depend only on prey type, not predators. At the end of this chapter, we explore some surprising dynamics that emerge from this simple two-prey, one-predator model.

The GLV description can be generalized further by using the machinery of matrix math to include n prey species and m different predators. The growth vector for each species can be partitioned into two subvectors, one for the prey species and one for the predator species.

$$\begin{bmatrix} \dfrac{dV_1}{V_1dt} \\ \dfrac{dV_2}{V_2dt} \\ \vdots \\ \dfrac{dV_n}{V_ndt} \\ \dfrac{dP_1}{P_1dt} \\ \dfrac{dP_2}{P_2dt} \\ \vdots \\ \dfrac{dP_m}{P_mdt} \end{bmatrix} = \begin{bmatrix} \dfrac{d\mathbf{V}}{\mathbf{V}\,dt} \\ \dfrac{d\mathbf{P}}{\mathbf{P}\,dt} \end{bmatrix}.$$

With this notation, we have

$$\begin{bmatrix} \dfrac{d\mathbf{V}}{\mathbf{V}\,dt} \\ \dfrac{d\mathbf{P}}{\mathbf{P}\,dt} \end{bmatrix} = \begin{bmatrix} \mathbf{r} \\ -\mathbf{d} \end{bmatrix} + \begin{bmatrix} \dfrac{-\mathbf{r}\alpha}{\mathbf{K}} & -\mathbf{a} \\ k\mathbf{a}^T\mathbf{w} & 0 \end{bmatrix} \begin{bmatrix} \mathbf{V} \\ \mathbf{P} \end{bmatrix}, \tag{15.6}$$

where $\mathbf{V}$ is an $n \times 1$ vector containing $V_1, V_2, \ldots, V_n$ and $\mathbf{P}$ is an $m \times 1$ vector containing $P_1, P_2, \ldots, P_m$. Similarly the vector $\mathbf{k}$ is composed of two subvectors: $\mathbf{r}$ is an $n \times 1$ vector holding the r_i term of each prey species, V_i, and $-\mathbf{d}$ is an $m \times 1$ vector holding the death rates, d_i, for each predator, P_i. The entire community matrix **A** has dimensions of $(n + m \times n + m)$. It has four blocks, of which the top-left block is

$$-\frac{\mathbf{r}\boldsymbol{\alpha}}{\mathbf{K}} = \begin{bmatrix} \dfrac{-r_1}{K_1} & \dfrac{-r_1\alpha_{12}}{K_1} & \cdots & \dfrac{-r_1\alpha_{1n}}{K_1} \\[2ex] \dfrac{-r_2\alpha_{21}}{K_2} & \dfrac{-r_2}{K_2} & \vdots & \dfrac{-r_2\alpha_{2n}}{K_2} \\[2ex] \vdots & \vdots & \vdots & \vdots \\[2ex] \dfrac{-r_n\alpha_{n1}}{K_n} & \dfrac{-r_n\alpha_{n2}}{K_n} & \vdots & \dfrac{-r_n}{K_n} \end{bmatrix}.$$

This block describes the interactions among the n different prey. Therefore it is $n \times n$, and both subscripts refer to prey species. The top-right block is

$$-\mathbf{a} = \begin{bmatrix} -a_{11} & -a_{12} & \cdots & -a_{1m} \\ -a_{21} & -a_{22} & \vdots & -a_{2m} \\ \vdots & \vdots & \vdots & \vdots \\ -a_{n1} & -a_{n2} & \vdots & -a_{nm} \end{bmatrix}. \tag{15.7}$$

Block $-\mathbf{a}$ describes the effects of each of the m predators on each of the n prey species. This block has dimension $n \times m$, where the first subscript is a prey species and the second subscript is a predator. If any two columns of this block are identical or are simple multiples of one another (i.e., two consumers utilize the resources in exactly the same relative proportions) or if there are more predator species then prey species (i.e., $n < m$), then the community matrix **A** will have a zero determinant and therefore the inverse $\mathbf{A}^{-1}$ will not exist. In this case, there can be no stable feasible equilibrium involving all the species.

The lower-left block is

$$\mathbf{ka}^T\mathbf{w} = \begin{bmatrix} k_1 & 0 & \cdots & 0 \\ 0 & k_2 & \cdots & 0 \\ \vdots & \vdots & \cdots & 0 \\ 0 & 0 & \cdots & k_m \end{bmatrix} \begin{bmatrix} a_{11} & a_{21} & \cdots & a_{n1} \\ a_{12} & a_{22} & \vdots & a_{n2} \\ \vdots & \vdots & \vdots & \vdots \\ a_{1m} & a_{2m} & \vdots & a_{nm} \end{bmatrix} \begin{bmatrix} w_1 & 0 & \cdots & 0 \\ 0 & w_2 & \cdots & 0 \\ \vdots & \vdots & \cdots & 0 \\ 0 & 0 & \cdots & w_n \end{bmatrix}, \tag{15.8}$$

and its dimensions are

$$m \times n = (m \times m) \quad \mathsf{X} \quad (m \times n) \quad \mathsf{X} \quad (n \times n).$$

The $\mathbf{ka}^T\mathbf{w}$ block describes the effects of each of the n prey *on* each of the m predators' growth rate. As shown, it is the product of three matrices: the first matrix gives the conversion rates, k_i, for each predator, $m \times m$; and the second matrix gives the consumption rates of each prey by each predator, $m \times n$, and is simply the transpose of the $-\mathbf{a}$ matrix but in positive form, since prey benefit predators. The first subscript refers to a specific predator and the second subscript to a particular prey. The different weights for each resource are contained in the diagonal matrix **w** on the right, $n \times n$. As usual with matrix arithmetic, the order of multiplication is important.

Problem: Verify that the matrix multiplication given by Eq. (15.8) produces a term for prey i and predator j that exactly matches that in Eq. (15.5), or $k_j w_i a_{ij}$.

The fourth and final submatrix lies in the bottom-right corner. It is an $m \times m$ matrix of zeros describing the direct interactions of predators on predators:

$$\mathbf{0} = \begin{bmatrix} 0 & 0 & \cdots & 0 \\ 0 & 0 & \cdots & 0 \\ \vdots & \vdots & \vdots & \vdots \\ 0 & 0 & \cdots & 0 \end{bmatrix}. \tag{15.9}$$

Naturally, if predators directly interfered with each other, then this block would contain negative terms describing those interference rates, which we labeled f_i in Chapter 12. With such interference, the **0** submatrix would become an **F** submatrix, where f_{ij} is the direct per capita effect of predator j on predator i's per capita growth rate.

We discuss some results for a two-prey, one-predator version of Eq. (15.6) at the end of this chapter.

Three Trophic Levels

An extension of the generalized Lotka–Volterra equations to describe interactions among three species, each at a different trophic level, is described by the following diagram.

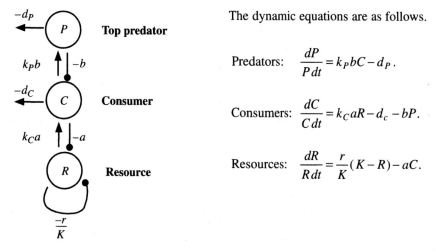

The dynamic equations are as follows.

Predators: $\dfrac{dP}{P\,dt} = k_P bC - d_P$.

Consumers: $\dfrac{dC}{C\,dt} = k_C aR - d_c - bP$.

Resources: $\dfrac{dR}{R\,dt} = \dfrac{r}{K}(K - R) - aC$.

Both consumers and predators experience some density independent mortality ($-d_C$ and $-d_P$), and resources and consumers have mortality due to consumption by a higher trophic level. These equations may be cast in matrix form as

$$\begin{bmatrix} \dfrac{dR}{R\,dt} \\ \dfrac{dC}{C\,dt} \\ \dfrac{dP}{P\,dt} \end{bmatrix} = \begin{bmatrix} r \\ -d_C \\ -d_P \end{bmatrix} + \begin{bmatrix} \dfrac{-r}{K} & -a & 0 \\ k_C a & 0 & -b \\ 0 & k_P b & 0 \end{bmatrix} \begin{bmatrix} R \\ C \\ P \end{bmatrix}. \tag{15.10}$$

The zero-isosurfaces for this system are shown in Figure 15.1.

Figure 15.1
The zero-isosurfaces for the three trophic level system of Eq. (15.10).

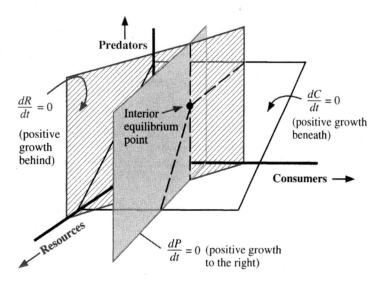

Figure 15.2
A cutaway view of Figure 15.1, showing the location of the boundary equilibria on each two-species phase plane.

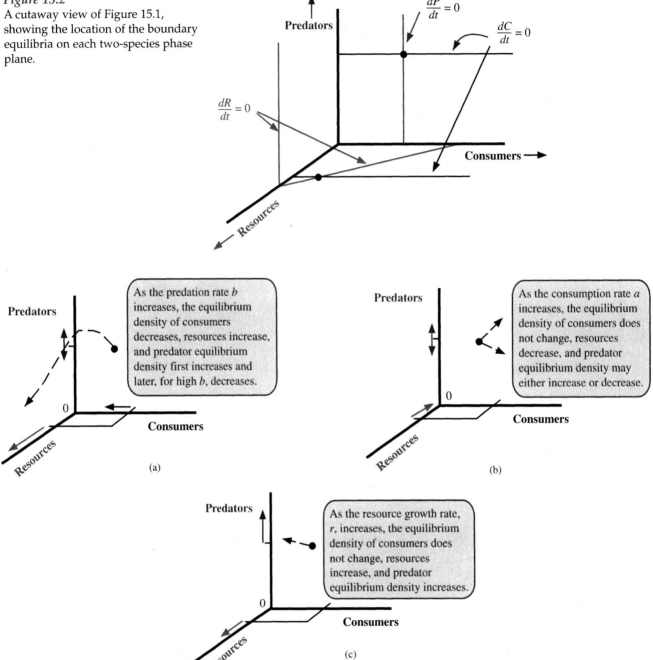

Figure 15.3
The arrows indicate the shift in the position of the three-species equilibrium point with qualitative changes in some of the parameters of the model given by Eq (15.10).

Figure 15.2 shows a cutaway view of the various boundary equilibria. Note that the zero-isoclines do not intersect in the Resource–Predator orthant because predators cannot eat resources; predators can only eat consumers. Multiple species at each trophic level could be added by putting Eq. (15.10) into block form like the two-trophic-level system of Eq (15.6).

An exploration of this three-trophic-level system is shown in Figure 15.3. The addition of a predator as a third trophic level will always result in greater resource abundance and lower consumer abundance. However, more complicated models of three-trophic-level systems may not have these same features, depending on the presence of self-damping and interference at higher trophic levels (Rosenzweig 1973).

STABILITY EVALUATION (ADVANCED)

For any generalized Lotka–Volterra system, the interior equilibrium point is found by applying Eq. (15.3),

$$\mathbf{N^*} = -\mathbf{A}^{-1}\mathbf{k}.$$

Depending on the actual terms in $\mathbf{A}$ and $\mathbf{k}$, this equilibrium point may not be feasible (i.e., $N_i^* < 0$ for some i), but assuming that it is feasible, the next step is to evaluate its local stability. The stability of this interior equilibrium is evaluated by forming the Jacobian matrix, $\mathbf{J}$ (Chapters 13 and 14). The Jacobian matrix contains the partial derivatives of the growth equations with respect to each variable (species). The form of the Jacobian matrix for GLV systems is a straightforward extension of that developed in Chapter 14 for LV competition communities. From Eq. (15.1a), the growth equation for the ith species is

$$\frac{dN_i}{dt} = N_i \left(k_i + \sum_{j=1}^{n} a_{ij} N_j \right).$$

The partial derivative of dN_i/dt with respect to N_j (for $j \neq i$) is therefore just

$$\frac{\partial \frac{dN_i}{dt}}{dN_j} = a_{ij} N_i , \tag{15.11}$$

and this partial derivative evaluated at the equilibrium point $\mathbf{N^*}$ is simply $a_{ij}N_i^*$. The partial derivative of dN_i/dt with respect to its own density, N_i, is

$$\frac{\partial \frac{dN_i}{dt}}{dN_i} = k_i + 2a_{ii}N_i + \sum_{\substack{j=1 \\ j \neq i}}^{n} a_{ij} N_j . \tag{15.12}$$

At equilibrium,

$$k_i = -\sum_{j=1}^{n} a_{ij} N_j^*. \tag{15.13}$$

When we substitute Eq. (15.13) into Eq. (15.12), we note that the sum on the right-hand side of Eq. (15.13) is over all species, including species i, but that the sum on the right-hand side of Eq. (15.12) excludes species i. Hence we can cancel k_i with the sum in Eq. (15.12) only if we also cancel an $a_{ii}N_i$ term. Thus Eq. (15.12) becomes

$$\frac{\partial \frac{dN_i}{dt}}{dN_i} \quad \text{(evaluated at } N^*) = (2a_{ii}N_i^* - a_{ii}N_i^*) \tag{15.14}$$

$$= a_{ii}N_i^*.$$

Thus for GLV systems we can write the Jacobian matrix, $\mathbf{J}$, as

$$\mathbf{J}(\text{interior}) = \mathbf{D^*A,} \tag{15.15}$$

where $\mathbf{D^*}$ is a square diagonal matrix containing the equilibrium densities on the diagonal and zeros elsewhere and $\mathbf{A}$ is the community matrix of Eq. (15.1). From Eq. (15.15), the elements of $\mathbf{J}$ will have the form

$$J_{ij} = N_i^* a_{ij}. \tag{15.16}$$

This is consistent with what we found earlier for the diagonal elements (Eqs. 15.11 and 15.14). For example, for LV competition the terms in the community matrix are $a_{ij} = -r_i\alpha_{ij}/K_i$ (Eq. 15.2), so the terms of $\mathbf{J}(\text{interior})$ are $J_{ij} = -N_i^*r_i\alpha_{ij}/K_i$. (Verify that this is the same result we reached in Chapter 14, using a somewhat different approach.)

The eigenvalues of **J**(interior) are indicative of the local stability of the interior equilibrium point **N***; this equilibrium point is locally stable if and only if all the eigenvalues of **J**(interior) have negative real parts.

> **Problem:** Form the Jacobian matrix for the predator–prey system described by Eq. (15.4).

All that you learned about stability evaluation in Chapters 13 and 14, applies directly to any generalized Lotka–Volterra system. The elements of the community matrix, **A,** may have different signs, depending on the type of direct interaction between each pair of species. If predator j eats prey species i, then $a_{ij} < 0$ and $a_{ji} > 0$. If the two species are mutualists, $a_{ij} > 0$ and $a_{ji} > 0$. Finally, two species that do not have any direct interactions will have $a_{ij} = a_{ji} = 0$.

> **Exercise:** Can you think of any biologically realistic cases where $a_{ji} > 0$ but $a_{ij} = 0$? Can you think of any biologically realistic cases where $a_{ji} < 0$ but $a_{ij} = 0$?
>
> **Exercise:** Show that, for Lotka–Volterra competition, Eq. (15.15) is equivalent to Eq. (14.13), that is, **J**(interior) = **D** α.
>
> **Exercise:** Form the Jacobian matrix to evaluate the stability of the interior equilibrium point for the three-trophic-level system of Eq (15.8). Algebraically solve for the equilibrium values of R, C, and P and use them to form the **D*** needed to evaluate the Jacobian matrix of Eq. (15.15).

NON-GLV SYSTEMS WILL BEHAVE AS GLV SYSTEMS IN A SMALL NEIGHBORHOOD AROUND AN EQUILIBRIUM

As we have demonstrated in Chapters 12 and 13, many predator–prey systems will not have linear zero-isoclines and thus cannot be cast in the framework of GLV. This is the case when predators have type 2 or 3 functional responses, when prey or predator recruitment comes, in part, from outside the system (open populations), or when species have Allee effects, refuges, or higher-order interference terms. For such systems, linear algebra cannot be used to solve for the equilibrium point. Some systems of nonlinear equations may still have analytical solutions for the equilibrium point, but others may not, typically forcing us to apply iterative numerical methods to find the equilibrium point(s).

Once an equilibrium point is found in a non-GLV system, its stability can be evaluated by making a linear approximation to the original dynamical equations using a Taylor's expansion (Chapters 5 and 13). In this way, a linear GLV system resembling Eq. (15.1) can still be formed, but these dynamics apply only to small perturbations in densities, n_i, around the focal equilibrium point, **N***. These perturbations are given by $n_i = N_i - N_i^*$, and over time the perturbation amount n_i changes approximately as

$$\frac{dn_i}{dt} = \sum_{j=1}^{n} J_{ij} n_i \,,$$

or in matrix form as

$$\frac{\mathbf{dn}}{\mathbf{dt}} = \mathbf{J}\,\mathbf{n}. \tag{15.17}$$

The J_{ij} are the partial derivative terms in the Jacobian matrix. As before, these partial derivatives are evaluated at the equilibrium point of interest, N^*. The stability of the equilibrium point is assessed as before, by determining the sign of the real parts of the eigenvalues of J. The only difference is that the Jacobian matrix for an interior equilibrium point in a GLV system had a particularly simple form given by Eq. (15.15) that did not require much mathematical manipulation to produce, whereas for non-GLV equations, it can be tedious indeed to find N^* to calculate all the partial derivatives that go into J and to evaluate them at N^*. In the next section we illustrate a simple example, using a slightly new twist on the old predator–prey model introduced in Chapter 13.

One Prey with Two Age Groups and a Predator That Eats Only Young Prey

In Chapter 13 we introduced the classical Lotka–Volterra predator-prey equations in which neither the prey nor the predator population is self-damped. This model produces neutrally stable cycles around the interior equilibrium point. The amplitude and period of the cycles is determined, in part, by the initial conditions. The cycles become larger and longer as the system is initialized farther from the equilibrium point. Only if the predator and prey are initialized precisely at the equilibrium point does the cycling disappear. We now modify this model to incorporate a simple age structure in the prey population. This modification should reinforce the notion that the state variables in predator–prey systems need not always be restricted to "species." Also we show that this simple modification drastically alters the nature of the dynamics. In this model, only the young of the prey are susceptible to predation. The model is shown in Figure 15.4.

Notice that this model cannot be put into the form of a GLV because the per capita equations for young and adults are nonlinear. Nevertheless, it is quite easy to

Figure 15.4
A predator–prey model with simple age structure in the prey.

$$\frac{dY}{dt} = bA - cPY - \gamma Y.$$

| The prey, P, are subdivided into young, Y, which are subject to predation, and adults, A, which are not | The change in the number of young | = | the births from adults | minus deaths from predators with capture rate c | minus losses due to the aging of young into adults. |

$$\frac{dA}{dt} = \gamma Y - d_A A.$$

| The change in the number of adults | = | the number of young turning into adults | minus the mortality of adults (not density dependent). |

For the Predator, P,

$$\frac{dP}{dt} = kcYP - d_P P.$$

| The change in the number of predators | = | the birth rate of predators, which is proportional to the number of young prey eaten, | minus a density independent mortality term. |

solve algebraically for the equilibrium point. Setting all three equations equal to zero, from the predator equation we have

$$Y^* = \frac{d_P}{kc}.\qquad(15.18)$$

Plugging this expression for Y^* into the growth equation for adults, we find that

$$A^* = \frac{\gamma d_P}{kcd_A}.\qquad(15.19)$$

Finally, putting the solutions for A^* and Y^* into the equation for young gives the predators' equilibrium abundance as

$$P^* = \frac{\gamma}{c}\left(\frac{b}{d_A} - 1\right).\qquad(15.20)$$

The next step is to form the Jacobian matrix to explore the dynamics of perturbations around the equilibrium point. We use the lowercase symbols y, a, and p to denote the perturbation amounts from the interior equilibrium point (Y^*, A^*, P^*). The first-order Taylor's expansion for the dynamics of these perturbation amounts is given by Eq. (15.17), which for this case is

$$\begin{bmatrix} \dfrac{dy}{dt} \\ \dfrac{da}{dt} \\ \dfrac{dp}{dt} \end{bmatrix} = \begin{bmatrix} \dfrac{\partial\left(\dfrac{dY}{dt}\right)}{\partial Y} & \dfrac{\partial\left(\dfrac{dY}{dt}\right)}{\partial A} & \dfrac{\partial\left(\dfrac{dY}{dt}\right)}{\partial P} \\ \dfrac{\partial\left(\dfrac{dA}{dt}\right)}{\partial Y} & \dfrac{\partial\left(\dfrac{dA}{dt}\right)}{\partial A} & \dfrac{\partial\left(\dfrac{dA}{dt}\right)}{\partial P} \\ \dfrac{\partial\left(\dfrac{dP}{dt}\right)}{\partial Y} & \dfrac{\partial\left(\dfrac{dP}{dt}\right)}{\partial A} & \dfrac{\partial\left(\dfrac{dP}{dt}\right)}{\partial P} \end{bmatrix} \begin{bmatrix} y \\ a \\ p \end{bmatrix}.$$

After evaluating these partial derivatives for the model shown in Figure 15.4, we reach

$$\mathbf{J} = \begin{bmatrix} -cP-\gamma & b & -cY \\ \gamma & -d_A & 0 \\ kcP & 0 & kcY-d_P \end{bmatrix}.\qquad(15.21)$$

The next step is to evaluate these partial derivatives at the interior equilibrium point $(Y = Y^*, A = A^*, P = P^*)$ as given by Eqs. (15.18)–(15.20). This gets a bit messy. However, even without going through the algebra, you can see that this matrix does not have zero elements on the diagonal as does the Jacobian of the classical LV predator–prey model. Although it is probably not immediately apparent, the eigenvalues of this Jacobian matrix always have negative real parts as long as the interior equilibrium is feasible.

Problem: Can you duplicate each of the partial derivatives in Eq. (15.21)?

In summary, the addition of a simple age structure to the prey population combined with differential vulnerability of these ages to predation causes the neutral stability of the classical Lotka–Volterra predation to be replaced by true stability. Fig 15.5 shows a typical time series.

Figure 15.5
The results of this predator–prey model when $b = 0.2$, $d_A = 0.1$, $\gamma = 0.2$, $d_P = 0.2$, $k = 0.3$, and $c = 0.04$. Any feasible interior equilibrium will be stable.

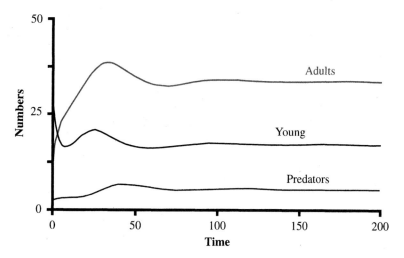

INDIRECT EFFECTS IN MULTISPECIES SYSTEMS

Note in the two-trophic-level model given by Eq. (15.6) that several consumer species are eating several different prey species. The **direct effects** between the predators, in the lower right-hand corner of **A,** were set to zero. Nevertheless, these different predators are exploitative competitors with one another to the extent that they share these limiting resources. These predators negatively affect the growth of one another through their effects on resource levels. This represents a subtle shift in modeling compared to our previous exploration of the Lotka–Volterra competition equations. There we considered two consumer species as competitors when α_{ij} and α_{ji} were greater than 0. By explicitly considering resource dynamics, we can show that two consumers, i and j, can still be exploitative competitors even though $a_{ij} = a_{ji} = 0$; the exploitation competition emerges as an indirect effect through their mutual exploitation of the same resources. **Indirect effects** are effects between two species mediated wholly in terms of changing population densities of intermediary species (Holt 1977, Bender et al. 1984, Roughgarden and Diamond 1986, Schoener 1993). Indirect effects pass from one species to another via the density changes in one or more intermediary species in the food chain. A colorful example of an indirect effect was offered by Charles Darwin in the *Origin of Species,* where he wrote concerning humblebees (an old name for bumblebees):

> *Humblebees alone visit red clover (Trifolium pratense), as other bees cannot reach their nectar. . . . The number of humblebees in any district depends in a great degree on the number of field-mice, which destroy their combs and nests. . . . Now the number of field mice is largely dependent, as every one knows, on the number of cats. . . . Hence it is quite credible that the presence of a feline animal in large numbers in a district might determine, through the intervention first of mice and then of bees, the frequency of certain flowers in that district!*

Indirect effects are also referred to as **interaction chains** (Wootton 1992). Different interaction chains in communities can produce different types of indirect effects between species. Figure 15.6 illustrates these effects for some simple multispecies communities. In Figure 15.6(a), two species that share the same predator, but otherwise have no direct interaction, may nevertheless negatively affect each others' density. As one prey increases, the predator's numbers may, in turn, increase. This increase in the predator, in turn, causes a decline in the density of the other prey species, and vice-versa. This type of indirect interaction is called **apparent competition** to distinguish it from true competition where the prey directly affect each other or compete for shared resources (Holt 1977).

Figure 15.6

Two simple food webs, showing direct and indirect effects. (a) Apparent competition between two noninteracting prey that share a common predator (or parasite). (b) Species 1 and 3 are indirect mutalists. The arrows on the direct effects (solid lines) pointing to a consumer indicate a positive effect flowing to the consumer; a negative effect occurs in the opposite direction on the prey. The sign of the indirect effect between a species pair (dashed line) may be positive or negative. For these two examples, the sign is the same in both directions.

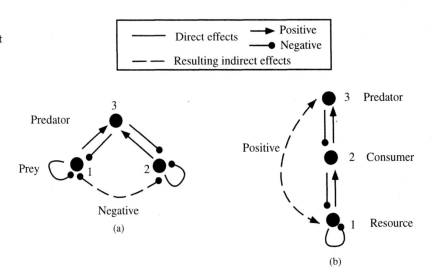

It is possible to evaluate the sign of an indirect interaction between two species by multiplying the elements of the community matrix connecting the species though an interaction chain. If this product is negative, the indirect effect between this species pair is negative; if it is positive, the two species are indirect mutualists, as shown in Figure 15.6(b). Apparent competition, as shown in Figure 15.6(a), has a community matrix with a sign structure as follows:

$$
\begin{array}{c@{}c}
 & \begin{array}{ccc} V_1 & V_2 & P \end{array} \\
\begin{array}{c} V_1 \\ V_2 \\ P \end{array} &
\left[\begin{array}{ccc}
- & 0 & - \\
0 & - & - \\
+ & + & 0
\end{array} \right].
\end{array}
\tag{15.22}
$$

Here we read the direct effect of the species in the column headings on the species in each row; for example, the direct effect of the predator on prey 1 is given in the first row, third column, as negative. To determine the sign of the indirect effect of prey 1 on prey 2 through the predator, we multiply the effect of prey 1 on the predator times the effect of the predator on prey 2. This gives $(a_{P1})(a_{2P}) = (+)(-) = (-)$. Thus prey 1 indirectly harms prey 2 through its positive effect on predator numbers.

For the three-trophic-level system (Figure 15.6b), we have

$$
\begin{array}{c@{}c}
 & \begin{array}{ccc} 1 & 2 & 3 \end{array} \\
\begin{array}{c} 1 \\ 2 \\ 3 \end{array} &
\left[\begin{array}{ccc}
- & - & 0 \\
+ & 0 & - \\
0 & + & 0
\end{array} \right].
\end{array}
$$

The indirect effect of species 1 on 3 through species 2 is $(a_{21})(a_{32}) = (+)(+) = (+)$. Thus species 1 indirectly benefits species 3 through its positive effect on species 2's numbers.

Exercise: Show for the following diagram that the indirect effect of species 3 on species 5 is positive and that the indirect effect of species 5 on species 3 is positive. Species 3 and 5 are indirect mutualists

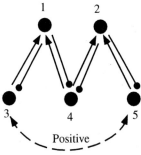

For complicated food webs with more than three species, several different interaction chains may connect a given pair of species through different other species. We may calculate the indirect effect of each possible path (i.e., interaction chain) as before. The total effect across all possible paths may be undetermined unless more detailed quantitative information about the relative strength of interactions in the different chains is available. (We deal with this issue in the next section.) For example, consider the following food web.

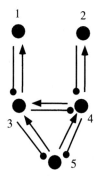

The indirect effect of 2 on 1 is

negative through the chain 4, 3, 1 and

positive through the chain 4, 5, 3, 1.

The necessary information can be determined, however, from all the strengths of the pairwise interactions in the community matrix, **A,** assuming, of course, that they are available.

The Inverse of the Community Matrix

The total effect of one species on another (direct plus indirect) can be measured by determining how that species responds to additions of the other. Consider this three-species example where the Press perturbation (see the next section) involves a continuous addition of species 2 at a rate I_2. If this was a removal experiment, then I_2 would be a negative quantity:

$$\frac{dN_1}{dt} = N_1\left(k_1 + \sum_{j=1}^{n} a_{1k}N_j\right);$$

$$\frac{dN_2}{dt} = N_2\left(k_2 + \sum_{j=1}^{n} a_{2j}N_j\right) + I_2;$$

$$\frac{dN_3}{dt} = N_3\left(k_3 + \sum_{j=1}^{n} a_{3j}N_j\right).$$

This system will come to equilibrium with some new densities for all species. These equilibrium densities, N_i^*, will be functions of the experimental perturbation I_2. At equilibrium, we may write

$$0 = k_1 + \sum_{j=1}^{n} a_{1j}N_j^*(I_2);$$

$$0 = k_2 + \sum_{j=1}^{n} a_{2j}N_j^*(I_2) + I_2;$$

$$0 = k_3 + \sum_{j=1}^{n} a_{3j}N_j^*(I_2).$$

Next, we differentiate these equations with respect to I_2, which is a value manipulated by the experimenter:

$$0 = \sum_{j=1}^{n} a_{1j} \frac{\partial N_j{}^*(I_2)}{\partial I_2};$$

$$0 = \sum_{j=1}^{n} a_{2j} \frac{\partial N_j{}^*(I_2)}{\partial I_2} + 1;$$

$$0 = \sum_{j=1}^{n} a_{3j} \frac{\partial N_j{}^*(I_2)}{\partial I_2}.$$

We evaluate these partial derivatives at the equilibrium point $(N_1{}^*, N_2{}^*, N_3{}^*)$. We solve these three linear equations in the three unknowns—the three partial derivatives with respect to I_2. Casting these three equations in matrix form, we get

$$-\begin{bmatrix} 0 \\ 1 \\ 0 \end{bmatrix} = \mathbf{A} \begin{bmatrix} \dfrac{\partial N_1{}^*(I_2)}{\partial I_2} \\[2mm] \dfrac{\partial N_2{}^*(I_2)}{\partial I_2} \\[2mm] \dfrac{\partial N_3{}^*(I_2)}{\partial I_2} \end{bmatrix},$$

which means that

$$\begin{bmatrix} \dfrac{\partial N_1{}^*(I_2)}{\partial I_2} \\[2mm] \dfrac{\partial N_2{}^*(I_2)}{\partial I_2} \\[2mm] \dfrac{\partial N_3{}^*(I_2)}{\partial I_2} \end{bmatrix} = -\mathbf{A}^{-1} \begin{bmatrix} 0 \\ 1 \\ 0 \end{bmatrix}. \tag{15.23}$$

This assumes that matrix $\mathbf{A}$ has a nonzero determinant; otherwise, the inverse of $\mathbf{A}$ does not exist. If the new equilibrium point is stable, then $\mathbf{A}$ must have an inverse. Equation (15.23) says that the effect of adding individuals of species 2 on species i's new resulting equilibrium density is given by

$$\frac{\partial N_1{}^*(I_2)}{\partial I_2} = -(a_{i2})^{-1},$$

where $(a_{i2})^{-1}$ is the element in position $(i, 2)$ of the inverse of the interaction matrix $\mathbf{A}^{-1}$. (*Note:* This is not the same as $1/a_{i2}$.) More generally, we could perturb any species j and look at the resulting response on the equilibrium densities of any other species i:

$$\frac{\partial N_i{}^*(I_j)}{\partial I_j} = -(a_{ij})^{-1}. \tag{15.24}$$

In this way the elements of the inverse of the community matrix, *following a sign reversal,* give the total effect (direct plus indirect effect) of one species on another within the context of the entire community. **In other words, if $(a_{ij})^{-1} > 0$, then $-(a_{ij})^{-1} < 0$, with the interpretation that the total effect of species j on species i is negative.**

Let's now reexamine one of the examples presented earlier, namely, the case of apparent competition in Figure 15.6(a). We use Matlab® to enter a community matrix with the sign structure shown in Eq. (15.22). (Boldface type denotes our input, and the computer prints out the rest.)

```
EDU» A = [-.2  0  -.3;
     0  -.2  -.4;
    .2  .3  0]
```

A =

-0.2000	0	-0.3000
0	-0.2000	-0.4000
0.2000	0.3000	0

EDU» **inv(A)**

ans =

-3.3333	2.5000	1.6667
2.2222	-1.6667	2.2222
-1.1111	-1.6667	-1.1111

This last matrix is the inverse of the community matrix, **A**. Note that the total effect of prey species 2 on prey species 1 given by $-\mathbf{A}^{-1}(1, 2)$ is –2.5. Similarly, the total effect of species 1 on species 2 is given by $-\mathbf{A}^{-1}(2, 1)$, which is –2.2222. Since the direct effects of these two prey species on each other are both zero, these total effects are simply due to the indirect effects—each of which is negative. This matches our earlier more qualitative interpretation for this community.

An interesting example of the importance of an indirect effect in pest control was provided by a study of birds feeding on larval insects on Eucalyptus trees in Australia (Loyn et al. 1983). One bird species, the bell miner, was extremely aggressive both intra- and interspecifically. It effectively kept many other birds from foraging in the trees. Since the bell miner spent so much time in territorial pursuits, its foraging rate on larval insects was lower than that of other less aggressive birds. The following diagram captures this situation (the thickness of the lines is roughly proportional to the strength of interaction). Here P_1 are the bell miners, P_2 are other insectivorous birds, and the larval insects are the resources, R.

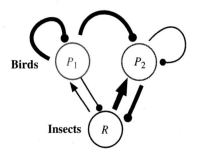

A community matrix qualitatively consistent with this arrangement is

$$
\begin{array}{c}
 \\
R \\
P_1 \\
P_2
\end{array}
\begin{array}{ccc}
R & P_1 & P_2 \\
\left[\begin{array}{ccc}
0 & -0.02 & -0.05 \\
0.002 & -0.04 & 0 \\
0.005 & -0.1 & -0.02
\end{array}\right].
\end{array}
\qquad (15.25)
$$

Using Matlab® to take the inverse of this matrix, we get

EDU» **inv(A)**

ans =

-1000.0	-5750.0	2500.0
-50.0	-312.5	125.0
0000.0	125.0	-50.0

You can see that the total effect of the bell miners, P_1, on insects, R, is positive, even though they eat these insects, since the $\mathbf{A}^{-1}(1, 2)$ element of the inverse of **A** is negative (–5750). According to this model, the removal of bell miners should increase the number of other birds while decreasing the numbers of insects. A simulation of this model, shown in Figure 15.7, quantifies this prediction.

Figure 15.7
A simulation of the bird/insect model. Prior to removal of the highly aggressive predator 1 (bell miners), they outnumber predator 2 (other insectivorous birds), and prey numbers are relatively high at equilibrium. The removal of predator 1 at time 40 leads to an increase in the number of predator 2 and a decrease in the number of prey (insects) at the new equilibrium. Note, however, that the prey's initial trend was briefly upward, following the removal. This reflects the negative direct effect of predator 1 on prey. Later, this direct effect is overwhelmed by the indirect effect as predator 2 increases, causing insect numbers to collapse. This model is based on Eq. (15.6) with community matrix *A* given in Eq. 15.25; other parameters are $r = 0.2$, $d_1 = 0.03$, and $d_2 = 0.05$.

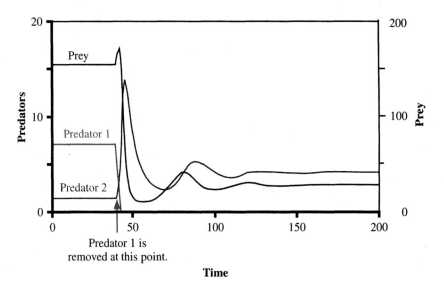

Figure 15.8
Birds and insects in two eucalyptus groves. (a) and (b) Bell miners were experimentally removed from the experimental plots on the date indicated by the arrows. (c) and (d) Control plots also with high insect numbers but no bell miner removal. The groves were about 3 ha in area. Insects were primarily lerps (a stage of a homopteran life history), which were sampled in trays on the forest floor. Numbers represent the mean number of lerps per tray and the number of birds per site. After Loyn et al. (1983).

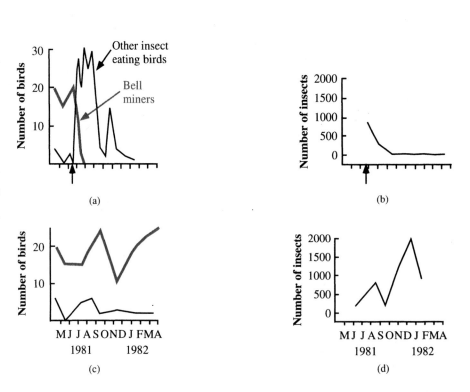

Loyn et al. (1983), in fact, undertook an experimental removal of bell miners. The results were broadly consistent with the model's prediction, as shown in Figure 15.8. Following the bell miner's experimental removal, insect numbers in the trees decreased as other birds moved into the now vacant territories to feed on them. In control plots, the other birds stayed at low numbers while insect numbers remained high. Case et al. (1979) and Holt and Polis (1997) generalized this phenomenon of joint resource exploitation and interference by using multispecies predator–prey models. Interference competition among the dominant predators can keep resources from being effectively exploited, with the result that the total numbers of all predators (summed over species) is not as high as it would be if the those predators with the higher interference rates were removed.

Press Experiments

Often it will be difficult experimentally to maintain and measure the I_j terms of Eq. (15.23). However, it will be relatively easy to set the perturbed species at some new equilibrium level (say, zero) by constantly removing all individuals of species j or excluding their entry. This is the type of experiment that Loyn et al. (1983) did when they removed the bell miners. In it one species (or set of species) is held at a new reduced level, while all the other species are allowed to reach new equilibrium abundances. It is called a **Press experiment** (Bender et al. 1984) since the experimenter is pressing the system in a sustained way. We examine the response in equilibrium density of some nonperturbed species i while we hold some other species (usually just one) at a fixed density. The perturbed species' change in equilibrium density is simply its old equilibrium density, N_j*, minus the new "equilibrium" density at which it is experimentally held. The latter is zero in the case of a complete removal of species j, so ΔN_j* is simply N_j*. The response of species i's equilibrium density following the removal of all individuals of species j then is

$$\frac{\Delta N_i*}{\Delta N_j*} = \frac{\Delta N_i*}{N_j*} = \frac{(a_{ij})^{-1}}{(a_{jj})^{-1}}. \tag{15.26}$$

The switch to deltas (Δ) from infinitesimal partials (∂) is justified for any perturbation of a generalized Lotka–Volterra system (as long as no species other than the perturbed species become extinct). However, is it valid only in a narrow region around the old equilibrium if the equations are not GLV. If we knew the values of $(a_{ij})^{-1}$ and $(a_{jj})^{-1}$, we could simply rearrange Eq. (15.26) to solve for change in equilibrium density of species i:

$$\Delta N_i* = \frac{(a_{ij})^{-1}}{(a_{jj})^{-1}} \Delta N_j*. \tag{15.27}$$

Therefore for a complete removal of species j the new equilibrium density of unmanipulated species i is

$$\text{New } N_i* = \text{old } N_i* - \frac{(a_{ij})^{-1}}{(a_{jj})^{-1}} \text{ old } N_j*.$$

Bender et al. (1984) made a slight modification to Eq. (15.26). They asked: What is the relationship between the change in equilibrium density of a species i relative to the change in another species j, following a press perturbation to a third species k? The answer, in the form of the resulting equation, is

$$\frac{\Delta N_{i,k}^*}{\Delta N_{j,k}^*} = \frac{(a_{ik})^{-1}}{(a_{jk})^{-1}}, \tag{15.28}$$

where $\Delta N_{i,k}^*$ is the change in equilibrium density of species i following a perturbation in species k. Note the subscripts in Eq. (15.28). Figures 15.9 and 15.10 illustrate the application of this equation to a three-species system. The three species' initial interior equilibrium point is the black dot in the center of phase space. The perturbation is to press species 3's density to zero and maintain its numbers there, allowing all the other species to adjust in density until they reach a new equilibrium (shown in the N_1–N_2 orthant).

It is important to remember that the long-term numerical response of species j to the removal of species i, is directly proportional to the ij element in the sign-reversed inverse of the community matrix, not the ij element in the community matrix itself. Thus the long-term response depends not only on the direct interaction between species i and j, but also on the direct interactions of these species with other species in the community through all the interaction chains connecting the species pair.

Figure 15.9
A diagram showing a Press
experiment for a three-species
community and the resulting
community response. Press
perturbation of species 3 forces its
density to zero, and the densities of
species 1 and 2 adjust to this
manipulation—eventually coming to a
new equilibrium in the $N_1 - N_2$
orthant.

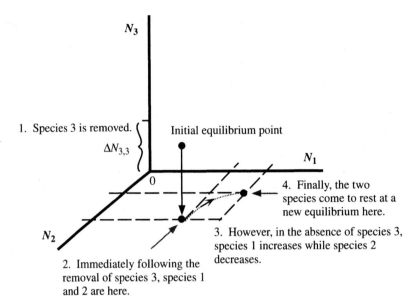

1. Species 3 is removed. $\left\{ \begin{array}{c} \\ \Delta N_{3,3} \end{array} \right\}$

Initial equilibrium point

4. Finally, the two
species come to rest at a
new equilibrium here.

3. However, in the absence of species 3,
species 1 increases while species 2
decreases.

2. Immediately following the
removal of species 3, species 1
and 2 are here.

Figure 15.10
The experiment depicted in Figure 15.9.
The community response following
the Press perturbation of species 3 can
be predicted from elements in the
inverse of the community matrix, A,
using Eq. (15.27).

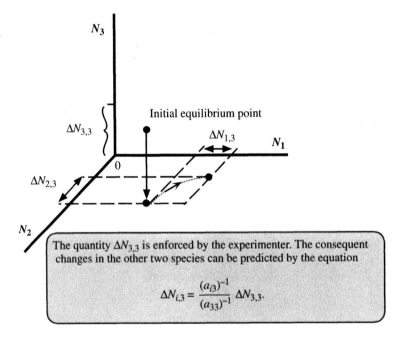

Initial equilibrium point

$\Delta N_{3,3}$

$\Delta N_{1,3}$

$\Delta N_{2,3}$

The quantity $\Delta N_{3,3}$ is enforced by the experimenter. The consequent
changes in the other two species can be predicted by the equation

$$\Delta N_{i,3} = \frac{(a_{i3})^{-1}}{(a_{33})^{-1}} \Delta N_{3,3}.$$

Because Press experiments integrate both direct and indirect effects, we can get
some counterintuitive results (Yodzis 1988). One such example, involving three
Lotka–Volterra competitors, is illustrated in Box 15.1.

Problem: Using the information provided in Box 15.1, calculate the equilibrium
population size of species 2 following the removal of species 3.

Problem: Using the information provided in Box 15.1, calculate the ΔN matrix
based on the complete removal of each of the three competitors in three successive experiments. Use Eq. (15.28) to verify that you can use ΔN to calculate the
community matrix, **A.**

Box 15.1 *Indirect Effects among Three Competitors*

Indirect effects among three competitors

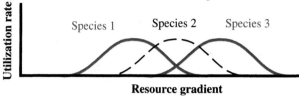

Niches of three competitors

In the diagram, the degree of overlap of the three species niches is indicative of the strength of pairwise competition. Species 1 and 2 and species 2 and 3 have strong competitive impacts on one another, but species 1 and 3 have weak competitive impacts on one another. An α matrix consistent with this diagram is

$$\alpha = \begin{bmatrix} 1 & 0.5 & 0.063 \\ 0.5 & 1 & 0.5 \\ 0.063 & 0.5 & 1 \end{bmatrix}.$$

If all three species have $K = 100$, then the equilibrium densities are $N_1^* = N_3^* = 88.81$ and $N_2^* = 11.19$. The community matrix is $\mathbf{A} = \mathbf{D}\,\alpha$, where $\mathbf{D}$ is a 3×3 diagonal matrix containing the terms $-r_i/K_i$ along the diagonal and zeros elsewhere. The inverse of $\mathbf{A}$ is

$$\mathbf{A}^{-1} = (\mathbf{D}\,\alpha)^{-1} = \alpha^{-1}\,\mathbf{D}^{-1}$$

Note the reversal of multiplication order in this last step (see Appendix 3). Obtaining α^{-1} via Matlab®, we get

$$\mathbf{A}^{-1} = \begin{bmatrix} 1.422 & -0.888 & 0.354 \\ -0.888 & 1.888 & -0.888 \\ 0.354 & -0.888 & 1.422 \end{bmatrix} \mathbf{D}^{-1}$$

Since $\mathbf{D}$ is a diagonal matrix, its inverse is just the matrix with 1 over each diagonal element of $\mathbf{D}$. These multiply each column of α^{-1} to yield

$$\mathbf{A}^{-1} = \begin{bmatrix} \dfrac{-K_1}{r_1}1.422 & \dfrac{K_2}{r_2}0.888 & \dfrac{-K_3}{r_3}0.354 \\[2ex] \dfrac{K_1}{r_1}0.888 & \dfrac{-K_2}{r_2}1.888 & \dfrac{K_3}{r_3}0.888 \\[2ex] \dfrac{-K_1}{r_1}0.354 & \dfrac{K_2}{r_2}0.888 & \dfrac{-K_3}{r_3}1.422 \end{bmatrix}.$$

Note that the total effect of species 1 on 3 is *positive,* since the (1, 3) element of the inverse matrix is $(-0.354K_3/r_3)$ is *negative* (see Eq. (15.24). Although species 3 competes directly with species 1 ($\alpha_{13} = 0.0625$), it competes much more with species 2 ($\alpha_{23} = 0.5$), which, in turn, is a strong competitor of species 1 ($\alpha_{12} = 0.5$) and depresses 1's density much more. Thus the indirect effect of species 1 on species 3 is beneficial through the depression of species 2—as in the saying, "My enemy's enemy is my ally."

The removal of species 3 will result in a decrease in species 1's new equilibrium density. The amount of this decrease can be quantified by applying Eq. (15.27):

$$\Delta N_{13}^* = \frac{(a_{13})^{-1}}{(a_{33})^{-1}}\,N_{33}^*$$

$$= \frac{-0.354\dfrac{K_3}{r_3}}{-1.422\dfrac{K_3}{r_3}}(-88.81) = -22.11. \qquad (15.29)$$

The new equilibrium density of species 1, following removal of species 3, is $88.81 - 22.11 = 66.70$.

Note how the K_3/r_3 terms canceled out in Eq (15.29). Because of this we could have gotten this same result by using just the inverse of the $-\alpha$ matrix instead of the $\mathbf{A}$ matrix to do the calculation with Eq. (15.29). The interaction matrix, $\mathbf{A},$ of a GLV can sometimes be written as the product of a diagonal matrix, $\mathbf{D},$ and a simpler matrix, $\mathbf{A}'$ (for example, for LV competition $\mathbf{A}' = -\alpha$). Since the $\mathbf{D}$ terms factor out anyway in the application of Eq. (15.27), we can predict density changes more readily by using the simpler matrix, $\mathbf{A}',$ and its inverse elements instead in Eq. (15.27).

The simulation depicted in Figure 15.11 verifies our predictions.

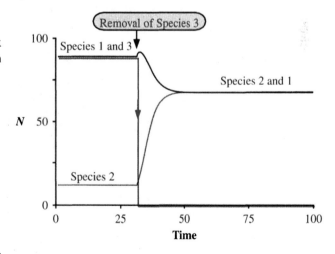

Figure 15.11
A simulation of the three-competitor model. Species 3 is removed at time 30. This results in a lower equilibrium density for species 1 (at 66.7) and a higher equilibrium density for species 2 (also at 66.7).

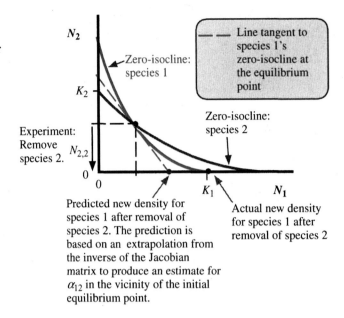

The other way to use Press experiments is to deduce the **A** matrix from the density changes in each experiment. This requires *n* different species removals to get the $\mathbf{A}^{-1}$ matrix for *n* interacting species (Bender et al. 1984). Then the experimentally derived $\mathbf{A}^{-1}$ matrix may be inverted to form the community matrix, **A.** If the species dynamics are not adequately described by GLV equations, we could still apply Press experiments and use the results to estimate the inverse, not of the community matrix, but of the Jacobian matrix evaluated at the initial equilibrium point. This Jacobian matrix could then be used to predict the response in species densities following more sustained manipulations. However, the perturbations must be kept relatively small for two reasons: first, the Jacobian matrix is a valid predictor of dN_i/dt only in a small neighborhood of the existing equilibrium; and second, if some species become extinct following the perturbation, then the analysis falls apart. The first sort of error is illustrated in Figure 15.12 for a simple two-species community whose zero-isoclines are like those of the *Drosophila* shown in Figure 14.25.

Pulse Experiments

Another type of ecological experiment is called a **Pulse experiment** (Bender et al. 1984). Theoretically a pulse experiment involves a quick one-time change in the numbers of one or more species. Then all the species, *including* the perturbed species, dynamically adjust. Figures 15.13 and 15.14 illustrate a Pulse experiment involving a three-species community. In this experiment only species 3 is perturbed from N_3* to a new density N_3. The initial growth rate of each species is measured immediately following this perturbation. As before, we let $N_{i,j}$ represent the abundance of species *i* following a perturbation, n_j, to species *j* ($n_j = N_j* - N_j$). We measure the change in $N_{i,j}$ over time, that is, $dN_{i,j}/dt$. In theory the terms in the community matrix can be calculated from these measurements of $dN_{i,j}/dt$, as illustrated in Figure 15.15.

By focusing on just the initial change in numbers—and if the perturbation is small—it is safe to assume that a linear differential equation model will approximate the true dynamics of the situation. (This is just the first-order Taylor's expansion again.) The immediate change in the numbers of species 1 following this Pulse perturbation to just species 3, then, is

$$\frac{dN_{1,3}}{dt} = n_3 \left. \frac{\partial f_1}{\partial N_3} \right|_{N_1^*, N_2^*, N_3} \tag{15.30}$$

Figure 15.13
An illustration of a Pulse experiment, involving perturbation of species 3. Three interacting species exist at an equilibrium point prior to the removal of some number, n_3, of species 3.

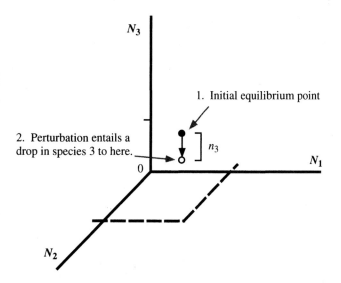

Figure 15.14
Measurement of the interaction terms of matrix **A** based on Figure 15.13 requires estimating the initial growth rate of each species following the perturbation of species 3.

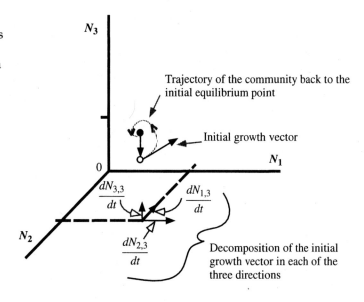

where f_1 is the growth equation for species 1 and n_3 is the perturbation amount. To keep things general, we assume that these f_i functions are *not necessarily* GLV. Also note where this partial derivative is evaluated—not at the original equilibrium point, (N_1^*, N_2^*, N_3^*), but at the new perturbed position, (N_1^*, N_2^*, N_3). The reason is that we are measuring the rates of change of each species from the perturbed position. For GLV equations the partial derivative in Eq. (15.30) is simply $N_1^* a_{13}$ (from Eq. 15.16), but for non-GLV it may be a more complicated function of the species densities. Similarly, the immediate growth rate of species 2 after the perturbation is

$$\frac{dN_{2,3}}{dt} = n_3 \left.\frac{\partial f_2}{\partial N_3}\right|_{N_1^*, N_2^*, N_3}$$

and for the perturbed species 3, we have

$$\frac{dN_{3,3}}{dt} = n_3 \left.\frac{\partial f_3}{\partial N_3}\right|_{N_1^*, N_2^*, N_3}$$

Figure 15.15
An example of the changes in population size of (a) an unperturbed species and (b) a perturbed species following a Pulse perturbation to species *j*.

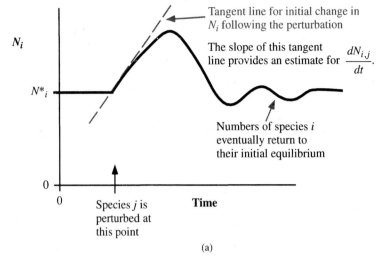

(a)

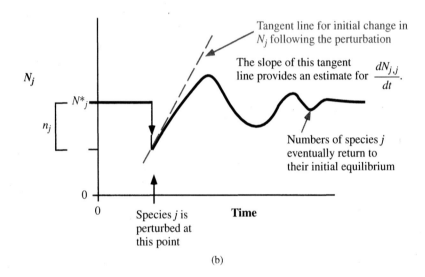

(b)

The elements of the Jacobian matrix can be solved as

$$J_{i3} = \frac{\partial f_i}{\partial N_3}\bigg|_{N_1^*, N_2^*, N_3}$$

$$= \frac{\dfrac{dN_{i,3}}{dt}}{n_3}. \tag{15.31}$$

The final term on the right-hand side of Eq. (15.31) contains two quantities, both measurable from the experiments.

As an example, suppose that we perturb a community of three competing species of *Paramecium.* Let's assume that all three species are at the same density of 100/ml. The perturbation is to add 20 individuals of species 3 so that $N_3 = 120$; thus, $n_3 = N_3 - N_3^* = 20$. We find that, immediately following this perturbation, species 1 decreases during the first few hours by 2 individuals/hour, species 2 decreases by 1.5 individuals/hour, and species 3 decreases at a rate of 3 individuals/hour. Then, with the growth equation for species 1, f_1, also expressed in per hour units, J_{i3} from Eq. (15.31) is

$$J_{1,3} = \frac{-2.0}{20} = -0.1.$$

If this is a GLV system, we can go one step further and rearrange Eq. (15.11) to solve for a_{13}:

$$a_{13} = \frac{J_{13}}{N_1^*}.$$

Hence

$$a_{13} = \frac{-0.1}{N_1^*} = -0.001. \tag{15.32}$$

Similarly,

$$a_{33} = \frac{-3}{(20)(120)} = -0.0015.$$

Still assuming that this is a GLV, we may find α_{13} but we require an additional experiment. To see this, first recall that for the Lotka–Volterra competition equations the Jacobian terms are

$$J_{13} = \frac{-r_1 \alpha_{13} N_1^*}{K_1} \quad \text{and} \quad J_{11} = \frac{-r_1 N_1^*}{K_1},$$

By taking the ratio of these two expressions, we can solve for α_{13}:

$$\frac{J_{13}}{J_{11}} = \frac{\dfrac{-r_1 \alpha_{13} N_1^*}{K_1}}{\dfrac{-r_1 N_1^*}{K_1}} = \frac{a_{13}}{a_{11}} = \alpha_{13}. \tag{15.33}$$

The problem though is that our single experiment of perturbing species 3, while giving us information on J_{13}, did not provide an estimate for J_{11}, the denominator of Eq. (15.33). To get α_{13}, a second experiment is necessary: we must perturb species 1. Let's imagine that we do so and find that the addition of 20 individuals of species 1 leads to a subsequent decline of species 1 equal to 3.5 individuals/hour immediately following the addition. From this result, we calculate a_{11} as

$$a_{11} = \frac{-3.5}{(20)(120)} = -0.00146. \tag{15.34}$$

Now, plugging in the values of a_{13} from Eq. (15.32) and a_{11} from Eq. (15.34) into Eq. (15.33), we find that

$$\alpha_{13} = \frac{a_{13}}{a_{11}} = \frac{-0.001}{-0.00146} = 0.685.$$

Solving for all the α's will take three separate perturbation experiments involving the manipulation of each of the three species. Even to estimate a single α in a Pulse experiment, it takes at least two perturbation experiments. Another problem with Pulse experiments is that the perturbation must be kept small. Consequently, the responses of the species growth rates, dn_i/dt, will also be small and thus easily obscured by experimental error.

APPLICATIONS TO NICHE THEORY

A species **niche** is the sum of its activities and relationships in the community. It is a difficult concept to measure, just as phenotype and genotype are abstract and difficult to quantify. Nevertheless these concepts have heuristic value. Hutchinson (1957) described niches by plotting species in an n-dimensional space where the dimensions are important environmental features affecting the species and one axis

Figure 15.16
A diagram of a two-dimensional niche of a hypothetical species and its activity level for different values of these niche dimensions.

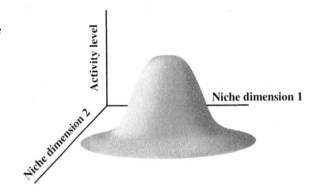

is the *activity level* of the species under those environmental conditions, as shown in Figure 15.16.

If a group of species compete for similar prey and if the only factor differentially influencing the competitors' success is the size of prey that they eat, then the relevant niche axis for these species would be prey size. Species that show greater diet overlap based on prey size, would be expected to compete more than species with less niche overlap in prey size. More typically several niche dimensions may influence population dynamics and competitive impacts since species may eat the same prey types but take them in different microhabitats or at different times.

> *The central tenet of niche theory is that niche overlap between two species, i and j, can be quantified into a metric o_{ij} such that the dynamic effects of competitors on one another, e.g. the α_{ij} terms, will be directly proportional to the degree of niche overlap between species pairs (i.e., $o_{ij} \approx \alpha_{ij}$).*

A corollary is that taxonomic relatedness will usually go hand in hand with morphological similarity and, in turn, niche overlap. Consequently interspecific competition is expected to be more severe among species in the same genus than in different genera or different families. For example, J. B. Steere (1894) investigating the biogeography of Philippine land birds wrote:

> *In 17 genera and 74 species each bird genus is represented in the Islands by several species, two or more of which may be found inhabiting the same island; but the species found together, with the same generic name, differ greatly in size, colouring or other characteristics and belong to different natural sections or subgenera. These sections or subgenera may each be represented in the archipelago by several species; but where this occurs each species is found isolated and separated for all the other species of the subgenus. . . . No two species structurally adapted to the same conditions will occupy the same area.*

Perhaps the most common metric for quantifying ecological similarity is based on a resource utilization matrix. Each row of this matrix represents a consumer species, *i*, and the elements in the row are the rates of consumption for each resource, *k*, by that consumer, a_{ik}. However, typically the resource use is converted into a proportional use by dividing a_{ik} by the total resource use for that consumer. The proportion of resource *k* used by consumer *i* then is

$$P_{ik} = \frac{a_{ik}}{\text{total resource use by consumer } i = \sum_{q=1}^{n} a_{iq}}.$$

These proportional uses are then used to produce a **niche overlap formula**:

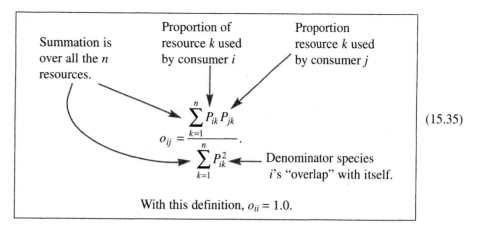

$$o_{ij} = \frac{\sum_{k=1}^{n} P_{ik} P_{jk}}{\sum_{k=1}^{n} P_{ik}^2}.$$ (15.35)

Summation is over all the n resources.

Proportion of resource k used by consumer i

Proportion resource k used by consumer j

Denominator species i's "overlap" with itself.

With this definition, $o_{ii} = 1.0$.

Another way to view this formula is to note that a convenient definition of the niche width of consumer species i is

$$NW_i = \frac{1}{\sum_{k=1}^{n} P_{ik}^2}.$$ (15.36)

As an example, contrast one species that takes equal proportions of two resources and another species that consumes primarily resource 1.

Consumer 1 takes equal amounts of two resources, R_1 and R_2:

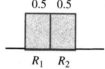

$$NW_1 = \frac{1}{0.5^2 + 0.5^2} = \frac{1}{0.5} = 2$$

Consumer 2 takes mostly R_1:

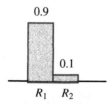

$$NW_2 = \frac{1}{0.9^2 + 0.1^2} = \frac{1}{0.82} = 1.22$$

Conclusion: By this formula the niche width of species 1 is larger than the niche width of species 2.

As it should, the niche width formula, Eq. (15.36), assigns a higher niche width to species 1. The magnitude of the niche width may be thought of as the number of "equally" consumed resources. Species 1 takes both resources equally; therefore NW_1 is 2. Species 2 takes the equivalent of 1.22 resources.

Now let's return to the niche overlap formula, Eq. (15.35). The o_{ij} term can be thought of as a measure of the shared niche space between species i and j multiplied by the niche width of species i, or

$$o_{ij} = NW_i \sum_{k=1}^{n} P_{ik} P_{jk}.$$ (15.37)

This formula does not necessarily produce a symmetric relationship (i.e., $o_{ij} \neq o_{ji}$) unless species i and j have the same niche width. Here's an example based on the two species depicted in the preceding box.

$$o_{12} = 2[(0.5)(0.9) + (0.5)(0.1)] = 2(0.5) = 1.0,$$

and

$$o_{21} = 1.22[(0.5)(0.9) + (0.5)(0.1)] = 1.22(0.5) = 0.61.$$

Conclusion: By Eq. (15.35), the niche overlap of species 1 (with the wider niche) on species 2, o_{21}, is less than the overlap of 2 on 1, o_{12}. Species with wider niches suffer more from competition by this formula than do narrower niche species. This is not a conclusion about the ecological world, however; it is only an assumption on which this particular formula is based. Some ecologists have argued that this particular aspect of the formula is unrealistic and have gone on to suggest alternative formulations. Ultimately this is an empirical question still to be settled by data.

Box 15.2 presents another example of the application of this niche overlap formula, showing that niche overlaps may be greater than one.

Extending these definitions to many competitors and many resources, we can rewrite the niche overlap formula, Eq (15.37), in matrix form as

$$\mathbf{O} = \mathbf{D}\,\mathbf{P}\,\mathbf{P}^T, \tag{15.38}$$

where $\mathbf{P}^T$ is the transpose of matrix $\mathbf{P}$, which means that the matrix $\mathbf{P}$ is flipped on its side by exchanging rows with columns. Since $\mathbf{P}$ had dimension $m \times n$, $\mathbf{P}^T$ has dimension $n \times m$ and the product $\mathbf{P}\,\mathbf{P}^T$ is $m \times m$. Matrix $\mathbf{D}$ is a diagonal matrix with the niche widths, NW_{ii}, along the diagonal and zeros elsewhere. Thus matrix $\mathbf{D}$ is a square matrix with dimensions $m \times m$.

Problem: Based on the example in Box 15.2, verify that the matrix in Eq. (15.38) gives the same values as calculated there.

Solution:

$$\begin{bmatrix} 1 & 6/5 \\ 2/3 & 1 \end{bmatrix} = \begin{bmatrix} 9/5 & 0 \\ 0 & 1 \end{bmatrix}\begin{bmatrix} 2/3 & 1/3 \\ 1 & 0 \end{bmatrix}\begin{bmatrix} 2/3 & 1 \\ 1/3 & 0 \end{bmatrix}$$

$$= \begin{bmatrix} 9/5 & 0 \\ 0 & 1 \end{bmatrix}\begin{bmatrix} 5/9 & 2/3 \\ 2/3 & 1 \end{bmatrix}$$

Finish this multiplication.

Strictly speaking, the overlap formula should not be based on proportional resource use, P_{ik}, since the central tenet of niche theory is that these overlap values are proportional to the dynamical effects that species have on one another. When we use proportions, the overlap of one species on another is independent of the total amount of resource consumption by those species. For example, if one species eats 10 times more resources per unit time than another species, then its impact on that species should be greater. But the formula in Eq. (15.35) does not capture this feature and therefore cannot be dynamically correct. MacArthur (1972) took up this issue and derived a niche theoretic formula for competitive impacts based on first principles. He used the predator–prey system of Eq. (15.6), without the resource–resource competition terms, to derive a formula for α_{ij} between consumer species. He assumed that, as consumer numbers changed, the resources adjusted to new equilibria instantaneously. With this assumption, MacArthur found that he could derive the consumer–consumer

Box 15.2   *Niche Width Calculations*

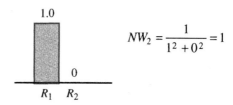

Consumer 1:

$$NW_1 = \frac{1}{(2/3)^2 + (1/3)^2} = 1.8$$

Consumer 2 takes only R_1:

$$NW_2 = \frac{1}{1^2 + 0^2} = 1$$

$$o_{12} = 1.8[(2/3)(1) + 0] = 1.8(2/3) = 1.2$$

and

$$o_{21} = 1[(2/3)(1) + 0] = 2/3$$

Conclusion: By this formula, the niche overlap of a narrow-niche species on a wide-niche species may exceed 1.

interactions in the form of the Lotka–Volterra competition equations where α_{ij} now had the form

$$\alpha_{ij} = \frac{\sum\limits_{k=1}^{n} \dfrac{K_k w_k}{r_k} a_{ik} a_{jk}}{\sum\limits_{k=1}^{n} \dfrac{K_k w_k}{r_k} a_{ik}^2}. \tag{15.39}$$

Equation (15.39) differs from the formula for o_{ij} based on niche overlap, Eq. (15.35), in two important ways. First Eq. (15.39) involves the use of raw consumption rates, a_{ij}, rather than proportional use of resources, p_{ij}. Second, it weights each resource, k, in terms of its importance to consumer–consumer impacts by a term $K_k w_k / r_k$. Resources that have very high r's are less important to consumer competition because once these resources are consumed they are quickly replaced. Shared consumption of resources that are highly nutritious (w_k is large) are more important to consumer competition than shared resources that are less nutritious. Finally, resources that are relatively common in the environment (high K) will also be common in the diets of the consumers and thus will be more important to consumer competition compared to relatively rare and therefore infrequently consumed resources.

Equation (15.39) also suggests a more mechanistic formulation for niche width:

$$\tilde{NW}_{ij} = \frac{1}{\sum\limits_{k=1}^{n} \dfrac{K_k w_k}{r_k} a_{ik}^2}. \tag{15.40}$$

The tilde is used to distinguish this measure of niche width from the earlier one of Eq. (15.36). It is nearly identical to the earlier formula. The exception is that it, too, is modified to include information about the relative importance of different resources to consumer growth rates. Finally MacArthur derived the consumer carrying capacities, $\tilde{K}_i$, based on these assumptions (the tilde over the K is added to distinguish the derived consumer carrying capacities from assumed resource carrying capacities):

$$\tilde{K}_i = \tilde{N}\tilde{W}_i \left(-\frac{d_i}{k_i} + \sum_{k=1}^{n} K_k w_k a_{ik} \right).$$ (15.41)

Thus by Eq. (15.41) a consumer's carrying capacity is directly proportional to its niche width as defined by Eq. (15.40), but becomes smaller as its death rate increases. The summation over resources in the parentheses can also be thought of as another measure of niche width, since it simply sums all the resource consumption rates, each weighted by an importance index for that resource ($K_k w_k$). One obvious problem in applying Eq. (15.41) to real communities is that it is difficult to enumerate all the resources used by consumers, let alone the r, K, and w for each.

Density Compensation

Support for some qualitative predictions made by niche theory comes from comparisons of island and mainland communities (MacArthur 1972, Cody 1974, Cox and Ricklefs 1977). Islands typically have only a subset of species that exist on comparable habitats on the mainland. If island habitats are perfectly matched to mainland habitats in other respects, the expected release from competition on an island can be predicted by using the framework developed for Press experiments.

The carrying capacities of competitors are very difficult to measure compared to their abundances. Carrying capacity is rarely observed, but abundances are measurable. If we assume that present abundances are at equilibrium, we could base estimates of K's on an independently derived α matrix. This is where the overlap matrix comes in. The hope is that detailed studies of foraging method, prey types, and other niche dimensions can lead to the estimation of an overlap $\mathbf{O}$ matrix that is proportional to α. For the moment, we make this very tenuous assumption and see where it leads us. Note, however, that niche overlap tells us nothing about how strong the interactions may be between these species and other trophic levels. Predators and parasites may be equally important in shaping patterns of abundance.

Let's work through a hypothetical example for a group of four competitors. We first estimate the 4×4 $\mathbf{O}$ matrix, using the niche metrics presented earlier. The relative densities of the four species on typical mainland sites can also be directly measured. Using the Lotka–Volterra competition equations, we can solve for the carrying capacities of the species:

$$\mathbf{K} = \mathbf{O}\,\mathbf{N}^*.$$

This last step invokes the important assumption that the species dynamics are following these competition equations. Now let's suppose that an island exists where some set of the mainland species are absent—for example, species 1 and 2. The relative abundances of species 3 and 4, then, should be

$$\mathbf{N}^*_I(1,2) = (\mathbf{O}_M(1,2)^{-1}\,\mathbf{K}_M(1,2),$$ (15.42)

where $\mathbf{N}^*_I(1,2)$ is the relative abundance vector on the island in the absence of species 1 and 2, $\mathbf{O}_M(1,2)$ is the overlap matrix estimated on the mainland with rows and columns 1 and 2 removed, and $\mathbf{K}_M(1,2)$ is the $\mathbf{K}$ vector for the mainland with rows 1 and 2 removed. By measuring the relative abundances of species 3 and 4 on the island we can check to see whether their actual densities match the values predicted from Eq (15.42).

Yeaton (1974) studied the passerine birds of chaparral habitats in 7-acre sites in southern California. He found 17 species in mainland Santa Monica and 12 species on Santa Cruz Island. He then calculated niche overlaps between bird species from field studies in the Santa Monica Mountains. Table 15.1 compares predicted and actual abundances of the commoner bird species on Santa Cruz Island.

The loss of some species from Santa Cruz Island has led to a net increase in the density of several species that are present. This phenomenon is called **density com-**

Table 15.1 **From Yeaton (1974).**

Species	Mainland density (pairs/acre)	Island density expected, based on formulas like Eq.(15.42)	Actual island density
Scrub jab	0.27	1.00	1.11
Orange crowned warbler	0.07	0.50	0.55
Huttons vireo	0.14	0.60	0.55
Bewicks wren	0.61	1.28	1.27
Total bird density	5.52	—	5.39

pensation. When it is common, the net result is that the total bird density is nearly equal in the two places despite the lowered species number on the island.

Density compensation is often accompanied by an expansion of a species niche and its habitat use. Collectively, density compensation and **niche and habitat expansion** are referred to as **competitive release.** Again, this pattern is suggestive but not definitive proof that competition occurs among these species, since the islands and mainland sites may not be replicates in all respects except for the presence or absence of certain competitors.

Limiting Similarity

For over a century naturalists have noticed striking regularities in the abundance, niche positions, and distribution of similar species (Vandermeer 1972, Cody 1974, Schoener 1974, Pianka 1976). Several of these regularities seem consistent with expectations from niche theory, and entire books have been written on the subject. One common theme is that related species that are similar in morphology, especially body size, and thus have similar diets often occupy different habitats or occur in different geographic regions (i.e., in allopatry). In contrast, closely related species that are sympatric and occur in the same habitat (i.e., syntopic) frequently have niches that segregate strongly in body size or other directions related to foraging mode. These observations spawned an analysis of how similar species niches can be, before they can no longer coexist.

We will work through one theoretical approach to this problem (MacArthur and Levins 1967). It assumes a single niche dimension and normal distributions for the niche of each species. Figure 15.17 presents a graphical description of the situation.

For the invasion to be successful, the invader's population must be able to increase when it is rare and in the presence of the two residents at their equilibrium densities with each other, that is, we require

$$\frac{dN_I}{dt} > 0.$$

We use niche theory to convert the static picture presented in Figure 15.17 to Lotka–Volterra dynamics. We then solve for the invader's growth rate when its population is rare:

$$\frac{dN_I}{dt} > 0 \quad \text{when} \quad K_I - \alpha N_1{}^*(I) - \alpha N_3{}^*(I) > 0. \tag{15.43}$$

K of the invader   Equilibrium densities of the two residents in the absence of the invader, species I.

We solve for the equilibrium densities of the two residents in the absence of the invader by applying Cramer's rule (see Appendix 3):

$$N_1{}^*(I) = \frac{\begin{vmatrix} K & \beta \\ K & 1 \end{vmatrix}}{\begin{vmatrix} 1 & \beta \\ \beta & 1 \end{vmatrix}} = \frac{K(1-\beta)}{1-\beta^2} = \frac{K(1-\beta)}{(1-\beta)(1+\beta)} = \frac{K}{1+\beta}.$$

Figure 15.17
A graphical depiction of the classical limiting similarity problem. Can an invading species insert itself between the niches of two resident species? All the niches have the same standard deviation, σ, and total area, A. Each species has the same carrying capacity, K, and the niches are evenly spaced d units apart, the distance between the peaks (means) of adjacent niches.

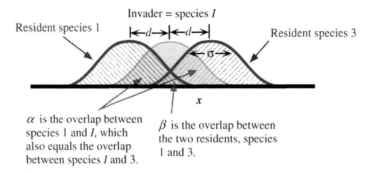

α is the overlap between species 1 and I, which also equals the overlap between species I and 3.

β is the overlap between the two residents, species 1 and 3.

From the symmetry of the niches depicted in Figure 15.17, the equilibrium density of species 3 in the absence of species 2 must be identical to that of species 1, or

$$N_1^*(I) = N_3^*(I) = \frac{K}{1+\beta}. \qquad (15.44)$$

We substitute Eq. (15.44) into the inequality of Eq. (15.43) to reach

$$K_I - \frac{2\alpha K}{1+\beta} > 0. \qquad (15.45)$$

Next, we want to find an expression for niche overlap as a continuous function of niche separation, d. We apply the continuous version of the niche overlap function, Eq. (15.35):

$$\beta = \frac{\int_{-\infty}^{\infty} p_1(x)\,p_2(x)\,dx}{\int_{-\infty}^{\infty} p_1(x)^2\,dx}. \qquad (15.46)$$

The niches p_1 and p_2 can be described by the gaussian curves with means $-d$ and d, standard deviation, σ, and area, A:

$$p_1(x) = \frac{A}{\sigma\sqrt{2\pi}} \exp\left(-\frac{(-d-x)^2}{2\sigma^2}\right) \qquad (15.47a)$$

and

$$p_2(x) = \frac{A}{\sigma\sqrt{2\pi}} \exp\left(-\frac{(d-x)^2}{2\sigma^2}\right) \qquad (15.47b)$$

We substitute Eqs. (15.47a) and (15.47b) into Eq. (15.46) and evaluate the integrals of Eq. (15.46) to reach the much simpler expression,

$$\beta = \exp\left(\frac{-s^2}{(2\sigma)^2}\right), \qquad (15.48)$$

where s = niche separation—in our case $s = 2d$ for the two outer species, 1 and 3. Equation (15.48) describes a gaussian curve with standard deviation of 2σ. When niche separation $s = 0$, niche overlap $\beta = 1$. β depends only on the ratio of s/σ. As s/σ increases, niche overlap decreases according to a gaussian curve, as shown in Figure 15.18.

Since in our case $s = 2d$ for species 1 and 3 and $s = d$ for species 1 and 2, as well as species 2 and 3, we have

$$\beta = \exp\left(\frac{-(2d)^2}{(2\sigma)^2}\right) \quad \text{and} \quad \alpha = \exp\left(\frac{-d^2}{(2\sigma)^2}\right).$$

Figure 15.18
Niche overlap, β, as a function of niche separation s/σ between two gaussian niches (as defined by Eq. 15.48); s is the distance between the two means, and σ is the standard deviation of the niche. Since negative niche separation is not defined, the dashed portion of the overlap curve is not observed.

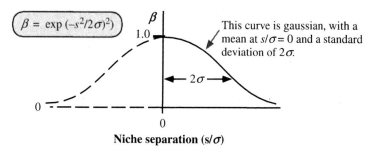

$$\beta = \exp\left(-s^2/2\sigma\right)^2$$

This curve is gaussian, with a mean at $s/\sigma = 0$ and a standard deviation of 2σ.

Niche separation (s/σ)

Thus $\beta = \alpha^4$ and the inequality, Eq. (15.45), can be rewritten as

$$K_I - 2\alpha\left(\frac{K}{1+\alpha^4}\right) > 0. \tag{15.49}$$

We can rearrange Eq. (15.49) to get

$$K_I > \frac{2\alpha K}{1+\alpha^4}. \tag{15.50}$$

If the invader has the same carrying capacity as the residents, then Eq. (15.50) simplifies further to

$$1 > \frac{2\alpha}{1+\alpha^4}.$$

Hence the critical α_c for invader success is obtained by solving

$$1 = \frac{2\alpha}{1+\alpha^4} \tag{15.51}$$

for α. There will be four roots, or solutions, for α_c from Eq. (15.51). Two are nonnegative real numbers:

1. $\alpha_c = 1.0$, which implies that $d/\sigma = 0$; and
2. $\alpha_c = 0.544$, which implies that $d/\sigma = 1.56$.

The second root is the most interesting. The invader will be successful if α, here defined as its niche overlap with the residents, is less than 0.544. Moreover, for this case of equal K's, if the invader can grow when rare, it will reach a positive equilibrium density with both residents present. The invader will not displace either resident. This limiting similarity paradigm may be extended in several directions. The K's may be made unequal, the number of resident species may be increased from two to several, the shape of the niche can be altered from gaussian to some other form, and the growth rate may include stochastic terms. Abrams (1975) and Roughgarden (1979) provide a review of some of these modifications.

Given enough time and a large species pool, we might expect communities to accumulate species until they become competitively saturated: any niche gaps will be invaded by new species, and competition-driven extinctions will eliminate species whose niches are too close. Thus **species-assortment** based on niche position may be one reason why some co-occurring competitors display a regularity in their niche positions or body sizes (Cody 1974, Schoener 1974). Another complementary explanation is that species may genetically coevolve over time to minimize competitive impacts with one another: This is generally referred to as **character displacement** when a pair of species is more divergent morphologically in sympatry than in allopatry. More complicated models combine these two processes to represent the long-term evolution of competition communities through species assortment and coevolution (see, e.g., Taper and Case 1992).

> **Problem:** Show that for a single-resident species, there is no limiting similarity for an invader (i.e., $\alpha_c = 1.0$) when the K's are equal.
>
> **Problem:** Prove that, for two residents (and equal K's for residents and invaders), both residents will be able to coexist with a successful invader.

LIMITING SIMILARITY AND PREDATION

How might a predator alter the limiting similarity of its prey? The answer depends on the way that the predator acts (Abrams 1977). If the predator indiscriminately eats prey in such a way that the carrying capacity, K, of each prey species is reduced by the same amount, then the predator has no impact on the limiting similarity of these prey species. Such a predator will not prevent the competitive exclusion of one prey by another if that would occur in the absence of predation. (Note that, when K's were equal, K canceled out of Eq. 15.50). As an example, mosquito larvae that filter-feed indiscriminately on protozoans living in rainwater collected in pitcher plants do not enhance the diversity of protozoan prey species in these communities. Addicott (1974) experimentally added varying numbers of mosquito larvae to pitcher plants and measured the impact to prey numbers and prey species diversity. Both measures decreased with increasing mosquito larvae.

Recall, however, that in Chapter 11 we introduced the situation where "switching" predators may take disproportionate numbers of prey that are relatively common. This situation was modeled by Roughgarden and Feldman (1975) in the context of the limiting similarity paradigm. They imagined that the invading prey, at very low numbers, would experience much less predation on a per capita basis then the resident prey. This would be the case, for example, if the predator had a type 3 functional response for all prey. In the limit when the invader experiences no predation, its growth when rare and thus the condition for it to invade, is from Eq. (15.49)

$$K_I - 2\alpha N^*(I) > 0,$$

where $N^*(I)$ is again the equilibrium density of the residents in the absence of the invader but now in the presence of the predator. As predation increases, $N^*(I)$ will become lower and thus it will be easier for the invader to insert itself in the community. In the absence of the predator, $N^*(I)$ was given by Eq. (15.44) as $K/(1 + \alpha^4)$, but with such a predator it will be a value less than this. Consequently, the critical limiting similarity, α_c, will be greater. In this way, a predator that feeds disproportionately on common prey items can enhance the coexistence of its several competing prey.

An example of a low-frequency advantage is seen in birds feeding on pine looper caterpillars. The older caterpillars exhibit a color polymorphism caused by a single genetic locus. The wild type aa genotypes are green and quite cryptic on their normal background of pine needles. The heterozygote genotypes Aa are yellow. The other homozygote AA is lethal and never found in nature. The color polymorphism is not expressed in adult moths or the youngest stages of the caterpillars. The typical frequency of the heterozygote yellow morphs in the wild is about 0.06 to 2.5%, which is rare but much higher than would be expected if only new mutations each generation were responsible for producing these yellow morphs. Boer (1971) could find no differences between adult moths from green and from yellow caterpillars with respect to mating preferences, mating success, fecundity, egg hatching, and larval growth rate. This caused Boer to look for factors that might favor the survival of heterozygote caterpillars in the wild. He placed caterpillars on green pine needles in an aviary and then introduced birds (great tits and coal tits) without previous exposure to either color morph. As expected yellow caterpillars are attacked much more frequently than green, as depicted in Figure 15.19(a).

Figure 15.19
Attacks by birds on equal numbers of both color morphs of pine looper caterpillars. (a) Naive birds. (b) Birds trained on green caterpillars.

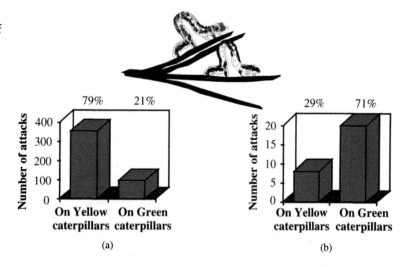

Figure 15.20
As the yellow morph increases in frequency from 0 to 1, the green morph must decrease in frequency since the frequency of both morphs always sums to 1. As yellow increases in frequency, its fitness declines. As green increases in frequency, its fitness declines. When the fitness of both types is equal, which occurs where the two fitness lines intersect, they will be at equilibrium. The equilibrium frequency of yellow, p^*, is less than that of green, $1 - p^*$, since green has a higher maximum fitness than yellow.

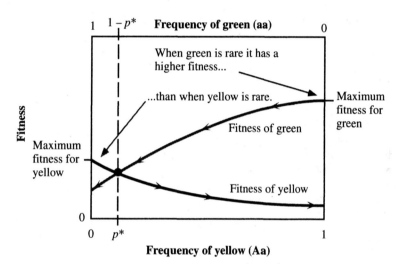

But, once birds have been exposed to the green caterpillars, they learn how to detect and find them, and they perfect a search image for these cryptic green morphs. If these experienced birds are now offered equal numbers of both color morphs in an aviary, the results are quite different, as shown in Figure 15.19(b): the green caterpillars are eaten more frequently than the yellow.

These results suggest that in nature both color morphs have a relative advantage when rare, since birds would not gain experience with them or be able to develop a search image. Yet, since the green morph is more cryptic than yellow on the normal green pine needle background, green's advantage when rare is greater than that of yellow when it is equally rare. Figure 15.20 puts these considerations together to explain the stable polymorphism and the relatively rare but persistent frequency of yellow in the population.

Another way that a predator can enhance the coexistence of competing prey types is by preferring to eat prey that happen to be competitively dominant, regardless of this prey's relative frequency. As an example, consider a situation with a single predator and two competing prey. The predator has a simple type 1 functional response for each prey, so there is no switching here. Let's imagine a case where the two prey also directly compete, as modeled by Eq. (15.5). Let's further suppose that, in the absence of the predator, prey 1 competitively excludes prey 2 (which means that $K_1 > K_2/\alpha_{21}$ and $K_2 < K_1/\alpha_{12}$, from Chapter 14). If the predator prefers to eat species 1, it may rescue species 2 from competitive exclusion. Moreover, if $\alpha_{12}\alpha_{21} > 1$, then a feasible

Figure 15.21
An example of spiral chaos involving two prey and a single predator. Prey 1 is competitively superior to prey 2, but the predator has a higher capture rate for prey 1 than for prey 2. The equations governing these dynamics are given by Eq. (15.5), with $r_1 = r_2 = d = k = K_1 = K_2 = 1$, $w_1 = w_2 = 0.5$, $\alpha_{12} = 1$, $\alpha_{21} = 1.5$, $a_1 = 10$, and $a_2 = 1$. The zero-isoclines for the two competing prey species are shown in the prey 1–prey 2 plane as thick straight lines. The trajectory begins at the red dot. The interior equilibrium point is at prey 1 = 0.1184, prey 2 = 0.8158, predator = 0.0066.

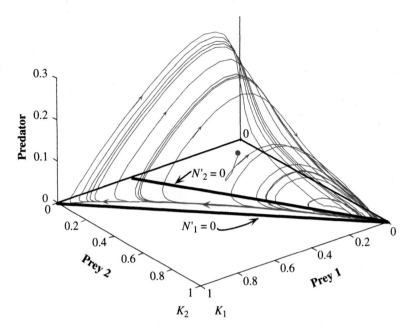

Figure 15.22
As in Figure 15.21, but now the predation rate on the competitively superior prey is lowered from $a_1 = 10$ to 5. All three species coexist at a stable equilibrium point. All other parameters are the same. The equilibrium point here is at prey 1 = 0.2963, prey 2 = 0.5185, predator = 0.0370.

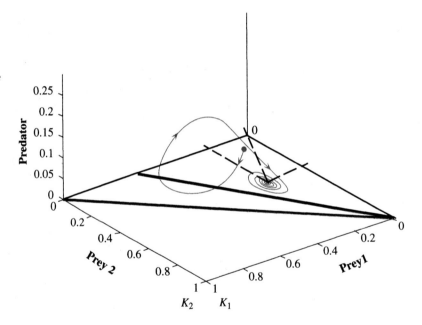

three-species equilibrium point may exist, which may be stable or unstable, depending on the other parameters. If it is unstable, all three species may still coexist via a limit cycle or on a chaotic attractor with the spiral shape illustrated in Figures 15.21 and 15.22 (Gilpin 1979). The trajectory never settles down, but continually roams around the spiral "horn."

Problem: Form the Jacobian matrix for the interior equilibrium point shown in the captions of Figures 15.21 and 15.22 and evaluate the eigenvalues. Verify that the interior equilibrium point is unstable in Figure 15.21 but is stable in Figure 15.22. This will require a computer and matrix software.

PROBLEMS

1. Suppose that you have a single-resident Lotka–Volterra competitor species with $K = 100$. Another competitor, also with $K = 100$, attempts to invade. The niches of the two competing species are gaussian with separation d/σ. How small can d/σ be for this invasion to still be successful?

2. Using Matlab® for the community matrix of Eq. (15.25) and the vector **k** given in the caption to Figure 15.7, verify that the equilibrium densities for this two-predator, one-prey model are:

Prey = 152.5000

Predator 1 = 6.8750

Predator 2 = 1.2500

Using the inverse of the community matrix in Eq. (15.25), follow the procedure in Box 15.1 to calculate the expected change in numbers of predator 2 following the removal of predator 1. Check your result with the simulation results of Figure 15.7. What would happen to the density of predator 1 if predator 2 was removed?

3. Using the niche overlap formula of Eq. (15.35), calculate the niche overlap for these two consumers for the proportionate use of the three resources shown.

	R_1	R_2	R_3
Consumer 1	0.5	0.5	0
Consumer 2	0	0.5	0.5

If the carrying capacities of the two competing consumer species are both $K = 100$—and assuming that these overlaps reflect α's in the Lotka–Volterra competition equations—what are the predicted equilibrium densities of the two consumers?

4. Suppose that the community matrix for a three species GLV community is

$$A = \begin{bmatrix} -1 & 0 & -0.5 \\ -0.5 & -1 & -0.5 \\ 1.5 & 1 & 0 \end{bmatrix} \quad \text{with} \quad k = \begin{bmatrix} 1 \\ 1 \\ -0.1 \end{bmatrix}.$$

Draw a food chain for these three species. What are the equilibrium densities of the three species? Suppose that species 3 is missing from an island, but everything else is equivalent. Then what are the predicted equilibrium densities for species 1 and 2 on the island?

5. Imagine a situation for three competitors like that shown in Figure 15.17, except that the outer tails of the niches of the two residents loop around in a circle so that they overlap to degree α. Let the niche separation between the means of adjacent niches be d for all three adjacent pairs on this ring. Form the α matrix for this community and determine the limiting similarity for these three species.

6. In desert scrub habitats on a mainland you determine that the niche overlap matrix for three lizard species is

$$O = \begin{bmatrix} 1 & 0.3 & 0.5 \\ 0.4 & 1 & 0.2 \\ 0.6 & 0.2 & 1 \end{bmatrix}.$$

The population densities you observe per acre are $N_1 = 10$, $N_2 = 20$, and $N_3 = 15$. Given the usual assumptions for niche theory (i.e., these overlaps represent competitive impacts, the community is at equilibrium, the Lotka–Volterra competition equations are valid, etc.), what are the carrying capacities for each of the three species on the mainland?

On some islands you find that species 2 is absent. What are the expected equilibrium densities of species 1 and 3 in the same desert scrub habitats on these islands (assuming all other conditions on the islands are the same)?

16

*Space, Islands, and Metapopulations**

The ecological processes that we have described in the previous chapters did not give much consideration to spatial position and real-world landscapes. Yet the patterns described by ecologists and conservation biologists are quite often spatially based: species A and B have a zone of competitive overlap at the edge of a forest; the range of species C is limited at the north by gene flow from midlatitude core populations; a nature reserve for species D must cover 100,000 hectares and include sufficient altitudinal diversity if species D is to persist; and so on. One of the outstanding intellectual challenges for ecologists is to meld the kinds of models considered previously in this book with realistic representations of one- and two-dimensional space (and the marine and aquatic people often have to add the third dimension).

Except for our brief excursion into movements in Chapter 2, the state variables we have considered have been quantities such as P, defined as the number or density of predators. We have left vague where these individuals are located: they are simply lumped together, irrespective of spatial position or their actual ability to interact with other members of the "population." What we really need are variables such as

$P(x)$ the number of predators at one-dimensional position x;

$P(x, y)$ the number of predators at two-dimensional position (x, y);

P_i the number of predators in spatial cell or patch i.

The challenge with spatially explicit dynamics is not just one of more complex descriptions and more difficult mathematics. Rather, it is that qualitatively new outcomes emerge. This chapter is only an introduction to this rich field—a sampling of a set of approaches to ecological modeling of spatially based population dynamics.

SPATIAL REPRESENTATIONS

We begin with a description of the commonly studied representations of space, as shown in Figure 16.1. The initial assumption illustrated is that the habitat is uniform over all the patches or cells. But remember, it is possible (even likely) that this is not true. For example, mean temperature declines as latitude or altitude increase or a subset of the available patches contains an exotic weed, parasite, competitor, and so on.

We may distinguish two different representations for species within spatial cells, on patches, or on islands. We can enumerate the actual population size (or density) or, alternatively, we might merely represent the presence or absence of each species (1 or 0). With this latter approach we lose considerable information, yet it is often a more consistent modeling representation when the available data are coarse. That is, a species

---

*Prepared jointly with Mike Gilpin, Professor of Biology, University of California at San Diego.

Figure 16.1
Some possible ways of including
spatial configuration in ecological
models.

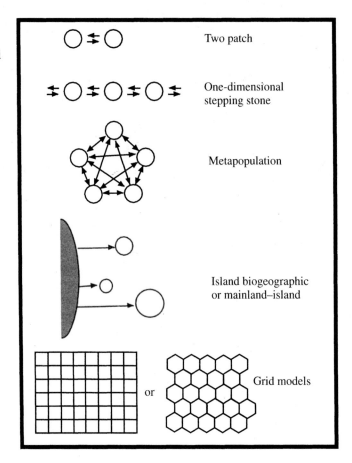

Two patch

One-dimensional
stepping stone

Metapopulation

Island biogeographic
or mainland–island

or   Grid models

list (presence or absence) for a patch is much more easily obtained than complete and accurate censuses of each of the species populations.

The simplest approach is to go from a spatially aggregated one-patch model to one with just two patches and movements of individuals between them. As described in Chapter 2, movements based on a random walk are a simple starting point. Here the probability of movement and the distance moved are independent of local population density and direction is random. Such movements exclude seasonal migrations and other movements that are more directed and sometimes to a predetermined endpoint.

We offer two examples where the introduction of the spatial dimension in simple two-patch models yields new emergent behavior. The first involves two competitors and shows that spatial pattern can arise even with complete equivalence of the two patches (except for initial conditions). The second involves a predator and its prey and shows that movements between two patches can lead to stable coexistence that was impossible when the patches were closed to interchange.

A Two-Patch Model for Competition

Imagine that we have two competing species, 1 and 2, each in two identical and adjacent patches, A and B. These two species are Lotka–Volterra competitors, and interspecific competition is stronger than intraspecific competition so that the interior equilibrium point is unstable and the species cannot coexist (see Chapter 14), as depicted in Figure 16.2.

Now suppose that one of the two adjacent patches is at one of the boundary equilibria while the other patch is at the other one, as shown in Figure 16.3. As long as individuals are restricted from moving between the patches, this configuration will be stable. However, suppose some gates are opened between the two patches so that individuals can move back and forth. What will happen? Will the species coexist, or will one species oust the other?

Figure 16.2
An example of two-species
Lotka–Volterra competition where the
internal equilibrium point is unstable.
There are two domains of attraction:
$(0, K_2)$ and $(K_1, 0)$. Trajectories will go
to one of the two boundary solutions,
each containing only one species (the
black dots).

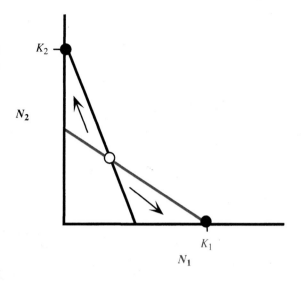

Figure 16.3
(a) Two adjacent and identical patches
exist at the two different domains of
attraction, K_1 and K_2. (b) We open
some gates between the two patches
and individuals begin to move back
and forth.

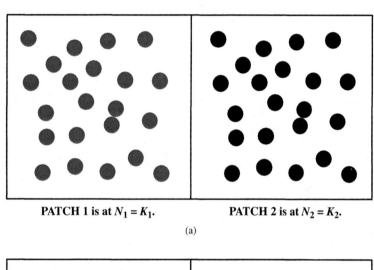

PATCH 1 is at $N_1 = K_1$.   **PATCH 2 is at $N_2 = K_2$.**

(a)

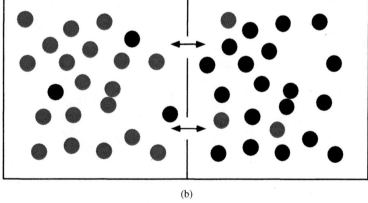

(b)

A full analysis involves writing four differential equations, one for each species in each of two patches (Levin 1974). The existence and stability of equilibria must be evaluated in four dimensions: $N_{1,A}$, $N_{1,B}$, $N_{2,A}$ $N_{2,B}$, where the first subscript is for species and the second is for patch. Even though we cannot plot isosurfaces in four dimensions, we can gain some insight about this system by approximating the four dimensional dynamics based on the special initial conditions—where each patch begins with just one species. An approximate zero-isocline depiction for just patch A is developed in Figure 16.4. We imagine that every individual of species 1 that leaves patch A dies from competition soon after it gets to patch B so that patch B stays at essentially all species 2.

Figure 16.4
Movements affect the position and shape of the zero-isoclines in patch A. The old zero-isoclines (i.e., without movement) for the two species are shown as dashed lines. The zero-isoclines with movements are solid lines. The assumption is that patch B has all species 2 and no species 1, while patch A is in the reverse configuration.

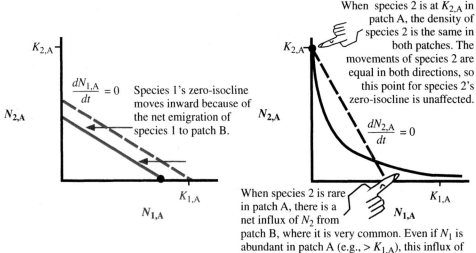

When species 2 is at $K_{2,A}$ in patch A, the density of species 2 is the same in both patches. The movements of species 2 are equal in both directions, so this point for species 2's zero-isocline is unaffected.

Species 1's zero-isocline moves inward because of the net emigration of species 1 to patch B.

When species 2 is rare in patch A, there is a net influx of N_2 from patch B, where it is very common. Even if N_1 is abundant in patch A (e.g., $> K_{1,A}$), this influx of species 2 from the neighboring patch B keeps species 2's growth rate positive. Hence species 2's zero-isocline does not hit the N_1 axis despite the intense local competition with species 1.

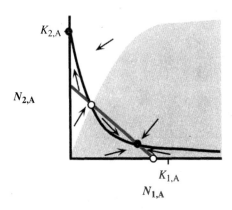

Figure 16.5
Approximate shapes for the zero-isoclines in patch A, as they would be if patch A began at $(K_{1,A}, 0)$ and patch B began at $(0, K_{2,B})$. The arrows show the qualitative direction of trajectories in different regions of state space. The gray area shows the approximate domain of attraction for this stable equilibrium. Open circles indicate unstable equilibrium points and filled circles are stable equilibrium points.

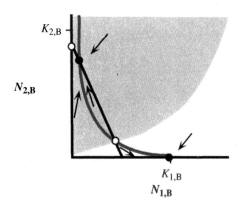

Figure 16.6
Approximate shapes for the zero-isoclines in patch B, as they would be if patch A began at $(K_{1,A}, 0)$ and patch B began at $(0, K_{2,B})$. The gray area shows the domain of attraction for this equilibrium.

Based on these arguments, the zero-isoclines for patch A look like those shown in Figure 16.5.

If the dispersal rate in both directions and for both species is μ with the gates open, then patch A receives an initial influx of approximately $\mu K_{2,B}$ immigrants of species 2 per unit time while it loses $\mu K_{1,A}$ emigrants of species 1 per unit time to patch B. Patch A reaches a new equilibrium with somewhat less than $K_{1,A}$ individuals of species 1 coexisting with a small number of individuals of species 2. Patch A at $K_{2,A}$ and no species 1 is also a stable equilibrium, but it would not be reached since the initial conditions $N_{1,A} = N_{1,A}$ and $N_{2,A} = 0$ are outside its domain of attraction. By an analogous argument the zero-isoclines for patch B have the form shown in Figure 16.6. With the gates open, patch B receives an initial influx of approximately $\mu K_{1,A}$ immigrants of species 1 per unit time (from patch A) and it is initially losing $\mu K_{2,B}$ emigrants of species 2 per unit time to patch A. Patch B reaches a new stable equilibrium with somewhat less than $K_{2,B}$ individuals of species 2 coexisting with a small number of individuals of species 1. Patch B at $K_{1,B}$ and no species 2 is also a stable equilibrium, but it would not be reached since the initial conditions $N_{2,B} = K_{2,B}$ and $N_{1,B} = 0$ are outside its domain of attraction.

Problem: Derive the form of species 2's zero-isocline in patch A. Begin with the equation

$$\frac{dN_{2,A}}{dt} = \frac{r_2 N_{2,A}}{K_2}(K_2 - N_{2,A} - \alpha_{21}N_{1,A}) - \mu N_{2,A} + \mu N_{2,B}.$$

The first subscript is for species and the second is for patch. We assume that $N_{2,B} = K_2$ and that $N_{1,B} = 0$. The parameters r, K, and α_{ij} are not different between the two patches, but they are potentially different between the two species. Show that the solution is the quadratic equation,

$$0 = N_{2,A}^2 - N_{2,A}\left(K_2 - \alpha_{21}N_{1,A} - \frac{\mu K_2}{r_2}\right) - \frac{\mu N_{2,B}K_2}{r_2}.$$

We can evaluate it numerically and plot the zero-isocline of species 2. If we let $r_2 = \mu$, and $K_2 = 1$, this equation simplifies to

$$0 = N_{2,A}^2 + N_{2,A}\alpha_{21}N_{1,A} - 1$$

The following is a plot of the solution of this equation in the positive orthant. Note the similarity with the isocline depiction in Figure 16.5.

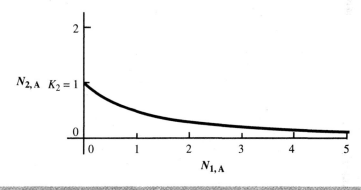

Figure 16.7 shows a simulation of this model for two different dispersal rates, μ. In this simulation, species 2 has a slight competitive edge because its carrying capacity is a tiny bit higher than species 1. The two patches are initiated with only species 1 on patch A and only species 2 on patch B. When dispersal is low, both species coexist in both patches, but when dispersal is doubled, species 1 becomes extinct everywhere.

Thus these two species, which cannot coexist in a single patch, can still coexist in the collection of two patches if the movement rates are not too great. This shows that even without spatial heterogeneity in the environment, when multiple domains of attraction exist, the landscape can contain a mosaic of species occupancies given sufficiently low dispersal rates (Yodzis 1978). With higher movements the two patches function as a single "well-mixed" patch. One or the other species would win out in both patches. Another way coexistence might be enhanced is when there is spatial heterogeneity in the patch characteristics. If species 1 were favored in patch A and species 2 were favored in patch B, the species might be able to coexist in both patches because each patch serves as a source of colonists of its favored species to the alternative patch.

Figure 16.7
Simulations of the two-patch
competition model for (a) low and
(b) high (doubled) movement rates.

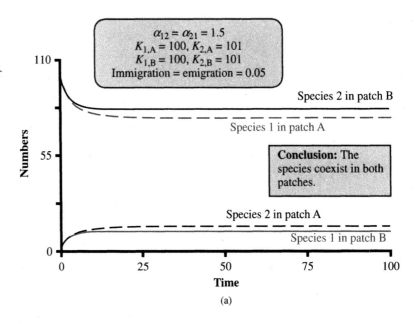

$\alpha_{12} = \alpha_{21} = 1.5$
$K_{1,A} = 100, K_{2,A} = 101$
$K_{1,B} = 100, K_{2,B} = 101$
Immigration = emigration = 0.05

Species 2 in patch B

Species 1 in patch A

Conclusion: The species coexist in both patches.

Species 2 in patch A

Species 1 in patch B

(a)

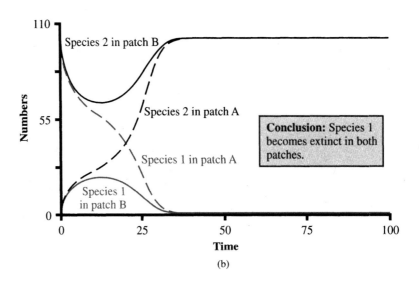

Species 2 in patch B

Species 2 in patch A

Conclusion: Species 1 becomes extinct in both patches.

Species 1 in patch A

Species 1 in patch B

(b)

Problem: Draw qualitative zero-isoclines for patch A that would be consistent with the situation illustrated in Figure 16.7(b).

A Two-Patch Predator–Prey Model

Recall from Chapter 13 the classical Lotka–Volterra predator–prey model with a neutrally stable equilibrium point:

$$\frac{dV}{dt} = V(b - aP) \tag{16.1a}$$

and

$$\frac{dP}{dt} = P(-d + kaV). \tag{16.1b}$$

Figure 16.8
The predator, *P*, and prey, *V*, zero-isoclines for the classical Lotka–Volterra equations. Regions of positive growth for each species are shaded and hatched. Two egg-shaped neutrally stable solutions are shown cycling around the equilibrium point (*V**, *P**). Neutrally stable cycles emerge, with their position dependent on the initial conditions.

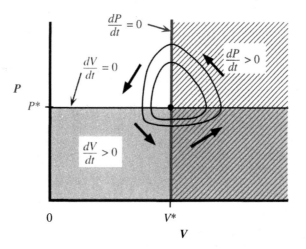

In the absence of the predator, the prey population, V, grows exponentially with an intrinsic growth rate, *b*. The prey death rate increases linearly with the number of predators. Predators have a type 1 functional response, *kaV*, with an encounter rate parameter, *a*. The zero-isocline is depicted in Figure 16.8. Trajectories spiral around the equilibrium point, but the cycle is neutrally stable and determined by the initial conditions. Now imagine two patches that are adjacent to one another. The predator and prey occur in each patch, and individuals of both species can potentially move between the two patches.

It again takes four differential equations to describe this system: one for each of the two species in each of the two patches. We label the two patches A and B and subscript prey and predator growth according to patch:

$$\frac{dV_A}{dt} = V_A (b_A - a_A P_A) - \mu_{V,A} V_A + \mu_{V,B} V_B ; \qquad (16.2a)$$

$$\frac{dV_B}{dt} = V_B (b_B - a_B P_B) - \mu_{V,B} V_B + \mu_{V,A} V_A ; \qquad (16.2b)$$

$$\frac{dP_A}{dt} = P_A (-d_A + k_A a_A V_A) - \mu_{P,A} P_A + \mu_{P,B} P_B ; \qquad (16.2c)$$

$$\frac{dP_B}{dt} = P_B (-d_B + k_B a_B V_B) - \mu_{P,B} P_B + \mu_{P,A} P_A . \qquad (16.2d)$$

Note in Eq. (16.2a) that some movement terms have been tacked onto Eq. (16.1a): prey emigrate from patch A with per capita rate $\mu_{V,A}$. The total number leaving patch A per unit time is then $\mu_{V,A}$ times the number of prey in patch A, $\mu_{V,A} V_A$. Prey also immigrate into patch A from patch B, so the number entering patch A per time period is $\mu_{V,B}$ times the numbers of prey in patch B (*not* in patch A). The other equations are similarly modified. To solve for the possible equilibrium points for this system, we set all four equations to zero. First, we focus on Eq. (16.2a):

$$0 = V_A*(b_A - a_A P_A*) - \mu_{V,A} V_A* + \mu_{V,B} V_B*.$$

Dividing each term by V_A* gives

$$0 = b_A - a_A P_A* - \mu_{V,A} + \mu_{V,B} \frac{V_B*}{V_A*}. \qquad (16.3)$$

We can see immediately that Eq. (16.3) does not yield a linear equation involving any of the equilibrium densities. That is, a change in P_A* is accompanied by a change in a *ratio*, V_B*/V_A*. This presents no fundamental problem but does make the math more difficult. We have four *nonlinear* equations to solve for four unknowns (V_A*, V_B*, P_A*, and P_B*) instead of four linear equations, which would be the case if we disal-

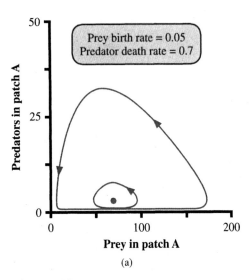

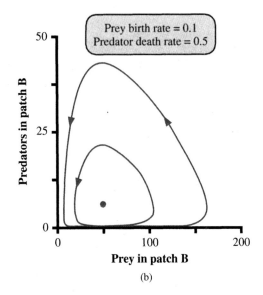

(a)   (b)

Figure 16.9
Lotka–Volterra predator–prey systems in two uncoupled patches of different quality: (a) patch A (poor) and (b) patch B (good). Neutrally stable cycles are shown for two different initial conditions. The equilibrium point in each patch is shown as a dot.

lowed movements between the patches. We also see that the equilibrium densities in each patch depend on the movement parameters as well as the birth and death rates.

If the movements in both cells are equal and if both cells are identical (have identical parameters), then $\mu V_A = \mu V_B$ and $V_B{}^* = V_A{}^*$ so that Eq. (16.3) simply becomes

$$0 = b - aP_A{}^*,$$

which is the same solution as without movements. Thus with movements the behavior of this system can be qualitatively altered, but only if movement rates differ across the two patches or the patches affect birth rates and death rates differently.

Once we have solved for the interior equilibrium point, we can determine the local stability of that point by solving for the eigenvalues of the Jacobian matrix evaluated around that point (see Chapters 13–15). However, this four-dimensional problem would be quite tedious to solve and, anyway, is beyond the scope of this primer. Instead, we simulate the system represented by Eq. (16.2), to illustrate the point. One patch is made better than the other for both predator and prey. Thus the prey have a higher birth rate, *b,* in patch B than in patch A, and the predators have a lower death rate, *d,* in patch B than in in patch A. First, we show the results without movements in Figure 16.9, obtaining the familiar neutrally stable cycles.

Figure 16.10 shows the results when the two patches are opened to movements. Note that the positions of the equilibrium points are in slightly different positions in Figures 16.10(a) and (b). Mathematically, the movements did not "stabilize" a previously neutrally stable equilibrium point; instead, they created a new equilibrium point in a four-dimensional state space. Similarly, if the two patches were identical in terms of birth and death rates for each species but movement rates differed between the patches, the neutrally stable equilibrium point could disappear, and a new stable equilibrium point could emerge. However, in either of these situations, if the movement rates are too high, then the two patches are so glued together by dispersal that they function as one large "well-mixed" patch, and we are back where we started with neutral stability. Similarly, if the movement rates are extremely low, then the patches are effectively isolated—each behaves independently yielding neutral stability. It is only for intermediate movement rates that a qualitatively new result emerges.

Imagine an experiment where species' dispersal rates between several patches are artificially enhanced. This would act to homogenize the species composition across patches. Some species may now become extinct because predators locally drive prey extinct or superior competitors outcompete inferior competitors. On the one hand, high

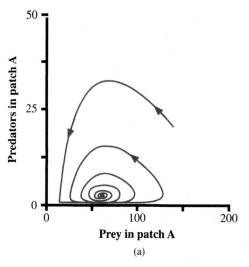

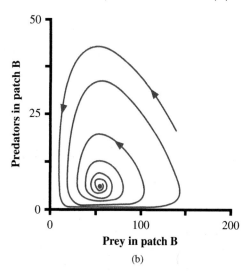

Figure 16.10

As in Figure 16.9, but the two patches are now coupled by movements of individuals between them. The neutrally stable cycles have been replaced by stable equilibrium points where prey and predator coexist in each patch but at different densities. In this simulation the movement rates are equal for both cells. The predators move with rate $\mu = 0.05$, and the prey move with rate $\mu = 0.1$. Also see Murdoch and Oaten (1975) for a discussion of these dynamics.

dispersal might bring efficient predators together with vulnerable prey, potentially causing extinction of both. On the other hand, if dispersal is too low, then subpopulations that decline cannot be "rescued" by immigrants. Environmental fluctuations could lead to both local and regional extinction. This suggests that there might be some intermediate level of dispersal that leads to the highest persistence and thus the greatest overall species diversity in patchy communities.

Vandermeer et al. (1980) performed an experiment to test these ideas. The "patches" were piles of compost. In the experimental piles, one-quarter of each compost pile was split off and the portions were transferred to each of the other piles twice a week. The control group was identical but no transfers took place. The result was that, while the average species number per patch was similar in both groups, the overall number of species in the entire collection of patches was about 50% less in the experimental group with enhanced dispersal. Thus species composition differed more from pile to pile in the control group than in the experimental "mixed-up" group of populations. Piles were not experimentally isolated, so it is unknown whether the diversity reduction that resulted from one-quarter transfers could be duplicated by experimentally enforcing less colonization than existed in the controls.

LARGER SPATIAL ARENAS

We want to move from two-patch models to many-patch models. One possible pathway is indicated by the following diagram. The habitat in each cell is assumed to remain the same; we're simply modeling more but smaller local populations, potentially exchanging individuals with immediate neighbors.

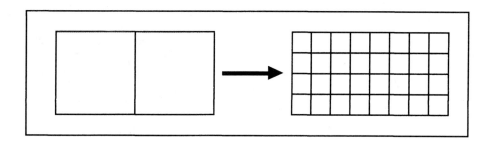

In the limit when patches are infinitely small and time is continuous, this model takes the form of a *reaction–diffusion equation*. For two-dimensional space (a plane) we would have equations like the following for species i:

$$\frac{\partial N_i(x,y,t)}{\partial t} = F_i\left[N_1(x,y,t), N_2(x,y,t), \ldots, N_n(x,y,t)\right] + D_i\left[\frac{\partial^2 N_i(x,y,t)}{\partial^2 x} + \frac{\partial^2 N_i(x,y,t)}{\partial^2 y}\right].$$

The growth rate of species i at position (x, y) and at time t $=$ a growth function for species i that depends on the densities of all the species at position (x, y) at time t. This is the **reaction** part of the equation. $+$ **diffusion** across two-dimensional space (x, y) with diffusion coefficient D_i.

We presented a single-species reaction–diffusion equation in Chapter 2 when we explored exponential growth (the reaction component) with diffusive movements. With several species the reaction–diffusion equations are coupled, and the reaction component contains both intra- and interspecific interactions. Further, the movement portion of the equations may be modified to represent different kinds of movements other than pure diffusion. The analysis and solution can be quite formidable and is far beyond the level of this primer. For such models, as we have shown for the two-patch models, the asymptotic pattern across space may be complex. Instead of the same equilibrium point at each position, different positions may take on different equilibrium values; these density differences between neighboring spatial locations are not smoothed out by diffusion but rather depend on it. Or a limit cycle may exist in both time and space—such that the abundance of a species oscillates in each single position, and at any single time species abundances fluctuate over space.

These bizarre dynamics are not simply theoretical constructs without bearing on the real world. Maron and Harrison (1997) described stable patches of tussock moth caterpillars feeding on their host plant, bush lupine, which is continuously distributed. Moth eggs and larvae that are experimentally introduced outside these patches do fine, so it seemed a mystery why the patches of caterpillars did not expand over time. The tussock moth females are wingless, so dispersal is limited. They are also predated and parasitized by several other species with greater mobility. Reaction–diffusion models of this interaction predict stable patches. This occurs because predators (or parasites) spill over the edges of prey patches, creating zones on the boundary where predator to prey ratios are elevated. Farther away from existing tussock moth patches, these ratios are lower and the prey population can increase (Brodmann et al. 1997). This suggests that new stable patches could be created if large enough experimental introductions are made. For those interested in pursuing these subjects in more detail, see the work by Tilman and Kareiva (1997).

A different modeling path is needed when space itself is patchy. We begin our analysis of such island biogeographic or metapopulation systems by restricting analysis to a single species on a single patch. This focal patch may be colonized either from a mainland or from a collection of similar patches nearby.

COLONIZATION AND EXTINCTION DYNAMICS OF A SINGLE PATCH

Often ecologists are willing to give up a detailed understanding of density differences across a wide region of space and settle for more qualitative information—simply the presence or absence of a species in a given area or quadrant. You may think that this is necessarily giving up a lot, but if the quadrants (or grid cells) are small, then not much is lost. In the limit where the quadrants are the same size as the average individual, we are back to a measure of absolute abundance. More typically, however, the landscape is patchy, with islands of livable space surrounded by a sea of more hostile terrain. The state variable in a patch–occupancy model is the fraction of patches that is occupied by a species, regardless of its density within. We assume that the location of the patch is irrelevant and that the densities in each patch are equivalent, which might be the case if the species quickly reached a carrying capacity that was approximately the same for all patches.

Let J be the fraction of patches occupied by some focal species. This **incidence** may change over time due to colonization of unoccupied patches and extinction events on presently occupied patches. Thus

$$\frac{dJ}{dt} = \text{gains through colonization} - \text{losses due to extinction}.$$

The fraction of islands experiencing an extinction in any time period is equal to the fraction of patches that are presently occupied, J, multiplied by the probability of an extinction, e, on such an occupied patch. Thus

Losses due to extinction = fraction of patches experiencing an extinction = eJ. (16.4)

At this point the development diverges, depending on the type of spatial configuration imagined and how it affects the "gains through colonization" expression. We will explore two approaches that differ in the geographic configuration of patches and the resulting colonization process. In the island biogeography framework, we imagine a group of islands that are colonized by individuals from a mainland (see Figure 16.1). The mainland presents an infinite shower of colonists, so the colonization rate does not depend on the number of islands or how many are already occupied. Instead, the colonization rate is simply a constant. Alternatively, a metapopulation is composed of patches, each loosely connected dynamically with one another through dispersal (see Figure 16.1). There is no mainland per se: Each patch or "island" serves as both a giver and receiver of colonists to/from the others. The colonization rate into any given vacant patch depends on how many other patches are already occupied and perhaps how close these patches are to the vacant patch. In both models, each patch has a finite chance of losing its population (an extinction) and, if this happens, then the patch may be recolonized by dispersal.

MAINLAND–ISLAND SYSTEMS

The physical configuration of the typical island biogeographic configuration is shown in Figure 16.1. It is assumed that a species under study comes from a single mainland, which is much larger than the island. The species also is presumed always to be present on the mainland, and thus ecological conditions on the mainland can be ignored.

Colonization can take place, by definition, only to empty islands. At any one time, the fraction of islands being colonized is therefore equal to the fraction of islands not presently occupied by the species, $1 - J$, multiplied by the probability of colonization, assumed to be a constant, c, for a given species. Thus for a single time interval, we have

Gains from colonization = fraction of all islands being colonized = $(1 - J)c$. (16.5)

At equilibrium, incidence J is not changing; the fraction of islands being colonized must equal the fraction of all islands suffering an extinction per unit of time.

If we call J^* the equilibrium incidence, then at J^* the gains through colonization equal the losses from extinction. Thus, using Eqs. (16.4) and (16.5), we have

$$J^* = \frac{c}{c+e} = \frac{1}{1+\dfrac{e}{c}}. \tag{16.6}$$

From Eq. (16.6), J^* approaches 1 as the colonization rate, c, grows large, and J^* approaches 0 as the extinction rate, e, becomes large.

Exercise Four species on an island have the following e's and c's:

$$e_1 = 0.1 \qquad c_1 = 0.1$$
$$e_2 = 0.01 \qquad c_2 = 0.01$$
$$e_3 = 0.001 \qquad c_3 = 0.001$$
$$e_4 = 0.0001 \qquad c_4 = 0.0001$$

What would their incidences be? What are you likely to see over a 500-year period of yearly presence/absence surveys on the island?

Answer: Since $e_i = c_i$ for all four species, all of their incidences are $J_i = 0.5$ (Eq. 16.6). Their turnover rates at equilibrium are $T_1 = 0.1$, $T_2 = 0.01$, $T_3 = 0.001$, and $T_4 = 0.0001$. Thus, over a 500-year period, there would be 50 extinctions and recolonizations of species 1, 5 of species 2, 1 of species 3, and probably none of species 4.

Exercise: A species has the following presence (1) and absence (0) pattern on an island over a 51-year period. Estimate its extinction probability, e, and its colonization probability, c.

00000000000000000011111111110000000000011100000000000

Solution: An extinction is a 1 followed by a 0. A colonization is a 0 followed by a 1. There are 11 "11"s and 2 "10"s; thus $e = 2/13$. There are 35 "00"s; and 2 "01"s; thus $c = 2/37$.

The Effects of Island Area and Isolation

To put this model in a geographic context, we allow the colonization and extinction curves to depend on island area and distance. One approach is to assume that colonization depends solely on the distance, d, that the island is from the mainland. A simple but still reasonable form would let colonization decline as an exponential with distance, or

$$c(d) = \omega \exp\left(\frac{-d}{d_0}\right), \tag{16.7}$$

where ω and d_0 are parameters that vary with species. Figure 16.11 illustrates the effect of d_0 on colonization rate. Changes in the parameter ω move these curves up or down on the y axis.

As we demonstrated in Chapters 2 and 6, the long term persistence of a population should increase with population size. An island with a larger area can support a larger absolute population size than a smaller island can. One reasonable guess for the extinction function therefore is

$$e(A) = \frac{\varepsilon}{A}, \tag{16.8}$$

Figure 16.11
Colonization rate decreases with an island's distance from the mainland according to an exponential function; $\omega = 1$.

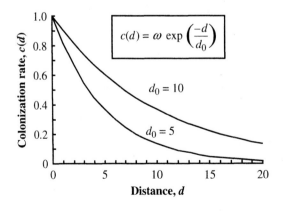

where ε is a constant to be determined from data. Now we substitute these expressions for $c(d)$ and $e(A)$ into Eq. (16.6) and solve for J^* as a function of area A and distance d:

$$J^* = \frac{1}{1 + \dfrac{\varepsilon}{A\omega\exp\left(\dfrac{-d}{d_0}\right)}} . \qquad (16.9)$$

Graphs of J^* versus d and A, based on Eq. (16.9), are shown in Figures 16.12 and 16.13.

Cole (1983) performed one of the few experimental tests of the relationship between island size and extinction rate by introducing ants of different species to very small mangrove islands off Florida. First, he found the smallest occupied islands for each of four species of ants. Then he introduced populations of ants to small islands where they were lacking. The results are shown in Table 16.1. Two ant species persisted on islands above but not below their empirical island-size threshold. However, introduced populations for two other species persisted even on islands smaller than their observed minimum. These later two species are competitive subordinates that are excluded by other ant species, and hence their absence is due more to the presence of competitors than island size per se.

Figure 16.12
Incidence on an island increases sigmoidally with island area. All else being equal, species with lower extinction rates will have higher incidence. Here, $d = 8$, $d_0 = 6$, and $\omega = 2$.

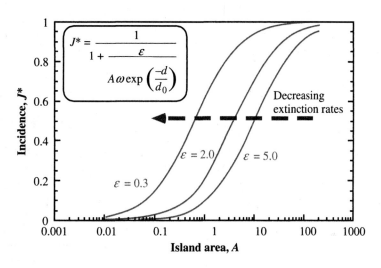

Figure 16.13
Incidence decreases on more isolated islands. All else being equal, the equilibrium incidence is higher for species with greater colonization abilities (higher ω). Here, $A = 10$, $d_0 = 6$, and $\varepsilon = 2$.

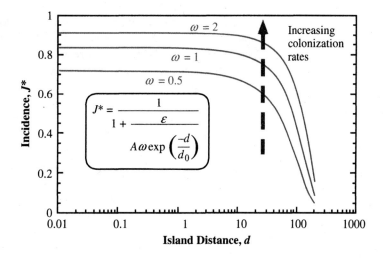

Table 16.1 Ants introduced to small mangrove islands off Florida (from Cole 1983). The number of populations persisting (for at least 27 weeks) out of the number of introductions. Failed introductions persisted an average of no more than 1.5 weeks.

Introduction species (minimum island size, m^3)	To islands smaller than minimum size	To islands larger than minimum size
Crematogaster ashmeadi **(0.3)**	0 of 3 persist	1 of 1 persist
Xenomyrmex floridanus **(1.2)**	0 of 3 persist	2 of 2 persist
Pseudomyrmex elongatus **(5.1)**	1 of 1 persist	—
Zacryptocerus varians **(12.6)**	2 of 2 persist	—

Figure 16.14
Colonization rates (new species per time period) decline with species number and with island distance. Extinction rates increase with species number and decrease with island area.

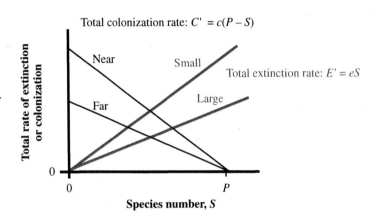

The Number of Species on Islands

The distribution of any single species will be clouded by chance factors. By studying all the species in some taxonomic group (a taxon), these factors should tend to cancel and the patterns expected due to island area and distance should stand out more clearly. Let's assume that the mainland contains P species that can possibly colonize the island and that S of these are already established on an island. Further, let's assume initially that all P species have the same colonization and extinction probabilities.

The total colonization rate in units of species per time period, denoted $C'(S)$, equals the colonization probability of individual species, $c(d)$, times the number of species not on the island, $P - S$. The total extinction rate of species, denoted $E'(S)$, equals the extinction probability per species, $e(A)$, times the number of species on the island, S. Extinction curves for larger and smaller islands and colonization curves for near and far islands are plotted in Figure 16.14. The total colonization rate must decline with species number because as more species occupy the island, fewer new species, $P - S$ are left to colonize. The extinction rate increases with species number simply because, if more species are present on an island, then a greater number of species exist that can potentially become extinct. The lines are linear because of the dual assumptions that all species have the same extinction and colonization rates, $e(A)$ and $c(d)$, and these per species rates are not influenced by species number, S. We relax these assumptions in the following sections.

An example of an intersection of $C'(S)$ and $E'(S)$, is shown in Figure 16.15. Where the two lines cross, extinction just equals colonization, so the number of species should reach an equilibrium. The expected equilibrium number of species on the island, S^*, and the expected rate of turnover of species, that is, the rate at which species replace each other over time, are labeled. This model, originally devised by MacArthur and Wilson (1967), predicts that in the face of ongoing colonization and extinction the number of species will not change greatly over time, once the equilibrium is reached.

Whether the number of species on an island reaches equilibrium depends critically on the rate of approach relative to the frequency of historical factors that may alter or

Figure 16.15
The intersection of the total extinction and colonization rates determines an equilibrium species number and an expected species turnover rate. This equilibrium species number, S^*, is stable.

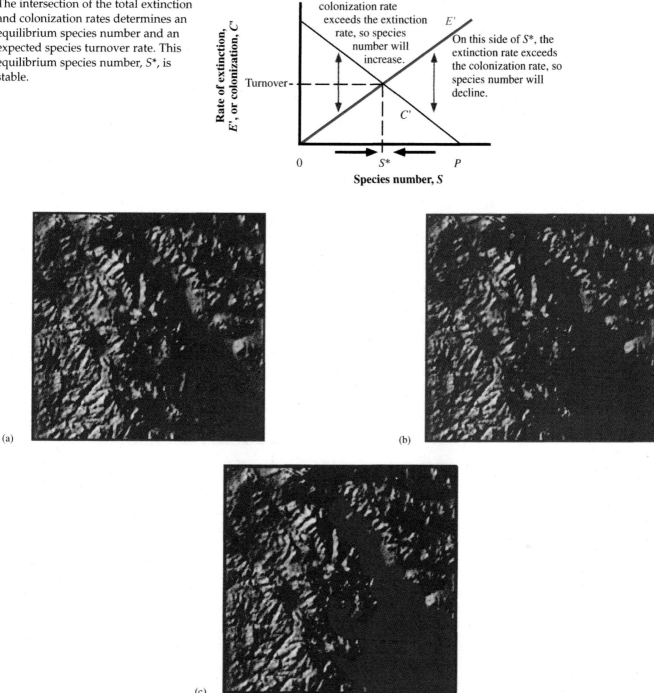

On this side of S^*, the colonization rate exceeds the extinction rate, so species number will increase.

On this side of S^*, the extinction rate exceeds the colonization rate, so species number will decline.

Rate of extinction, E', or colonization, C

Turnover

E'

C'

0　　S^*　　P

Species number, S

(a)

(b)

(c)

Figure 16.16
As the sea level (red) rises on this hypothetical landscape, high parts of the terrain become islands. Such islands are called landbridge islands.

perturb equilibrium species number, island area, and distance. Volcanic and plate tectonic upheavals can move islands and change their size. On a shorter time scale human activities like logging, urban development, and overhunting can change the island environment. For birds on the California Channel Islands, the rate of approach is apparently fast relative to the perturbations due to human disturbance (Jones and Diamond 1976). In other instances the approach to equilibrium is slow. One notable example is the apparent excess of species on post-Pleistocene landbridge islands. These islands became isolated as the glaciers melted about 12,000 years ago. Sea levels rose, pinching off mountain tops from the rest of the mainland as isolated islands, as shown in Figure 16.16.

Figure 16.17
The approach to the equilibrium species number, S^*, is in opposite directions for landbridge islands compared to islands that arise from the ocean floor.

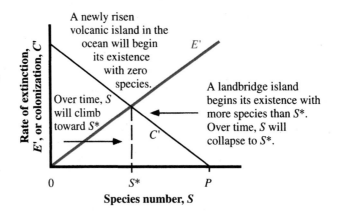

These islands initially hold more species than they can support at equilibrium and experience a net loss of species as they relax toward a lower equilibrium species number, as depicted in Figure 16.17. A number of studies have detected faunal relaxation on real and habitat islands (Brown 1971, Case 1975, Terborgh 1976, Gilpin and Diamond 1976, Wilcox 1978).

Exercise: Accurately draw the S versus $\log A$ curve for the following nine species with these incidence curves. Also indicate the approximate bounds of the 95% confidence intervals for the number of total species on islands of different sizes.

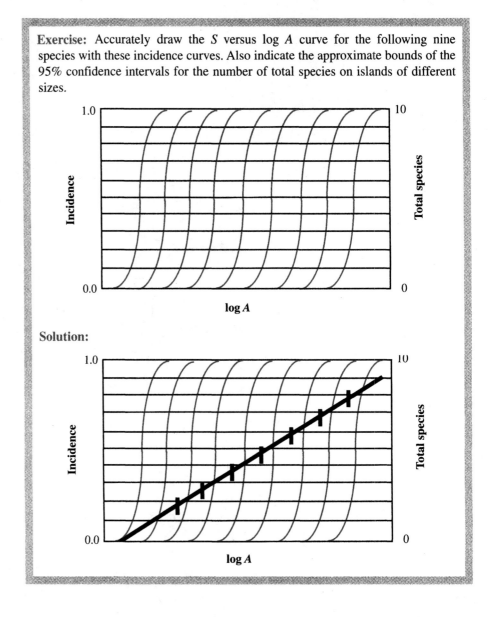

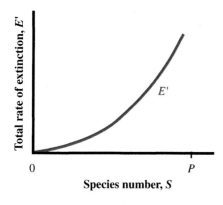

Figure 16.18
A modification of the total rate of extinction curve, E', to include potential interspecific competition. This extinction curve is based on Eq. (16.11).

Potential Effects of Competition on Extinction Rates

The presence of a competing species might reduce another species' access to physical and biotic resources and might thereby increase its probability of extinction. That is, the extinction probability per species, $e(A)$, should increase with the number of species on the island. For example, the extinction rate per species could be described by

$$e(A, S) = \frac{\varepsilon S^x}{A}, \qquad (16.10)$$

where x and ε are parameters, presumably positive, that must be determined by the analysis of actual data. This means that the total extinction function for all species will be

$$E'(S) = \frac{\varepsilon S^{x+1}}{A}, \qquad (16.11)$$

which has the form shown in Figure 16.18.

Interspecific competition could also effect the shape of the colonization curve. Case (1990) simulated colonists arriving in model Lotka–Volterra competition communities of different species numbers. The result was that the success of these invaders, evaluated as their ability to increase when rare in the presence of the residents (see Chapter 15), declined, on average, with increases in species number (Figure 16.19). As we described earlier, Cole (1983) introduced populations of ants to very small mangrove islands off Florida. While introduced species persisted on islands that were vacant and of sufficient size, all the introductions failed on islands already occupied by another ant species.

Problem: Let the per species colonization rate be given by the linear equation $c(S) = 0.5 - 0.01S$. Write an expression for the total colonization rate, $C'(S)$, as a function of species number, S. Plot this function for S from 0 to $P = 100$. Do you see why $C'(S)$ can go to zero before reaching $S = P$ in Figure 16.19?

Differences among Species

A second important consideration is that species will not be identical in their colonization and extinction probabilities, even on the same island. A top-trophic-level carnivore is certain to have a relatively small population size on the island and should therefore have a relatively higher probability of extinction. Species will also differ in their dispersal capabilities and their capacity to survive an overwater journey. These differences

Figure 16.19
Interspecific competition may also decrease the colonization success of colonizing species. (a) In simulation models this decline is linear (Case 1990). (b) Consequently, the total colonization success $C'(S) = c(S)(P - S)$, would decline with a concave form.

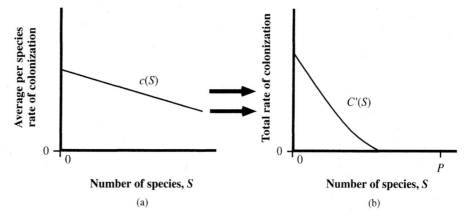

will be reflected in the parameters c and d_0. Thus (leaving out interspecific competition) we may write individual biogeographic rates for a particular species i as

$$c_i(d) = \omega_i \exp\left(\frac{-d}{d_{oi}}\right) \tag{16.12}$$

and

$$e_i(A) = \frac{\varepsilon_i}{A}. \tag{16.13}$$

The distributions for these three specific biogeographic rate parameters, ω, d_0, and ε, are usually not known. However, the mere fact that they differ from species to species allows for some partial conclusions based on the following logic.

Some species will be good colonists; they will tend to be the first to arrive on islands and the most likely to be present when S is low. Other species will be poor colonists and will tend to arrive late; they will thus usually be restricted to islands with high S. These factors will qualitatively affect the curves $C'(S)$ and $E'(S)$, as shown in Figure 16.20.

Thus mere heterogeneity among the mainland pool species (even without interspecific competition) can produce the same kind of curvature in the extinction curve as did competition among resident island species. Consequently, if we have a concave extinction curve, it will be difficult to tell which of these two features is the chief contributor. Figure 16.21 shows some colonization and extinction curves that have been extracted for land birds on islands in the Solomon Archipelago of the South Pacific. The concavity of both the colonization and extinction curves is extreme. It is not likely that interspecies competition could cause this concavity because these species are able to coexist on the source island of New Guinea. This result suggests that the species are very different (by orders of magnitude) in their colonization ability and their proneness to extinction.

Figure 16.20
When species differ in terms of colonization and extinction rates, the total colonization and the total extinction curves will be concave.

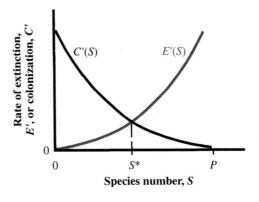

Figure 16.21
Examples of total colonization (black dash) and extinction (red) curves for landbirds on some Solomon islands. The intersections of the extinction and colonization curves for the same islands are marked by dots, which give S^* on the x axis. (a) All three islands share the same immigration curve. The area of these islands is in the order Shortland>Fauro>Vatilau.
(b) Nissan is larger and closer to the mainland than Ongong Java (from Gilpin and Diamond 1976).

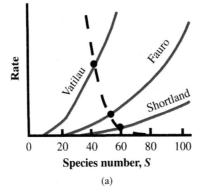

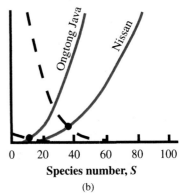

Species groups with rapid dispersal, good colonization, and high extinction rates should respond quickly to physical changes in the environment. A new equilibrium species number will be achieved quickly following physical and climatic changes. Hence we would expect this equilibrium theory to work relatively well in explaining differences in species number and composition across islands. Probably the clearest experimental support for the equilibrium theory comes from the studies of Simberloff and Wilson (1969, 1970). Small mangrove islands off the Florida Keys were defaunated with insecticide and the ensuing colonization of the islands by arthropods was studied over the next 4 years and compared to control islands. On manipulated islands, species number quickly climbed back to a level similar to that before defaunation. During both the colonization and plateau stages, species extinction and species turnover were evident (Simberloff 1976).

However, for species groups with low extinction rates and limited dispersal ability, changes in species number over time will be slow. Species number may not be able to keep pace with changes in the physical and climatic regime. For these organisms, and according to the distance and area of the islands, the biota will reflect more of history's past imprint than the ongoing dynamics of colonization and extinction as described in this equilibrium model. In such circumstances, an equilibrium species number will be a meaningless concept because it can never be achieved in the time span between major physical changes.

Nearly all theories of island biogeography predict changes in species number and composition with alterations in the physical and climatic setting. Equilibrium theory, however, predicts species turnover even in the absence of such physical changes. In theory this essential difference provides a way of testing the various theories. In practice, though, it is next to impossible to find a set of natural islands without environmental change, physical change, or the more recent impact of human activities. The mere presence of shifts in species number in the presence of environmental change is usually equivocal evidence, since it can be interpreted either as the system tracking a moving equilibrium or as the absence of any equilibrium whatsoever.

METAPOPULATION DYNAMICS

A metapopulation is simply a population of populations (Levins 1970). Habitat destruction associated with human activities inevitably leads to a patchwork of natural habitats surrounded by a sea of human-modified habitats. Consequently, metapopulation dynamics is relevant to the field of conservation biology. In a metapopulation model we assume that already occupied patches, rather than a mainland, are the source of colonists to vacant patches. Furthermore, in the simplest case, we invoke another very restrictive assumption: the spatial arrangement of occupied and empty patches makes no difference to the colonization rate for empty patches, as illustrated in Figure 16.22. With these assumptions, instead of Eq. (16.5), we have

$$\begin{bmatrix} \text{Gains from} \\ \text{colonization} \end{bmatrix} = \begin{bmatrix} \text{colonization} \\ \text{rate/empty} \\ \text{path/source} \\ \text{patch} \end{bmatrix} \begin{bmatrix} \text{number of} \\ \text{empty} \\ \text{patches} \end{bmatrix} \begin{bmatrix} \text{number of} \\ \text{source} \\ \text{patches} \end{bmatrix},$$

or in symbols,

$$\text{Gains from colonization} = c(1 - J)J. \qquad (16.14)$$

Thus in total we have

$$\frac{dJ}{dt} = c(1 - J)J - eJ. \qquad (16.15)$$

At equilibrium, J^*, the colonization rate must equal the extinction rate, or

$$c(1 - J^*)J^* = eJ^*$$

Figure 16.22
Patch occupancy models. After Levins (1969).

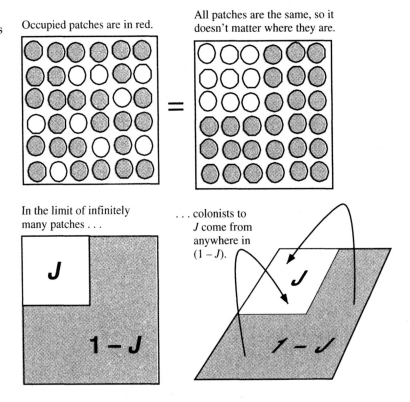

Occupied patches are in red.

All patches are the same, so it doesn't matter where they are.

In the limit of infinitely many patches . . .

. . . colonists to *J* come from anywhere in (1 − *J*).

Figure 16.23
The colonization curve from Eq. (16.14) and the extinction curve determine the equilibrium incidence, *J**. When *J* is below *J**, colonization exceeds extinction and thus *J* grows. When *J* is above *J**, extinction exceeds colonization and thus *J* declines. Therefore *J** is stable.

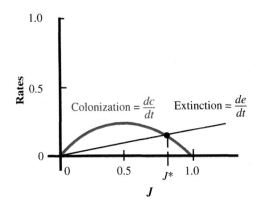

This yields (as the counterpart to Eq. 16.6) for the Mainland–island situation)

$$J^* = \frac{c - e}{c}.$$

(16.16)

We can conclude immediately that, unless $c > e$, a positive equilibrium incidence, J^*, cannot exist. Is this equilibrium, J^*, stable? Figure 16.23 shows a plot of the colonization and extinction rates as a function of incidence, J. The shapes of these curves match the situation of the continuous logistic curve with some superimposed density independent mortality (see Chapter 6), which yields a stable equilibrium point.

The equilibrium incidence, Eq. (16.16), is graphed as a function of colonization rate in Figure 16.24.

Figure 16.24
The equilibrium proportion of occupied patches as a function of the colonization rate for three different extinction rates. The amount of open space is given by the distance above each curve. For example, if $c = 1$ and $e = 0.2$, 80% of the patches will be occupied and 20% will be vacant.

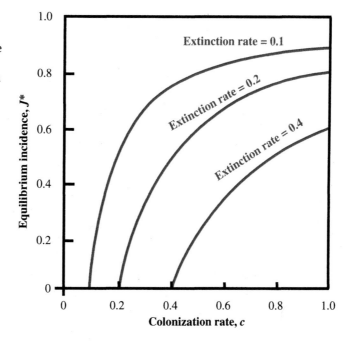

Figure 16.25
As in Figure 16.23, but a rescue effect is incorporated into the extinction rate. When the parameter ε is greater than the colonization rate c, it is possible to get an unstable equilibrium point in addition to a stable equilibrium point plus a stable trivial equilibrium point at $J = 0$. Here, $a = 3$, $c = 1$, and $\varepsilon = 1.5$.

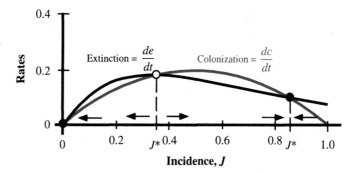

MODIFICATIONS TO THE BASIC METAPOPULATION MODEL

A Rescue Effect

Brown and Kodric-Brown (1977) argued that colonization and extinction will often not be independent. An island or patch that receives a high rate of colonization will also have a high rate of immigration when it is already occupied. Thus it will have a higher population size and therefore less likelihood of extinction. They called this the **rescue effect** since a population is rescued from extinction by continual immigration. In terms of the equilibrium theory of island biogeography, they reasoned that extinction rates would therefore increase with distance (as well as decrease with area). In a metapopulation setting, this notion can be incorporated into the model by assuming that the more filled the metapopulation (i.e., the higher the J), the greater is the rate of immigration and thus the lower is the extinction rate. The following functional form (Hanski 1991) is one way to incorporate these considerations:

$$\text{Losses from extinction, } e(J) = eJ^{-aJ}. \tag{16.17}$$

Figure 16.25 shows a plot of Eq. (16.17), along with the colonization curve from Figure 16.23. Note that for some parameter values, $\varepsilon > c$, an unstable equilibrium point appears.

Multispecies Extensions

An important question in ecology and conservation is whether species that cannot coexist locally (i.e., within a single patch) can nevertheless coexist within a network of patches (i.e., within a metapopulation). Such might be the case, for example, if for two competitors one is superior to the other and competitively excludes it wherever they co-occur, yet the inferior competitor has better dispersal capabilities and thus arrives first in vacant patches. In the ecological literature, an inferior competitor with a higher colonization rate is called a **fugitive species.** Clearly for such a fugitive species to persist, local extinction rates must be high enough that vacant patches are a relatively common occurrence. Let's now try to model this situation within the framework that we have already developed.

For the superior competitor, species 1, we may write

$$\frac{dJ_1}{dt} = c_1(1 - J_1)J_1 - e_1 J_1.$$

(16.18)

Equation (16.18) is the same Eq. (16.15) because the superior competitor's colonization and extinction probabilities are unaffected by the inferior competitor. But, for the inferior competitor, we have

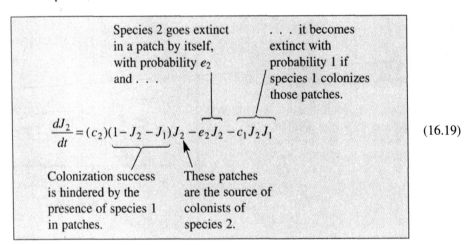

(16.19)

> **Problem:** Write a single equation for species i that is the analog to Eq. (16.19) and that applies to n competing species. Rank the species by their competitive ability such that species 1 is the superior competitor, then species 2, then species 3, and so on. (*Hint:* You will need to use the summation sign, Σ, twice in this equation.)

What is required for species 2 to coexist with species 1 at equilibrium: this will occur when $J_2^* > 0$ and $J_1^* > 0$. Setting Eq. (16.19) to zero and solving for J_2^* gives

$$c_2(1 - J_2^* - J_1^*) = e_2 + c_1 J_1^*$$

or

$$J_2^* = \left(1 - \frac{e_2}{c_2}\right) - J_1^*\left(1 + \frac{c_1}{c_2}\right).$$

(16.20)

Because

$$J_1^* = \frac{c_1 - e_1}{c_1},$$

(16.21)

we find that for $J_2^* > 0$ we must have

$$c_2 > \frac{c_1(c_1 + e_2 - e_1)}{e_1}.$$ (16.22)

We can see immediately that, for the condition of Eq. (16.22) to be true, the extinction rate of species 1, e_1 must be nonzero. Moreover, if extinction rates are equal, $e_2 = e_1$, then the condition of Eq. (16.22) simplifies further to

$$\frac{c_2}{c_1} > \frac{c_1}{e}.$$ (16.23)

Finally, if species 1 has a positive incidence to begin with (i.e., without species 2 present, $J_1^* > 0$), then c_1 must be greater than e_1. This implies that the right-hand side of Eq. (16.23) is greater than 1. So a necessary (but not sufficient condition) for invasion of species 2 is

$$\frac{c_2}{c_1} > 1.$$

Thus for coexistence we have a condition that the colonization rate of the inferior competitor must be greater than that of the superior competitor, $c_2 > c_1$, if extinction rates are equal. There will always be some magnitude of colonization advantage for species 2 that will be sufficient to override the competitive superiority of species 1—we just have to find a value of c_2 that fulfills Eq. (16.23). If the inferior competitor has a lower extinction rate than the superior competitor, it is possible for it to coexist even if its colonization rate is lower than that of the superior competitor.

Another useful generalization is to consider the criterion for coexistence in terms of the amount of open space left in the metapopulation with just the superior competitor present:

$$S_1 = 1 - J_1^* = 1 - \left(\frac{c_1 - e_1}{c_1}\right)$$

$$= \frac{e_1}{c_1}.$$

And again if extinction rates are equal, we can use the condition of Eq. (16.23) to find that species 2 can coexist if

$$J_2^* > 0,$$

which occurs when

$$S_1 > \frac{c_1}{c_2}.$$ (16.24)

In words, the inferior competitor can coexist if the proportion of open space when it invades exceeds the ratio of the colonization rates of the two species. This condition is plotted in Figure 16.26 for different levels of open space, S_1. The greater the amount of open space and the greater the colonization rate of species 2 relative to species 1, the easier it will be for species 2 to invade. Open space can be created by extinctions in occupied patches, but it can also be created by geological processes—landslides, floods, fires, and hurricanes can open up new habitat patches, thereby favoring fugitive species. Fugitive species make efficient use of these open disturbed habitats. They are adapted for high dispersal and colonization ability. However, if disturbances are too frequent—too much habitat upheaval—then even the fugitive species ultimately are eliminated. The **intermediate disturbance hypothesis** posits that species diversity should be highest at intermediate disturbance levels (Connell 1978). With too little disturbance the fugitive species become extinct, and only the competitive dominant species survive;

Figure 16.26
The region of parameter space that will support the coexistence of competitively inferior species 2 with species 1 for different levels of open space. The regions above the lines provide for a positive incidence for species 2 under different degrees of open space, S_1.

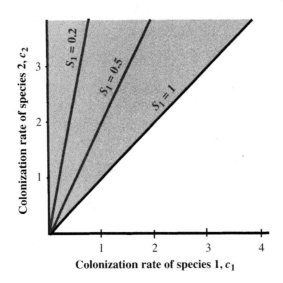

with too much disturbance, all species become extinct. This theory seems to work best for sessile, space-limited organisms like plants, corals, and intertidal invertebrates—all with indeterminate growth. One example is the sea palm, a large brown algae that grows in exposed waters with heavy wave action off Washington state. In more protected bays, it is competitively overgrown by mussels, which are frequently dislodged by wave action in the less protected sites where sea palms thrive (Paine 1979). If mussels are experimentally removed from protected bays, then sea palms grow well in these habitats too. Thus it is competition with mussels that normally prevents sea palms from dominating protected shores.

Exercise: Are there any conditions in this model that yield a situation such that the inferior competitor can exist in the metapopulation but the superior competitor cannot? If so, what are they?

Sousa (1979a, b) studied a diverse assortment of macroscopic algae and barnacles growing on submerged boulders in the near-shore ocean off California. These boulders get pushed around by storms and waves. He noticed that medium-sized boulders seemed to have a higher species diversity than either larger or smaller boulders, as shown in Figure 16.27(a). On the one hand, he found that competitive dominants monopolized space on the large boulders, which were resistant to physical disturbance by wave and tidal action. On the other hand, on the smallest boulders, disturbance was so frequent and catastrophic that most species were absent and open space was high, as shown in Figure 16.27(b). In a clever test of the hypothesis, Sousa secured the light-weight boulders to the ocean floor with strong adhesive. He reasoned that this experimental manipulation should cause an increase in diversity (and a decrease in open space) compared to unstabilized small boulders, and this is just the result that he found. Unstabilized rocks were dominated by sea lettuce (*Ulva*), a green alga that is a pioneer species and inhibits further colonization. However, even *Ulva* is eliminated by severe storms, setting the succession back to zero. Stabilized rocks were first dominated by *Ulva* too, but with time the rocks were colonized by several tougher red algae species and barnacles, which were more resistant to herbivores, particularly crabs (Sousa 1979a, b). After a year and half, the stabilized small rocks had over three times more species, on average, than the unstabilized ones.

It is possible to extend the two-competitor metapopulation model to a larger number of competitors; they all can coexist as long as there is a sufficient trade-off between their colonization rates and competitive ability (plus some additional bounds on the extinction rates; Tilman 1994). An example for three species is shown in Figure 16.28.

Figure 16.27
(a) The mean number of species of marine algae and barnacles (and the standard error) found growing on boulders of different sizes in near-shore communities off California. Two of four total samples are shown. Intermediate-sized boulders (which require a force of 50–294 Newtons to move them) consistently had the highest diversity. (b) The percent open space was highest for the smallest boulders, which are scoured most heavily by wave action. After Sousa (1979b).

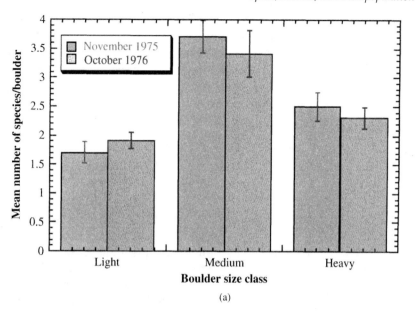

(a)

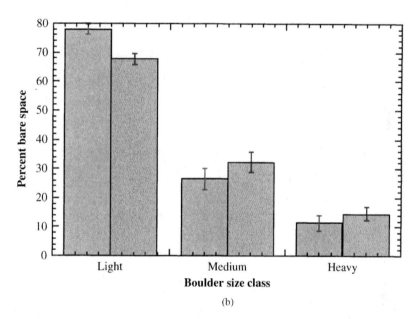

(b)

Figure 16.28
The dynamics of three competitors whose competitive abilities are in the rank order species 1 > species 2 > species 3. All three competitors have the same local extinction rate, $e = 0.1$, but differ in their colonization rates, c. All species begin at $J = 0.1$. The worst competitor, species 3, has the highest colonization rate and is the first species to dominate space but ultimately becomes the least common. The most dominant competitor, species 1, is the last to become common, but it ultimately gains the highest frequency. In this model, because of the trade-off between competitive dominance and colonizing ability, all three competitors can coexist at a stable equilibrium.

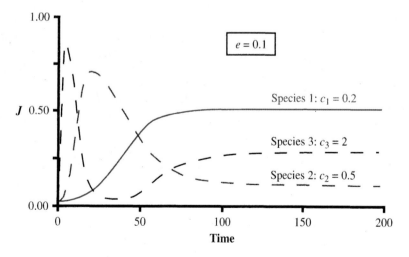

In this model the number of competitors that can coexist is theoretically unlimited, even though the best competitor displaces all other species locally wherever it occurs. However, as the number of species increases, the constraints on the parameters of extinction and colonization rates increase, leaving an increasingly small region of parameter space that will allow the next species to "fit in" as open space declines (see Figure 16.27). Also, it is important to realize that this model assumes an infinite number of patches. In the real world, where space is finite, rare species might still decline to extinction simply by chance, so it is doubtful that this model alone can explain infinite species coexistence. Also in the real world, the dispersal capabilities of species are more limited. Fugitive species may not be able to find vacant patches. In the next section we consider this problem.

Adding Neighborhood Dispersal

So far, these metapopulation models dealing simply with presence or absence have assumed that any patch can be colonized from any other patch with equal probability, an assumption that implies an infinite dispersal distance. This simplification might be a reasonable approximation for marine organisms with planktonic larvae or plants with windborne seeds, but more typically individual movements are limited. Unfortunately, attempts to restrict movements to smaller neighborhoods inevitably leads to the need for simulation for most realistic situations. The ecological software package Nemesis© (by Mike Gilpin) allows for a large uniform spatial habitat of square cells, as shown in Figure 16.1 (bottom). Movements of individuals from a focal cell can either be to just the four bordering cells (i.e., the nearest neighbors with direction chosen randomly) or to any cell in the entire grid with equal probability, as assumed in the classical Levins formulation. As before, a colonization event occurs only when an individual moves into a vacant cell. A comparison of the results of these two movement assumptions is shown in Figure 16.29.

One consequence of limited dispersal distance is immediately obvious: the equilibrium patch occupancy declines compared to the analytical Levins model with infinite dispersal. With limited dispersal there is a greater chance that several individuals leaving one set of cells will land at the same group of cells—leading to clumping. As clumping develops, more individuals will fall in or near the clumps and thus experience a much fuller habitat than is actually the case—leading to a lower equilibrium occupancy. As is evident in Figure 16.29, this can lead to extinctions that would not have resulted if movement distances were longer or habitat patches were closer (Lande 1987).

Another spatial simulation is to configure habitat patches in some realistic setting—with a "sea" of intervening habitat hostile to population growth. An example is shown in Figure 16.30. A simulation model developed by Gilpin (1987) represents an advance over so-called space-free models since the probability of colonization of a particular patch is determined by the sizes and distances of neighboring patches, rather than being the same for all patches regardless of their spatial distribution. The dynamical components in the model are the processes of colonization and extinction of individual patches. Within-patch dynamics from birth and death are not considered; a patch is either occupied or it is not. The dispersal between any pair of patches is assumed to be an exponentially declining function of their distance apart. Also since bigger patches hold more individuals and thus throw out dispersers at a greater rate, emigration from bigger patches should be higher than for smaller patches. The following function captures these features and is used in simulations:

$$c_j = k \sum_{i=1}^{n} A_i e^{-d_{ij}/d_0}, \tag{16.25}$$

where c_j is the colonization rate to patch j, d_{ij} is the distance from patch i to patch j, and d_0 and k are constants. The summation is over all the n patches of the metapopulation.

Figure 16.29
The frequency, *J*, of occupied patches over time based on (a) global dispersal or (b) dispersal only to the immediate bordering cells in the grid. Three simulations are shown for each case. The predicted equilibrium incidence, *J** is shown as a red horizontal line at 0.0909. All patches are occupied at the beginning of each simulation.

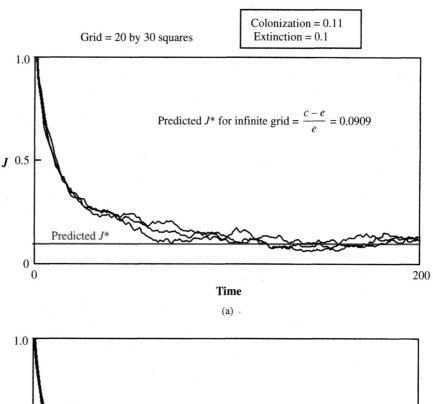

Grid = 20 by 30 squares

Colonization = 0.11
Extinction = 0.1

Predicted J^* for infinite grid $= \dfrac{c - e}{e} = 0.0909$

Predicted *J**

(a)

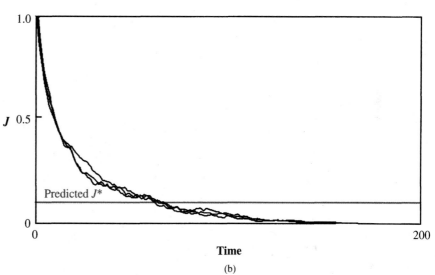

Predicted *J**

(b)

Figure 16.30
A possible metapopulation with (a) four patches and (b) the occupancy of each patch over time during 500 time steps. Occupancy is indicated by a black bar.

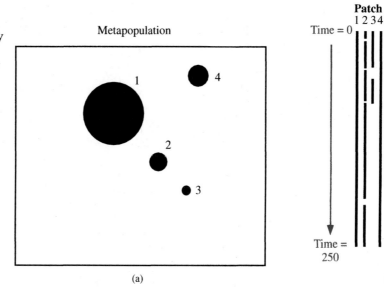

Metapopulation

(a)

(b)

Figure 16.31
Colonization and extinction rates for the patches shown in Figure 16.30. The last column shows the observed incidence for an extended run of 1500 time periods for each patch.

To Patch	From Patch 1	2	3	4	Sum, C
1	—	0.009	0.000	0.001	0.010
2	0.126	—	0.002	0.000	0.128
3	0.001	0.008	—	0.000	0.009
4	0.006	0.000	0.000	—	0.006

Patch	Extinction probability of single patch, e	Predicted approximate incidence, $(1 + e/c)^{-1}$	Observed incidence over 1500 time periods
1	0.001	0.909	1.00
2	0.010	0.928	0.921
3	0.040	0.184	0.170
4	0.007	0.461	0.589

The probability of extinction, e_j for a patch, j, is calculated as the inverse of the area of the patch, or

$$e_j = \frac{1}{A_j}. \qquad (16.26)$$

At each time step, a random number is compared to the probability of extinction for each patch. For empty patches (those that have gone extinct in some preceding time step) a random number is compared to the probability of colonization of that patch. Figure 16.30(b) shows the occupancy of each patch over 500 time steps of the simulation. All patches are initialized as occupied. The big patch, 1, stays occupied during the entire time while the smaller patches blink on and off to varying degrees. The smallest patch, 3, has the highest extinction rate, but since it is close to patch 2, which is itself close to patch 1, it is also frequently recolonized. Patch 4, is reasonably large and thus populations persist, on average, a long time; however, if they do become extinct, the isolation of patch 4 prevents it from being readily recolonized.

The parameters of extinction and pairwise colonization for this metapopulation based on Eqs. (16.25) and (16.26) are shown in Figure 16.31. Also shown is the predicted incidence of each patch from Eq. (16.6). This predicted incidence is only approximate because colonization is calculated as the sum of each pairwise colonization probability. Since neighboring patches are not always occupied they cannot always be a source for colonists; thus colonization is overestimated by this formula. Nevertheless, the predicted and observed incidences over 1500 time steps of a single simulation show reasonable agreement.

Exercise: In Figure 16.31, why not use the Levins formula, $J = (c - e)/c$ (Eq. 16.16) to calculate predicted incidence?

How important is each patch to the persistence of the entire metapopulation? An answer can be obtained by comparing the persistence time of the metapopulation with and without each patch. This is very time consuming since multiple replicates must be

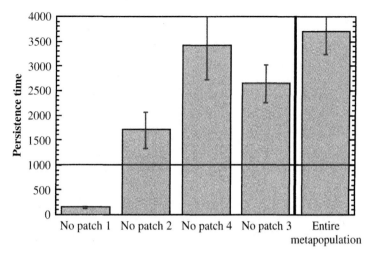

Figure 16.32
The average persistence time (plus or minus one standard error) for 25 simulations of each of five metapopulation scenarios. The entire four-patch metapopulation of Figure 16.30 is shown at the far right. The other bars represent different three-patch configurations based on the removal of a single patch, as indicated. The horizontal line shows the persistence time that the largest patch, 1, would have if it were alone.

performed for each scenario. An example based on 25 simulations for each case is shown in Figure 16.32.

Clearly, the largest patch, 1, is most critical: following its removal, the rest of the metapopulation collapses in short order. Patch 2 is also relatively important despite its small size, since it serves as a source of colonists for patch 1, in the rare event that patch 1 becomes extinct. This "subsidy" effect can be seen in Figure 16.32 by comparing the persistence time of all the metapopulations containing patch 1 (all but the far left bar) with the expected persistence time of patch 1 if it were alone (the black horizontal line). While patch 1 is important, the persistence time of each of these metapopulations containing patch 1 is substantially higher than that of patch 1 alone. Patches 3 and 4 are about equal in importance. While patch 3 is much smaller than patch 4, it is also closer to the center of mass of the metapopulation. This increases its importance as a source of colonists.

Of course, there are an infinite number of patch arrangements that could be explored with this model. The most relevant are those based on real situations—actual species occupying a real landscape. To apply this type of model to practical conservation issues, it is necessary to fit the parameters of the model to express each species' capabilities and to adjust the equations for extinction and colonization dynamics (Eqs. 16.25 and 16.26) to ensure that they are reasonable for the landscape.

Finally, it should be obvious that real-world landscapes differ from the abstract ones considered here because they contain patches of differing quality and movements across a real landscape are often more frequent and longer in some directions than others, due to winds, currents, elevational gradients, and the like. The incorporation of such realistic details into predictive models for the management of sensitive species is an active area of ecological research (Kareiva 1990).

Exercise: How would including population size and birth and death processes within patches affect the models presented in this last section? How might the presence of other species (e.g., competitors and predators) influence the likelihood of establishment and persistence of a "focal" species in a patch?

PROBLEMS

1. Which metapopulation probably has the lowest extinction probability?

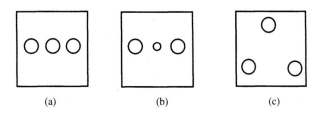

2. Carefully draw the S versus $\log A$ curve, using the coordinates at the right, given the four incidence versus $\log A$ curves shown.

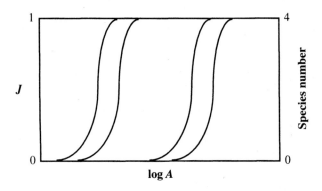

3. Three species can be on a patch. Their e's and c's are $e_1 = 0.001$, $c_1 = 0.1$, $e_2 = 0.01$, $c_2 = 0.1$, $e_3 = 0.1$, and $c_3 = 0.1$. What are their expected equilibrium incidences? Sketch the total $C'(S)$ and $E'(S)$ curves for the three-species pool, using the coordinate system shown.

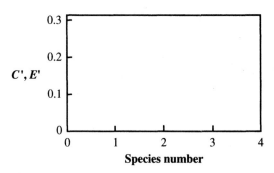

Appendix 1
Preparation

PART 1—VISUALIZING EQUATIONS

One goal of this primer is to develop analytical tools to understand population dynamics. We hope to predict events in the future and better understand the past by developing models and contrasting their behavior with empirical phenomenon. To ensure that a model's extrapolations are logically consistent with our understanding of the nature of the processes, we often use the precise language of mathematics. This preparatory section explains some of the common language in mathematics and makes connections between the algebraic expressions and their graphical forms.

Visualizing Equations

When you see an equation like the continuous logistic growth equation,

$$\frac{dN}{dt} = rN\left(\frac{K-N}{K}\right), \tag{A.1}$$

which is a very simple equation for population growth, you can apply some simple mental tricks to visualize the shape of its graph. Begin by arranging the equation in standard form: $y = f(x_1, x_2, x_3, \ldots)$. The y, or **dependent** variable, is generally placed on the left-hand side of the equals sign. The normal Cartesian plot has this y variable plotted on the **ordinate**, or vertical axis, of the graph. On the right-hand side of the equation are placed all the x terms, or **independent** variables. These can be plotted one at a time on the **abscissa**, or horizontal axis, of a graph. In the case of the continuous logistic equation we have one x variable, N, so the general form is $y = dN/dt = f(N)$.

Basically There Are Three Approaches to Visualizing the Appearance of f(x)

1. *The brute force approach.* For any equation $y = f(x)$ you can simply plug in various numeric values of x, calculate the corresponding values of y numerically, and then plot them on graph paper. In Eq. (A.1) the y variable is dN/dt and the x variable is N. Plug in various values of N and perform the arithmetic as prescribed in the equation to get the particular value of dN/dt. To do this you must choose some values for the two parameters, r and K. You can do the calculation more easily by choosing values that make the arithmetic simple (e.g., $r = 1$ and $K = 100$). You might want to get a feel for how this choice of r and K affects the resulting graph by making a **family** of different graphs, one each for a particular choice of a parameter, as follows.

$r = 1, K = 100$

$$\frac{dN}{dt} = rN\left(\frac{K-N}{K}\right)$$

N	dN/dt
0	0
10	9
30	21
50	25
80	16
100	0
150	−75

2. *Qualitative guess.* This method requires some thinking. You look at the equation and notice immediately that dN/dt will be zero at $N = 0$ and at $N = K$, but dN/dt will be bigger than 0 for any value of N between 0 and K. So a good guess at this point is that, since f starts at 0 when $N = 0$, then f gets bigger as N gets large, and eventually f returns to 0 when $N = K$, the expression will be humped-shaped, as shown.

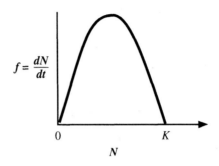

3. *Recalling similar forms from memory.* This technique will only work as well as your memory and your past experience with other mathematical relationships allow. Here are a few common forms that you should learn now.

   a. The equation of a **straight line:**

$$y = mx + b. \tag{A.2}$$

Memorize the equation of a straight line and you will be able to recognize immediately about 50% of the equations in population dynamics, simply by interpreting what is the x variable and what is the y variable in the equation. There are two parameters, m and b. The parameter m is the slope of the line, and b is the y intercept (i.e., the point on the y axis where the line intersects it). A straight line's slope is always a constant value by definition.

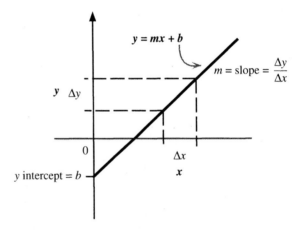

A linear equation in two dimensions, x and y, describes a plane.

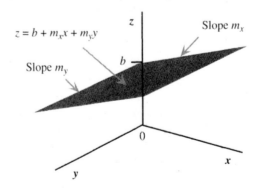

   b. A **quadratic equation:**

$$y = a + bx + cx^2. \tag{A.3}$$

Any equation with an x and an x^2 term will produce a **parabola.** When the sign of the c parameter is negative, it will lie with its hump pointing up so that it looks like a hill. When $c > 0$ it will lie upside down so that it looks like a valley. Note that the continuous logistic equation falls in the hill category: $dN/dt = rN - r/kN^2$. Here $c = -r/k$, so the hump points up. For the quadratic equation and for $f(x)$ other than straight lines, the slope of the curve is different at different places along x. We can define the slope at a particular point at x as the tangent line to curve y at point x.

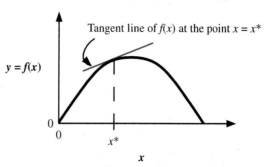

A quadratic equation that is a function of two variables, x and y, may bend up or down in either the x or y direction, depending on the sign of the coefficient for the squared terms, x^2 and y^2. For example, in the following case the coefficient of x^2 is positive, and the coefficient of y^2 is negative.

$$z = 0.5x + 0.25y + 0.5x^2 - 0.3y^2 + 1$$

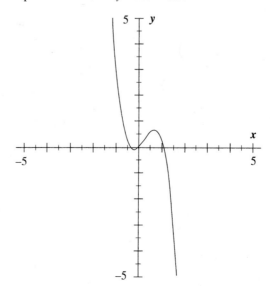

The curve in the x direction points down, and the hump in the y direction points up.

  c. A **polynomial equation:**

$$y = a + bx + cx^2 + dx^3 + ex^4 + \cdots . \qquad (A.4)$$

In general, this expression is going to yield a squiggly curve with multiple peaks and valleys. The total number of peaks plus valleys equals the highest exponent minus 1. Thus if you have a polynomial equation of maximum exponent 2, the graph of the parabola will have the number of peaks + valleys = $2 - 1 = 1$. As you add more higher-order terms, you get more and more bends in the curve. Here is a plot of the function $y = 0.5x^2 - 3x^3$.

You see one hill and one valley for this third-order polynomial. Note also that the curve crosses the *x* axis at three places (i.e., three places where *y* = 0). These are called the **roots** of an equation. An ***n*-order polynomial** has **n** roots and, if all the roots are real numbers, then the function crosses the *x* axis *n* times. If some of the roots are complex numbers (i.e., numbers with both a real and an imaginary part), the function cannot cross the axis in plots like the preceding one, since the *x* axis is made up only of real numbers. For the special case of a quadratic equation, $a + bx + cx^2$, which is a second order polynomial, there is a simple analytic formula for the two roots:

$$x = \frac{-b \pm \sqrt{b^2 - 4ac}}{2a}.$$

A discussion of finding complex roots for higher-order polynomials is complicated and beyond the scope of this primer.

    d. Sin and cosine functions: These functions have the form

$$y = a + b\ \sin(cx). \tag{A.5}$$

They oscillate up and down smoothly. In Eq. (A.5), the average value is *a,* the amplitude (or the height of the waves) is *b,* and the period of the cycle is 2*c,* the distance from peak to peak. The bigger *c* is, the wider are the waves. Here is the graph of *y* = 2 + 0.5 sin(2*x*).

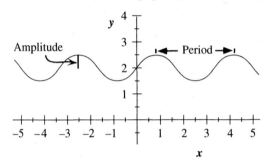

This curve has an amplitude of 0.5 and a period of 4, and the mean of the curve is at *y* = 2.

    e. Logarithms of *x.* These are functions of the form *y* = *a* + *b* log *x.* In the graphs of these functions, as *x* gets larger, so does *y* but at an ever decreasing rate. Usually we write log *x* when we mean log to the base 10, and ln *x* or $\log_e x$ to imply log to the base *e,* where *e* is the base of the natural logarithms, or about 2.7138. . . . Here's a normal plot of ln *x* over the range 0.01 to 100.

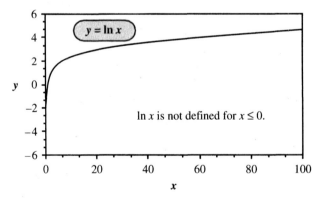

Now look what happens when *x* is plotted on a log scale. Here the scale is $\log_{10}$.

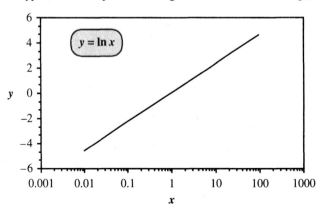

The curve is turned into a straight line because we're plotting one log function against another.

f. A **normal,** or **gaussian, curve** arises frequently when you're dealing with processes that have a random component. The curve is specified in one dimension by two parameters: a mean, $\bar{x}$, and a standard deviation, σ. Hence we write $\mathcal{N}(x; 10, 2)$ to describe the normal curve with mean = 10 and standard deviation = 2. The equation for a normal curve of one variable x is

$$\mathcal{N}(x; \bar{x}, \sigma) = \frac{1}{\sigma\sqrt{2\pi}} \exp\left(\frac{-(x - \bar{x})^2}{2\sigma^2} \right).$$

The first term, with π, simply scales the distribution to have an area of 1. Without this term, the value of $\mathcal{N}(x)$ at the mean would be 1, since $e^0 = 1$. Note that the value of $\mathcal{N}(x)$ can never become negative because of the squared term, but declines asymptotically to 0 as x gets farther from the mean. A plot of this curve for mean = 10 and standard deviation = 2 is shown. As σ becomes larger the bell-shaped normal distribution curve flattens out and becomes wider.

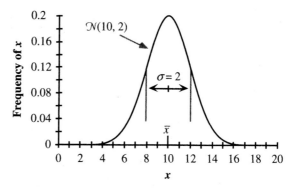

A normal distribution in two uncorrelated variables, x and y, would have this bell-shaped appearance in both directions if the standard deviations were equal.

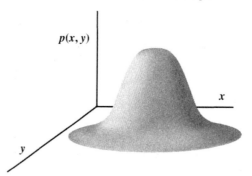

PART 2—TERMS AND METHODS OF MODEL BUILDING IN POPULATION BIOLOGY

A **model** is a way of describing the behavior of some process in order to predict its future or understand its past. In order to make a model logically rigorous and avoid ambiguities, we usually cast the model in concrete mathematical terms or as a computer program, although it could be made of clay, balsa wood, or simply words. If the behavior to be modeled is one in which variables change over time, it is called a **dynamical model.** It is often easier to understand a dynamical system by describing the forces acting on the system rather than the time course of the system itself. Think of the movements of all the heavenly bodies. Some comets have exceedingly complicated orbits; the position of a comet can be quite difficult to describe, yet these orbits are exactly predicted by the simple forces of gravity. Hence it is often easier to start with the forces and derive the consequences. Forces are typically expressed as rates of change. By integrating the rates of change, you can find functions that describe the position of the system variables over time. The rate of change is usually expressed as some function of the variables and parameters.

For example, here again is the continuous logistic equation Eq. (A.1), which is a dynamical model for single species population growth:

$$\frac{dN(t)}{dt} = rN(t)\left(\frac{K - N(t)}{K} \right).$$

This model expresses the instantaneous rate of change of a population's numbers, $dN(t)/dt$, as a function of those numbers $N(t)$ and two parameters, r and K. Once it is clear which symbols represent variables and which represent parameters, the explicit notation for time may be removed, leaving a slightly less cluttered notation:

$$\frac{dN}{dt} = rN\left(\frac{K-N}{K}\right),$$

which means exactly the same thing. When $dN/dt = 0$, the population size is no longer changing. We say that an **equilibrium** has been reached. Since the equilibrium is reached at population size $N = K$, K has a biological meaning: it is the carrying capacity of the population. The other parameter of this model is r, the intrinsic growth rate or the maximum instantaneous rate of change. The parameter r tells you how quickly K will be reached. The higher the value of r, the swifter the approach to K will be.

If a lot of different variables are changing in magnitude over time and if these variables affect each other's dynamics, then such a model is called a **dynamical system.** In this dynamical system the numbers of a prey species, V (for victim), and a predator, P, affect one another's growth rates:

$$\frac{dV}{dt} = bV(t) - aP(t)V(t) \quad \text{and} \quad \frac{dP}{dt} = aP(t)V(t) - dP(t). \qquad (A.6)$$

Don't worry now about the details of this model except to note that each prey gives birth with a per capita rate b and also has a death rate proportional to the numbers of predators at time $P(t)$. The exact death rate per prey is the product of the number of predators at the time and a constant a that expresses the predation rate *per* predator. The higher the predation rate, a, and the more predators around at time t, $P(t)$, the greater the prey death rate. Predators also die with a constant per capita death rate d. The model is developed in more detail in Chapter 13.

Variables or, more precisely, **state variables** are those elements of the model that are changing over time and whose ultimate behavior is to be predicted by the model. In Eq. (A.1) there is one state variable, population size, N. In Eqs. (A.6) there are two state variables, V and P. Sometimes, to avoid any confusion, the variables are written **explicitly** as functions of time, such as $N(t)$ and $P(t)$, $V(t)$, as in Eqs. (A.6). However, usually the model's structure makes it clear **implicitly** which terms are the variables. The left-hand side of Eqs. (A.6) state that V and P are changing over time. Thus we could rewrite Eq. (A.6) more compactly as

$$\frac{dV}{dt} = bV - aPV \quad \text{and} \quad \frac{dP}{dt} = aPV - dP. \qquad (A.7)$$

Parameters are the terms in the model whose magnitudes are preordained by the modeler; usually they are constants, but they could also vary in time or in space according to some specified arrangement. In Eq. (A.1) there are two parameters: r and K, both of which are constants. In Eqs. (A.6) and (A.7) there are three parameters: b, d, and a, all of which are constants.

We could suppose that the environment is seasonal and specify K as a sine function over time: $K(t) = K_{mean} + A \sin t$. While $K(t)$'s magnitude is now "variable" as it affects the state variable N, $K(t)$'s magnitude is not, in turn, affected by N. When parameters vary in a prescribed manner, they are called **exogenous variables** because they are not a part of the feedback structure of the system—they are outside it. We have specified their behavior to begin with. Exogenous variables act *on* the state variables, but they are not acted *upon* by the state variables. In this example we have used two new parameters to describe $K(t)$: K_{mean} is the mean value of K, and A is the amplitude of its fluctuations. Exogenous variables may have regular fluctuations, as in the example of $K(t)$ as a sine function, or they be described by unpredictable or **stochastic** fluctuations such as the mortality due to fires or storms.

Initial conditions are the magnitudes of the state *variables* at the beginning of time in the model. To solve Eq. (A.1) you will need to start the population growing at some specified initial population size (e.g., 10 individuals). To be more general, you may give these initial numbers a symbol, say, N_0. Furthermore, you must specify the time units (months, years, etc.) and the beginning time, say, August 3, 1999, or give it a more general symbol, t_0. Once units have been specified, you have to be careful to cast parameters in equivalent units; that is, if time is to be measured in years in Eq (A.1), then you will need to express r on a per year basis. It's easy to be careless about such things, and inconsistency in units can cause big numerical problems in the model's predictions.

Dynamical models may not always be cast in terms of **differential equations** (like Eqs. A.1 and A.6). For many systems it might be more realistic or more useful to cast these changes in terms of variables that change at discrete time intervals. For example, Eq. (A.6) could be modified to become the **difference equation:**

$$\frac{\Delta N}{\Delta t} = RN_t \, \frac{(K - N_t)}{K}.$$

It might also be desirable to restrict the variables so that they only take on integer values (e.g., 1, 2, 3, . . .) not fractional values. A differential equation would not be suited for this purpose, since you must assume that all the variables and time vary continuously.

Appendix 2
Some Matrix Operations

Matrices are boxes that hold numbers in rows and columns. Generally in this book, we use the boldface type to denote matrices and vectors.

MATRIX MULTIPLICATION

Matrix multiplication can be illustrated with an example from age-structured population growth. In Chapter 3, you learn how to form Leslie matrices: $n_1(0)$ is the number of individuals in age class 1 at time zero, s_1 is the survival rate of age class 1 to age class 2, and F_1 is the fecundity of females in age class 1.

Let's plug in some actual numbers: $n_1(0) = 30$, $n_2(0) = 20$, $n_3(0) = 10$, $s_1 = 0.6$, $s_2 = 0.9$, $F_1 = 0$, $F_2 = 1$, and $F_3 = 2$. One time step into the future we have

$$n_1(1) = n_1(0)\,(F_1) + n_2(0)(F_2) + n_3(0)(F_3)$$

$$= 30(0) + 20(1) + 10(2) + 40;$$

$$n_2(1) = n_1(0)(s_1)$$

$$= 30(0.6) = 18;$$

$$n_3(1) = n_2(0)(s_2)$$

$$= 20(0.9) = 18.$$

This can all be put in a much more concise format by using a Leslie matrix equation:

$$\begin{bmatrix} n_1(t+1) \\ n_2(t+1) \\ n_3(t+1) \end{bmatrix} = \begin{bmatrix} F_1 & F_2 & F_3 \\ s_1 & 0 & 0 \\ 0 & s_2 & 0 \end{bmatrix} \begin{bmatrix} n_1(t) \\ n_2(t) \\ n_3(t) \end{bmatrix}.$$

Using boldface to designate matrices and vectors, we can write the last expression as

$$\mathbf{n}(t + 1) = \mathbf{L}\,\mathbf{n}(t).$$

Note that this Leslie matrix is a square matrix: the number of rows equals the number of columns. To use the matrix to multiply the population vector, successively take each row of the matrix and multiply that row by the population vector. This vector by vector multiplication yields a scalar; these scalars stack up in $\mathbf{n}$ to become the new population vector at $t + 1$:

$$F_1 n_1(t) + F_2 n_2(t) + F_3 n_3(t) = n_{1(t} + 1);$$

$$s_1 n_1(t) + 0 n_2(t) + 0 n_3(t) = n_2(t + 1);$$

$$0 n_1(t) + s_2 n_2(t) + 0 n_3(t) = n_3(t + 1).$$

So for our example we have for $t = 1$:

$$\begin{bmatrix} 0 & 1 & 2 \\ 0.6 & 0 & 0 \\ 0 & 0.9 & 0 \end{bmatrix} \begin{bmatrix} 30 \\ 20 \\ 10 \end{bmatrix} \begin{matrix} = 0(30)+1(20)+2(10) = 40 \\ = 0.6(30)+0(20)+0(10) = 18 \\ = 0(30)+0.9(20)+0(10) = 18. \end{matrix}$$

The next time step, $t = 2$, yields

$$\begin{bmatrix} 0 & 1 & 2 \\ 0.6 & 0 & 0 \\ 0 & 0.9 & 0 \end{bmatrix} \begin{bmatrix} 40 \\ 18 \\ 18 \end{bmatrix} = \begin{matrix} 0(40) + 1(18) + 2(18) = 54 \\ 0.6(40) + 0(18) + 0(18) = 24 \\ 0(40) + 0.9(18) + 0(18) = 16.2. \end{matrix}$$

Remember this answer because you are now going to learn how to get it again by using a slightly different approach. The double iteration we just did to reach $\mathbf{n}(t + 2)$ can be described by

$$\mathbf{n}(t + 2) = \mathbf{L}\, \mathbf{n}(t + 1)$$

$$= \mathbf{L}\, \mathbf{L}\, \mathbf{n}(t).$$

You just learned how to multiply a matrix times a vector, but how do you multiply two matrices (e.g., $\mathbf{L}$ times $\mathbf{L}$)? Let's be more general and consider the product

$$\mathbf{C} = \mathbf{A}\, \mathbf{B}.$$

One rule of matrix–matrix multiplication is that, for the multiplication to even be defined, the number of columns of $\mathbf{A}$ must equal the number of rows in $\mathbf{B}$. So if $\mathbf{A}$ is of size $n \times m$ (i.e., n rows and m columns) and $\mathbf{B}$ is $m \times s$, then the product $\mathbf{A}\, \mathbf{B}$ is possible. However, If $\mathbf{B}$, were, say, of size $n \times m$, then the product $\mathbf{A}\, \mathbf{B}$ would not be defined. As a simple device we have the following matrix multiplication rule:

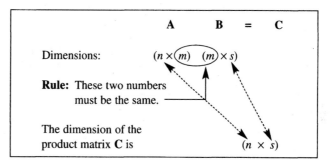

Before, we multiplied $\mathbf{L}$ (size 3×3) times $\mathbf{n}$ (size 3×1), so this is a compatible product. The result we found was a 3×1 vector. Note that, by this rule, in general the multiplication $\mathbf{A}\, \mathbf{B}$ may be permitted while the multiplication $\mathbf{B}\, \mathbf{A}$ may not be. Unlike the situation for scalar multiplication, the product of two matrices is different, depending on the order of multiplication. In other words $\mathbf{A}\, \mathbf{B} \neq \mathbf{B}\, \mathbf{A}$ except under restricted circumstances (e.g., $\mathbf{A} = 0$).

The multiplication of two matrices is performed similarly to that of a matrix times a vector, which you have already learned. The element c_{ij} of the matrix $\mathbf{C}$ is produced by multiplying the ith row of $\mathbf{A}$ times the jth column of $\mathbf{B}$. An example will illustrate.

Find $\mathbf{C} = \mathbf{A}\, \mathbf{B}$.

$$\overset{\mathbf{A}}{\begin{bmatrix} 1 & 2 \\ 3 & 4 \end{bmatrix}} \quad \overset{\mathbf{B}}{\begin{bmatrix} 1 & 2 & 2 \\ 3 & 1 & 2 \end{bmatrix}}$$

$$(2 \times 2) \quad (2 \times 3)$$

Therefore $\mathbf{C}$ will be of size (2×3). Then

$$c_{11} = \begin{bmatrix} \boxed{1 \quad 2} \\ 3 \quad 4 \end{bmatrix} \begin{bmatrix} \boxed{1} & 2 & 2 \\ \boxed{3} & 1 & 2 \end{bmatrix}$$

$$= (1)(1) + (2)(3) = 7,$$

$$c_{12} = \begin{bmatrix} \boxed{1 \quad 2} \\ 3 \quad 4 \end{bmatrix} \begin{bmatrix} 1 & \boxed{2} & 2 \\ 3 & \boxed{1} & 2 \end{bmatrix}$$

$$= (1)(2) + (2)(1) = 4,$$

$$c_{13} = \begin{bmatrix} \boxed{1 \quad 2} \\ 3 \quad 4 \end{bmatrix} \begin{bmatrix} 1 & 2 & \boxed{2} \\ 3 & 1 & \boxed{2} \end{bmatrix}$$

$$= (1)(2) + (2)(2) = 6,$$

and so on, to reach

$$\mathbf{C} = \begin{bmatrix} 7 & 4 & 6 \\ 15 & 10 & 14 \end{bmatrix}.$$

Returning to our age-structure example, since $\mathbf{L}$ is a square matrix (size 3×3), multiplying $\mathbf{L}$ by itself is defined and this product is a matrix also with dimension 3×3. Thus

$$\mathbf{L}^2 = \mathbf{L}\,\mathbf{L} = \begin{bmatrix} 0 & 1 & 2 \\ 0.6 & 0 & 0 \\ 0 & 0.9 & 0 \end{bmatrix} \begin{bmatrix} 0 & 1 & 2 \\ 0.6 & 0 & 0 \\ 0 & 0.9 & 0 \end{bmatrix} = \begin{bmatrix} 0.6 & 1.8 & 0 \\ 0 & 0.6 & 1.2 \\ 0.54 & 0 & 0 \end{bmatrix}.$$

Now we multiply the matrix $\mathbf{L}^2$, which we just found, times $\mathbf{n}(t)$ to see if we get the same answer for $\mathbf{n}(t + 2)$ that we got previously by doing the double iteration:

$$\mathbf{L}^2\mathbf{n}(t) = \mathbf{n}(t + 2)$$

or

$$\begin{bmatrix} 0.6 & 1.8 & 0 \\ 0 & 0.6 & 1.2 \\ 0.54 & 0 & 0 \end{bmatrix} \begin{bmatrix} 30 \\ 20 \\ 10 \end{bmatrix} = \begin{bmatrix} 54 \\ 24 \\ 16.2 \end{bmatrix}.$$

As expected, we do.

TRANSPOSE OF A MATRIX

The transpose operation is accomplished by simply flipping a matrix on its side—by exchanging rows and columns. That is, the first row becomes the first column; the second row becomes the second column, and so on. We usually use the symbol T as a superscript to denote a transpose. Here are two examples. If

$$\mathbf{A} = \begin{bmatrix} 0 & 1 & 2 \\ 0.6 & 0 & 0 \\ 0 & 0.9 & 0 \end{bmatrix}, \quad \text{then} \quad \mathbf{A}^T = \begin{bmatrix} 0 & 0.6 & 0 \\ 1 & 0 & 0.9 \\ 2 & 0 & 0 \end{bmatrix},$$

and if

$$\mathbf{B} = \begin{bmatrix} 0 & 8 \\ 1 & 3 \\ 2 & 4 \end{bmatrix} \quad \text{then} \quad \mathbf{B}^T = \begin{bmatrix} 0 & 1 & 2 \\ 8 & 3 & 4 \end{bmatrix}.$$

Note that the transpose is defined for matrices (or vectors) of any dimension, not just square ones).

DETERMINANT OF A MATRIX

Think of a 2×2 matrix as two columns of numbers (i.e., two column vectors x_1 and x_2). Each vector expresses a point in two-dimensional space, as in

$$\mathbf{A} = [x_1 \ x_2] = \begin{bmatrix} 3 & -1 \\ 1 & 4 \end{bmatrix}.$$

Think of this matrix as representing the two vectors in this picture.

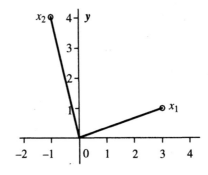

The determinant of a square matrix is a number (i.e., a scalar, not a matrix) whose absolute value is equal to the area of a parallelogram encompassed by these two vectors, as shown.

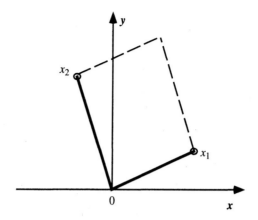

Taking the determinant of a square matrix for n = 2 Suppose that for $\mathbf{A}$ =

$$\mathbf{A} = \begin{bmatrix} a_{11} & a_{12} \\ a_{21} & a_{22} \end{bmatrix}.$$

The determinant equals the difference between the two cross products: $a_{11}a_{22} - a_{21}a_{12}$. If the two vectors were superimposed on each other, then the area would be zero, as depicted.

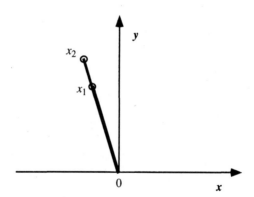

In this way a determinant tells you whether the vectors making up a matrix are all **linearly independent** or whether some duplicate others' directions.

When we extend this idea to three dimensions, the picture for a matrix with a nonzero determinant (the determinant is the volume of the dashed three-dimensional box) looks like this.

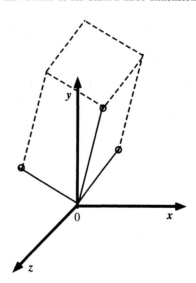

If two of the vectors were linearly dependent, then the picture would look like a flattened cardboard box in three-dimensional space.

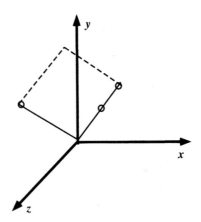

This box is only two-dimensional, so the determinant is zero; in this case we we say that this matrix has **rank** 2. If all three vectors were linearly dependent, then we'd only have a line in three-dimensional space; and we'd say that the matrix has rank 1. But, if the determinant of the original matrix were nonzero, as in the diagram showing the three dimensional box, it would have rank 3.

The determinant of a matrix with a single element is simply that element; that is, the determinant of 2 is 2. Calculating the determinant of a square matrix for $n = 3$ or higher can be tedious. We illustrate the process for a 3×3 matrix. Suppose that

$$\mathbf{A} = \begin{bmatrix} a_{11} & a_{12} & a_{13} \\ a_{21} & a_{22} & a_{23} \\ a_{31} & a_{32} & a_{33} \end{bmatrix} = \begin{bmatrix} 1 & 2 & 3 \\ 4 & 5 & 6 \\ 7 & 8 & 9 \end{bmatrix}.$$

To find the determinant of matrix **A** choose any single row or column. It doesn't make any difference which one you pick, except that calculations will be easier if you pick a row or column with lots of zeros. In this case, there are no zeros, so let's just choose row 1.

1. For that row or column, begin with the first element and strike out the row and column to which it belongs. For row 1 you would begin with a_{11} (= 1). After striking out the first row and first column you get a 2×2 matrix:

$$\begin{bmatrix} 5 & 6 \\ 8 & 9 \end{bmatrix}.$$

Then take the determinant of the resulting 2×2 matrix: $(5)(9) - (6)(8) = -3$. The determinant of this submatrix is called the **minor** of a_{11}, written as **minor**(a_{11}).

2. Now add the indexes (i.e., subscripts) of the original element chosen—in this case a_{11}—and add $1 + 1 = 2$. If this sum is an even number, as it is here, multiply the minor by $(+1)(a_{ij})$; if it is odd, multiply the minor by $(-1)(a_{ij})$. For this example you have in total $(1)(1)(-3) = -3$. (Do you see where the two 1's come from?) The result of this multiplication gives the **cofactor** of a_{11} often written as cof(a_{11}).

3. Repeat this process for the next element in the chosen row or column of matrix **A**. In this example we have decided to "expand" along the first row, so the next element is $a_{12} = 2$. Strike out the first row and second column to get

$$\begin{bmatrix} \cancel{1} & \cancel{2} & \cancel{3} \\ 4 & \cancel{5} & 6 \\ 7 & \cancel{8} & 9 \end{bmatrix} = \begin{bmatrix} 4 & 6 \\ 7 & 9 \end{bmatrix}.$$

The determinant of this submatrix is $(4)(9) - (6)(7) = -6 = \text{minor}(a_{12})$. Multiply this value by $(-1)(2)$ to get cof(a_{12}) = 12.

Keep doing this until you've finished all the calculations for the chosen row or column and then add all the cofactors. This sum gives you the determinant. For this example, the determinant of $\mathbf{A} = \text{cof}(a_{11}) + \text{cof}(a_{12}) + \text{cof}(a_{13}) = (-3) + (12) + (-9) = 0$. The process is the same for larger matrices. For example, for a 4×4 matrix, you would need to take the determinant of four matrices, each of size 3×3.

Appendix 3
Solving for the Equilibrium Points of Dynamical Systems and Finding the Inverse of a Square Matrix

In ecological dynamical systems the state variables are often species densities or resource levels, but they can be subcategories like different ages of a single species or males and females. Let n_i be the density of species i. Suppose that its population growth is given by the differential equation

$$\frac{dn_i}{dt} = n_1 f_i(\mathbf{n}, t), \tag{A.8}$$

where $\mathbf{n}$ is the vector $(n_1, n_2, \cdots n_m)$. We may have several (let's say m) species with the per capita growth of each described by as yet unspecified equations called the f functions. (Remember: there's nothing sacred about choosing the symbol f for this; you could use g or σ, if you like. The Greek letters look more impressive, but they're harder to find on the keyboard.)

These m different per capita growth equations given by the f_i functions can potentially be complicated functions involving nonlinear terms; not all f_i's might have terms involving every species, but in general they could.

At equilibrium each species population has stopped growing by definition. Thus

$$\frac{dn_i}{dt} = 0 \quad \text{for all the species } (i = 1 \text{ to } m).$$

Our goal is to find the values of $\mathbf{n}$ where $f_i(\mathbf{n}, t) = 0$ for all m species. At what densities will all species population sizes stop changing? There may be several; each is an equilibrium point and is denoted $\mathbf{n}^*$, meaning the vector $(n_1^*, n_2^*, n_3^*, \ldots, n_m^*)$. The asterisk indicates the particular values of $\mathbf{n}$ that are equilibrium values. Because of the form of Eq. (A.8), there are two basic ways that dn_i/dt may equal zero. The first is when n_i equals zero, and the second is when $f_i(\mathbf{n}, t)$ equals zero. For closed biological populations, population growth will always be zero when $n_i = 0$; if there are no individuals, the population can't possibly grow.

One way to find a multispecies equilibrium is to send some or all of the species to zero and let the others reach an equilibrium in their absence. That is, for a four-species system, we might set species 1 and 2 to zero and then for species 3 and 4 solve

$$f_3(\mathbf{n}, t) = 0 \quad \text{and} \quad f_4(\mathbf{n}, t) = 0.$$

These two equations can be solved for equilibrium values n_3^* and n_4^* in the absence of species 1 and 2. We took this approach in Chapter 14 on competition when we tried to find boundary solutions for three Lotka–Volterra competitors. Extending that case to four species, we get two equations:

$$0 = (K_3 - n_3{}^* - \alpha_{34}n_4{}^*] \quad \text{and} \quad 0 = (K_4 - n_4{}^* - \alpha_{43}n_3{}^*).$$

After some algebra, this pair of equations can be solved to yield

$$\mathbf{n}^* = \begin{bmatrix} 0 \\ 0 \\ \dfrac{(K_3 - a_{34}K_4)}{(1 - a_{34}a_{43})} \\ \dfrac{(K_4 - a_{43}K_3)}{(1 - a_{34}a_{43})} \end{bmatrix}$$

for the boundary solution corresponding to $n_1 = n_2 = 0$.

Many subsets of species can be set to "zero"; the possibilities are numerous and include the *trivial* equilibrium where all species have zero density. For example for $m = 3$ species, you have the following potential equilibriums to evaluate: $(0, 0, 0)$, $(n_1, 0, 0)$, $(0, n_2, 0)$, $(0, 0, n_3)$, $(n_1, n_2, 0)$, $(n_1, 0, n_2)$, and $(0, n_2, n_3)$. Sometimes, of course, when you set one species to zero, the others may not even be able to reach a positive equilibrium. If the prey is at zero in a one-predator, one-prey system, the only possible equilibrium for the predator is also zero.

The *interior* equilibrium point is that particular equilibrium point $\mathbf{n}^*$ that satisfies

$$f_i(\mathbf{n}, t) = 0 \quad \text{for } all \ i \tag{A.9}$$

Note the emphasis on the word *all*. You can solve for the interior equilibrium if you have m independent equations (one for each i) in m unknowns (the m values of $n_i{}^*$).

If the solution to Eq. (A.9) is strictly positive (i.e., every $n_i{}^*$ is greater than zero), the interior equilibrium point is then said to be a *feasible interior equilibrium*. A separate question is whether that equilibrium point is *stable*. Stability takes very different logic and math to evaluate, as you learned in Chapters 5 and 12. Usually, use of the term *stable* is restricted to the meaning that the system returns to the equilibrium following small perturbations away from it, although other definitions are possible.

Sometimes the algebraic solution of Eq. (A.9) yields an equilibrium point with the biologically nonsensical result that one or more of the species is at a negative density! Since, in reality, negative densities cannot be reached in a community, we say that the interior point is *unfeasible*. In this case the community will not reach the unfeasible equilibrium; instead, populations grow until a boundary equilibrium is reached (assuming that one of the boundaries is stable). The confusing part of multispecies dynamics, however, is that it's not necessarily true that if, say, species 2 and 6 have negative $n_i{}^*$, the boundary equilibrium involving all species but 2 and 6 will be a feasible or stable boundary equilibrium. We show this in Chapter 14 for two competitors (see Figure 14.26). It's also not necessarily true that the boundary equilibrium involving the absence of just species 2 (or just species 6) is itself necessarily unfeasible. Basically an unfeasible interior equilibrium point is not very helpful for determining which sets of species will be able to stably coexist and where the trajectory will end up starting from some initial conditions.

When the f_i are linear equations (e.g., the Lotka–Volterra competition equations), it is, of course, easier to solve for the interior equilibrium point than when these equations are nonlinear. However, in general, things can still be complicated since, even with linear per-capita systems, a number of possible equilibrium points exist and more than one of them may be locally stable. If this occurs, there are *alternative domains of attraction* or *alternative stable points*. For such a system, you might end up at one equilibrium point for some initial starting densities and another equilibrium point for other initial conditions. You can even see this in Lotka–Volterra competition involving just two species when $(\alpha_{12})(\alpha_{21}) > 1$ (see Chapter 14).

In the case of nonlinear f's, as in Chapter 12 on predator–prey dynamics, the equilibrium point can be feasible but unstable, and instead the equilibrium behavior of the system is a fixed cycle (i.e., not a point). In other words, the species densities keep oscillating up and down. You can even get limit cycles as the solution of these equations when the f's are linear if there are many interacting species. You will see this in Chapter 14 for three Lotka–Volterra competitors. It's still an active research problem in mathematics to determine these cyclic solutions when the number of interacting state variables (species) is large and the equations are nonlinear. Often researchers have to resort to simulation on a computer. Finally, the asymptotic behavior of some nonlinear systems may be neither a point or a cycle but rather a strange attractor of varying size and dimension; this is what we called chaos in Chapter 5.

Consider the more manageable case where the f's are linear. Now you can use some matrix tricks. Imagine a simple linear case of m Lotka–Volterra competitors and use linear algebra to solve for the *interior* equilibrium point. In this case after setting all the growth equations (i.e., the f_i) to zero, you're left with the linear matrix equation

$$\mathbf{K} - \mathbf{A}\, \mathbf{n}^* = \mathbf{0},$$

where $\mathbf{K}$ is the m by 1 vector of carrying capacities, $\mathbf{A}$ is the α matrix, $\mathbf{n}^*$ is the m by 1 vector of equilibrium densities and $\mathbf{0}$ is an m by 1 vector of zeros. (Remember, $\alpha_{ii} = 1$.) Rearranging yields

$$\mathbf{K} = \mathbf{A}\, \mathbf{n}^*. \tag{A.10}$$

To solve for $\mathbf{n}^*$ you now have two choices.

1. Apply **Cramer's rule** to Eq. (A.10). Cramer's rule works like this: To solve for the equilibrium level of species 1, n_1^*, take the ratio of the determinants of two different matrices: in the numerator form the matrix $\mathbf{A}\,(1)$, which is simply the matrix $\mathbf{A}$ with the first column replaced by the column of K's. In the denominator place the matrix of $\mathbf{A}$. Hence for a three-species system,

$$\mathbf{n}_1^* = \frac{|\mathbf{A}(1)|}{|\mathbf{A}|} = \frac{\begin{vmatrix} K_1 & a_{12} & a_{13} \\ K_2 & a_{22} & a_{23} \\ K_3 & a_{32} & a_{33} \end{vmatrix}}{\begin{vmatrix} a_{11} & a_{12} & a_{13} \\ a_{21} & a_{22} & a_{23} \\ a_{31} & a_{32} & a_{33} \end{vmatrix}}.$$

Similarly, the equilibrium density of species 2 is $n_2^* = \det(\mathbf{A}(2))/\det(\mathbf{A})$, where now, the matrix $\mathbf{A}(2)$ is formed by replacing the second column of matrix $\mathbf{A}$ with the $\mathbf{K}$ vector. In this manner, all n_i^* can be found.

2. Rearrange Eq. (A.10) to get $\mathbf{n}^* = \mathbf{A}^{-1} \mathbf{K}$. This involves finding the matrix $\mathbf{A}^{-1}$ called the *inverse of* $\mathbf{A}$. It's not at all like scalar arithmetic in the sense that the inverse of a matrix is not simply the inverse of each element in it. The concept, however, is similar: We want to find a matrix $\mathbf{A}^{-1}$ such that $\mathbf{A}\mathbf{A}^{-1} = \mathbf{I}$, where $\mathbf{I}$ is the **identity matrix** . This matrix has 1's down the diagonal and 0's everywhere else. Thus

$$\mathbf{I} = \begin{bmatrix} 1 & 0 & 0 & \cdots & 0 \\ 0 & 1 & 0 & \cdots & 0 \\ 0 & 0 & 1 & \cdots & 0 \\ \vdots & \vdots & \vdots & \vdots & \vdots \\ 0 & 0 & 0 & \cdots & 1 \end{bmatrix}$$

HOW TO FIND THE INVERSE OF A MATRIX, A^{-1}

Only square matrices have inverses. There are basically four steps in the calculation of an inverse matrix for a 2×2 matrix.

1. Form a new matrix $\mathbf{B}$, where each element b_{ij} is found by taking the determinant of the $\mathbf{A}$ matrix that results from striking out the ith row and jth column. The determinants of these submatrices are the same minors that we developed earlier. Here is an example.

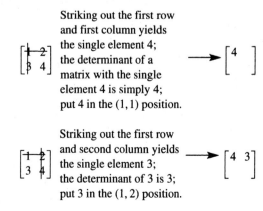

Striking out the first row and first column yields the single element 4; the determinant of a matrix with the single element 4 is simply 4; put 4 in the (1, 1) position.

Striking out the first row and second column yields the single element 3; the determinant of 3 is 3; put 3 in the (1, 2) position.

$$\begin{bmatrix} 1 & 2 \\ 3 & 4 \end{bmatrix}$$ Striking out the second row and first column yields the single element 2; the determinant of 2 is 2; put 2 in the (2, 1) position. $\longrightarrow$ $$\begin{bmatrix} 4 & 3 \\ 2 & \end{bmatrix}$$

$$\begin{bmatrix} 1 & 2 \\ 3 & 4 \end{bmatrix}$$ Striking out the second row and second column yields the single element 1; put 1 in the (2, 2) position. $\longrightarrow$ $$\begin{bmatrix} 4 & 3 \\ 2 & 1 \end{bmatrix}$$

Thus $\mathbf{B} = \begin{bmatrix} 4 & 3 \\ 2 & 1 \end{bmatrix}$.

2. Change the signs of the elements of $\mathbf{B}$ in the following manner: for each minor associated with the a_{ij} element of $\mathbf{A}$, if $i + j$ is an even number, then multiply the corresponding b_{ij} element of $\mathbf{B}$ by +1; if the sign of $i + j$ is an odd number, then multiply b_{ij} by –1. Matrix $\mathbf{B}$ becomes

$$\begin{bmatrix} 4 & -3 \\ -2 & 1 \end{bmatrix}.$$

3. Next divide each element of matrix $\mathbf{B}$ by the determinant of $\mathbf{A}$. For example, in this example the determinant of $\mathbf{A}$ is $4 - 6 = -2$ so we get

$$\frac{\mathbf{B}}{\det(\mathbf{A})} = \begin{bmatrix} -2 & 1.5 \\ 1 & -0.5 \end{bmatrix}. \tag{A.11}$$

4. Finally, take the **transpose** of the matrix in Eq. (A.11); this now is the inverse matrix $\mathbf{A}^{-1}$. Recall that the transpose operation involves flipping the matrix on its side so that the *i*th row now becomes the *i*th column. Thus

$$\mathbf{A}^{-1} = \begin{bmatrix} -2 & 1 \\ 1.5 & -0.5 \end{bmatrix}.$$

To verify that this answer is correct, multiply $\mathbf{A}\,\mathbf{A}^{-1}$.

$$\begin{bmatrix} 1 & 2 \\ 3 & 4 \end{bmatrix}\begin{bmatrix} -2 & 1 \\ 1.5 & -0.5 \end{bmatrix} = \begin{bmatrix} 1 & 0 \\ 0 & 1 \end{bmatrix}.$$

Since the product $\mathbf{A}\,\mathbf{A}^{-1}$ equals the identity matrix, you know that you did the inverse operation correctly. In practice, taking the inverse of a large matrix can be a tedious process and it's usually easier to let computers do the calculations.

Exercise: What is the inverse of matrix $\mathbf{A}$:

$$\mathbf{A} = \begin{bmatrix} 1 & 0 & -2 \\ 0 & 2 & 3 \\ -1 & 3 & 4 \end{bmatrix}.$$

Partial Solution: Expanding along the first row of $\mathbf{A}$ gives

$$\begin{bmatrix} 1 & 0 & -2 \\ 0 & 2 & 3 \\ -1 & 3 & 4 \end{bmatrix}$$ The determinant of this 2×2 submatrix is $(2)(4) - (3)(3) = 1$. Therefore put 1 in the (1, 1) position of $\mathbf{B}$. $\longrightarrow$ $$\begin{bmatrix} 1 & & \\ & & \\ & & \end{bmatrix}$$

$$\begin{bmatrix} 1 & 0 & -2 \\ 0 & 2 & 3 \\ -1 & 3 & 4 \end{bmatrix}$$ The determinant of this 2×2 submatrix is $(0)(4) - (3)(-1) = 3$. Therefore put 3 in the (1, 2) position of $\mathbf{B}$. $\longrightarrow$ $$\begin{bmatrix} 1 & 3 & \\ & & \\ & & \end{bmatrix}$$

Finish this process and verify that you have found the correct inverse by testing to see whether $\mathbf{A}\,\mathbf{A}^{-1} = \mathbf{I}$.

Some Useful Facts from Linear Algebra

- For two square matrices **A** and **B,** the inverse of their product is

$$(\mathbf{A}\,\mathbf{B})^{-1} = \mathbf{B}^{-1}\,\mathbf{A}^{-1}.$$

Note the reversal of multiplication order.

- The transpose of **A B** is

$$(\mathbf{A}\,\mathbf{B})^{T} = \mathbf{B}^{T}\,\mathbf{A}^{T}.$$

Note the reversal of multiplication order.

- If **D** is a diagonal matrix with elements d_{ii}, the result of multiplying **D A** is to multiply each element in the ith *row* of **A** by the ith element of **D,** $d_{ii.}$ For example, if

$$\mathbf{A} = \begin{bmatrix} a_{11} & a_{12} \\ a_{11} & a_{22} \end{bmatrix} \quad \text{and} \quad \mathbf{D} = \begin{bmatrix} d_{11} & 0 \\ 0 & d_{22} \end{bmatrix},$$

then

$$\mathbf{DA} = \begin{bmatrix} d_{11}a_{11} & d_{11}a_{12} \\ d_{22}a_{11} & d_{22}a_{22} \end{bmatrix}.$$

The result of multiplying **A D** is to multiply each element in the ith *column* of **A** by the ith element of **D,** d_{ii}, or

$$\mathbf{A}\,\mathbf{D} = \begin{bmatrix} d_{11}a_{11} & d_{22}a_{12} \\ d_{11}a_{11} & d_{22}a_{22} \end{bmatrix}.$$

- A matrix **A** may be written in block form as

$$\mathbf{A} = \begin{bmatrix} \mathbf{A}_{11} & \mathbf{A}_{12} \\ \mathbf{A}_{21} & \mathbf{A}_{22} \end{bmatrix},$$

such that blocks $\mathbf{A}_{11}$ and $\mathbf{A}_{22}$ are square. A theorem in linear algebra gives the determinant of **A** as

$$|\mathbf{A}| = |\mathbf{A}_{22}|\,|\mathbf{A}_{11} - \mathbf{A}_{12}\,\mathbf{A}_{22}^{-1}\,\mathbf{A}_{21}|.$$

When either $\mathbf{A}_{12}$ or $\mathbf{A}_{21}$ contain all zeros, then this simplifies further to

$$|\mathbf{A}| = |\mathbf{A}_{22}|\,|\mathbf{A}_{11}|.$$

For example, if

$$\mathbf{A} = \begin{bmatrix} 1 & 2 & 4 \\ 0 & 1 & 2 \\ 0 & 1/4 & 1 \end{bmatrix},$$

then

$$|\mathbf{A}| = 1(1 - 0.5) = 0.5.$$

- The inverse of a diagonal matrix **D** has elements $1/d_{ii}$ on the diagonal and zeros elsewhere. For example,

$$\mathbf{D}^{-1} = \begin{bmatrix} \dfrac{1}{d_{11}} & 0 \\ 0 & \dfrac{1}{d_{22}} \end{bmatrix}.$$

- The eigenvalues of a triangular matrix are simply the elements on the diagonal. For example, if

$$\mathbf{A} = \begin{bmatrix} 1 & 2 & 4 \\ 0 & 1 & 2 \\ 0 & 0 & 4 \end{bmatrix},$$

then the eigenvalues of **A** are 1, 1, and 4.

Appendix 4
Some Useful Mathematical Identities and Approximations

1. e is the base of the natural logarithms = 2.7138. . . .

2. $\ln 1 = 0$

3. As $a \to 0$, $\ln a \to -\infty$.

4. $\ln ab = \ln a + \ln b$

5. $\ln(a + b) \neq \ln a + \ln b$

6. $e^{z \ln a} = a^z$

7. $\ln(1 + x) \approx x$ when $|x| \ll 1$

8. $e^x \approx 1 + x$ when $|x| \ll 1$

9. $\dfrac{a_1}{a_2} - \dfrac{a_3}{a_4} = \dfrac{a_1 a_4 - a_2 a_3}{a_2 a_4}$

10. The two roots of the quadratic equation

$$ax^2 + bx + c = 0$$

are

$$x = \frac{-b \pm \sqrt{b^2 - 4ac}}{2a}.$$

11. $e^{iz} = \cos z + i \sin z$ where $i = \sqrt{-1}$

12. $e^{-iz} = \cos z - i \sin z$

13. $e^{i\pi} = -1$ (Think about it!)

14. The **absolute value** (also called the magnitude or modulus) of a **complex** number, $z = a + bi$, is

$$|z| = \sqrt{a^2 + b^2}$$

and

$$|z_1 + z_2| \leq |z_1| + |z_2| \quad \text{for any numbers } z_1 \text{ and } z_2.$$

15. $1° = (\pi/180)$ radians = 0.01745 radians.

16. 1 radian = $(180/\pi)° = 57.296°$.

SOME SERIES, SUMS, AND APPROXIMATIONS

1. $\dfrac{a}{(1-x)} = \displaystyle\sum_{i=0}^{\infty} ax^i$ when $-1 < x < 1$

2. $\dfrac{a(1-x^n)}{(1-x)} = \displaystyle\sum_{i=0}^{n} ax^i$ when $-1 < x < 1$

3. $\dfrac{1}{(1-x)^2} = \displaystyle\sum_{i=0}^{\infty} ix^{(i-1)}$ when $-1 < x < 1$

4. $\dfrac{1}{(1-x)^2} = \displaystyle\sum_{i=0}^{\infty} (i+1)x^i$ when $-1 < x < 1$

5. $e^x = 1 + x + \dfrac{x^2}{2!} + \dfrac{x^3}{3!} + \ldots = \displaystyle\sum_{i=0}^{\infty} \dfrac{x^i}{1!}$ for all real x,

where $n! = n$ factorial $= n(n-1)(n-2) \cdots 1$. For example $3! = (3)(2)(1) = 6$.

6. $(x+1)^n = \displaystyle\sum_{i=0}^{n} \binom{n}{i} x^i$ when $-1 < x < 1$,

where $\dbinom{n}{i} = \dfrac{n!}{i!(n-i)!}$,

e.g., $(0.5+1)^3 = 1 + (3)(0.5) + (3)(0.5)^2 + (1)(0.5)^3 = 3.375$.

7. $\dfrac{1}{1+x} = \displaystyle\sum_{i=0}^{\infty} (-1)^i x^i$ for $-1 < x < 1$

8. $\ln(1+x) = \displaystyle\sum_{i=1}^{\infty} (-1)^{i+1} \left(\dfrac{x^i}{i} \right)$ for $-1 < x \leq 1$

9. $\displaystyle\sum_i \sum_j a_i b_j \neq \sum_i a_i \sum_j b_j$

10. $\displaystyle\sum_i \sum_j a_{ij} b_j = \sum_i \left(\left(\sum_i a_{ij} \right) b_j \right)$

11. Using 5: $e^x \cong 1 + x$ when $|x| \ll 1$.

12. Using 8: $\ln(1-x) \approx -x$ when $|x| \ll 1$.

13. Taylor's series approximation for a continuous function $F(x)$ at point x^* is

$$F(x) \approx F(x^*) + \frac{F'(x^*)(x-x^*)}{1!} + \frac{F''(x^*)(x-x^*)^2}{2!} + \frac{F'''(x^*)(x-x^*)^3}{3!} + \ldots,$$

where F' is the first derivative, F'' is the second derivative, and so on.

14. Taylor's series approximation for a continuous function $F(x, y)$ at point (x^*, y^*) is

$$F(x,y) \approx F(x^*,y^*) + \frac{1}{1!}\left[(x-x^*)\frac{\partial F(x^*,y^*)}{\partial x} + \left(y-y^*\right)\frac{\partial F(x^*,y^*)}{\partial y} \right] +$$

$$\frac{1}{2!}\left[\left((x-x^*)\frac{\partial}{\partial x} + (y-y^*)\frac{\partial}{\partial y} \right)^2 F(x^*,y^*) \right] + \ldots.$$

Appendix 5
Calculus

DERIVATIVES

Function	Derivative
1. $y = \text{constant}$	$\dfrac{dy}{dx} = 0$
2. $y = x^n$	$\dfrac{dy}{dx} = nx^{n-1}$
3. $y = \sin x$	$\dfrac{dy}{dx} = \cos x$
4. $y = \cos x$	$\dfrac{dy}{dx} = -\sin x$
5. $y = \ln x$	$\dfrac{dy}{dx} = \dfrac{1}{x}$
6. $y = e^x$	$\dfrac{dy}{dx} = e^x$

Reduction formulas for functions of x: in the following, y, u, and v are arbitrary functions of x; f is a function of u; and a is a constant.

Function	Derivative
7. au	$a\left(\dfrac{du}{dx}\right)$
8. y	$\dfrac{dy}{dx}$
9. $u + v$	$\dfrac{du}{dx} + \dfrac{dv}{dx}$
10. $\sum_i u_i$	$\sum_i \dfrac{du_i}{dx}$
11. uv	$u\dfrac{dv}{dx} + v\dfrac{du}{dx}$
12. $\dfrac{u}{v}$	$\dfrac{v\left(\dfrac{du}{dx}\right) - u\left(\dfrac{dv}{dx}\right)}{v^2}$
13. $f(u(x))$	$\dfrac{df}{dx} = \dfrac{df}{du}\dfrac{du}{dx}.$

14. $y = y(x)$ $$\frac{dx}{dy} = \frac{1}{\frac{dy}{dx}}.$$

Differential $$dy = \frac{dy}{dx}(dx).$$

Alternative notations for a derivative:

$$\frac{dy}{dx}, \quad y', \quad f'(x), \quad \frac{df}{dx}, \quad \frac{d}{dx}f(x), \quad D_x y$$

INTEGRATION

The indefinite integral is given without the constant of integration because the limits of integration are not specified. The inclusion of a multiplicative constant involves only a simple transformation (a and b are constants).

$$\int f(ax)dx = \frac{1}{a}\int f(x)dx$$

$$\int bf(x)dx = b\int f(x)dx$$

An integral that is the sum of several terms may be broken up, as in

$$\int [f_1(x) + f_2(x)]dx = \int f_1(x)dx + \int f_2(x)dx.$$

The following are general forms for some easily integrable functions.

1. $\int x^n dx = \dfrac{x^{n+1}}{n+1}$ (except for $n = -1$)

2. $\int \dfrac{1}{x}dx = \ln x$

3. $\int e^x dx = e^x$

4. $\int (\ln x)dx = x\ln x - x$

5. $\int \dfrac{dx}{a^2 + x^2} = \dfrac{1}{a}\tan^{-1}\left(\dfrac{x}{a}\right)$

6. $\int \sin x \, dx = -\cos x$

7. $\int \cos x \, dx = \sin x$

8. $\int \sin^2 x \, dx = \dfrac{1}{2}x - \dfrac{1}{4}\sin^2 x$

9. $\int \dfrac{dx}{\sqrt{a^2 + x^2}} = \sin^{-1}\left(\dfrac{x}{a}\right)$

10. $\int e^{-ax}dx = \dfrac{-1}{a}e^{-ax}$

Appendix 6
Functions of Random Variables

If x is a continuous random variable, then its expected value, $E(x)$, is defined as

$$E(x) = \int f(x)x\,dx.$$

Where $f(x)$ is the density function (or probability distribution) of x.

The expected value is a measure of the center of the probability distribution. For discrete variables, like a string of n numbers, the expected value, or **mean,** is

$$E(x) = \frac{1}{n}\sum_{i=1}^{n} x_i.$$

The **variance** is a measure of the spread of the distribution of x:

$$\mathrm{Var}(x) = \int (x - E(x))^2\, f(x)dx$$

$$= \int x^2 f(x)dx - \left(\int xf(x)dx\right)^2$$

$$= E(x^2) - E(x)^2.$$

The variance is often written as σ^2. The **standard deviation** is defined as the square root of the variance is denoted as σ. If x and y are independent random variables and C is any constant, then

Function	Expected Value, E( )	Variance, var
$x + y$	$E(x) + E(y)$	var x + var y
$x - y$	$E(x - E(y))$	var x + var y
Cx	$CE(x)$	C^2 var x
xy	$E(x)E(y)$	var x var y + $E(x)^2$ var y + $E(y)^2$ var x

If x and y are random variables with expected values for $x^2 = E(x^2)$ and for $y^2 = E(y^2)$, then by the **Schwarz inequality,**

$$[E(xy)]^2 \leq E(x^2)E(y^2).$$

SOME PROBABILITY DISTRIBUTIONS AND THEIR PROPERTIES

Binomial distribution: The probability of k successes in n Bernoulli trials (e.g., coin tosses). The probability of one outcome, say, a head, is p.

432

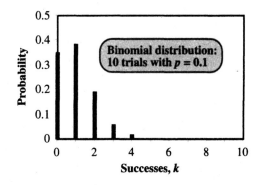

The probability of k heads appearing in n flips gives the density function as

$$b(x;k,n) = \binom{n}{k} p^k q^{n-k},$$

where $q = 1 - p$ and $\binom{n}{k}$ is the binomial coefficient $\dfrac{n!}{(n-k)!k!}$. The mean of the binomial distribution is $\bar{x} = np$, and the variance is npq. The range of the binomial is the integers from 0 to n.

Poisson distribution: The probability of k successes in n Bernoulli trials (when n is very large and the probability of success, p, for each is very small) is given by the Poisson distribution.

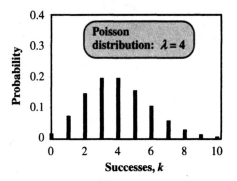

$$\text{Density function} = p(k;\lambda) = \frac{\lambda^k e^{-\lambda}}{k!}, \text{ where } \lambda = np.$$

$$\text{Mean, } \bar{x} = \lambda.$$

$$\text{Variance} = \lambda.$$

The range of the Poisson distribution is the integers from 0 to ∞.

Normal (or Gaussian) Distribution

$$\text{Density function} = \mathcal{N}(x;\bar{x},\sigma) = \frac{1}{\sigma\sqrt{2\pi}} \exp\left(\frac{-(x-\bar{x})^2}{2\sigma^2}\right)$$

where $\bar{x}$ is the expected value and σ^2 is the variance. This distribution is bell-shaped, and the range extends from $-\infty$ to ∞. The *standard* normal distribution has mean $\bar{x} = 0$ and $\sigma = 1$.

$$\mathcal{N}(y;\bar{x},\sigma) = \frac{1}{\sqrt{2\pi}} \exp\left(-\frac{1}{2}y^2\right)$$

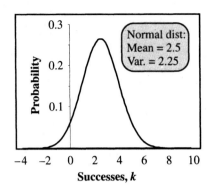

Estimate of the mean for a sample of size n:

$$\bar{x} = \frac{1}{n}\sum_{1}^{n} y_i .$$

Estimate of the variance for a sample of size n:

$$\mathrm{var} = \frac{1}{(n-1)}\sum_{i=1}^{n}(y_i - \bar{x})^2 .$$

Rectangular Distribution (or Uniform Distribution)

The density on the interval (a, b) with $a < b$ is given by the density function:

$$p(x) = \frac{1}{b-a} .$$

$$\mathrm{Mean} = \bar{x} = \frac{b+a}{2} .$$

$$\mathrm{var}\, x = E(x^2) - (E(x))^2 = \frac{1}{b-a}\int_a^b x^2 dx - \left(\frac{(b+a)^2}{4}\right)$$

$$= \frac{1}{b-a}\left(\left(\frac{b^3}{3} - \frac{a^3}{3}\right) - \left(\frac{(b+a)^2}{4}\right)\right)$$

$$= \frac{(b-a)^2}{12} .$$

Exponential Distribution

The exponential distribution has one parameter, u. The probability of x is given by the density function

$$p(x) = \begin{cases} ue^{-ux}, & x > 0 \\ 0, & x \leq 0 \end{cases} .$$

The mean is $\bar{x} = \frac{1}{u}$, and $\mathrm{var}\, x = \frac{1}{u^2}$.

The range extends from 0 to ∞.

A useful compendium of these and many other statistical distributions is Hale et al. (1993).

Literature Cited

Abrams, P. A. 1975. Limiting similarity and the form of the competition coefficient. *Theoretical Population Biology* 8: 356–375.

Abrams, P. A. 1977. Density-independent mortality and interspecific competition: A test of Pianka's niche overlap hypothesis. *American Naturalist* 111: 539–552.

Abrams, P. A. 1991. Strengths of indirect effects generated by optimal foraging. *Oikos* 62: 167–176.

Addicott, J. F. 1974. Predation and prey community structure: An experimental study of the effect of mosquito larvae on the protozoan communities of pitcher plants. *Ecology* 55: 475–492.

Allee, W. C. 1931. *Animal aggregations: A study in general sociology.* University of Chicago Press, Chicago.

Andow, D. A., P. M. Kareiva, S. A. Levin, and A. Okubo. 1990. Spread of an invading organism. *Landscape Ecology* 4: 177–188.

Armbruster, P., and R. Lande. 1993. A population viability analysis for African elephant (*Loxodonta africana*): How big should reserves be? *Conservation Biology* 7: 602–610.

Armstrong, R. A., and M. E. Gilpin. 1977. Evolution in a time-varying environment. *Science* 195: 591–592.

Atwood, J. L. 1980. Breeding biology of the Santa Cruz Island scrub jay, pp. 675–688. In D. M. Power (ed.), *The California Islands: Proceedings of a Multidisciplinary Symposium.* Santa Barbara Natural History Museum, Santa Barbara.

Atwood, J. L., M. J. Elpers, and C. T. Collins. 1990. Survival of breeders in Santa Cruz island and mainland California scrub jay populations. *Condor* 92: 783–788.

Beckerman, A. P., M. Uriarte, and O. J. Schmitz. 1997. Experimental evidence for a behavior-mediated trophic cascade in a terrestrial food web. *Proceedings of the National Academy of Science, USA* 94: 10735–10738.

Bender, E. A., T. J. Case, and M. E. Gilpin. 1984. Perturbation experiments in community ecology: Theory and practice. *Ecology* 65: 1–13.

Berg, H. C. 1983. *Random walks in biology.* Princeton University Press, Princeton.

Billick, I., and T. J. Case. 1994. Higher order interactions in ecology: What are they and how can they be detected? *Ecology* 75: 1529–1543.

Birch, L. C. 1948. The intrinsic rate of natural increase of an insect population. *Journal of Animal Ecology* 17: 15–26.

Boer, M. H. D. 1971. A colour polymorphism in caterpillars of *Bupalus piniarius* (L.) (Lepidoptera: Geometridae). *Netherlands Journal of Zoology* 21: 61–116.

Borema, L. K. and J. A. Gulland. 1973. Stock assessment of the Peruvian anchovy (*Engraulis ringens*) and management of the fishery. *Journal of the Fisheries Research Board of Canada* 39: 2226–2235.

Boutin, S., C. J. Krebs, R. Boonstra, M. R. T. Dale, S. J. Hannon, K. Martin, A. R. E. Sinclair, J. Smith N. M., R. Turkington, M. Blower, A. Byrom, F. I. Doyle, D. Hik, L. Hofer, A. Hubbs, T. Karels, D. L. Murray, V. Nams, M. O'Donoghue, C. Rohner, and S. Schweiger. 1995. Population changes of the vertebrate community during a showshoe hare cycle in Canada's boreal forest. *Oikos* 74: 69–80.

Boyce, M. S., and C. M. Perrins. 1987. Optimizing great tit clutch size in a fluctuating environment. *Ecology* 68: 142–153.

Broadmann, P. A., C. V. Wilcox, and S. Harrison. 1997. Mobile parasitoids may restrict the spatial spread of an insect outbreak. *Journal of Animal Ecology* 66: 65–72.

Brown, J. H. 1971. Mammals on mountain tops: nonequilibrium insular biogeography. *American Naturalist* 105: 467–478.

Brown, J. H., and A. Kodric-Brown. 1977. Turnover rates in insular biogeography: Effect of immigration on extinction. *Ecology* 58: 445–449.

Burnett, T. 1956. Effects of natural temperatures on oviposition of various numbers of an insect parasite (Hymenoptera, Chacidadae, Tehthredinidae). *Annals of the Entomological Society of America* 49: 55–59.

Buss, L. W., and J. B. C. Jackson. 1979. Competitive networks: Nontransitive competitive relationships in cryptic coral reef environments. *American Naturalist* 113: 223–234.

Case, T. J. 1975. Species numbers, density compensation, and colonizing ability of lizards on islands in the Gulf of California. *Ecology* 56: 3–18.

Case, T. J. 1978. Body size, endothermy, and parental care in the terrestrial vertebrates. *American Naturalist* 112: 861–874.

Case, T. J. 1990. Invasion resistance arises in strongly interacting species-rich model competition communities. *Proceedings of the National Academy of Science, USA* 87: 9610–9614.

Case, T. J., and R. Casten. 1979. Global stability and multiple domains of attraction in ecological systems. *American Naturalist* 113: 705–714.

Case, T. J., J. M. Diamond, and M. E. Gilpin. 1979. Interference competition, overexploitation and excess density compensation. *American. Naturalist* 113:843–854.

Case, T. J., and M. E. Gilpin. 1974. Interference competition and niche theory. *Proceedings of the National Academy of Science, USA* 71: 3073–3077.

Caswell, H. 1989. *Matrix population models: Construction, analysis, and interpretation.* Sinauer Associates, Sunderland, MA.

Charlesworth, B. 1971. Selection in density regulated populations. *Ecology* 52: 469–474.

Charlesworth, B. 1994. *Evolution in age-structured populations* (Second edition). Cambridge University Press, Cambridge.

Charnov, E. L. 1976. Optimal foraging: The marginal value theorem. *Theoretical Population Biology* 9: 129–136.

Charnov, E. L., and W. M. Schaffer. 1973. Life history consequences of natural selection: Cole's result revisited. *American Naturalist* 107: 791–793.

Chesson, P. L., and N. Huntley. 1997. The roles of harsh and fluctuating conditions in the dynamics of ecological communities. *American Naturalist* 150: 519–553.

Chesson, P. L. 1986. Environmental variation and the coexistence of species, pp. 240–256. In J. M. Diamond and T. J. Case (eds.), *Community Ecology.* Harper & Row, New York.

Cody, M. L. 1966. A general theory of clutch size. *Evolution* 20: 174–184.

Cody, M. L. 1971. Ecological aspects of reproduction, pp. 461–512. In D. S. Farner and J. R. King (eds.), *Avian Biology.* Vol. 1 Academic Press, New York.

Cody, M. L. 1974. *Competition and the structure of bird communities.* Princeton University Press, Princeton.

Cole, B. J. 1983. Assembly of mangrove ant communities: Patterns of geographic communities. *Journal of Animal Ecology* 52: 349–355.

Cole, L. C. 1954. The population consequences of life history phenomena. *Quarterly Review of Biology* 29: 103–137.

Connell, J. H. 1978. Diversity in tropical rainforests and coral reefs. *Science* 199: 1302–1310.

Cox, G. W., and R. E. Ricklefs. 1977. Species diversity and ecological release in Caribbean land bird faunas. *Oikos* 28: 113–122.

Crombie, A. C. 1946. Further experiments on insect competition. *Proceedings of the Royal Society of London, B* 133: 76–109.

Crow, J. F. 1986. *Basic concepts in population, quantitative, and evolutionary genetics.* W. H. Freeman, New York.

Crowe, D. 1975. Model for exploited bobcat populations in Wyoming. *Journal of Wildlife Management* 39: 408–415.

Dale, B. W., L. G. Adams, and R. T. Bowyer. 1994. Functional response of wolves preying on barren-ground caribou in a multiple-prey ecosystem. *Journal of Animal Ecology* 63: 644–652.

Darwin, C. 1859. *On the origin of species by means of natural selection, or preservation of favored races in the struggle for life* (First edition). Murray, London.

Darwin, C. 1872. *On the origin of species by means of natural selection, or preservation of favored races in the struggle for life* (Sixth edition). Murray, London.

Davidson, J. 1938. On the growth of the sheep population in Tasmania. *Transactions of the Royal Society of South Australia* 62: 342–346.

Deevey, E. S. J. 1947. Life tables for natural populations of animals. *Quarterly Review of Biology* 22: 283–314.

Dennis, B., and M. L. Taper. 1994. Density dependence in time series observations of natural populations: Estimation and testing. *Ecological Monographs* 64: 205–224.

Diamond, J. M. 1970. Ecological consequences of island colonization by southwest Pacific birds. II. The effect of species diversity on total population density. *Proceedings of the National Academy of Science* 67: 1715–1721.

Dobzhansky, T. 1967. Adaptedness and fitness, pp. 109–121. In R. C. Lewontin (ed.), *Population biology and evolution.* Syracuse University Press, Syracuse.

Elton, C. S., and M. Nicholson. 1942. The ten-year cycle in numbers of the lynx in Canada. *Journal of Animal Ecology* 11: 215–244.

Elton, C. S. 1958. *The ecology of invasions by animals and plants.* Methuen, London.

Erickson, J. M. 1971. The displacement of native ant species by the introduced Argentine ant *Iridomyrmex humilis* Mayr. *Psyche* 78: 257–266.

Evans, M., N. Hastings, and B. Peacock. 1993. *Statistical Distributions* (Second edition). John Wiley & Sons, New York.

Fagen, R. 1987. A generalized habitat matching rule. *Evolutionary Ecology* 1: 5–10.

Finnerty, J. P. 1980. *The population ecology of cycles in small mammals.* Yale University Press, New Haven.

Fisher, R. A. 1958. *The genetical theory of natural selection* (Second edition). Dover, New York.

Flack, J. D. D. 1976. Aspects of the ecology of the New Zealand Robin (*Petroica australis*). *Emu* (Supplement) 74: 286.

Fons, R., F. Poitevin, J. Catalan, and H. Croset. 1997. Decrease in litter size in the shrews *Crocidura suaveolens* (Mammalia, Insectivora) from Corsica (France): Evolutionary response to insularity? *Canadian Journal of Zoology* 75: 954–958.

Fowler, C. W. 1988. Population dynamics as related to rate of increase per generation. *Evolutionary Ecology* 2: 197–204.

Fretwell, S. D., and H. L. Lucas. 1970. On territorial behaviour and other factors influencing habitat distribution in birds. *Acta Biotheoretica* 19: 16–36.

Fryar, S., 1998. *Competitive interactions between wood decay Basidiomycetes.* Ph.D., Flinders University of South Australia.

Fujii, K. 1968. Studies on interspecies competition between the azuki bean weevil and the southern cowpea weevil: III, some considerations of strains of two species. *Research in Population Ecology* 10: 87–98.

Gadgil, M., and W. H. Bossert. 1970. Life historical consequences of natural selection. *American Naturalist* 104: 1–24.

Gause, G. F. 1934. *The struggle for existence.* Hafner, New York.

Gause, G. F. 1935. Experimental demonstration of Volterra's periodic oscillation in the numbers on animals. *Journal of Experimental Biology* 12: 44–48.

Gause, G. F. 1935. La théorie mathématique de la lutte pour la vie. *Actualités Scientifiques et Industrielles* 277: 1–63.

George, T. L. 1987. Greater land bird densities on island vs. mainland: Relation to nest predation level. *Ecology* 68: 1393–1400.

Gilpin, M. E. 1973. Do hares eat lynx? *American Naturalist* 107: 727–730.

Gilpin, M. E. 1975a. *Group selection in predator–prey communities.* Princeton University Press, Princeton.

Gilpin, M. E. 1975b. Limit cycles in competition communities. *American Naturalist* 109: 51–60.

Gilpin, M. E. 1979. Spiral chaos in a predator–prey model. *American Naturalist* 113: 306–308.

Gilpin, M. E. 1987. Spatial structure and population vulnerability, pp. 125–139. In M. E. Soulé (ed.), *Viable populations for conservation,* Cambridge University Press, Cambridge.

Gilpin, M. E., and J. M. Diamond. 1976. Calculation of immigration and extinction curves for the species-area-distance relation. *Proceedings of the National Academy of Science, USA* 73: 4130–4134.

Gilpin, M. E., and K. E. Justice. 1972. Reinterpretation of the invalidation of the principle of competitive exclusion. *Nature* 236: 273–301.

Goodman, D. 1987. The demography of chance extinction, pp. 11–34. In M. Soulé (ed.), *Viable populations for conservation.* Cambridge University Press, Cambridge.

Grant, P. R., and B. R. Grant. 1987. The extraordinary El Niño event of 1982–83: Effects on Darwin's Finches on Isla Genovesa, Galápagos. *Oikos* 49: 55–66.

Grime, J. P. 1979. *Plant strategies and vegetation processes.* John Wiley & Sons, New York.

Grosholz, E. D. 1996. Contrasting rates of spread for introduced species in terrestrial and marine systems. *Ecology* 77: 1680–1686.

Halbach, U. 1979. Introductory remarks: strategies in population research exemplified by rotifer populations. *Fortschritte der Zoologie* 25: 1–27.

Hamilton, W. D. 1966. The molding of senescence by natural selection. *Journal of Theoretical Biology* 12: 12–45.

Hanski, I. 1991. Single species metapopulation dynamics: Concepts, models and observations. *Biological Journal of the Linnean Society* 42: 17–38.

Harte, J. 1988. *Consider a spherical cow. A course in environmental problem solving.* University Science Books, Mill Valley, CA.

Hastings, A., and H. Caswell. 1979. Role of environmental variability in the evolution of life histories. *Proceedings of the National Academy of Science, USA* 76: 4700–4703.

Hengeveld, R. 1989. *Dynamics of biological invasions.* Chapman and Hall, London and New York.

Higuchi, H. 1976. Comparative study of the breeding of island and mainland subspecies of the varied tit, *Parus varius. Tori* 25: 11–20.

Holling, C. S. 1959. The components of predation as revealed by a study of small mammal predation of the European pine sawfly. *The Canadian Entomologist* 91: 293–320.

Holt, R. D. 1977. Predation, apparent competition, and the structure of prey communities. *Theoretical Population Biology* 12: 197–229.

Holt, R. D. 1997. On the evolutionary stability of sink populations. *Evolutionary Ecology* 11: 723–731.

Holt, R. D., and G. A. Polis. 1997. A theoretical framework for intraguild predation. *American Naturalist* 149: 745–764.

Hornfeldt, B. 1994. Delayed density dependence as a determinant of vole cycles. *Ecology* 75: 791–806.

Houston, D. B., E. G. Schreiner, and B. B. Moorhead. 1994. *Mountain goats in Olympic National Park: Biology and management of an introduced species.* National Park Service, Natural Resources Publication Office, Denver.

Hutchinson, E. W., and M. R. Rose. 1990. Quantitative genetic analysis of postponed aging in *Drosophila melanogaster,* pp. 66–87. In D. E. Harrison (ed.), *Genetic effects on aging II.* Telford Press, Caldwell, NJ.

Hutchinson, G. E. 1948. Circular causal systems in ecology. *New York Academy of Science* 50: 221–246.

Hutchinson, G. E. 1957. Concluding remarks. *Cold Spring Harbor Symposium on Quantitative Biology* 22: 415–427.

Ives, T. R. 1995. Measuring resilience in stochastic systems. *Ecological Monographs* 65: 217–233.

Johnson, T. E. 1990. Increased life-span of age-1 mutants in *Caenorhabditis elegans* and lower Gompertz rate of aging. *Science* 249: 908–912.

Johnston, J. P., W. J. Peach, R. D. Gregory, and S. A. White. 1997. Survival rates of tropical and temperate passerines: A Trinidadian perspective. *American Naturalist* 150: 771–789.

Jones, H. L., and J. M. Diamond. 1976. Short-time-base studies of turnover in breeding bird populations on the California Channel Islands. *Condor* 78: 526–549.

Joshi, N. V., M. Gadgil, and S. Patil. 1996. Correlates of the desired family size among Indian communities. *Proceedings of the National Academy of Science, USA* 93: 6387–6392.

Kareiva, P. M. 1990. Population dynamics in spatially complex environments—theory and data. *Philosophical Transactions of the Royal Society of London B* 330: 175–190.

Kendell, M. G., and A. Stuart. 1969. *The advanced theory of statistics.* Griffin, London.

Keyfitz, N. 1985. *Applied mathematical demography* (Second edition). Springer-Verlag, New York.

Keyfitz, N., and W. Flieger. 1968. *World population: an analysis of vital data.* University of Chicago Press, Chicago.

King, J. R. 1973. Energetics of reproduction in birds, pp. 78–107. In D. S. Farner (ed.), *Breeding biology of birds.* National Academy of Sciences, Washington, D.C.

Krebs, C. J. 1986. Are lagomorphs similar to other small mammals in their population ecology? *Mammal Reviews* 16: 187–194.

Krebs, C. J. 1988. *The message of ecology.* Harper & Row, New York.

Krebs, C. J. 1994. *Ecology: The experimental analysis of distribution and abundance* (Fourth edition). HarperCollins, New York.

Krebs, C. J., S. Boutin, R. Boostra, A. R. E. Sinclair, J. N. M. Smith, M. R. Dale, K. Martin, and R. Turkington. 1995. Impact of food and predation on the snowshoe hare cycle. *Science* 269: 1112–1115.

Lack, D. 1947. The significance of clutch size. *Ibis* 89: 302–352.

Lack, D. 1948. Natural selection and family size in the starling. *Evolution* 2: 95–110.

Lack, D. 1954. *The natural regulation of animal numbers.* Oxford University Press, Oxford.

Lack, D. 1968. *Ecological adaptations for breeding in birds.* Methuen, London.

Lande, R. 1987. Extinction thresholds in demographic models of territorial populations. *American Naturalist* 130: 624–635.

Lande, R. 1993. Risks of population extinction from demographic and environmental stochasticity and random catastrophes. *American Naturalist* 142: 911–927.

Law, R., A. D. Bradshaw, and P. D. Putwain. 1977. Life history variation in *Poa annua. Evolution* 31: 233–246.

Law, R. N. 1975. *Colonization and the evolution of life histories in Poa annua.* Ph.D., University of Liverpool.

Lawton, J. H., J. R. Beddington, and R. Bonser. 1974. Switching in invertebrate predators, pp. 141–158. In M. B. Usher and M. H. Williamson (eds.), *Ecological stability.* Chapman and Hall, London.

Lawton, J. H., and M. P. Hassell. 1981. Asymmetrical competition in insects. *Nature* 289: 793–795.

Lebreton, J. D., K. P. Burnham, J. Clobert, and D. R. Anderson. 1992. Modeling survival and testing biological hypotheses using marked animals: A unified approach with case studies. *Ecological Monographs* 62: 67–118.

Lefkovitch, L. P. 1965. The study of population growth in organisms grouped by stages. *Biometrics* 21: 1–18.

Leslie, P. H. 1945. On the use of matrices in certain population mathematics. *Biometrika* 35: 183–212.

Levin, S. A. 1974. Dispersion and population interactions. *American Naturalist* 108: 207–225.

Levins, R. 1968. *Evolution in changing environments: Some theoretical explorations.* Princeton University Press, Princeton.

Levins, R. 1969. Some demographic and genetic consequences of environmental heterogeneity for biological control. *Bulletin of the Entomological Society of America* 15: 237–240.

Levins, R. 1970. Extinction, pp. 77–107. In M. Gerstenhaber (ed.), *Some mathematical problems in biology.* American Mathematical Society, Providence, RI.

Lewontin, R. C. 1965. Selection for colonizing ability, pp. 77–94. In H. G. Baker and G. L. Stebbins (eds.), *The genetics of colonizing species.* Academic Press, New York.

Lewontin, R. C., and D. Cohen. 1969. On population growth in a randomly varying environment. *Proceedings of the National Academy of Science, USA* 62: 1056–1060.

Lithgow, G. J., and T. B. Kirkwood. 1996. Mechanisms and evolution of aging. *Science* 273: 80.

Lotka, A. J. 1920. Analytical note on certain rhythmic relations in organic systems. *Proceedings of the National Academy of Science (USA)* 6: 410.

Lotka, A. J. 1924. *Elements of physical biology.* Williams and Wilkins, New York.

Lowe, V. P. W. 1969. Population dynamics of the red deer (*Cervus elaphus* L.) on Rhum. *Journal of Animal Ecology* 38: 425–457.

Loyn, R. H., R. G. Runnalls, G. Y. Forward, and J. Tyers. 1983. Territorial bell miners and other birds affecting populations of insect prey. *Science* 221: 1411–1413.

Luckinbill, L. S. 1973. Coexistence in laboratory populations of *Paramecium aurelia* and its predator *Didinium nasutum.* *Ecology* 54: 1320–1327.

Luckinbill, L. S. 1978. *r* and *K* selection in experimental populations of *Escherichia coli.* *Science* 202: 1201–1203.

Luckinbill, L. S. 1979. Selection and the *r/K* continuum in experimental populations of protozoa. *American Naturalist* 113: 427–437.

MacArthur, R. H. 1972. *Geographical ecology.* Harper & Row, New York.

MacArthur, R. H., J. H. Diamond, and J. R. Karr. 1972. Density compensation in island faunas. *Ecology* 53: 330–342.

MacArthur, R. H., and R. Levins. 1967. The limiting similarity, convergence, and divergence of coexisting species. *American Naturalist* 101: 377–385.

MacArthur, R. H., and E. O. Wilson. 1967. *The theory of island biogeography.* Princeton University Press, Princeton.

Maher, W. J. 1970. The pomarine jaeger as a brown lemming predator in northern Alaska. *Wilson Bulletin* 82: 130–137.

Malthus, R. T. 1798. *An Essay on the Principles of Population as it Affects the Future Improvement of Society.* Johnson, London.

Maly, E. J. 1969. A laboratory study of the interaction between the predatory rotifer *Asplanchna* and *Paramecium.* *Ecology* 50: 59–73.

Maron, J. L., and S. Harrison. 1997. Spatial pattern formation in an insect host–parasitoid system. Science 278: 1619–1621.

May, R. M. 1973. Time delay versus stability in population models with two and three trophic levels. *Ecology* 54: 315–325.

May, R. M. 1975. *Stability and complexity in model ecosystems* (Second edition). Princeton University Press, Princeton.

May, R. M. 1976. Simple mathematical models with very complicated dynamics. *Nature* 261: 439–467.

May, R. M. 1981. Models for single populations, pp. 5–29. In R. N. May (ed.), *Theoretical ecology: principles and applications* (Second edition). Sinauer, Sunderland, MA.

May, R. M., J. R. Beddington, C. W. Clark, S. J. Holt, and R. M. Laws. 1979. Management of multispecies fisheries. *Science* 205: 267–277.

McCauley, D. E. 1978. Demographic and genetic responses of two strains of *Tribolium castaneum* to a novel environment. *Evolution* 32: 398–415.

Middleton, D. A. J., A. R. Veitch, and R. M. Nisbet. 1995. The effect of an upper limit to population size on persistence time. *Theoretical Population Biology* 48: 277–305.

Milinski, M. 1979. An evolutionarily stable feeding strategy in sticklebacks. *Zeitschrift für Tierpsychologie* 51: 36–40.

Mook, L. J. 1963. Birds and the spruce budworm, pp. 268–271. In R. F. Morris (ed.), The dynamics of epidemic spruce budworm populations. *Memoirs of the Ecological Society of Canada,* no. 31.

Moore, F. R. 1978. Interspecific aggression: Toward whom should a mockingbird be aggressive? *Behavioral Ecology and Sociobiology* 3: 173–176.

Mousseau, T. A., and D. A. Roff. 1987. Natural selection and the heritability of fitness components. *Heredity* 59: 181–197.

Murdoch, W. W., and A. Oaten. 1975. Predation and population stability. *Advances in Ecological Research* 9: 1–131.

Myers, J. H. 1988. Can a general hypothesis explain population cycles of forest Lepidoptera? *Advances in Ecological Research* 18: 179–242.

Newton, I., and P. Rothery. 1997. Senescence and reproductive value in sparrowhawks. *Ecology* 78: 1000–1008.

Neyman, J., T. Park, and E. L. Scott. 1956. Struggle for existence. The *Tribolium* model: Biological and statistical aspects. *Proceedings of the Third Berkeley Symposium on mathematical statistics and probability* 4: 41–79.

Nicholson, A. J. 1958. The self-adjustment of populations to change. *Cold Springs Harbor Symposium of Quantitative Biology* 22: 153–173.

Nicholson, A. J., and V. P. Bailey. 1935. The balance of animal populations. *Proceedings of the Royal Society of London* 3: 551–598.

Nisbet, R. M. 1997. Delay-differential equations for structured populations, pp. 89–118. In S. Tuljapurkar and H. Caswell (eds.), *Structured population models.* Chapman and Hall, New York.

Nisbet, R. M., and W. S. C. Gurney. 1982. *Modelling fluctuating populations.* John Wiley & Sons, New York.

Noon, B. R., and K. S. McKelvey. 1996. Management of the spotted owl: A case history in conservation biology. *Annual Review of Ecology and Systematics* 27: 135–162.

Noordwijk, A. J. van., A. J. van Balen, and W. Scharloo. 1981. Genetic and environmental variation in clutch size of the great tit. *Netherlands Journal of Zoology* 31: 342–372.

Olsen, L. F., and H. Degn. 1985. Chaos in biological systems. *Quarterly Review of Biophysics* 18: 165–225.

Paine, R. T. 1979. Disaster, catastrophe, and local persistence of the sea palm (*Postelsia palmaeformis*). *Science* 205: 685–687.

Park, T. 1954. Experimental studies of interspecific competition. II. Temperature, humidity, and competition in two species of *Tribolium.* *Physiological Zoology* 27: 177–238.

Park, T., D. B. Mertz, W. Grodinski, and T. Prus. 1965. Cannibalistic predation in populations of flour beetles. *Physiological Zoology* 38: 289–321.

Peacor, S. D., and E. E. Werner. 1997. Trait-mediated indirect interactions in a simple aquatic food web. *Ecology* 78: 1146–1156.

Pearl, R. 1928. *The rate of living.* Knopf, New York.

Pianka, E. R. 1976. Competition and niche theory, pp. 167–196. In R. M. May (ed.), *Theoretical Ecology* (Second edition). Sinauer, Sunderland, MA.

Pianka, E. R. 1994. *Evolutionary ecology* (Fifth edition). HarperCollins, New York.

Pielou, E. C. 1969. *An introduction to mathematical ecology.* Wiley-Interscience, New York.

Pimm, S. L., and A. Redfearn. 1988. The variability of population densities. *Nature* 334: 613–614.

Pollock, K. H., J. D. Nichols, C. Browne, and J.E. Hines. 1990. Statistical inference for capture–recapture experiments. *Wildlife Monographs* 107: 1–97.

Pomerantz, M. J., W. R. Thomas, and M. E. Gilpin. 1980. Asymmetries in population growth regulated by intraspecific competition: Empirical studies and model tests. *Oecologia* 47: 311–322.

Pratt, D. M. 1943. Analysis of population development in *Daphnia* at different temperatures. *Biological Bulletin* 85: 116–140.

Price, T. 1998. Maternal and paternal effects in birds: Effects on offspring fitness, pp. 202–226. In T. A. Mousseau and C. W. Fox (eds.), *Maternal effects as adaptations.* Oxford University Press, Oxford.

Price, T., and L. Liou. 1989. Selection on clutch size in birds. *American Naturalist* 134: 950–959.

Primack, R. B. 1978. Regulation of seed yield in *Plantago. Journal of Ecology* 66: 835–847.

Ranta, E., V. Kaitala, and P. Lundberg. 1997. The spatial dimension in population fluctuations. *Science* 278: 1621–1623.

Reese, J. G. 1975. Productivity and management of feral mute swans in Chesapeake Bay. *Journal of Wildlife Management* 39: 280–286.

Reznick, D. N. 1982. The impact of predation on life history evolution in Trinidadian guppies: Genetic basis of observed life history patterns. *Evolution* 36: 1236–1250.

Reznick, D. N., and J. A. Endler. 1982. The impact of predation of life history evolution in Trinidadian guppies (*Poecilia reticulata*). *Evolution* 36: 160–177.

Reznick, D. N., F. H. Shaw, H. F. Rodd, and R. G. Shaw. 1997. Evaluation of the rate of evolution in natural populations of guppies (*Poecilia reticulata*). *Science* 275: 1934–1937.

Ricker, W. E. 1952. Stock and recruitment. *Journal of the Fisheries Research Board of Canada* 11: 559–623.

Ricklefs, R. E. 1969. An analysis of nesting mortality in birds. *Smithsonian Contributions in Zoology* 9: 1–48.

Ricklefs, R. E. 1979. *Ecology* (Second edition). Chiron, New York.

Robinson, S. K., and J. Terborgh. 1995. Interspecific aggression and habitat selection by Amazonian birds. *Journal of Animal Ecology* 64: 1–11.

Romanoff, A. L., and A. J. Romanoff. 1949. *The avian egg.* John Wiley & Sons, New York.

Rose, M. R. 1991. *Evolutionary biology of aging.* Oxford University Press, New York.

Rosenzweig, M. L. 1973. Exploitation in three trophic levels. *American Naturalist* 107: 275–294.

Rosenzweig, M. L., and R. H. MacArthur. 1963. Graphical representation and stability conditions of predator–prey interactions. *The American Naturalist* 47: 209–223.

Roughgarden, J. 1971. Density-dependent natural selection. *Ecology* 52: 453–468.

Roughgarden, J. 1979. *Theory of population genetics and evolutionary ecology: An introduction.* MacMillan, New York.

Roughgarden, J. and J. M. Diamond. 1986. Overview: The role of species interactions in community ecology, pp. 333–343. In J. M. Diamond and T. J. Case (eds.), *Community ecology.* Harper and Row, New York.

Roughgarden, J., and M. Feldman. 1975. Species packing and predation pressure. *Ecology* 56: 489–492.

Schaffer, W. M. 1974. Optimal reproductive effort in fluctuating environments. *American Naturalist* 108: 783–790.

Schaffer, W. M. 1984. Stretching and folding in lynx fur returns: evidence for a strange attractor. *American Naturalist* 124: 798–820.

Schaffer, W. M., and P. F. Elson. 1975. The adaptive significance of variations in life history among local populations of Atlantic salmon in North America. *Ecology* 56: 577–590.

Scheffer, V. B. 1951. The rise and fall of a reindeer herd. *Scientific Monthly,* 73: 356–362.

Schluter, D., T. D. Price and P. R. Grant. 1985. Ecological character displacement in Darwin's finches. *Science* 227:1056–1059.

Schoener, T. W. 1974. Resource partitioning in ecological communities. *Science* 185: 27–39.

Schoener, T. W. 1993. On the relative importance of direct versus indirect effects in ecological communities, pp. 365–411. In H. Kawanabi, J. Cohen, and I. Iwasaki (eds.), *Mutualism and Community Organization.* Oxford University Press, Oxford.

Shelford, V. E. 1943. The relation of snowy owl migration to the abundance of collared lemming. *Auk* 62: 592–594.

Shigesada, N., and K. Kawasaki. 1997. *Biological invasions: Theory and practice.* Oxford, Oxford University Press.

Simberloff, D. 1976. Species turnover and equilibrium island biogeography. *Science* 194: 572–578.

Simberloff, D., and E. O. Wilson. 1969. Experimental island zoogeography of islands: the colonization of empty islands. *Ecology* 50: 278–296.

Simberloff, D., and E. O. Wilson. 1970. Experimental zoogeography of islands: A two year record of colonization. *Ecology* 51: 934–937.

Sinclair, A. R. E., J. M. Gosline, G. Holdsworth, D. J. Krebs, S. Boutin, J. N. M. Smith, R. Boonstra, and M. Dale. 1993. Can the solar cycle and climate synchronize the snowshoe hare cycle in Canada? Evidence for tree rings and ice cores. *American Naturalist* 141: 173–198.

Skellam, J. G. 1951. Random dispersal in theoretical populations. *Biometrica* 38: 196–218.

Solomon, M. E. 1949. The natural control of animal populations. *Journal of Animal Ecology* 18: 1–35.

Sousa, W .P. 1979a. Experimental investigation of disturbance and ecological succession in a rocky intertidal algal community. *Ecological Monographs* 49: 227–254.

Sousa, W. P. 1979b. Disturbance in marine intertidal boulder fields: The nonequilibrium maintenance of species diversity. *Ecology* 60: 1225–1239.

Southwood, T. R. E., R. M. May, M. P. Hassell, and G. R. Conway. 1974. Ecological strategies and population parameters. *American Naturalist* 108: 791–804.

Stearns, S. C. 1992. *The evolution of life histories.* Oxford University Press, New York.

Steere, J. B. 1894. On the distribution of genera and species of non-migratory land birds in the Philippines. *Ibis* 1894: 411–420.

Strobeck, C. 1973. *N* species competition. *Ecology* 54: 650–654.

Strong, D. R. 1986. Density vagueness: Abiding the variance in the demography of real populations, pp. 257–268. In J. Diamond and T. J. Case (eds.), *CommunityEcology*. Harper & Row, New York.

Tanner, J. T. 1966. Effects of population density on growth rates of animal populations. *Ecology* 47: 733–745.

Taper, M. L., and T. J. Case. 1992. Coevolution among competitors, pp. 63–109. In J. Antonovics and D. J. Futuyma (eds.), *Oxford Surveys in Evolutionary Biology*. Oxford University Press, Oxford.

Terborgh, J. W. 1976. Island biogeography and conservation: Strategies and limitations. *Science* 193: 1029–1030.

Tilman, D. 1986. Evolution and differentiation in terrestrial plant communities: The importance of the soil resource: Light gradient, pp. 359–380. In J. M. Diamond and T. J. Case (eds.), *Community Ecology*. Harper & Row, New York.

Tilman, D. 1994. Competition and biodiversity in spatially structured habitats. *Ecology* 75: 2–16.

Tilman, D., and P. M. Kareiva. 1997. *Spatial ecology: the role of space in population dynamics and interspecific interactions*. Princeton University Press, Princeton.

Utida, S. 1957. Population fluctuation, an experimental and theoretical approach. *Cold Springs Harbor Symposium on Quantitative Biology* 22: 139–151.

Vandermeer, J. H. 1969. The competitive structure of communities: An experimental approach with protozoa. *Ecology* 50: 362–371.

Vandermeer, J. H. 1972. Niche theory. *Annual Reviews of Ecology and Systematics* 3: 107–132.

Vandermeer, J. H., J. Lazarus, C. Ludwig, J. Lyon, B. Schultz, and K. Yih. 1980. Migration as a factor in the community structure of a macroarthropod litter fauna. *American Naturalist* 115: 606–612.

Volterra, V. 1926. Variazione e fluttuazini del numero d'individuals in specie animali conviventi. *Mem. Accad. Nazionale Linceri* (Ser. VI) 2: 31–113.

Wade, M. J. 1977. An experimental study of group selection. *Evolution* 31: 134–153.

Wade, M. J. 1979. The primary characteristics of *Tribolium* populations group selected for increased and decreased population size. *Evolution* 33: 749–764.

Wade, M. J. 1980. Group selection, population growth rate, and competitive ability in the flour beetles, *Tribolium* spp. *Ecology* 61: 1056–1064.

Wangersky, P. J., and W. J., Cunningham. 1957. Time lag in prey–predator population models. *Ecology* 38: 136–139.

Wilcox, B. A. 1978. Supersaturated island faunas. A species–age relationship for lizards on post-Pleistocene land-bridge islands. *Science* 199: 996–998.

Wootton, J. T. 1993. Indirect effects and habitat use in an intertidal community: interaction chains and interaction modifications. *American Naturalist* 141: 71–89.

Wootton, J. T. 1994. Putting the pieces together: testing the independence of interactions among organisms. *Ecology* 75: 1544–1551.

Yeaton, R. I. 1974. An ecological analysis of chaparral and pine forest bird communities on Santa Cruz Island and mainland California. *Ecology* 55: 959–973.

Yodzis, P. 1978. *Competition for space and the structure of ecological communities*. Springer-Verlag, Berlin.

Yodzis, P. 1988. The indeterminacy of ecological interactions as perceived through perturbation experiments. *Ecology* 69: 508–515.

Zwarts, L., and R. H. Drent. 1981. Prey depletion and regulation of predator density: Oystercatchers (*Haematopus ostralegus*) feeding on mussels (*Mytilus edulis*), pp. 193–216. In N. V. Jones and W. J. Wolff (eds.), *Feeding and survival strategies of esturine organisms*. Plenum, London.

Index